T0320929

Magnetic Small-Angle Neutron Scattering

OXFORD SERIES ON NEUTRON SCATTERING IN
CONDENSED MATTER

Magnetic Small-Angle Neutron Scattering

A Probe for Mesoscale Magnetism Analysis

Andreas Michels

University of Luxembourg

OXFORD

UNIVERSITY PRESS

OXFORD

UNIVERSITY PRESS

Great Clarendon Street, Oxford, OX2 6DP,
United Kingdom

Oxford University Press is a department of the University of Oxford.
It furthers the University's objective of excellence in research, scholarship,
and education by publishing worldwide. Oxford is a registered trade mark of
Oxford University Press in the UK and in certain other countries

First Edition published in 2021

Impression: 1

Published in the United States of America by Oxford University Press
198 Madison Avenue, New York, NY 10016, United States of America

British Library Cataloguing in Publication Data
Data available

Library of Congress Control Number: 2021932525

ISBN 978–0–19–885517–0

DOI: 10.1093/oso/9780198855170.001.0001

Printed and bound by
CPI Group (UK) Ltd, Croydon, CR0 4YY

This book is dedicated to the memory of my parents, and to Anna and Niki.

Preface

One of the very first publications on magnetic small-angle neutron scattering (SANS) is "Depolarisation und Kleinwinkelstreuung von Neutronen durch Gitterfehler in ferromagnetischen Kristallen" by Helmut Kronmüller, Alfred Seeger, and Manfred Wilkens from 1963 [1]. The paper is written in German (English translation: "Depolarization and small-angle scattering of neutrons by lattice imperfections in ferromagnetic crystals") and was dedicated to Max Born on the occasion of his 80th birthday. In their theoretical study, the authors have pioneered the use of the continuum theory of micromagnetics for calculating the magnetic SANS cross section of magnetic materials. Specifically, Kronmüller, Seeger, and Wilkens studied the spin disorder that is related to the strain fields of dislocations: in mechanically deformed metals, the magnetization is highly inhomogeneous in the vicinity of dislocations, which is due to the presence of magnetoelastic coupling. The associated static long-wavelength magnetization fluctuations represent a contrast for elastic magnetic SANS, and the scattering cross section can be computed by means of micromagnetic theory for samples close to magnetic saturation. This type of magnetic SANS, denoted as spin-misalignment scattering, is related to spatial variations in the orientation and magnitude of the magnetization. It has been predicted to be about $10-100$ times larger than the nuclear SANS that is related to the volume dilatations of dislocations. With the advent of nuclear research reactors and the concomitant construction and development of the first dedicated SANS instruments, e.g., at Jülich [2] and Grenoble [3], started the exploration of magnetism and superconductivity on a mesoscopic length scale using neutrons as a probe. The main ideas of [1] were then verified and extended only later in SANS experiments on cold-worked single crystals, as summarized in [2, 4–6].

Besides being a pioneering study that is full of insights regarding the complexity of the interaction between lattice imperfections and spin structure, the work by Kronmüller, Seeger, and Wilkens has undoubtly demonstrated the fact that magnetic SANS is in many respects different than the (nonmagnetic) small-angle scattering by structural and chemical inhomogeneities. The latter statement may be considered as the paradigm for this book. It certainly served as the central motivation for the author to write it. Nuclear SANS and small-angle x-ray scattering are largely based on the particle-matrix concept, with particle form factors and structure factors being the basic quantities. Magnetic SANS is about the magnetization distribution. The particle form factor is obtained as the solution of a volume integral, while the structure-factor problem, related to the arrangement of and the interaction between the particles, may be solved by using the methods of statistical mechanics. On the other hand, the continuous vectorial magnetization distribution of a magnetic material is obtained by solving a set of nonlinear partial differential equations (known as Brown's equations [7]), which in the context of magnetic SANS—and despite the early work [1]—is an often overlooked fact. Still, many magnetic SANS studies analyze their data based on the particle-matrix concept, with the underlying shortcoming assumption of homogeneously magnetized domains, and neglect the important and often even dominant magnetic scattering contribution due to misaligned magnetic moments.

As one may anticipate from the previous considerations, the origin of magnetic SANS is very closely related to the presence of lattice defects in the microstructure of magnetic materials (e.g., vacancies, dislocations, grain boundaries, pores). This viewpoint has also been emphasized in the early SANS review by Springer and Schmatz [8]. On the mesoscopic length scale that is probed by conventional SANS ($\sim 1-300\,\mathrm{nm}$), the defects are locally decorated by nanoscale spin disorder, which is generated by (i) spatial variations in the magnetic anisotropy field, and by (ii) spatial variations in the magnetic materials parameters, most notably the local saturation magnetization. To be more specific, forces due to the distortion of the crystal lattice in the vicinity of a microstructural defect tend to rotate the local magnetization vector field along the main axes of the system of internal stresses (magnetoelastic coupling), while magnetocrystalline anisotropy tries to pull the magnetic moments along the principal axes of the crystal [7]. Likewise, nanoscale spatial variations of the saturation magnetization, exchange, or anisotropy constants (e.g., at internal interfaces in a magnetic nanocomposite or in a nanoporous ferromagnet) give rise to inhomogeneous magnetization states, which represent a contrast for magnetic SANS. It is of decisive importance to emphasize that the adjustment of the magnetization along the respective local easy axes does not occur abruptly, i.e., on a scale of the interatomic spacing, but takes place over a more extended range. This is a consequence of the quantum-mechanical exchange interaction, which spreads local perturbations in the magnetization over larger distances. The size of such spin inhomogeneities is characterized by the micromagnetic exchange length l_{H}, which varies continuously with the applied field and takes on values between about $1-100\,\mathrm{nm}$. The ensuing magnetic neutron scattering appears at scattering angles $\psi \cong \lambda/l_{\mathrm{H}}$ (with λ the neutron wavelength), a regime which is routinely accessible by the SANS technique. We also emphasize that the observation that lattice-defect-induced magnetization nonuniformities are continuous functions of the position does not imply the absence of sharp features in the nuclear grain microstructure; for instance, there may well exist sharp particle-matrix interfaces (e.g., in the chemical composition) in a magnetic material, but the corresponding spin distribution (which decorates these interfaces) is continuous over the defects. This is nothing more than saying that the magnetic microstructure in real space corresponds to the convolution of the nuclear grain microstructure with micromagnetic response functions which vary with position and field. Therefore, particle form factors may also appear within the micromagnetic description of magnetic SANS, but they naturally emerge in the course of a calculation, by specifying the geometry of the defect.

The richness and the complex character of magnetic SANS can be grasped by looking at the figure at the end of this preface, which depicts a selection of computed spin structures and experimental magnetic SANS cross sections of various polycrystalline magnetic materials. While the angular anisotropies which are visible in some of the cross-section images are exclusively related to the saturated magnetization state, other patterns have their origin in the nonuniform magnetization distribution of the material, which depends on the magnetic interactions and on the characteristics of the underlying microstructure. The displayed anisotropies obviously go beyond the well-known $\sin^2 \alpha$ dependency of magnetic neutron scattering, which epitomizes the homogeneously magnetized single-domain state. They can only be understood by analyzing the spin distribution of the material, i.e., by carrying out micromagnetic calculations of the SANS cross section. Since the standard textbooks on small-angle scattering [9–13] do not cover the subject of magnetic SANS in sufficient depth, we believe that there is the necessity to fill this gap with the present book.

The central aim of the book is to provide an introduction into the theoretical background that is required to compute SANS cross sections and correlation functions related to long-wavelength magnetization structures; and to scrutinize these concepts based on the discussion of experimental unpolarized and polarized neutron data. The book is primarily about the technique of magnetic SANS, not about materials. Its writing style and diction may be described as a mixture between monograph, textbook, and review article (in particular Chapter 5). Regarding prior background knowledge, some familiarity with the basic magnetic interactions and phenomena as well as scattering theory is desired. The target audience consists of Ph.D. students and postdoctoral and senior researchers working in the field of magnetism and magnetic materials who wish to make efficient use of the magnetic SANS method. The principles and methods that are laid out in this book will hopefully enable them to analyze and interpret their SANS experiments.

Besides exposing the different origins of magnetic SANS (Chapter 1), and furnishing the basics of the magnetic SANS technique (Chapter 2), a large part of the book is devoted to a comprehensive treatment of the continuum theory of micromagnetics (Chapter 3), as it is relevant for the study of the elastic magnetic SANS cross section. Analytical expressions for the magnetization Fourier components allow one to highlight the essential features of magnetic SANS and to analyze experimental data both in reciprocal (Chapter 4) as well as in real space (Chapter 6). Chapter 5 provides an overview on the magnetic SANS of nanoparticles and so-called complex systems (e.g., ferrofluids, magnetic steels, spin glasses and amorphous magnets). It is this subfield where we expect a major progress to be made in the coming years, mainly via the increased usage of numerical micromagnetic simulations (Chapter 7), which is a very promising approach for the understanding of the magnetic SANS from systems exhibiting nanoscale spin inhomogeneity.

Being the result of more than two decades of research, the book contains the contributions of many people in one or other form. I would like to express my sincere gratitude to my former and present master and Ph.D. students, postdocs, and to the many collaborators and colleagues worldwide. Very special thanks are due to: Michael Adams, Natalie Baddour, Philipp Bender, Frank Bergner, Dmitry Berkov, Salvino Ciccariello, Sergey Erokhin, Artem Feoktystov, Luis Fernández Barquín, Arsen Goukassov, Sergey Grigoriev, Patrick Hautle, Dirk Honecker, Joachim Kohlbrecher, Artem Malyeyev, José Luis Martínez, Konstantin Metlov, Denis Mettus, Sebastian Mühlbauer, Yojiro Oba, Ivan Titov, and Andrew Wildes. These researchers took over the most important task of critically reading various parts of the book, and their most valuable comments and suggestions have without doubt improved the final result. Stephen Lovesey is thanked for his help in establishing the initial contact to Oxford University Press and for his continuous interest in the progression of the book. It is also a pleasure to acknowledge the contribution of Rainer Birringer, who has relentlessly supported me throughout my scientific career. Last, but not least, I would like to thank my Ph.D. adviser, Jörg Weißmüller, who introduced me to the magnetic SANS technique.

Luxembourg   Andreas Michels
February 2021

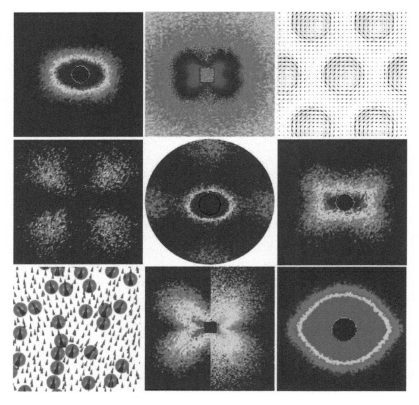

Selection of computed spin structures and experimental magnetic SANS cross sections.

Acknowledgements

I am indebted to the following colleagues for providing original data and figures of their own research, which have formed the basis for many illustrations in the book: Diego Alba Venero, Frank Bergner, Dmitry Berkov, Sabrina Disch, Sergey Erokhin, Patrick Hautle, André Heinemann, Dirk Honecker, Joachim Kohlbrecher, Christian Kübel, Dina Mergia, Sebastian Mühlbauer, Feodor Ogrin, Roger Pynn, Kiyonori Suzuki, Dmitri Svergun, Laura Vivas, and Albrecht Wiedenmann. I am also grateful to the following publishers, organizations, and learned societies for permission to reuse many figures in the book: EDP Sciences, Elsevier, Springer, Taylor & Francis, Wiley, American Institute of Physics, IEEE Magnetics Society, Institute of Physics Publishing, American Physical Society, Deutsche Physikalische Gesellschaft, and the International Union of Crystallography. The University of Luxembourg, the National Research Fund of Luxembourg, and the Deutsche Forschungsgemeinschaft are acknowledged for providing the financial means to carry out large parts of this research.

Contents

1

INTRODUCTION

1.1 Outline of the book

Small-angle neutron scattering (SANS) is a powerful and unique technique, which allows one to investigate microstructural (density and compositional fluctuations) as well as magnetic inhomogeneities in the volume of materials and on a mesoscopic length scale between a few and a few hundred nanometers. A further advantage of SANS, compared e.g., to electron-microscopy-based imaging methods, is that it provides statistically averaged information about a large number of scattering objects. Figure 1.1 gives an overview on the microstructural size regimes which are accessible by various observational methods. When conventional SANS is supplemented by ultra or very small-angle neutron scattering (USANS or VSANS) the spatial resolution can be extended up to the micrometer range [14, 15]. This is an important size regime in which many macroscopic material properties are realized. Since SANS and its x-ray counterpart, small-angle x-ray scattering (SAXS), are utilized in diverse fields of science such as materials science, physics, chemistry, biology, and in the derived interdisciplinary research areas, there exists an enormous body of research literature. The standard references for nuclear (i.e., nonmagnetic) SANS and SAXS are the well-known textbooks by Guinier and Fournet [9], Glatter and Kratky [10], Feigin and Svergun [11], Svergun, Koch, Timmins, and May [12], and by Gille [13]. For a selection of reviews on various materials classes and aspects of small-angle scattering, for instance, on biological structures, polymers, ferrofluids, magnetic materials, superconductors, disordered and porous materials such as coal, fractal systems, colloidal suspensions, ceramics, steels, precipitates in metallic alloys and composites, defect clusters, or dislocations we refer the reader to [2,5,16–68]. The proceedings of the International Small Angle Scattering Conference, which are usually published in the *Journal of Applied Crystallography*, are also an excellent source for following the latest trends and developments in the field.

From the historical point of view, experimental and theoretical progress in the domain of small-angle scattering is closely connected to the development of laboratory SAXS methods [9]. Perhaps related to this historical perspective is the fact that the theoretical concepts and foundations of nuclear SANS and SAXS are relatively well developed—in contrast to magnetic SANS—and widely acknowledged and applied in experimental studies [9–13]. This book is devoted to "diffuse" magnetic SANS, i.e., magnetic neutron scattering at small scattering angles around the forward direction arising from quasi-non-periodic, continuous long-wavelength magnetization fluctuations [compare panel (A1) in Fig. 1.2]. The term diffuse magnetic SANS is used here to distinguish it from magnetic small-angle diffraction, which is the method of choice for investigating long-range-ordered periodic structures such as helical spin systems,

Magnetic Small-Angle Neutron Scattering: A Probe for Mesoscale Magnetism Analysis. Andreas Michels, Oxford University Press (2021). © Andreas Michels. DOI: 10.1093/oso/ 9780198855170.003.0001

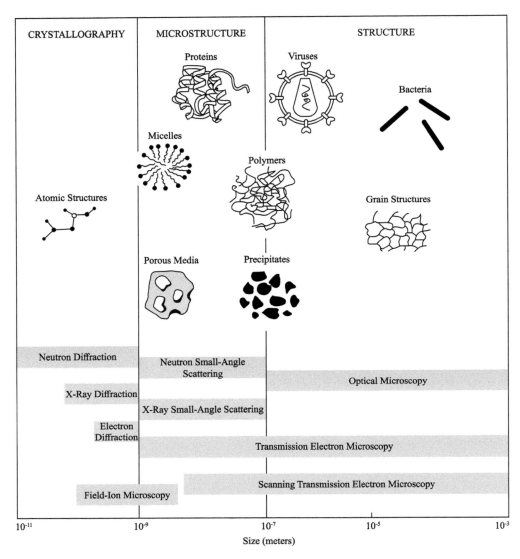

Fig. 1.1: Microstructural size regimes accessible by different observational methods. The techniques range from neutron diffraction, which is used to investigate atomic structures, to optical microscopy, which can be employed to image macroscopic objects such as bacteria or crystalline grain structures. Image courtesy of Roger Pynn, Indiana University, Bloomington, USA. After [69].

spin-density waves, flux-line lattices in superconductors, or the recently discovered skyrmion crystals [compare panel (A2) in Fig. 1.2]. Many of the latter problems can be dealt with by making use of the well-known cross-section expressions derived in the context of magnetic single-crystal neutron diffraction; for instance, SANS studies of

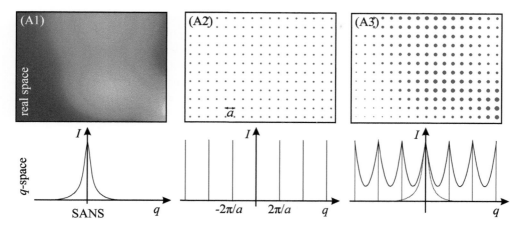

Fig. 1.2: Schematic illustration of the transformation from real space to reciprocal space for a smooth texture (A1), a discrete lattice with lattice spacing a (A2), and a smooth incommensurate modulation on top of a discrete lattice (A3). After [64].

the vortex lattice can be considered as crystallography of a two-dimensional system of lines (vortices), providing information about the lattice structure and correlations, as well as the internal structure of the individual scatterers [64]. Therefore, these topics will not be covered in this book, and likewise subjects such as neutron instrumentation, data treatment, or multiple scattering.

The book is organized as follows. In Chapter 1 we introduce the basic expressions for the neutron scattering cross sections and, in particular, for the elastic magnetic differential SANS cross section. Special emphasis is put on the discussion of the origins of magnetic SANS and on its distinctness from conventional particle scattering. Chapter 2 is concerned with general aspects of SANS such as the experimental setup, instrumental resolution, the influence of inelastic contributions due to phonon and magnon scattering to the elastic SANS cross section, the basics of nuclear SANS, the magnetic SANS cross sections for the two most often used scattering geometries, and their relation to the particle-matrix model. Chapter 3 introduces the continuum expressions for the magnetic energy contributions, which are employed for describing the mesoscale magnetic microstructure of magnetic materials. It is then shown how the static equations of micromagnetics, the so-called Brown's equations, can be solved in the high-field regime and how the Fourier components of the magnetization are related to the magnetic SANS cross section. In Chapter 4 we use the results for the Fourier components to compute the unpolarized and spin-polarized SANS cross sections. These expressions serve to highlight certain features of magnetic SANS such as the role of the magnetodipolar and Dzyaloshinskii–Moriya interaction, and are then applied to analyze experimental SANS data on various bulk magnetic materials such as soft and hard magnetic nanocomposites. Furthermore, Chapter 4 contains discussions on the magnetic Guinier law and on the asymptotic power-law exponents found in magnetic SANS experiments, as well as two sections summarizing magnetic SANS results on nanocrystalline rare-earth metals in the paramagnetic temperature regime and on dislocations. Chapter 5 provides an overview on the magnetic SANS of nanoparticles and complex systems, which include ferrofluids, magnetic steels, and spin glasses and amorphous magnets. We discuss the underlying assumptions of the conventional particle-matrix-based model of magnetic SANS, which assumes uniformly magnetized domains, characteristic e.g., for superparamagnets, and we provide a complete specification of the micromagnetic boundary-value problem. First attempts to provide analytical expressions for the vortex-state-related magnetic SANS of thin circular discs are considered. Spin-misalignment correlations in real space are the subject of Chapter 6. The correlation function and correlation length of the spin-misalignment SANS cross section are introduced, their properties are discussed within the context of micromagnetic theory, and selected experimental data on Nd–Fe–B-based permanent magnets and nanocrystalline elemental soft (Co and Ni) and hard (Gd and Tb) magnets are reviewed. Finally, in Chapter 7 we report on the progress made in using full-scale micromagnetic simulations for the understanding of the fundamentals of magnetic SANS. These studies take into account the nonlinearity of Brown's static equations of micromagnetics. We discuss prototypical sample microstructures, the implementation of the different energy contributions, and the state-of-the-art regarding simulations on multiphase nanocomposites and nanoparticle assemblies.

1.2 Basic properties of the neutron and numerical relations

The neutron is an elementary particle, which was discovered by James Chadwick in 1932 [70]. The basic properties of the neutron are its mass of $m_n = 1.675 \times 10^{-27}$ kg, its zero net electrical charge, and its spin angular momentum of $S = \pm \frac{1}{2} \hbar$ and associated magnetic dipole moment of $\mu_n = -1.913 \mu_N$ ($\hbar$ is the Planck constant h divided by 2π and μ_N is the nuclear magneton). These properties, together with the fact that the average lifetime of a free neutron is $\sim$886 s [71], render neutrons extremely attractive for research purposes; in particular, for investigating the structure and dynamics of matter on a wide range of length and timescales; more specifically:

- The value of the neutron mass results in a de Broglie wavelength of research neutrons which is of the order of interatomic distances in many crystalline and liquid materials. This allows one to access the structure of matter. Likewise, since the typical energy of cold and thermal neutrons is of the order of the elementary excitations in solids, dynamic features can also be explored by studying the inelastic scattering of neutrons.

- The zero net electrical charge implies that neutrons interact only very weakly with matter, in contrast to e.g., electrons, which are subject to a Coulomb barrier. As a consequence, neutrons are able to penetrate deeply into materials and they interact with the atomic nuclei of the material. From a theoretical point of view, this entails that the neutron-nucleus scattering process can be analyzed within first-order perturbation theory, which in the context of scattering formalism is known as the Born approximation [72]. Multiple-scattering effects can usually be ignored. This circumstance facilitates the analysis of neutron scattering data, since well-known expressions for the nuclear (and magnetic) scattering cross sections have been derived for this case.

- The scattering length for the neutron-nucleus interaction varies in a non-systematic manner from one element to the other across the periodic table of elements, and even between the isotopes of the same element (see Fig. 2.11). This brings with it an important advantage of neutron scattering over x-ray scattering: light elements such as hydrogen or oxygen strongly scatter neutrons, whereas they only weakly scatter x-rays, which is a consequence of the fact that the x-ray scattering length is proportional to the atomic number. The non-systematic variation of the nuclear scattering length between isotopes of the same element creates the possibility to tune the neutron scattering cross section by varying the isotopic concentration; for instance, light water, H_2O, has a theoretical macroscopic coherent scattering cross section per molecule of $\Sigma_{\mathrm{coh}} = 3.92 \times 10^{-3}\,\mathrm{cm}^{-1}$, whereas for heavy water, D_2O, one finds $\Sigma_{\mathrm{coh}} = 0.512\,\mathrm{cm}^{-1}$ (see, e.g., chapter 9 in [73]). The related so-called contrast-variation technique is of great importance in soft matter science, where compounds containing many hydrogen molecules are usually studied.
- The magnetic moment of the neutron interacts with unpaired electrons of atoms. Hence, magnetic structure and dynamics can be investigated by means of this interaction. Magnetic neutron scattering may be as strong as nuclear scattering. In fact, we will see in this book that magnetic SANS due to long-wavelength spin-misalignment fluctuations can be several orders of magnitude larger than the nuclear SANS.

Highly energetic neutrons are produced by either nuclear fission of heavy nuclei or by spallation of heavy metal targets [74]. By means of a moderator (e.g., light water, heavy water, graphite), the neutrons are slowed down to the energy of the moderator, where the neutron-velocity spectrum is to a very good approximation described by a Maxwellian distribution [75]. The following equations summarize the basic relations for the de Broglie wavelength λ, wave vector $\mathbf{k}_0$, momentum $\mathbf{p}_n$, and kinetic energy E_0 of a neutron moving with velocity $\mathbf{v}_n$:

$$\lambda = \frac{h}{m_n v_n}, \tag{1.1}$$

$$k_0 = |\mathbf{k}_0| = \frac{2\pi}{\lambda}, \tag{1.2}$$

$$\mathbf{p}_n = m_n \mathbf{v}_n = \hbar \mathbf{k}_0, \tag{1.3}$$

$$E_0 = kT = \frac{1}{2} m_n v_n^2 = \frac{\hbar^2 k_0^2}{2m_n} = \frac{h^2}{2m_n \lambda^2}. \tag{1.4}$$

By comparing the kinetic energy E_0 of the moderated neutrons to the thermal energy kT, neutrons for research purposes are commonly classified as hot, thermal, and cold neutrons (see Table 1.1). Using the values for the physical constants m_n, h, and k (see

Table 1.1 Classification scheme for ranges of energy, temperature, wavelength, and velocity of research neutrons. The indicated values correspond to the cold, thermal, and hot spectrum of the ILL (taken from [77]).

	E_0 (meV)	T (K)	λ (Å)	v_n (km/s)
cold neutrons	0.1–10	1–120	3–30	0.13–1.3
thermal neutrons	10–100	120–1200	1–3	1.3–4
hot neutrons	100–500	1200–6000	0.4–1	4–10

Appendix G), eqns (1.1)–(1.4) suggest the following numerical relations between the involved quantities [76]:

$$\lambda = 6.283/k_0 = 3.956/v_n = 9.045/\sqrt{E_0} = 30.811/\sqrt{T}; \tag{1.5}$$

$$E_0 = 0.08618\,T = 5.227\,v_n^2 = 81.805/\lambda^2 = 2.072\,k_0^2, \tag{1.6}$$

where λ is expressed in units of Å, k_0 in $10^{10}\,\mathrm{m^{-1}}$, v_n in km/s, E_0 in meV, and T is in degrees of Kelvin.

1.3 Definitions of scattering cross sections

The neutron scattering cross sections are defined with reference to Fig. 1.3, which depicts a sketch of the general scattering geometry. In the following considerations the spin degree of freedom of the neutron is ignored. The incoming neutrons are characterized by the wave vector $\mathbf{k}_0$ and the energy E_0. The sample (in a crystalline, amorphous, liquid, or gaseous state) is composed of N scattering atoms in the sample volume V. It is positioned such that the origin of the coordinate system is at some arbitrary point within the sample, e.g., at the center of mass of the system. The neutrons are scattered into a direction which is specified by the angles ψ and θ. Moreover, one commonly assumes that the characteristic size of the sample is much smaller than the distance between it and the detector. After interaction with the sample, the wave vector and energy of the neutrons are, respectively, given by $\mathbf{k}_1$ and $E_1 = \hbar^2 k_1^2/(2m_n)$. The change in the wave vector of the neutrons defines the momentum-transfer or scattering vector

$$\mathbf{q} = \mathbf{k}_0 - \mathbf{k}_1, \tag{1.7}$$

whereas the neutron-energy change is given by

$$\Delta E = E_0 - E_1 = \frac{\hbar^2}{2m_n}\left(k_0^2 - k_1^2\right). \tag{1.8}$$

The momentum which is transferred to the target in the scattering event equals $\hbar\mathbf{q}$. With reference to Fig. 1.3 one easily finds that [74]

$$q^2 = \frac{2m_n}{\hbar^2}\left(E_0 + E_1 - 2\sqrt{E_0 E_1}\cos\psi\right). \tag{1.9}$$

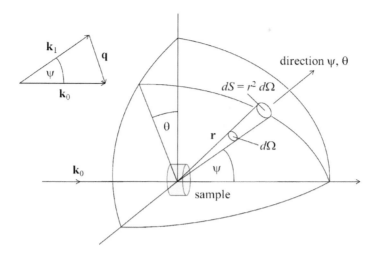

Fig. 1.3: Sketch of the general neutron scattering geometry.

For elastic scattering ($k_0 = k_1 = \frac{2\pi}{\lambda}$ and $\Delta E = 0$), the case which is most relevant in this book, it is readily verified that the magnitude of **q** is given by:

$$q = |\mathbf{q}| = 2k_0 \sin(\psi/2) = \frac{4\pi}{\lambda} \sin(\psi/2) \cong k_0 \psi, \qquad (1.10)$$

where the last approximation becomes valid in the small-angle regime ($\psi \ll 1$). The relation $q \cong k_0 \psi$ for small-angle scattering can be connected to a well-known result from classical optics [78]: an object with a linear dimension of Z gives rise to a Fraunhofer diffraction peak along the forward direction with an angular width of $\psi \cong \lambda/Z$; in other words, most of the diffracted intensity is concentrated at scattering angles $\psi \lesssim \lambda/Z$, which translates into $q \lesssim 2\pi/Z$. For typical values of $\lambda = 0.6\,\text{nm}$ and $Z = 10\,\text{nm}$, we find $\psi \cong 3.4°$.

The partial or double differential neutron scattering cross section $\frac{d^2\sigma}{d\Omega dE_1}$ is defined as:

$$\frac{d^2\sigma}{d\Omega dE_1} = \frac{n_1}{\Phi d\Omega dE_1}, \qquad (1.11)$$

where n_1 denotes the number of neutrons which are scattered per second into the element of solid angle $d\Omega = \sin\psi d\theta d\psi$ and which have a final energy between E_1 and $E_1 + dE_1$, and Φ denotes the incident beam flux, i.e., the number of neutrons per second through a unit area, where the area is assumed to be perpendicular to the direction of the incident neutron beam. The dimensions of n_1 and Φ are, respectively, time^{-1} and time^{-1}area^{-1}, so that $\frac{d^2\sigma}{d\Omega dE_1}$ is commonly expressed in units of barn eV^{-1}, where 1 barn $= 10^{-24}\,\text{cm}^2$ and $1\,\text{eV} = 1.602 \times 10^{-19}\,\text{J}$. Integrating $\frac{d^2\sigma}{d\Omega dE_1}$ over all final energies, one obtains the differential scattering cross section:

$$\frac{d\sigma}{d\Omega} = \frac{n_2}{\Phi d\Omega} = \int_0^\infty \frac{d^2\sigma}{d\Omega dE_1} dE_1, \tag{1.12}$$

where n_2 equals the total number of neutrons which are scattered per second into $d\Omega$. The total scattering cross section

$$\sigma = \frac{n_3}{\Phi} = \int_{4\pi} \frac{d\sigma}{d\Omega} d\Omega \tag{1.13}$$

is obtained from $d\sigma/d\Omega$ by integration over all directions, where n_3 denotes the total number of neutrons scattered per second at all energies and in all directions. In many experimental situations—however, generally not in magnetic SANS—$d\sigma/d\Omega$ depends only on the angle ψ, so that

$$\sigma = 2\pi \int_0^\pi \frac{d\sigma}{d\Omega} \sin\psi d\psi. \tag{1.14}$$

Both $d\sigma/d\Omega$ and σ are given in the unit of the barn. In addition to elastic and inelastic scattering, neutrons may be removed from the incident beam by absorption within the nuclei of the material under study. This process is characterized by an absorption cross section σ_a, so that the total collision or extinction cross section (taking into account scattering and absorption) is given by

$$\sigma_{tot} = \sigma + \sigma_a. \tag{1.15}$$

The quantity σ_{tot} is related via the optical theorem, also known as the Bohr–Peierls–Placzek relation, to the imaginary part of the nuclear scattering amplitude $f_N(\psi)$ along the forward direction ($\psi = 0$) [72, 79, 80]:

$$\sigma_{tot} = \frac{4\pi}{k_0} Im[f_N(0)]. \tag{1.16}$$

The optical theorem embodies the conservation of particle number, i.e., the number of incoming neutrons equals the number of outgoing neutrons plus those which are captured [75].*

The total collision cross section also determines the neutron transmission. Assuming perpendicular neutron incidence on a planar slab-shaped sample, the sample transmission T is related to the atomic number density $\rho_a = N/V$ (number N of nuclei in the sample volume V), σ_{tot}, and the sample thickness t by the well-known Lambert–Beer law [82]:

$$T = \exp(-\rho_a \sigma_{tot} t) = \exp(-\mu t), \tag{1.17}$$

where $\mu = \rho_a \sigma_{tot}$ denotes the attenuation coefficient (in cm^{-1}). The mean free path of the neutron between collisions is given by $L_{mfp} = \mu^{-1}$. For $L_{mfp} \gg t$ one has

---

*In this context we refer to Toperverg et al. [81] for an interesting polarized SANS study on a ferrofluid, which utilizes the optical theorem for analyzing the whole range of the scattering pattern including the zone of the interference between scattered and transmitted neutron waves.

T $\cong 1$ and the sample is transparent to neutrons with a particular energy, while T $\cong 0$ for $L_{\mathrm{mfp}} \ll t$ and the medium is opaque. Values for L_{mfp} depend on the neutron wavelength and, for not-too-strongly absorbing and scattering materials, are typically in the $0.1-10\,$cm regime (see table 12.3 in [83]). As a rule of thumb, in order to avoid multiple scattering one should aim for T $\gtrsim 90\%$ [84, 85] (neglecting absorption).

Finally, we remark that in the domain of small-angle scattering it is customary to display the macroscopic differential SANS cross section $d\Sigma/d\Omega$ per unit volume (in cm^{-1}), which is related to the microscopic $d\sigma/d\Omega$ per nucleus [2] via

$$\frac{d\Sigma}{d\Omega} = \frac{N}{V}\frac{d\sigma}{d\Omega}. \tag{1.18}$$

1.4 Elastic differential SANS cross section

In a SANS experiment, the energy of the scattered neutrons is usually not analyzed and one measures an energy-integrated cross section, i.e., the detector counts neutrons that have been scattered elastically as well as inelastically. However, it can be shown that inelastic coherent scattering contributions (e.g., due to phonons and/or magnons) play only a minor role in the small-angle regime, so that it is justified in many cases to consider SANS as elastic scattering (see Section 2.3). The assumption of elasticity implies that the internal states of the sample are not changed by the scattering event. Following Lovesey [86] and Squires [76] and ignoring the spin of the neutron, the elastic differential cross section can be computed by summing up all scattering processes in which the state of the neutron changes from $\mathbf{k}_0$ to $\mathbf{k}_1$, according to:

$$\frac{d\Sigma}{d\Omega} = \frac{1}{V}\frac{1}{\Phi}\frac{1}{d\Omega}\sum_{\mathbf{k}_1 \text{ in } d\Omega} W_{\mathbf{k}_0 \to \mathbf{k}_1}, \tag{1.19}$$

where $W_{\mathbf{k}_0 \to \mathbf{k}_1}$ denotes the number of transitions per second from the state $\mathbf{k}_0$ to the state $\mathbf{k}_1$. The sum in eqn (1.19) runs over all values of the final wave vector $\mathbf{k}_1$ that lie within the element of solid angle $d\Omega$, keeping the values of $\mathbf{k}_0$ constant. The transition rate can be evaluated by using Fermi's golden rule [72, 77]:

$$\sum_{\mathbf{k}_1 \text{ in } d\Omega} W_{\mathbf{k}_0 \to \mathbf{k}_1} = \frac{2\pi}{\hbar}\rho_{\mathbf{k}_1}\left|\langle\mathbf{k}_1|V_{\mathrm{int}}|\mathbf{k}_0\rangle\right|^2, \tag{1.20}$$

where $\rho_{\mathbf{k}_1}$ is the number of final momentum states in $d\Omega$ per unit energy range, and V_{int} denotes the interaction potential between the neutron and the scatterer (inducing the transition from $\mathbf{k}_0$ to $\mathbf{k}_1$). By assuming that the incoming and outgoing neutron waves are in a plane-wave state, respectively, $\mathbf{A}_{\mathrm{in}} \propto \exp\left(i\mathbf{k}_0 \cdot \mathbf{r}\right)$ and $\mathbf{A}_{\mathrm{out}} \propto \exp\left(i\mathbf{k}_1 \cdot \mathbf{r}\right)$, the matrix element eqn (1.20) can be evaluated. Within this so-called Born approximation, and using the definition of the scattering vector $\mathbf{q}$ [eqn (1.7)], the transition rate $W_{\mathbf{k}_0 \to \mathbf{k}_1}$ is then proportional to the magnitude square of the Fourier transform of the neutron-matter interaction potential. The derivation of the elastic nuclear and magnetic neutron scattering cross sections can be found in many textbooks on neutron scattering (e.g., [76, 86]) and it is therefore not repeated in detail here. Two interactions are most relevant for the scattering of thermal neutrons from matter, the nuclear and the magnetic interaction, which are briefly discussed in the following.

1.4.1 Nuclear SANS cross section

The short-range and isotropic neutron-nucleus interaction is modeled by Fermi's pseudopotential [87]:

$$V_{\text{int}}^{\text{nuc}}(\mathbf{r}) = \frac{2\pi\hbar^2}{m_{\text{n}}} b\, \delta(\mathbf{r}), \tag{1.21}$$

where b is the atomic scattering length of a strongly bound nucleus sitting at the origin $\mathbf{r} = 0$, and $\delta(\mathbf{r})$ is Dirac's delta function. It should be emphasized that Fermi's pseudopotential does not reflect the actual nuclear interaction potential and that the assumptions underlying the Born approximation, in particular, the approximation of the incoming neutron wave function as a plane wave within the interaction range of the nuclear potential (~ 1 fm; 1 fm $= 10^{-15}$ m), are not valid for the very strong nuclear potential (see the article by Schober [77] for a discussion of this point). However, eqn (1.21) is the only potential which, in the Born approximation, yields the required isotropic (s-wave) scattering behavior [76, 86].

The scattering length is generally a complex quantity which can be written as [88]:

$$b = b' - ib'', \tag{1.22}$$

where the real part b' takes on positive as well as negative values and is of the order of a few fm (see Fig. 2.11). The imaginary part $b'' > 0$ describes the absorption of neutrons. In fact, it can be shown that for all practical purposes the absorption cross section is given by [82]

$$\sigma_{\text{a}} = \frac{4\pi}{k_0} b''. \tag{1.23}$$

Note that σ_{a} is inversely proportional to k_0 and, hence, to the neutron velocity v_{n}. Equation (1.23) is called Fermi's $1/v_{\text{n}}$-law. Nuclear scattering lengths are usually taken from experiment. They can be measured with a range of techniques; for instance by means of mirror reflection [89–93], with a gravity refractometer [94–97], or by neutron interferometry [98–100] (see the article by Rauch and Waschkowski [101] and the book by Sears [82] for further details). Absorption is negligible for most elements and their isotopes ($b'' \ll |b'|$), however, there exist a few isotopes (e.g., ^{3}He, ^{10}B, ^{113}Cd, ^{157}Gd) which do possess a strong neutron capture cross section. Values for b and for the coherent σ_{coh}, incoherent σ_{inc}, and absorption σ_{a} cross sections can be found in [102].

By inserting the interaction potential eqn (1.21) into the eqns (1.19) and (1.20) it can be shown that the elastic nuclear differential scattering cross section of a collection of N nuclei at positions $\mathbf{r}_k$ equals [76, 103]:

$$\frac{d\Sigma_{\text{nuc}}}{d\Omega}(\mathbf{q}) = \frac{1}{V} \left| \sum_{k=1}^{N} b_k \exp\left(-i\mathbf{q}\cdot\mathbf{r}_k\right) \right|^2 = \frac{1}{V} \sum_{k,l} b_k b_l^* \exp\left(-i\mathbf{q}\cdot[\mathbf{r}_k - \mathbf{r}_l]\right), \tag{1.24}$$

where the b_k denote the nuclear atomic scattering lengths, $i^2 = -1$, the asterisk ($*$) marks the complex-conjugated quantity, and it is understood that $d\Sigma_{\text{nuc}}/d\Omega$ represents an ensemble-averaged quantity; more specifically, an average over the atomic configuration as well as an average over both the isotope distribution and the nuclear-spin orientations, which are random for not-too-large fields and low temperatures (see Section 2.4 for further details).

At this point it is important to realize that for small-angle scattering the discrete atomic structure of many crystalline materials is generally of no importance. Often, the probing neutron wavelength is larger than twice the maximum lattice-plane distance for allowed reflections from the crystal structure (so-called Bragg cutoff). The resolution range of the conventional SANS technique ranges between a few and a few hundred nanometers. SANS can therefore be described within a coarse-grained picture of the actual discrete atomic structure of matter. For nuclear SANS, the relevant coarse-grained variable is the scalar scattering-length density $N(\mathbf{r})$ (in m^{-2}, and not to be confused with the number N of nuclei or magnetic moments in a sample). This quantity is assumed to be a smooth and continuous function of the position $\mathbf{r} = \{x, y, z\}$ inside the material. It is therefore permissible to replace the discrete sum in eqn (1.24) by an integral over $N(\mathbf{r})$. The final result for the macroscopic nuclear SANS cross section then reads (e.g., [2]):

$$\frac{d\Sigma_{\mathrm{nuc}}}{d\Omega}(\mathbf{q}) = \frac{1}{V} \left| \int_V N(\mathbf{r}) \exp\left(-i\mathbf{q} \cdot \mathbf{r}\right) d^3 r \right|^2 = \frac{8\pi^3}{V} |\widetilde{N}(\mathbf{q})|^2, \qquad (1.25)$$

where $\widetilde{N}(\mathbf{q})$ denotes the Fourier transform of $N(\mathbf{r})$, and the integral is taken over the volume V of the sample. In Section 2.4.3, we will see how $d\Sigma_{\mathrm{nuc}}/d\Omega$ can be related to the Fourier transform of the autocorrelation function of $N(\mathbf{r})$.

Nuclear SANS (and SAXS) is not further considered here, and we refer the reader to Section 2.4 for a summary of the basics. Before moving on to the discussion of the magnetic SANS cross section, and with respect to polarized neutron scattering, we emphasize that the nuclear SANS contribution which is related to the spins of the atomic nuclei gives rise to a q-independent incoherent background contribution. It is ignored in this book, simply because the related scattering signal is usually small when compared to the magnetic SANS that is due to the electronic spins of nonhydrogenated magnetic materials.

1.4.2 Magnetic SANS cross section

Magnetic neutron scattering originates from the interaction between the magnetic moment of the neutron and the magnetic field that is created by the spin and the orbital motion of atomic electrons. In contrast to the nuclear interaction [eqn (1.21)], magnetic scattering is anisotropic. The interaction potential for magnetic neutron scattering is given by [76]:

$$V_{\mathrm{int}}^{\mathrm{mag}}(\mathbf{r}) = -\boldsymbol{\mu}_{\mathrm{n}} \cdot \mathbf{B}(\mathbf{r}), \qquad (1.26)$$

where

$$\boldsymbol{\mu}_{\mathrm{n}} = -\gamma_{\mathrm{n}} \mu_{\mathrm{N}} \boldsymbol{\sigma}_{\mathrm{P}} \qquad (1.27)$$

denotes the magnetic dipole moment of the neutron, $\gamma_{\mathrm{n}} = 1.913$ is the magnetic moment of the neutron expressed in units of the nuclear magneton [104], and $\boldsymbol{\sigma}_{\mathrm{P}}$ is the Pauli spin operator for the neutron. The quantity

$$\mathbf{B} = \mathbf{B}_{\mathrm{S}} + \mathbf{B}_{\mathrm{L}} = \frac{\mu_0}{4\pi}\left(\nabla \times \frac{\boldsymbol{\mu}_{\mathrm{e}} \times \mathbf{r}}{r^3} - \frac{2\mu_{\mathrm{B}}}{\hbar}\frac{\mathbf{p} \times \mathbf{r}}{r^3}\right) \tag{1.28}$$

is the magnetic field at the position $\mathbf{r}$ that is produced by the spin ($\mathbf{B}_{\mathrm{S}}$) and the orbital ($\mathbf{B}_{\mathrm{L}}$) motion of an electron possessing a magnetic moment $\boldsymbol{\mu}_{\mathrm{e}} = -2\mu_{\mathrm{B}}\mathbf{S}$ and a linear momentum $\mathbf{p}$ (μ_0: permeability of free space; μ_{B}: Bohr magneton; $\mathbf{S}$: electron-spin operator in units of $\hbar$). Note that eqn (1.26) represents only the leading term in an expansion of the electromagnetic neutron-sample interaction potential in powers of $m_{\mathrm{e}}/m_{\mathrm{n}}$, where m_{e} denotes the electron mass [76]. Higher-order contributions, e.g., due to the scattering by the atomic electric field [105], are several orders of magnitude smaller than the $-\boldsymbol{\mu}_{\mathrm{n}} \cdot \mathbf{B}$ term and can be ignored for the purpose of this book [75].

Based on the work by Bloch [106,107], Schwinger [108], and in particular by Halpern and Johnson [109], the theory of polarized thermal neutron scattering on condensed matter has been largely developed by Maleev et al. [110–112] and Blume [113] in the early 1960s. These so-called Blume–Maleev equations are commonly formulated in terms of discrete atomic spins rather than in the continuum picture, as is appropriate for magnetic SANS. We therefore write down first the general expression for the elastic magnetic differential scattering cross section and discuss subsequently the approximations necessary in order to obtain the magnetic SANS cross section. In addition to the standard references [76, 86, 103], we recommend the papers by Izyumov, de Gennes, and Maleev [114–116] for detailed accounts of polarized magnetic neutron scattering.

For an unpolarized incident neutron beam, the macroscopic elastic magnetic differential scattering cross section at scattering vector $\mathbf{q}$ due to an arrangement of N atomic magnetic moments $\boldsymbol{\mu}_{\mathrm{a}}$ at positions $\mathbf{r}_k$ is given by [76, 104, 117]:

$$\frac{d\Sigma_{\mathrm{M}}}{d\Omega}(\mathbf{q}) = \frac{1}{V}\left|\sum_{k=1}^{N} b_{\mathrm{m},k}\mathbf{Q}_k \exp\left(-i\mathbf{q} \cdot \mathbf{r}_k\right)\right|^2$$
$$= \frac{1}{V}\sum_{k,l} b_{\mathrm{m},k}b_{\mathrm{m},l}\mathbf{Q}_k\mathbf{Q}_l^* \exp\left(-i\mathbf{q} \cdot [\mathbf{r}_k - \mathbf{r}_l]\right), \tag{1.29}$$

where b_{m} is the atomic magnetic scattering length, which is sometimes denoted with the symbol p [104, 117], and

$$\mathbf{Q}_k = \hat{\mathbf{q}} \times (\hat{\mathbf{q}} \times \hat{\mathbf{m}}_k) = \hat{\mathbf{q}}(\hat{\mathbf{q}} \cdot \hat{\mathbf{m}}_k) - \hat{\mathbf{m}}_k \tag{1.30}$$

is the Halpern–Johnson vector [109] (also denoted as the magnetic interaction or magnetic scattering vector). This quantity depends on the unit scattering vector $\hat{\mathbf{q}} = \mathbf{q}/q = \{\hat{q}_x, \hat{q}_y, \hat{q}_z\}$ and on the unit vector $\hat{\mathbf{m}}_k = \boldsymbol{\mu}_{\mathrm{a},k}/\mu_{\mathrm{a},k}$ in the direction of a magnetic moment. The special dependency of $\mathbf{Q}_k$ on $\hat{\mathbf{q}}$ and on $\hat{\mathbf{m}}_k$ implies that for magnetic neutron scattering only the component of the magnetization which is perpendicular to the scattering vector is of relevance (compare Fig. 2.17). This reflects the dipolar nature of the neutron-magnetic interaction. The atomic magnetic scattering length is given by [117, 118]:

$$b_{\mathrm{m}} = \frac{\gamma_{\mathrm{n}}r_{\mathrm{e}}}{2}\frac{\mu_{\mathrm{a}}}{\mu_{\mathrm{B}}}f(\mathbf{q}) \cong 2.70 \times 10^{-15}\mathrm{m}\frac{\mu_{\mathrm{a}}}{\mu_{\mathrm{B}}}f(\mathbf{q}) \cong b_{\mathrm{H}}\mu_{\mathrm{a}}, \tag{1.31}$$

Table 1.2 Atomic magnetic scattering lengths b_m in the SANS regime (computed using eqn (1.31) with $f(\mathbf{q}) = 1$), Curie temperatures T_C, and atomic magnetic moments of various elements exhibiting ferromagnetism in bulk form. T_C and μ_a/μ_B-values taken from [119, 120].

Element	Crystal Structure (at RT)	T_C (K)	μ_a/μ_B	b_m (10^{-15} m)
Fe	bcc	1043	2.217	5.98
Co	hcp	1403	1.721	4.64
Ni	fcc	627	0.616	1.66
Gd	hcp	293	7.63	20.57
Tb	hcp	220	9.34	25.18
Dy	hcp	89	10.33	27.84
Ho	hcp	20	10.34	27.87
Er	hcp	20	9.1	24.53
Tm	hcp	32	7.14	19.25

where $r_e = 2.818 \times 10^{-15}$ m denotes the classical radius of the electron, and $f(\mathbf{q})$ is the normalized atomic magnetic form factor. Note that

$$f(\mathbf{q}) \cong 1 \tag{1.32}$$

in the small-angle region, which is the approximation which we will adopt throughout this book. Table 1.2 provides the b_m-values of some of the most important elemental magnetic materials. As can be seen, b_m has the same order of magnitude as the nuclear scattering length and can be quite large for the heavy rare-earth metals (see also Fig. 2.11). Equation (1.31) also defines the constant [104]

$$b_H = 2.70 \times 10^{-15} \, \text{m} \, \mu_B^{-1} = 2.91 \times 10^8 \, \text{A}^{-1} \text{m}^{-1}. \tag{1.33}$$

The magnitude of $\mathbf{Q}_k$ for a single magnetic moment equals

$$|\mathbf{Q}_k|^2 = \sin^2 \alpha_k, \tag{1.34}$$

where α_k is the angle enclosed between $\hat{\mathbf{m}}_k$ and $\hat{\mathbf{q}}$. This simple result highlights the anisotropic character of magnetic neutron scattering and demonstrates that, for a single uniformly magnetized domain, the magnetic scattering vanishes for $\hat{\mathbf{q}} \parallel \hat{\mathbf{m}}_k$ and is at maximum for $\hat{\mathbf{q}} \perp \hat{\mathbf{m}}_k$ [see Fig. 1.12(b)].

As with nuclear SANS, eqn (1.29) needs to be transformed into a continuum picture. The quantity of interest for magnetic SANS is the magnetization vector field of the sample, $\mathbf{M} = \mathbf{M}(\mathbf{r})$, which, similarly to $N(\mathbf{r})$, is assumed to be a smooth and continuous function of $\mathbf{r}$. The definition of $\mathbf{M}$ (see Section 3.1) entails that this quantity represents an average over the thermal fluctuations of many individual atomic magnetic moments $\boldsymbol{\mu}_{\mathrm{a}}$ within a domain. Moreover, since the magnetic moment of an atom is due to the spin and the orbital motions of the electrons, these contributions are included in the definition of the magnetization. The magnitude of $\mathbf{M}$ is called the saturation magnetization $M_{\mathrm{s}} = |\mathbf{M}|$, which is assumed to be only a function of temperature, i.e., $M_{\mathrm{s}} = M_{\mathrm{s}}(T)$. Inserting eqns (1.30)−(1.33) into eqn (1.29), replacing the sum by an integral over the sample volume and the discrete terms $\mu_{\mathrm{a},k}\mathbf{Q}_k = \hat{\mathbf{q}}(\hat{\mathbf{q}} \cdot \boldsymbol{\mu}_{\mathrm{a},k}) - \boldsymbol{\mu}_{\mathrm{a},k}$ by the corresponding continuum expression $\mathbf{Q}(\mathbf{r}) = \hat{\mathbf{q}}[\hat{\mathbf{q}} \cdot \mathbf{M}(\mathbf{r})] - \mathbf{M}(\mathbf{r})$, the unpolarized elastic magnetic differential SANS cross section reads [76]:

$$
\begin{aligned}
\frac{d\Sigma_{\mathrm{M}}}{d\Omega}(\mathbf{q}) &= \frac{1}{V}b_{\mathrm{H}}^2 \left| \int_V \mathbf{Q}(\mathbf{r}) \exp\left(-i\mathbf{q}\cdot\mathbf{r}\right) d^3r \right|^2 \\
&= \frac{8\pi^3}{V}b_{\mathrm{H}}^2 |\widetilde{\mathbf{Q}}|^2 \\
&= \frac{8\pi^3}{V}b_{\mathrm{H}}^2 \left| \hat{\mathbf{q}} \times \left(\hat{\mathbf{q}} \times \widetilde{\mathbf{M}}(\mathbf{q}) \right) \right|^2 \\
&= \frac{8\pi^3}{V}b_{\mathrm{H}}^2 \sum_{\alpha,\beta} \left(\delta_{\alpha\beta} - \hat{q}_\alpha\hat{q}_\beta \right) \widetilde{M}_\alpha \widetilde{M}_\beta,
\end{aligned} \tag{1.35}
$$

where α and β stand for the Cartesian coordinates x, y, z, and $\delta_{\alpha\beta}$ is the Kronecker delta function. The three-dimensional Fourier-transform pair of the magnetization is given by:

$$
\begin{aligned}
\widetilde{\mathbf{M}}(\mathbf{q}) &= \left\{ \widetilde{M}_x(\mathbf{q}), \widetilde{M}_y(\mathbf{q}), \widetilde{M}_z(\mathbf{q}) \right\} \\
&= \frac{1}{(2\pi)^{3/2}} \int_{-\infty}^{+\infty}\int_{-\infty}^{+\infty}\int_{-\infty}^{+\infty} \mathbf{M}(\mathbf{r}) \exp\left(-i\mathbf{q}\cdot\mathbf{r}\right) d^3r,
\end{aligned} \tag{1.36}
$$

$$
\begin{aligned}
\mathbf{M}(\mathbf{r}) &= \left\{ M_x(\mathbf{r}), M_y(\mathbf{r}), M_z(\mathbf{r}) \right\} \\
&= \frac{1}{(2\pi)^{3/2}} \int_{-\infty}^{+\infty}\int_{-\infty}^{+\infty}\int_{-\infty}^{+\infty} \widetilde{\mathbf{M}}(\mathbf{q}) \exp\left(i\mathbf{q}\cdot\mathbf{r}\right) d^3q,
\end{aligned} \tag{1.37}
$$

where $i^2 = -1$, and $\mathbf{q} = \{q_x, q_y, q_z\}$ is the wave vector.[†] $\widetilde{\mathbf{Q}} = \hat{\mathbf{q}} \times (\hat{\mathbf{q}} \times \widetilde{\mathbf{M}})$ is a linear vector function of $\widetilde{\mathbf{M}}$. Both $\widetilde{\mathbf{Q}}(\mathbf{q})$ and $\widetilde{\mathbf{M}}(\mathbf{q})$ are in general complex vectors. Regarding the dimensions of the involved quantities in eqn (1.35): $\mathbf{M}(\mathbf{r})$ and $\mathbf{Q}(\mathbf{r})$ have dimensions

---

[†]The Fourier transform of the magnetization is a function of the wave vector $\mathbf{q}$, whereas the magnetic SANS cross section $d\Sigma_{\mathrm{M}}/d\Omega$ is a function of the scattering vector, say $\mathbf{q}'$, which is the difference between the incident and scattered wave vectors of the neutron (compare Fig. 2.1). Contrary to $\mathbf{q}$, the quantity $\mathbf{q}'$ is not a wave vector, since its magnitude for elastic scattering is not given by $2\pi/\lambda$. However, $d\Sigma_{\mathrm{M}}/d\Omega$ at scattering vector $\mathbf{q}'$ depends exclusively on the Fourier components of the magnetization at wave vector $\mathbf{q} = \mathbf{q}'$, and a consistent separate handling of the symbols $\mathbf{q}$ and $\mathbf{q}'$ would unnecessarily encumber the discussion. In agreement with common usage in the neutron scattering literature, we ignore the distinction between the two quantities and use the symbol $\mathbf{q}$ to denote both the wave vector and the scattering vector.

of A/m, so that $\widetilde{\mathbf{M}}(\mathbf{q})$ and $\widetilde{\mathbf{Q}}(\mathbf{q})$ are in Am^2, with the consequence that $d\Sigma_{\mathrm{M}}/d\Omega$ has dimensions of m^{-1}. We also mention that in a small-angle scattering experiment only correlations in the plane perpendicular to the incident neutron beam (with wave vector $\mathbf{k}_0$) are probed. This implies that the component of the scattering vector $\hat{\mathbf{q}}$ along $\mathbf{k}_0$ is ignored, i.e., the three-dimensional $\hat{\mathbf{q}}$ is approximated by a two-dimensional one (see Section 2.1).

Unpolarized SANS cross section. The nuclear SANS cross section [eqn (1.25)] adds to $d\Sigma_{\mathrm{M}}/d\Omega$, so that the total (nuclear and magnetic) unpolarized SANS cross section is given by:

$$\frac{d\Sigma}{d\Omega}(\mathbf{q}) = \frac{d\Sigma_{\mathrm{nuc}}}{d\Omega}(\mathbf{q}) + \frac{d\Sigma_{\mathrm{M}}}{d\Omega}(\mathbf{q}) = \frac{8\pi^3}{V}\left(|\widetilde{N}|^2 + b_{\mathrm{H}}^2|\widetilde{\mathbf{Q}}|^2\right). \qquad (1.38)$$

With reference to eqn (1.34) one can then see that for a saturated single-domain magnetic microstructure ($|\widetilde{\mathbf{Q}}|^2 \propto \sin^2\alpha$), the unpolarized scattering along the direction of magnetization ($\mathbf{q} \parallel \mathbf{M}$) is exclusively nuclear in origin, while $d\Sigma/d\Omega$ for $\mathbf{q} \perp \mathbf{M}$ contains both the nuclear and magnetic scattering. Therefore, if it becomes possible by the application of a strong external magnetic field $\mathbf{H}_0$ perpendicular to the beam to completely saturate a magnetic material, then this provides a means to separate nuclear and magnetic scattering. However, we emphasize that when $\mathbf{H}_0$ is not large enough to completely saturate the sample, then the scattering of unpolarized neutrons for $\mathbf{q} \parallel \mathbf{H}_0$ does not represent the pure nuclear SANS, but also contains the magnetic SANS due to the misaligned transversal spins [121]. This is a very delicate issue and we recommend to carry out experiments with increasing values of the "supposed" saturation field (see Appendix A). The mere visual inspection of a measured hysteresis loop and the assertion that the horizontal saturation regime has been reached might be misleading in this context (see also the discussion in [122]). We also refer the reader to Section 6.5.2, where the field-dependent unpolarized SANS of nanocrystalline cobalt is discussed. In Fig. 6.16, the field-dependent SANS is compared to the magnetization curve and one can see that, even in the saturation regime, $d\Sigma/d\Omega$ exhibits an extraordinarily large field variation.

Polarized SANS cross section. When the incident neutron beam is polarized, additional nuclear-magnetic and magnetic-magnetic interference terms may appear in the SANS cross section. We refer to Fig. 2.1 in Section 2.1, which depicts the typical setup of a polarized SANS experiment. The polarization $\mathbf{P}$ of a neutron beam containing N_{s} spins can be defined as the average over the individual polarizations $\mathbf{P}_j$ of the neutrons as [104]:

$$\mathbf{P} = \frac{1}{N_{\mathrm{s}}}\sum_{j=1}^{N_{\mathrm{s}}}\mathbf{P}_j, \qquad (1.39)$$

where $0 \leq |\mathbf{P}| \leq 1$. In experimental SANS studies the beam is usually partially polarized along a certain guide-field direction (quantization axis), which we take here, unless otherwise stated, as the z-direction. Assuming that the expectation values of the perpendicular polarization components vanish, i.e., $P_x = P_y = 0$, and that $P_z = P$, one can then introduce the fractions

$$p^+ = \frac{1}{2}(1 + P) \quad \text{and} \quad p^- = \frac{1}{2}(1 - P) \tag{1.40}$$

of neutrons in the spin-up $(+)$ and spin-down $(-)$ state, with

$$p^+ + p^- = 1 \quad \text{and} \quad p^+ - p^- = P. \tag{1.41}$$

Obviously, for an unpolarized beam $p^+ = p^- = 0.5$ and $P = 0$, while $P = +1$ $(p^+ = 1)$ and $P = -1$ $(p^- = 1)$ for a fully polarized beam.

When there is an additional analyzer behind the sample, configured such that it selects only neutrons with spins either parallel or antiparallel to the initial polarization, then one can distinguish four scattering cross sections (scattering processes) [104,113, 117]: two of which conserve the neutron-spin direction $(++$ and $--)$, the so-called non-spin-flip cross sections

$$\frac{d\Sigma^{++}}{d\Omega} = \frac{8\pi^3}{V}b_{\mathrm{H}}^2\left[b_{\mathrm{H}}^{-2}|\widetilde{N}|^2 + b_{\mathrm{H}}^{-1}(\widetilde{N}\widetilde{Q}_z^* + \widetilde{N}^*\widetilde{Q}_z) + |\widetilde{Q}_z|^2\right],$$

$$\frac{d\Sigma^{--}}{d\Omega} = \frac{8\pi^3}{V}b_{\mathrm{H}}^2\left[b_{\mathrm{H}}^{-2}|\widetilde{N}|^2 - b_{\mathrm{H}}^{-1}(\widetilde{N}\widetilde{Q}_z^* + \widetilde{N}^*\widetilde{Q}_z) + |\widetilde{Q}_z|^2\right], \tag{1.42}$$

and two cross sections which reverse the neutron spin $(+-$ and $-+)$, the so-called spin-flip cross sections

$$\frac{d\Sigma^{+-}}{d\Omega} = \frac{8\pi^3}{V}b_{\mathrm{H}}^2\left[|\widetilde{Q}_x|^2 + |\widetilde{Q}_y|^2 - i\mathbf{e}_z \cdot (\widetilde{\mathbf{Q}} \times \widetilde{\mathbf{Q}}^*)\right],$$

$$\frac{d\Sigma^{-+}}{d\Omega} = \frac{8\pi^3}{V}b_{\mathrm{H}}^2\left[|\widetilde{Q}_x|^2 + |\widetilde{Q}_y|^2 + i\mathbf{e}_z \cdot (\widetilde{\mathbf{Q}} \times \widetilde{\mathbf{Q}}^*)\right]. \tag{1.43}$$

These expressions, written here in terms of the Cartesian components of $\widetilde{\mathbf{Q}}$, will be discussed in more detail in Section 2.5, where they are explicitly displayed in terms of the Cartesian components of $\widetilde{\mathbf{M}}$. The latter representation is more intuitive for comparing the cross sections to experimental data and to the outcome of micromagnetic computations of the real-space spin structure. We also emphasize that nuclear-spin-dependent SANS is not taken into account in this book, so that the corresponding scattering contributions do not show up in eqns (1.42) and (1.43). The particular experimental arrangement when the neutrons are polarized and analyzed along the same guide-field direction is denoted as uniaxial (sometimes also as one-dimensional or longitudinal) polarization analysis. As emphasized by Moon, Riste, and Koehler [117], in uniaxial polarization-analysis experiments, one measures these four scattering cross sections which connect the two neutron-spin states. From these measurements the polarization of the scattered beam may be obtained. Three-dimensional polarization analysis on a SANS instrument, which generally demands for a zero magnetic field environment, has been demonstrated a long time ago (e.g., [123–126]). The technique has been used to characterize the magnetic domain structure and to study the critical spin dynamics in ferromagnets. However, the state-of-the-art in "routine" SANS polarization-analysis experiments is the above-sketched uniaxial case, where a large field can be applied to the sample. The progress made with the development of efficient ^{3}He spin filters [127] has decisively contributed to this (see Section 2.5 for further details).

The total SANS cross section $d\Sigma/d\Omega$ can be expressed in terms of the initial spin populations $p^\pm$ as [104, 113, 117]

$$\frac{d\Sigma}{d\Omega} = p^+ \left(\frac{d\Sigma^{++}}{d\Omega} + \frac{d\Sigma^{+-}}{d\Omega} \right) + p^- \left(\frac{d\Sigma^{--}}{d\Omega} + \frac{d\Sigma^{-+}}{d\Omega} \right). \tag{1.44}$$

Inserting the above expressions for p^+ and p^- [eqns (1.40) and (1.41)] and for the partial SANS cross sections $d\Sigma^{\pm\pm}/d\Omega$ and $d\Sigma^{\pm\mp}/d\Omega$ [eqns (1.42) and (1.43)], eqn (1.44) evaluates to:

$$\frac{d\Sigma}{d\Omega} = \frac{8\pi^3}{V} b_{\mathrm{H}}^2 \left[b_{\mathrm{H}}^{-2} |\widetilde{N}|^2 + |\widetilde{\mathbf{Q}}|^2 + \mathbf{P} \cdot b_{\mathrm{H}}^{-1} (\widetilde{N}\widetilde{\mathbf{Q}}^* + \widetilde{N}^*\widetilde{\mathbf{Q}}) - i\mathbf{P} \cdot (\widetilde{\mathbf{Q}} \times \widetilde{\mathbf{Q}}^*) \right], \tag{1.45}$$

which, using $\mathbf{P} = \{0, 0, P_z = \pm P\}$, can be rewritten as:

$$\frac{d\Sigma^\pm}{d\Omega} = \frac{8\pi^3}{V} b_{\mathrm{H}}^2 \left[b_{\mathrm{H}}^{-2} |\widetilde{N}|^2 + |\widetilde{\mathbf{Q}}|^2 \pm P b_{\mathrm{H}}^{-1} (\widetilde{N}\widetilde{Q}_z^* + \widetilde{N}^*\widetilde{Q}_z) \mp i P \mathbf{e}_z \cdot (\widetilde{\mathbf{Q}} \times \widetilde{\mathbf{Q}}^*) \right]. \tag{1.46}$$

Since the cross section is a scalar quantity and the polarization is an axial vector (or pseudovector), eqn (1.45) shows that the system under study must itself contain an axial vector. As emphasized by Maleev [116], examples for such built-in pseudovectors are related to the interaction of a polycrystalline sample with an external magnetic field (inducing an average magnetization directed along the applied field), the existence of a spontaneous magnetization in a ferromagnetic single crystal, the antisymmetric Dzyaloshinskii–Moriya interaction, mechanical (torsional) deformation, or the presence of spin spirals. If on the other hand there is no preferred axis in the system, then $d\Sigma/d\Omega$ is independent of $\mathbf{P}$. Examples include a collection of randomly oriented non-interacting nuclear (electronic) spins, which describe the general case of nuclear (paramagnetic) scattering at not-too-low temperatures and large applied fields, or a multi-domain ferromagnet with a random distribution of the domains. The same condition—existence of an axial system vector—applies for neutrons to be polarized in the scattering process [compare the last two terms in eqn (1.50)]. We also refer to Section 2.8.3, where spin-polarized SANS data on a ferromagnetic nanocrystalline alloy are discussed.

In the domain of magnetic SANS it is customary to denote experiments with a polarized incident beam only, and no spin analysis of the scattered neutrons, with the acronym SANSPOL. For spin-resolved SANS experiments (uniaxial polarization analysis) the term POLARIS is commonly used. The two SANSPOL cross sections $d\Sigma^+/d\Omega$ and $d\Sigma^-/d\Omega$ (also sometimes denoted as the half-polarized SANS cross sections) combine non-spin-flip and spin-flip scattering contributions, according to ($p^\pm = 1$):

$$\frac{d\Sigma^+}{d\Omega} = \frac{d\Sigma^{++}}{d\Omega} + \frac{d\Sigma^{+-}}{d\Omega}, \tag{1.47}$$

$$\frac{d\Sigma^-}{d\Omega} = \frac{d\Sigma^{--}}{d\Omega} + \frac{d\Sigma^{-+}}{d\Omega}. \tag{1.48}$$

Finally, noting that an unpolarized beam can be viewed as consisting of 50% spin-up and 50% spin-down neutrons [compare eqns (1.40) and (1.41)], the unpolarized SANS cross section is obtained as [compare eqn (1.44) and see Section 2.6]:

$$\frac{d\Sigma}{d\Omega} = \frac{1}{2}\left(\frac{d\Sigma^{++}}{d\Omega} + \frac{d\Sigma^{+-}}{d\Omega} + \frac{d\Sigma^{--}}{d\Omega} + \frac{d\Sigma^{-+}}{d\Omega}\right) = \frac{1}{2}\left(\frac{d\Sigma^{+}}{d\Omega} + \frac{d\Sigma^{-}}{d\Omega}\right). \tag{1.49}$$

Polarization of the scattered beam. Although not further used in this book, we display for completeness also the equation for the polarization of the scattered neutrons (uniaxial case). The polarization P' of the scattered beam along the direction of the incident neutron polarization P is obtained from the following relation [104, 113, 117]:

$$
\begin{aligned}
P'\frac{d\Sigma}{d\Omega} = {} & p^{+}\frac{d\Sigma^{++}}{d\Omega} + p^{-}\frac{d\Sigma^{-+}}{d\Omega} - p^{-}\frac{d\Sigma^{--}}{d\Omega} - p^{+}\frac{d\Sigma^{+-}}{d\Omega} \\
= {} & \frac{8\pi^3}{V}b_H^2 P\left[b_H^{-2}|\tilde{N}|^2 + |\tilde{Q}_z|^2 - |\tilde{Q}_x|^2 - |\tilde{Q}_y|^2\right] \\
& + \frac{8\pi^3}{V}b_H^2\left[b_H^{-1}(\tilde{N}\tilde{Q}_z^* + \tilde{N}^*\tilde{Q}_z) + i\mathbf{e}_z\cdot(\tilde{\mathbf{Q}}\times\tilde{\mathbf{Q}}^*)\right],
\end{aligned}
\tag{1.50}
$$

where $d\Sigma/d\Omega$ is given by eqn (1.44), and the polarization fractions and partial cross sections are again expressed using eqns (1.40)–(1.43). The first four terms in the second line on the right-hand side of eqn (1.50) demonstrate that nuclear scattering (to be more precise, the nuclear coherent scattering, the isotopic disorder scattering, and 1/3 of the nuclear-spin-dependent scattering) and scattering due to the longitudinal component $\tilde{Q}_z$ of the magnetic scattering vector $\tilde{\mathbf{Q}}$ do not reverse the initial polarization, while the two transversal components $\tilde{Q}_x$ and $\tilde{Q}_y$ give rise to spin-flip scattering. The last two terms in eqn (1.50) do create polarization: these are the familiar nuclear-magnetic interference terms $(\tilde{N}\tilde{Q}_z^* + \tilde{N}^*\tilde{Q}_z)$, which are commonly used to polarize beams, and the chiral term $i\mathbf{e}_z\cdot(\tilde{\mathbf{Q}}\times\tilde{\mathbf{Q}}^*)$, which is of relevance in spin-wave scattering, in elastic scattering on spiral structures, or in the presence of the Dzyaloshinskii–Moriya interaction in microstructural-defect-rich magnets (see Section 4.3.3). We remind that nuclear-spin-dependent scattering is not taken into account in the expressions for the magnetic SANS cross sections, although Section 2.4.2 features some basic results combining SANS with the technique of dynamic nuclear polarization. In the general expression for the polarization of the scattered neutrons, a term $i\mathbf{P}\times(\tilde{N}\tilde{\mathbf{Q}}^* - \tilde{N}^*\tilde{\mathbf{Q}})$ appears [104], which is ignored in eqn (1.50). This term rotates the polarization perpendicular to the initial polarization and cannot be observed in the uniaxial setup. We emphasize that in linear neutron polarimetry it is not possible to distinguish between a rotation of the polarization vector and a change of its length [116, 117].

1.5   Diffraction versus refraction

In the context of the validity of the Born approximation for SANS studies it is worth referring to the early investigation of Weiss [128], who computed the differential and the total scattering cross section for the scattering of slow neutrons on a single spherical particle with a diameter very much larger than the neutron wavelength. Weiss's treatment reveals that for $\phi_{ps} \ll 1$ the result for the cross section agrees with the solution obtained using the Born approximation (diffraction regime), while for $\phi_{ps} \gg 1$ the result agrees with the one obtained using the methods of geometrical optics (refraction regime). The parameter ϕ_{ps} denotes the difference between the neutron phase shift in traversing the particle diameter D and the phase shift in traversing the same distance in vacuum. For normal neutron incidence, it is given by the following relation:

$$\phi_{ps} = kD \left(n - 1 \right), \tag{1.51}$$

where $k = 2\pi/\lambda$ is the neutron wave vector outside the sample (vacuum), $K = nk$ is the wave vector inside the medium, and

$$n^2 = \left(\frac{K}{k} \right)^2 = 1 - \frac{\lambda^2}{\pi} \rho_a \bar{b} \tag{1.52}$$

is the square of the index of refraction, with ρ_a the number density of nuclei and $\bar{b}$ the mean-bound coherent scattering length [82]; $\langle N \rangle = \rho_a \bar{b}$ is also denoted as the mean nuclear scattering-length density. As discussed in the paper by Weiss [128], van de Hulst [129] has shown that the cases $\phi_{ps} \ll 1$ and $\phi_{ps} \gg 1$ correspond to different limiting forms of the same scattering problem: $\phi_{ps} \ll 1$ corresponds to the Rayleigh–Gans theory of diffraction [130, 131], while $\phi_{ps} \gg 1$ corresponds to the refraction theory of von Nardroff [132]. For most materials $\bar{b}$ is real and positive, so that $n < 1$, suggesting that the phenomenon of total reflection can occur if the neutrons enter the medium from vacuum. The deviation of n from unity, $|n - 1|$, is for thermal and cold neutrons of the order of 10^{-5}–10^{-6} [118]. Figure 1.4 displays $\phi_{ps}(D)$ for $\lambda = 6$ Å and for several values of $1 - n$. It is seen that for particle sizes below a few 100 nm the condition $\phi_{ps} \ll 1$ is fulfilled for cold neutrons.

For ferromagnetic materials the index of refraction is double-valued (birefringant). As shown e.g., by Sears [82], it can be obtained by replacing $\bar{b}$ in eqn (1.52) with $\bar{b} \pm b_m$, i.e.,

$$n_{\pm}^2 = 1 - \frac{\lambda^2}{\pi} \rho_a \left(\bar{b} \pm b_m \right), \tag{1.53}$$

where b_m denotes the atomic magnetic scattering length [eqn (1.31)], and the "+" and "−" signs refer, respectively, to neutrons with spin parallel or antiparallel to the magnetization. Equation (1.53) is the basis for the production of polarized neutron beams and for the experimental determination of scattering lengths; see, e.g., [76, 103, 118] for a further discussion of this aspect of eqn (1.53). We also refer to the treatment by Schaerpf [133], who has analyzed two ways for solving the problem of the dependence of the refractive index on the direction of magnetization on both sides of the refractive boundary, one employing the Halpern–Johnson vector, the other by applying the dynamical theory of diffraction. Since b_m is of the same order of magnitude as $\bar{b}$ for many magnetic materials, similar conclusions as in the nuclear case are expected to hold [128]: as a rule of thumb, for magnetic domain sizes $D = D_{ds}$ below a few 100 nm, diffraction is generally the dominant process ($\phi_{ps} \ll 1$), while for micron-sized domains refraction may occur ($\phi_{ps} \gg 1$).

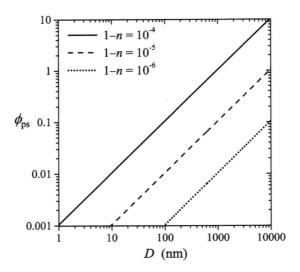

Fig. 1.4: The parameter ϕ_{ps} [eqn (1.51)] as a function of D for cold neutrons ($\lambda = 6\,\text{Å}$) and for several values of $1 - n$ (see inset) (log-log scale).

The question which remains to be discussed is the meaning of D_{ds} for a ferromagnetic material. This depends e.g., on the specific microstructure (single crystalline, polycrystalline coarse-grained, or nanocrystalline), on the magnetic interactions and the anisotropy energy (hard or soft magnetic), and on the value of the applied magnetic field. It is well known that many ferromagnets at small applied fields exhibit a multi-domain structure with domain sizes in the μm–mm range (see Fig. 1.5). In an idealized picture the domains are uniformly magnetized to saturation and domain walls with a thickness of the order of $l_K = \sqrt{A/K}$ separate the domains (A: exchange-stiffness constant; K: anisotropy constant). For nickel $l_K \cong 40\,\text{nm}$, while $l_K \cong 1\,\text{nm}$ for $\text{Nd}_2\text{Fe}_{14}\text{B}$ [134]. On top of the above two length scales, the domain size D_{ds} and the domain-wall width l_K, there is at least the average crystallite (grain) or particle size D_{gs} in a polycrystalline material. This quantity decorates random spatial variations in the magnetic anisotropy field due to spatial variations in the set of crystallographic easy axes at internal interfaces (e.g., grain boundaries). Such variations in the anisotropy field may give rise to deviations of the magnetization from the mean magnetization direction within a domain; in other words, and more generally speaking, due to the inevitable presence of microstructural defects (e.g., vacancies, dislocations, grain boundaries, pores) in polycrystalline magnets, the individual domains may not be uniformly magnetized (see the discussion in Section 1.6). The size of the associated spin-misaligned regions is characterized by the exchange length l_H, which is a function of the applied field; typically, $l_H \sim 1$–$100\,\text{nm}$ [compare Fig. 1.7 and Fig. 3.5(a)].

When a finely collimated neutron beam is scattered off a macroscopic domain structure at not-too-large applied fields, multiple refraction at the magnetic domain boundaries occurs and one observes a concomitant broadening of the incident neutron beam. This feature, along with probably the first experimental observation of magnetic SANS,

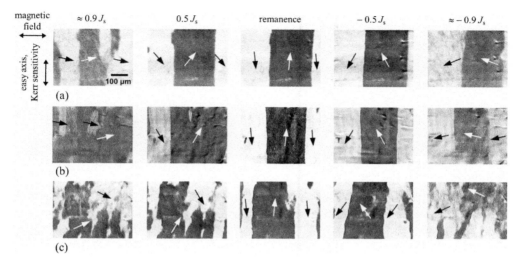

Fig. 1.5: Kerr microscopy images illustrating the magnetization process in soft magnetic transverse-field-annealed alloys on a $100\,\mu$m length scale: (a) an amorphous material with an induced magnetic anisotropy of $K_{\mathrm{u}} \approx 0.6\,\mathrm{J/m^3}$; (b) a nanocrystalline material with a high induced anisotropy ($K_{\mathrm{u}} \approx 30\,\mathrm{J/m^3}$); (c) a nanocrystalline alloy with a low induced anisotropy ($K_{\mathrm{u}} \approx 3\,\mathrm{J/m^3}$). Arrows indicate the magnetization direction, which is estimated from inductive magnetization measurements in the dependence of the applied magnetic field. The magnetization is given in fractions of the respective saturation polarization $J_{\mathrm{s}} = \mu_0 M_{\mathrm{s}}$. After [135].

has been discovered early on in neutron research by Hughes et al. [136, 137]. These authors have performed neutron experiments at small scattering angles on an unmagnetized and a magnetized polycrystalline iron block and demonstrated the presence of two indices of refraction for neutrons. It was suggested that the magnetic refraction of neutrons at domain boundaries may be used to investigate the magnetic domain structure.‡ Along with refraction, the intra-domain nanoscale spin disorder may result in a magnetic neutron scattering signal along the forward direction, i.e., the conventional diffuse magnetic SANS. Increasing the external field gradually transforms the multidomain structure into a quasi-single-domain state. Within the approach-to-saturation regime one then considers small spin deviations (caused e.g., by defects) relative to the applied-field direction with a characteristic wavelength given by $l_{\mathrm{H}} \sim 1-100\,\mathrm{nm}$. The dominant scattering process in this case is magnetic small-angle scattering. We will now proceed to the discussion of the origins of magnetic SANS.

---

‡Nowadays, neutron depolarization is the standard technique for determining the average domain size and domain magnetization, and the mean-square-direction cosines of the domains. When a polarized neutron beam is traversing a magnetic material, then the beam may be depolarized due to the Larmor precession of the neutron spin in the changing magnetic induction field of the domain structure. Three-dimensional neutron depolarization is an important method for studying the static and dynamic properties of magnetic structures in the micrometer and submicrometer regime, including superconductors and critical phenomena (e.g., [103, 123, 124, 138–151]).

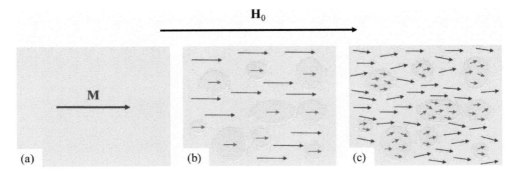

Fig. 1.6: Simplified sketches of the magnetization distribution. (a) A homogeneous and uniformly magnetized ferromagnet; (b) an inhomogeneous and uniformly magnetized material; (c) an inhomogeneous and nonuniformly magnetized magnet. The external magnetic field $\mathbf{H}_0$ is applied along the $\mathbf{e}_z$-direction of a Cartesian laboratory coordinate system. Effects due the external sample surface are ignored. Note that in (b) the magnetization vectors of the particle and matrix phase have a different length but the same orientation. Case (c) reduces to (b)-type for complete magnetic saturation (infinite applied field H_0).

1.6   Origin of magnetic SANS: relation to magnetic microstructure

The magnetization vector field $\mathbf{M}$ is the quantity which is most relevant for magnetic SANS. The central part of this book deals with the computation and the experimental investigation of the elastic magnetic SANS cross section, eqn (1.35), which in the Born approximation exclusively depends on the Fourier components of $\mathbf{M}$. These functions carry the information on the underlying physics of the problem, since they depend on the relevant parameters such as the momentum-transfer vector $\mathbf{q}$, the applied magnetic field $\mathbf{H}_0$, the magnetic interactions [symmetric (A) and antisymmetric (D) exchange, magnetocrystalline (K) and magnetoelastic ($\boldsymbol{\sigma}$) anisotropy, magnetostatics M_s], the particle-size or crystallite-size distribution $f(R)$, or the crystallographic texture; in other words,

$$\widetilde{M}_{x,y,z} = \widetilde{M}_{x,y,z}(\mathbf{q}; \mathbf{H}_0, A, D, K, \boldsymbol{\sigma}, M_s, f(R), \ldots). \tag{1.54}$$

In the following, we discuss the various origins of magnetic SANS and we highlight some important differences between magnetic SANS and conventional particle scattering. Some of the ideas which are presented in this section require some basic knowledge about magnetism as well as about nuclear SANS. We refer the reader to Sections 2.4 and 3.1 for a quick reference, or to the standard magnetism (e.g., [152–155]) and small-angle scattering [9–13] textbooks for more detailed information. We aim to keep the discussion on a qualitative level and guided by physical intuition.

Magnetic SANS has its origin in nanometer-scale variations in both the orientation and/or magnitude of the magnetization vector $\mathbf{M}(\mathbf{r})$. Figure 1.6 illustrates this statement for several magnetic microstructures. The simplest case is the homogeneous and uniformly magnetized ferromagnet [Fig. 1.6(a)]. Here, both the magnitude $M_s = |\mathbf{M}|$ and the direction of $\mathbf{M} \parallel \mathbf{H}_0 \parallel \mathbf{e}_z$ are the same at each point $\mathbf{r}$ inside the material.

The magnetization distribution is given by $\mathbf{M} = \{0, 0, M_s\}$. Consequently, there exists no nanometer-scale contrast for magnetic SANS (no deviation from a mean value regarding both magnitude and direction of $\mathbf{M}$) and the magnetic SANS cross section reduces to a delta function at the origin of reciprocal space,

$$\frac{d\Sigma_M}{d\Omega}(\mathbf{q}) \propto |\delta(\mathbf{q})|^2. \tag{1.55}$$

For a fully saturated but inhomogeneous system, such as the two-phase structure which is sketched in Fig. 1.6(b), we have $\mathbf{M} = \{0, 0, M_z = M_s(\mathbf{r})\}$, and the magnetic SANS is exclusively due to spatial variations in the saturation magnetization. Hence, $d\Sigma_M/d\Omega$ is determined by the Fourier transform $\widetilde{M}_s(\mathbf{q})$ of $M_s(\mathbf{r})$ [compare eqn (1.35)],

$$\frac{d\Sigma_M}{d\Omega}(\mathbf{q}) \propto |\widetilde{M}_s(\mathbf{q})|^2. \tag{1.56}$$

Prototypical microstructures for this scenario are e.g., a distribution of saturated or single-domain nanoparticles in a nonmagnetic or saturated matrix (of different magnetization), or pores in a saturated matrix. Analysis of $d\Sigma_M/d\Omega$ then provides information on the size and shape of the particles, on the magnetic contrast between particle and matrix (difference in M_s), and on the arrangement and interaction potential between the particles. Figure 1.6(c) depicts the most general case of an inhomogeneous and nonuniformly magnetized magnetic material: the magnetization exhibits spatial variations in the magnitude and direction, and the magnetic SANS cross section is given by eqn (1.35). Examples for such materials are polycrystalline elemental magnets with a nanometer crystallite size, heavily deformed (cold-worked) metals, nanoporous ferromagnets, or multiphase magnetic nanocomposites, including magnetic steels, shape-memory alloys, and permanent magnets. Only in the completely saturated state is the $d\Sigma_M/d\Omega$ of these materials described by eqn (1.56).

As far as the analysis procedure is concerned, there is an important difference between the cases (b) and (c). As we will see in Chapter 3 and in Section 5.1.3, the problem of finding the vectorial magnetization distribution

$$\mathbf{M}(\mathbf{r}) = \left\{ \begin{array}{l} M_x(x, y, z) \\ M_y(x, y, z) \\ M_z(x, y, z) \end{array} \right\}, \tag{1.57}$$

which is subject to the constraint of constant magnitude $|\mathbf{M}| = M_s$, consists of the solution of a set of nonlinear partial differential equations with complex boundary conditions. By contrast, the problem which is embodied in case (b) amounts to finding the scalar function $M_s(\mathbf{r})$ [eqn (1.56)], which is determined by the geometry of the microstructure; for instance, for a dilute scattering system consisting of saturated nanoparticles in a homogeneous nonmagnetic matrix or in a homogeneous magnetically saturated matrix, $d\Sigma_M/d\Omega$ is determined by the form factor of the particle. Problem (b) is therefore fully analogous to the usual problem, encountered in nuclear SANS and SAXS, of determining the scalar scattering-length density distribution (see Section 2.4.3); in other words, the well-known procedures of nuclear SANS can be applied in order to analyze the idealized situation of a saturated microstructure.

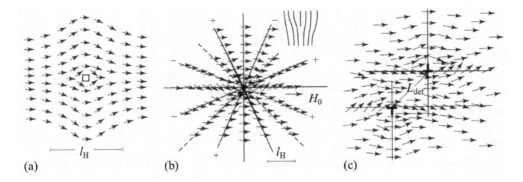

(a)   (b)   (c)

Fig. 1.7: Magnetization distributions around (a) a vacancy-type defect, (b) an isolated quasi-dislocation, and (c) a quasi-dislocation dipole. At a particular value of the externally applied magnetic field H_0, the micromagnetic exchange length l_H specifies the range or size of the perturbed region around the defect core. L_{def} in (c) denotes the spatial separation between defect cores. Image courtesy of Dagmar Goll, Aalen University, Aalen, Germany. After [156].

1.6.1   Origins of inhomogeneous magnetization states

After these elementary considerations we would like to briefly discuss in the following the origins of inhomogeneous magnetization states which appear on the length scale that is probed by conventional SANS. In fact, this is a complicated and challenging subject and a large part of this book is devoted to the discussion of this issue. Many of the magnetic materials which are studied using the SANS technique are polycrystalline in nature, implying that their macroscopic properties, as well as their magnetic SANS signal, are largely determined by crystalline lattice imperfections (e.g., pores, grain or phase boundaries, dislocations, point defects). The local magnetization $\mathbf{M}(\mathbf{r})$ in the material is linked to the defect via the magnetoelastic coupling energy (see Section 3.1.4): the stress field related to the defect couples to $\mathbf{M}$, which results in a deviation from the mean magnetization and, hence, in a contrast for magnetic SANS. As an example for this mechanism, Fig. 1.7 displays the inhomogeneous spin distribution in the vicinity of point and line defects. The spatial extension of the perturbed region around a defect is quantified by the magnetic exchange length l_H, which depends on the value of the external field H_0 and on the exchange interaction. Increasing H_0 results in a decrease of l_H, i.e., in a suppression of fluctuations. The ferromagnetic exchange interaction (see Section 3.1.1) plays a special role here, since this energy contribution aims to avoid gradients in $\mathbf{M}(\mathbf{r})$ and, therefore, a perturbed magnetization state spreads over many lattice sites into the surrounding crystal lattice.

Figure 1.7 nicely serves to highlight one of the central aspects of this book. The magnetic microstructure of many materials is characterized by the continuous "flow" of magnetization around defects, and a picture in terms of uniform and discontinuous domains is not adequate for the description of magnetic SANS. Naively, one may argue that replacing sharp discontinuous changes in the magnetization by a smooth profile,

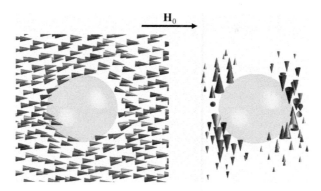

Fig. 1.8: Result of a micromagnetic simulation for the magnetization distribution around a spherical pore (diameter: 10 nm) in a ferromagnetic iron matrix ($\mu_0 H_0 = 0.6$ T). Shown is a two-dimensional projection of the spin structure out of a three-dimensional simulation. The jump of the magnetization magnitude at the pore-matrix interface is $\mu_0 M_s = 2.2$ T. The left image shows the magnetization distribution (arrows) in the iron matrix. In order to highlight the spin misalignment in the iron phase, the right image displays the magnetization component $\mathbf{M}_\perp$ perpendicular to $\mathbf{H}_0 \parallel \mathbf{e}_z$. Size of arrows is proportional to the magnitude of $\mathbf{M}_\perp$. After [59].

i.e., going from $\rightarrow\uparrow\rightarrow$ to $\curlywedge$, is only a mathematical detail which does not affect the main physics of a problem. However, as we will see in the course of this book and as is already well known from nuclear small-angle scattering theory (see the chapter by Porod in [10]), this change of paradigm has important consequences for the magnetic SANS cross section and correlation function.

On top of the magnetoelastic defect-related mechanism one has to take into account that at a defect site (e.g., at an internal interphase in a magnetic nanocomposite) the values of the magnetic materials parameters (saturation magnetization, exchange, or anisotropy constants) may change, in this way giving rise to inhomogeneous magnetization states, which contribute significantly to the magnetic SANS signal. Figure 1.8 depicts the numerically computed magnetization distribution $\mathbf{M}(\mathbf{r})$ around a spherical pore in a ferromagnetic iron matrix; see the papers by Kronmüller [157] and by Dietze and Schröder [158–160] for analytical calculations. The jump of the magnetization magnitude at the pore-matrix interface amounts to $\mu_0 M_s = 2.2$ T. This discontinuity in M_s and the related magnetic surface charges give rise to an inhomogeneous magnetic stray field $\mathbf{H}_{\text{stray}}^{\text{pore}}(\mathbf{r})$, which in turn exerts a torque $\mathbf{H}_{\text{stray}}^{\text{pore}} \times \mathbf{M}$ on the magnetic moments of the iron matrix [134]. The perpendicular (canted) moments which are produced by $\mathbf{H}_{\text{stray}}^{\text{pore}}$ represent a contrast for magnetic SANS (right image in Fig. 1.8).

The discussion can be made more quantitative by considering the closely related problem of a uniformly magnetized spherical particle of radius R and magnetization M_s^{p}, which is embedded in a saturated matrix of magnetization M_s^{m}. For this situation (see also Section 3.1.6), the Cartesian components of the stray field (outside of the particle at $r \geq R$) equal [161]:

$$\mathbf{H}_{\text{stray}}(r, \vartheta, \zeta) = \Delta M \left(\frac{R}{r}\right)^3 \left\{ \begin{array}{c} \sin\vartheta\cos\vartheta\cos\zeta \\ \sin\vartheta\cos\vartheta\sin\zeta \\ \cos^2\vartheta - \frac{1}{3} \end{array} \right\}, \tag{1.58}$$

where $\Delta M = M_{\text{s}}^{\text{p}} - M_{\text{s}}^{\text{m}}$ is the jump of the magnetization magnitude, the polar angle ϑ is measured between the magnetization axis ($\parallel \mathbf{e}_z$) and the position vector $\mathbf{r}$ with $|\mathbf{r}| = r$, and ζ denotes the azimuthal angle. As one can see [Fig. 1.9(a)], the dipole field eqn (1.58) is highly anisotropic and decays as r^{-3} with the distance from the surface of the particle. The components of $\mathbf{H}_{\text{stray}}$ which are perpendicular to the magnetization of the matrix (initially assumed to be saturated along $\mathbf{e}_z$) give rise to a torque and produce spin disorder. The orientation average of the square of the normal component of $\mathbf{H}_{\text{stray}}$ can be computed as follows:

$$\langle \mathbf{H}_{\text{stray},\perp}^2 \rangle_\Omega = \frac{1}{4\pi} \int\limits_0^{4\pi} \mathbf{H}_{\text{stray},\perp}^2 \sin\vartheta d\vartheta d\zeta = \frac{2}{15}(\Delta M)^2 \left(\frac{R}{r}\right)^6. \tag{1.59}$$

The square root of $\langle \mathbf{H}_{\text{stray},\perp}^2 \rangle_\Omega$ is plotted in Fig. 1.9(b) as a function of the reduced distance parameter $\hat{r} = R/r$. It is seen that the magnitude of the average dipole field can take on quite large values in the immediate vicinity of the particle/pore-matrix interface (e.g., $154\,\text{mT}$ at $\hat{r} = 0.75$ and for $\mu_0\Delta M = 1.0\,\text{T}$). This field also competes with the externally applied magnetic field $\mathbf{H}_0$, and it implies that—under realistic experimental conditions, where $\mu_0 H_0^{\max} \sim 1\text{--}10\,\text{T}$—the related torques do result in a significant spin canting of the matrix magnetic moments; in other words, for strongly inhomogeneous microstructures (large ΔM) and for not-too-large applied fields, it may be a too-idealized assumption to consider the matrix as being uniformly magnetized.

Besides vacancies, dislocations, pores, and second-phase particles, grain boundaries may also give rise to significant spin disorder, so that, even for a nearly defect-free and homogeneous polycrystalline ferromagnet, the magnetic SANS signal may be quite strong and dominate over the nuclear contribution; compare, e.g., the field-dependent SANS cross section of nanocrystalline Co (Fig. 6.16). Consider a polycrystalline single-phase elemental ferromagnet with a nanometer-sized average crystallite size D_{gs}, with uniform exchange interaction, saturation magnetization, and magnitude of the magnetic anisotropy, and with atomically sharp grain boundaries. For such an idealized system, the crystallographic set of easy axes for the magnetization changes randomly at each grain boundary, giving rise to nanometer-scale spatial variations in the orientation of the magnetocrystalline anisotropy field $\mathbf{H}_{\text{K}}$ (see Section 3.1.3). The magnetocrystalline anisotropy energy tries to pull the magnetic moments along certain crystallographic axes within each single-crystalline nanocrystal. Therefore, if $\mathbf{H}_{\text{K}}$ varies randomly on a scale of $D_{\text{gs}} \sim 10\text{--}20\,\text{nm}$, the magnetization is being perturbed by means of the related torques $\mathbf{H}_{\text{K}} \times \mathbf{M}$, which results in a strong magnetic SANS response.

Up to this point we have implicitly based the discussion on infinitely extended bulk ferromagnets, where all the phases of the material are magnetic. However, the above-described mechanisms giving rise to a magnetic SANS contrast are also of relevance for isolated finite-sized magnetic nanoparticles, which are embedded in a nonmagnetic

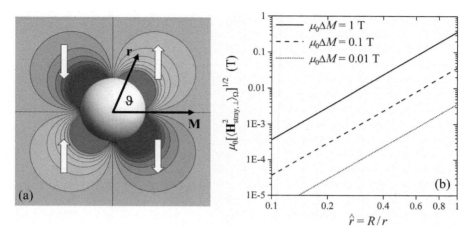

Fig. 1.9: (a) Sketch illustrating the geometry of the dipole field [eqn (1.58)]. Shown is the projection of the normal component of $\mathbf{H}_{\text{stray}}(r, \vartheta)$ into a plane containing the magnetization $\mathbf{M}$ of the particle ($\Delta M > 0$). Note that this projected field changes sign at the border between quadrants. (b) Root-mean-square dipolar stray field around a spherical particle as a function of $\hat{r} = R/r$ [eqn (1.59)] (log-log scale).

matrix. For nanoparticles, vacancies or the presence of surface anisotropy are potential sources of inhomogeneous magnetization states. Moreover, it is important to realize that, even if the nanoparticles are defect-free and homogeneous, it is the competition between the magnetic energy contributions which decides on the spin configuration, irrespective of whether the shape of the particle is ellipsoidal (e.g., a sphere) or not (see Section 1.6.2). Following Aharoni [154], the main difference between a homogeneously magnetized ellipsoid and any other homogeneously magnetized body is that the demagnetizing field $\mathbf{H}_d$ inside an ellipsoid is uniform, i.e., it is the same at every point inside the ellipsoid, which is not true for any non-ellipsoidal shape, where $\mathbf{H}_d = \mathbf{H}_d(\mathbf{r})$ (see Section 3.1.6 for a more-thorough discussion of magnetostatics). As a result, deviations from ellipsoidal particle shape may result in an inhomogeneous spin texture due to the associated nonuniformity of the individual particle's demagnetizing field, which is due to the magnetodipolar interaction.

This is illustrated in Fig. 1.10, where the results of micromagnetic simulations for the remanent spin structures of a long cobalt cylinder ($L_{\text{cyl}} = 500$ nm) are presented for diameters D_{cyl} ranging between $30-90$ nm [162, 163]. The uniaxial magnetocrystalline anisotropy K_u of Co was neglected in these simulations. We refer to [162] for results with a nonzero K_u directed along the cylinder's diameter (see also Fig. 7.15 in Section 7.4.2). The data in Fig. 1.10 reflect the competition between the exchange interaction, which tries to keep the ferromagnetic spin alignment, and the magnetodipolar interaction, which aims to avoid surface and volume charges (trying to align the magnetic moments parallel to the surface of the nanorod's ends). While the $D_{\text{cyl}} = 30$ nm cylinder is in a quasi-single-domain state (magnetized along the long x-axis), localized transversal spin inhomogeneities at the end faces of the rod, caused by torques

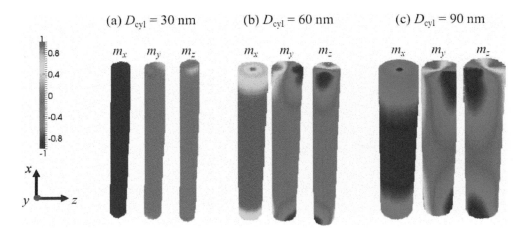

Fig. 1.10: Color-coded maps of the computed real-space spin structures of a Co nanorod with a length of $L_{cyl} = 500\,\mathrm{nm}$ and diameters of (a) $D_{cyl} = 30\,\mathrm{nm}$, (b) $D_{cyl} = 60\,\mathrm{nm}$, and (c) $D_{cyl} = 90\,\mathrm{nm}$ ($K_u = 0$) (remanent state). The cylinders were first saturated by an external field $\mathbf{H}_0 \parallel \mathbf{e}_z$ applied perpendicular to the wire axis (along the diameter). The magnetization distributions $\mathbf{m}(x, y, z)$ are identical in all three images in (a)−(c), respectively. The color scale encodes the different Cartesian components of $\mathbf{m}$. After [162].

$\mathbf{H}_d \times \mathbf{M}$, appear when the diameter is increased. The growing internal spin disorder results in a magnetic SANS cross section that is different from the uniform particle case (see Section 7.4.2), as described by the Stoner–Wohlfarth model [164].

The preceding considerations have clearly emphasized the pivotal role of microstructural defects and the related spatial variations of the magnetic materials parameters (e.g., saturation magnetization, direction and/or magnitude of magnetic anisotropy, exchange constant) for the understanding of the magnetic microstructure and the ensuing magnetic SANS cross section $d\Sigma_M/d\Omega$. In the following, we find it instructive to continue the discussion by comparing the $d\Sigma_M/d\Omega$ which are related to a uniformly and to a nonuniformly magnetized defect-free spherical nanoparticle.

1.6.2   Example: uniform versus nonuniform sphere

The discussion in this section may be perceived as a bit premature, since we already rely on some of the basic concepts of SANS. We emphasize, however, that the focus here is on grasping the differences between conventional particle scattering (assuming uniformly magnetized domains) and the magnetic scattering from nonuniform magnetization textures. Therefore, it is advisable that the reader returns to this section, once some of the other chapters have been worked through.

On a length scale of the interatomic spacing, strong quantum-mechanical exchange forces hold all atomic spins in parallel, which is true for many ferromagnetic materials [154]. Therefore, very small magnetic particles may be assumed to be uniformly

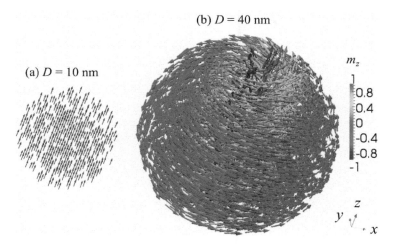

(b) $D = 40$ nm

(a) $D = 10$ nm

m_z

1
0.8
0.4
0
-0.4
-0.8
-1

z

y

x

Fig. 1.11: Numerically computed magnetization distributions (using the software package MuMax3 [166, 167]) of a defect-free iron sphere with a diameter of (a) $D = 10$ nm and (b) $D = 40$ nm. Shown are the spin structures in the remanent state after prior saturation along the z-direction. After [163].

magnetized and are said to be in a single-domain state. Increasing the size of nanoparticles leads to an increase of the magnetostatic stray-field energy (see Section 3.1.6), and eventually it may become energetically advantageous to introduce a domain wall into the particle. The critical single-domain size D_c marks the transition between the single-domain and the two-domain state. For a single spherical particle, D_c is approximately given by [119, 165]:

$$D_c \cong \frac{72\sqrt{AK_1}}{\mu_0 M_s^2},$$ (1.60)

where A denotes the exchange-stiffness constant, and K_1 is the first-order magnetocrystalline anisotropy constant; see Table B.1 in Appendix B for D_c-values of some selected magnetic materials.

Figure 1.11 depicts the remanent spin structures of defect-free iron spheres with diameters of $D = 10$ nm and $D = 40$ nm. Note that the critical single-domain size for iron is $D_c \cong 10$ nm (see Table B.1). As one can see, the magnetization distribution of the $D = 10$ nm sphere is quasi-uniform, while the one of the $D = 40$ nm sphere is highly inhomogeneous [i.e., $\mathbf{M} = \mathbf{M}(x, y, z)$] and exhibits a vortex-type spin configuration.

Before proceeding with the discussion of the magnetic SANS of the iron sphere, we will briefly furnish some technical information regarding the micromagnetic simulations. For the computation of the spin structures of the iron sphere, the GPU-based open-source software package MuMax3 [166, 167] was employed, which uses a finite-difference discretization scheme of space. MuMax3 can take into account all the standard contributions to the total magnetic Gibbs free energy, i.e., energy in the external magnetic field, magnetodipolar interaction energy, energy of the magnetocrystalline anisotropy, antisymmetric Dzyaloshinskii–Moriya interaction, as well as the symmetric and isotropic Heisenberg exchange energy. For the examples shown in Fig. 1.11, the Dzyaloshinskii–Moriya interaction was ignored, and likewise thermal fluctuations, which may become relevant for small nanoparticles. The following materials parameters were used: saturation magnetization $M_s = 1700\,\text{kA/m}$; first-order cubic anisotropy constant $K_1 = 4.7 \times 10^4\,\text{J/m}^3$; exchange-stiffness constant $A = 1.0 \times 10^{-11}\,\text{J/m}$.

We will now compare the magnetic SANS cross section of the $D = 40\,\text{nm}$ particle to the one of a uniformly magnetized sphere of the same size. The general question of what is the actual ground-state spin configuration of a nanoparticle depends on the microstructure and on the related magnetic energy contributions which are taken into account. As discussed in the previous Section 1.6.1, lattice defects (e.g., vacancies in nanoparticles), surface anisotropy, or strong deviations from ellipsoidal particle shape in otherwise perfect particles may give rise to a nonuniform magnetization (see, e.g., [162, 163, 168–179] and references therein).

When the incident neutron beam is along the x-direction and the scattering vector varies within the y-z-plane, we have $\hat{\mathbf{q}} \cong \{0, \sin\theta, \cos\theta\}$ in small-angle approximation ($\hat{q}_x \cong 0$), where θ denotes the angle between $\mathbf{q}$ and the z-direction (parallel to $\mathbf{H}_0$). The elastic magnetic SANS cross section [eqn (1.35)] is then explicitly given in terms of the Cartesian magnetization Fourier components as:

$$\frac{d\Sigma_M}{d\Omega} \sim |\widetilde{m}_x|^2 + |\widetilde{m}_y|^2 \cos^2\theta + |\widetilde{m}_z|^2 \sin^2\theta - (\widetilde{m}_y \widetilde{m}_z^* + \widetilde{m}_y^* \widetilde{m}_z) \sin\theta \cos\theta, \quad (1.61)$$

where the $\widetilde{m}_{x,y,z}$ represent the Fourier transforms of the $m_{x,y,z}$. Note that $\mathbf{m} = \mathbf{M}/M_s$ denotes the unit magnetization vector field. In the most-general case, $m_{x,y,z} = m_{x,y,z}(x,y,z)$ so that the $\widetilde{m}_{x,y,z}$ depend on the three Cartesian components of the wave vector $\mathbf{q} = \{q_x, q_y, q_z\}$, i.e.,

$$\widetilde{m}_{x,y,z}(q_x, q_y, q_z) = \int_V m_{x,y,z}(x,y,z) \exp(-i\mathbf{q}\cdot\mathbf{r})\,dxdydz. \quad (1.62)$$

The integration in eqn (1.62) generally extends over the sample volume V, but in the present case, where we consider only a single particle, it equals the particle volume V_p. Equation (1.62) can in most cases only be solved numerically, since the magnetization is—within the continuum theory of micromagnetics—the solution of a set of nonlinear partial differential equations (see Section 5.1.3), i.e., no closed-form expression for $m_{x,y,z}$ exists. Moreover, since the magnetic SANS cross section is experimentally accessible only in the plane perpendicular to the incoming neutron beam, it becomes necessary to project the $\widetilde{m}_{x,y,z}$ into the detector plane. For the above-described scattering geometry, this implies that $\widetilde{m}_{x,y,z} = \widetilde{m}_{x,y,z}(q_x = 0, q_y, q_z)$.

The uniformly magnetized sphere is assumed to have its atomic magnetic moments all aligned along the $\mathbf{e}_z$-direction of a Cartesian laboratory coordinate system, so that in a continuum description the magnetization vector field is given by:

$$\mathbf{m} = \frac{\mathbf{M}}{M_s} = \begin{Bmatrix} m_x \\ m_y \\ m_z \end{Bmatrix} = \begin{Bmatrix} 0 \\ 0 \\ 1 \end{Bmatrix}. \quad (1.63)$$

Note that the above magnetization distribution has a constant length, $|\mathbf{m}|^2 = m_x^2 + m_y^2 + m_z^2 = 1$, which in physical terms means that the spontaneous magnetization is constant and only a function of temperature, i.e., $M_s = M_s(T)$. As we will discuss in Chapter 3, this constraint implies that only two components of $\mathbf{m}$ are independent. Note that the $m_{x,y,z}$ are here dimensionless, so that the $\widetilde{m}_{x,y,z}$ and $d\Sigma_M/d\Omega$ have, respectively, dimensions of m^3 and m^6 [compare eqns (1.61) and (1.62)]. Inserting eqn (1.63) into eqn (1.62) yields the well-known sphere form factor

$$\widetilde{m}_z(q_x, q_y, q_z) = 3V_p \frac{j_1(qR)}{qR}, \tag{1.64}$$

where $j_1(z = qR)$ is the spherical Bessel function of the first order, and $V_p = \frac{4\pi}{3}R^3$. The zeros (z_0) and maxima (z_{max}) of the function $(3j_1(z)/z)^2$ are approximately found for:

$$z_0 = \pi \times (1.43; 2.46; 3.47; 4.48; 5.48; 6.48; \ldots),$$
$$z_{max} = \pi \times (1.83; 2.90; 3.92; 4.94; 5.95; 6.96; \ldots). \tag{1.65}$$

The magnetic SANS cross section of a uniformly magnetized particle reduces to [compare eqn (1.61)]:

$$\frac{d\Sigma_M}{d\Omega}(\mathbf{q}) = |\widetilde{m}_z|^2 \sin^2\theta. \tag{1.66}$$

Figure 1.12 depicts the results for $|\widetilde{m}_z|^2$, $d\Sigma_M/d\Omega$, as well as for the azimuthally averaged $d\Sigma_M/d\Omega$. The thin solid line in Fig. 1.12(c) is the analytical result, based on eqns (1.64) and (1.66), while the data points are the result of the micromagnetic simulation. As expected for a saturated sphere, $|\widetilde{m}_z|^2$ [Fig. 1.12(a)] is isotropic and $d\Sigma_M/d\Omega$ [Fig. 1.12(b)] is highly anisotropic in the detector plane ($q_x = 0$), exhibiting two maxima along the directions perpendicular ($\theta = 90°$ and $\theta = 270°$) to the horizontal saturation direction. The azimuthally averaged data [Fig. 1.12(c)] exhibit some of the characteristic and well-known features of particle scattering (see Section 2.4.3 or [9–13] for further details); namely, at low q, the SANS cross section of the uniformly magnetized particle exhibits the so-called Guinier behavior [compare Fig. 1.12(c)], i.e.,

$$\frac{d\Sigma_M}{d\Omega} \cong \frac{d\Sigma_M}{d\Omega}(q = 0) \exp\left(-\frac{q^2 R_G^2}{3}\right), \tag{1.67}$$

where R_G denotes the radius of gyration of the particle. For a uniform sphere, $R_G^2 = \frac{3}{5}R^2$. As shown by Feigin and Svergun [11], for a homogeneous particle, the Guinier law is valid for $q < 1.3/R_G$. At the larger q, $d\Sigma_M/d\Omega$ follows the Porod law, i.e.,

$$\frac{d\Sigma_M}{d\Omega} \cong 2\pi(\Delta\rho)_{mag}^2 \frac{S}{V}\frac{1}{q^4}, \tag{1.68}$$

where $(\Delta\rho)_{mag}^2 = b_H^2 (\Delta M)^2$ is the magnetic scattering-length density contrast, the squared difference between the saturation magnetizations of the iron particle and vacuum, and S/V denotes the surface-to-volume ratio. The Porod q^{-4}-law is a direct

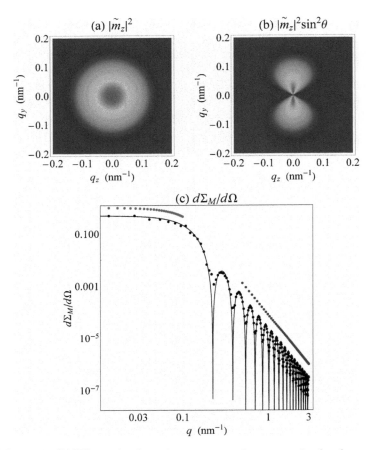

Fig. 1.12: Magnetic SANS results for a homogeneously magnetized sphere of diameter $D = 40\,\text{nm}$. (a) $|\widetilde{m}_z|^2$ [square of eqn (1.64)]; (b) $|\widetilde{m}_z|^2 \sin^2\theta$ [eqn (1.66)]; (c) (solid line) azimuthally averaged $d\Sigma_M/d\Omega = \frac{1}{2\pi}\int_0^{2\pi}|\widetilde{m}_z|^2 \sin^2\theta d\theta = \frac{9}{2}[j_1(qR)/(qR)]^2$ (log-log scale). (●) Result of the micromagnetic simulation. The vertically shifted data points in (c) at small and large momentum transfers represent, respectively, the Guinier and the Porod approximation (see text). Saturation direction in (a) and (b) is horizontal in the plane. Note that all expressions have been normalized by V_p^2 [compare eqn (1.64)], so that $\frac{d\Sigma_M}{d\Omega}(q=0) = \frac{1}{2}$. After [163].

consequence of the discontinuous jump of the magnetization at the surface of the spherical particle.

When the particle's internal spin distribution is inhomogeneous [see Fig. 1.11(b)] all the Fourier components, and not just $|\widetilde{m}_z|^2$ as in the homogeneous case (Fig. 1.12), contribute to $d\Sigma_M/d\Omega$. This is demonstrated in Fig. 1.13, where $d\Sigma_M/d\Omega$ is decrypted into its individual Fourier contributions according to eqn (1.61). The Fourier components are now also dependent on the angle θ in the detector plane. It is also seen that $d\Sigma_M/d\Omega$ is for the particular case of the $D = 40\,\text{nm}$ iron sphere dominated by

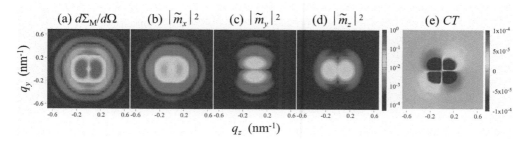

Fig. 1.13: Magnetic SANS results for an inhomogeneously magnetized iron sphere of diameter $D = 40\,\text{nm}$ (remanent state). (a) $d\Sigma_M/d\Omega$; (b) $|\widetilde{m}_x|^2$; (c) $|\widetilde{m}_y|^2$; (d) $|\widetilde{m}_z|^2$; (e) $CT = -(\widetilde{m}_y\widetilde{m}_z^* + \widetilde{m}_y^*\widetilde{m}_z)$. Note that the cross term (CT) in (e) can take on positive as well as negative values (red color: $CT > 0$; blue color: $CT < 0$) (compare also Fig. 7.9). After [163].

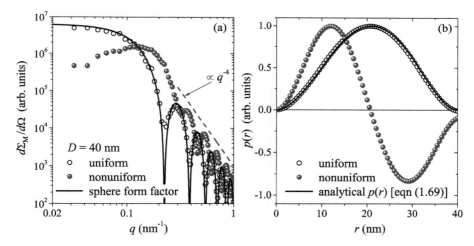

Fig. 1.14: (a) Comparison of the azimuthally averaged magnetic SANS cross sections $d\Sigma_M/d\Omega$ of a uniformly (open circles ○) and a nonuniformly (closed circles ●) magnetized iron sphere (remanent state) with a diameter of $D = 40\,\text{nm}$ (log-log scale). Both cross sections have been numerically computed using the continuum theory of micromagnetics. Solid line: analytical solution for the homogeneous case using the sphere form factor, $d\Sigma_M/d\Omega \propto [j_1(qR)/(qR)]^2$. Dashed line: $d\Sigma_M/d\Omega \propto q^{-4}$. (b) Corresponding distance distribution functions $p(r)$. Solid line: equation (1.69). After [163].

the $|\widetilde{m}_x|^2$ Fourier component (in the remanent state), which is different than $|\widetilde{m}_y|^2$. Correspondingly, the azimuthally averaged cross section can no longer be described by the sphere form factor [see Fig. 1.14(a)]. Compared to the homogeneous case, the maxima in $d\Sigma_M/d\Omega$ for the inhomogeneous sphere are shifted to larger q-values, since structures smaller than D appear in the magnetization distribution. If the $d\Sigma_M/d\Omega$ data in the remanent state would be fitted using the sphere form factor, assuming a uniform magnetization, then an erroneous value for the particle size may result. A Guinier region can also not be discerned for the inhomogeneous sphere. Its reduced magnetization in the remanent state manifests as a lower value of $d\Sigma_M/d\Omega$ as $q \to 0$. The Porod law [eqn (1.68)] is however also found for the inhomogeneous particle, since the spin structure is still confined by a sharp interface. We emphasize that the Porod behavior naturally emerges here from the micromagnetic computations without a priori assumptions. Figure 1.14(b) shows the computed distance distribution functions $p(r)$ for the two cases. While the homogeneously magnetized sphere reveals the well-known bell-shaped form (compare Section 2.4.3)

$$p(r) = r^2 \left(1 - \frac{3r}{4R} + \frac{r^3}{16R^3}\right),\tag{1.69}$$

the distance distribution function of the nonuniform sphere has an oscillatory character with a zero at $r \cong 20.7\,\text{nm}$. Currently, no analytic description for the $d\Sigma_M/d\Omega$ and $p(r)$ of an inhomogeneously magnetized sphere is available, which represents a challenge for future studies.

The main message of Section 1.6 is to emphasize the importance of microstructural defects for the understanding of magnetic SANS: lattice imperfections may result in nanometer-scale spatial variations in the orientation and magnitude of the magnetization, which in turn give rise to a large contribution to the magnetic SANS cross section, in particular for bulk ferromagnets. Secondly, as the micromagnetic simulation results for the iron sphere have demonstrated, even if nanoparticles are defect-free and homogeneous, nonuniform spin structures may emerge and render their magnetic SANS response different as compared to the uniform case. The usage of analytical and numerical micromagnetic calculations is of vital importance for connecting the nanoscale spin inhomogeneity to the experimentally measured quantity.

2
BASICS OF SANS

This chapter is concerned with the discussion of the basics of nuclear and magnetic SANS. Sections 2.1−2.4 provide, respectively, a brief description of a typical SANS setup, elementary considerations regarding the q-resolution and the coherence properties of the neutron beam, a discussion of the influence of inelastic scattering contributions (due to phonons and magnons) to the energy-integrated SANS signal, and a summary of the basics of elastic nuclear SANS. Expressions for the various unpolarized and spin-polarized elastic SANS cross sections $d\Sigma/d\Omega$ will then be displayed and discussed in Sections 2.5−2.8. We focus on the two most relevant scattering geometries which have the applied magnetic field $\mathbf{H}_0$ either perpendicular or parallel to the wave vector $\mathbf{k}_0$ of the incoming neutron beam. In the Born approximation, the $d\Sigma/d\Omega$ are fully determined by the three Cartesian Fourier components $\widetilde{M}_{x,y,z}(\mathbf{q})$ of the magnetization vector field $M_{x,y,z}(\mathbf{r})$ of the sample. Displaying the $d\Sigma/d\Omega$ in terms of the $\widetilde{M}_{x,y,z}$ allows for a direct comparison to experimental SANS data and to the results of micromagnetic computations. Finally, Section 2.9 discusses the SANS cross sections at magnetic saturation, in this way providing the connection to the customary particle-matrix concept.

2.1 Description of the SANS setup

Figure 2.1 depicts schematically the typical SANS setup along with sketches of the two most commonly used scattering geometries, while Fig. 2.2 displays as an example the layout of the SANS-1 instrument at the Heinz Maier-Leibnitz Zentrum (MLZ), Garching, Germany. Cold neutrons that emerge from a nuclear reactor can be monochromatized by means of a mechanical velocity selector. Depending on the rotational speed and tilting angle of the selector drum relative to the incident neutron-beam direction, a mean wavelength between about 5−20 Å and with a wavelength resolution $\Delta\lambda/\lambda$ between 5−30% (FWHM) can be selected. As an alternative to a velocity selector one may employ the time-of-flight method for carrying out the velocity (wavelength) discrimination, in particular, at a spallation neutron source. Such a system, in addition to a velocity selector, is e.g., installed at the instrument D33 at the Institut Laue-Langevin, Grenoble, France [180], or at BILBY at the Australian Centre for Neutron Scattering, Lucas Heights, Australia [181]. Here, a chopper system pulses the incoming beam having a broad wavelength distribution and the detector electronics records the time of arrival of the neutrons at the detector. However, the usage of the time-of-flight mode of operation is still rare in magnetic SANS research, and many experiments rely on the monochromatic mode using a velocity selector. In the evacuated pre-sample flight path (source-to-sample distance: ∼1−20 m), a set of apertures

Magnetic Small-Angle Neutron Scattering: A Probe for Mesoscale Magnetism Analysis. Andreas Michels, Oxford University Press (2021). © Andreas Michels. DOI: 10.1093/oso/ 9780198855170.003.0002

collimates the monochromatized beam, resulting in an angular divergence of the beam of $\Delta\psi \sim 1-10\,\mathrm{mrad}$ [182]. A particular strength of the SANS technique is that experiments can be conducted under rather flexible conditions and under different sample environments (e.g., temperature, electric and magnetic field, pressure, neutron polarization, time-resolved data acquisition). The typical size of the irradiated area of sample is of the order of $1\,\mathrm{cm}^2$.

A two-dimensional position-sensitive detector, moving along rails in an evacuated post-sample flight path (sample-to-detector distance: $L_{\mathrm{SD}} \sim 1-20\,\mathrm{m}$), counts the scattered neutrons during acquisition times ranging between a few minutes and a few hours. The recorded neutron counts (in each pixel element of the detector) are corrected for detector dead time, normalized to incident-beam monitor counts, and a solid-angle correction is applied to the data which corrects for the planar geometry of the detector. Further corrections relate to sample transmission, background scattering, detector dark current, and detector efficiency. The size of an individual pixel element of the detector is typically $\lesssim 10\,\mathrm{mm} \times 10\,\mathrm{mm}$, so that the related resolution effects become negligible. The scattering cross section of the sample is obtained by comparing the corrected signal to that of a reference sample (e.g., water, polystyrene, porous silica, vanadium single crystal) of known cross section. The data-reduction procedure provides the macroscopic differential scattering cross section $d\Sigma/d\Omega$ of the sample in absolute units (typically cm^{-1}) and as a function of the magnitude and orientation of the momentum-transfer or scattering vector $\mathbf{q}$ (see Fig. 2.1). In order to conveniently present the neutron data, one often carries out a so-called azimuthal averaging procedure, whereby the data at a constant magnitude of $\mathbf{q}$ are summed up (integrated) within a certain angular range (e.g., over a full circle of $360°$). This yields $d\Sigma/d\Omega$ as a function of $|\mathbf{q}| = q$. As quoted in [183, 184], the uncertainty in the cross sections determined by the above-sketched procedure is estimated to be $5-10\%$.

In the case when the incident neutrons travel e.g., along $\mathbf{k}_0 \parallel \mathbf{e}_x$, the scattering angle is denoted with ψ, and the angle θ is used to specify the orientation of $\mathbf{q}$ [with $q = 2k_0 \sin(\psi/2)$] on the two-dimensional detector, the scattering vector is found to be [compare Fig. 2.1(a)]:

$$\mathbf{q} = \begin{Bmatrix} q_x \\ q_y \\ q_z \end{Bmatrix} = q \begin{Bmatrix} -\sin(\psi/2) \\ \cos(\psi/2)\sin\theta \\ \cos(\psi/2)\cos\theta \end{Bmatrix} = k_0 \begin{Bmatrix} \cos\psi - 1 \\ \sin\psi\sin\theta \\ \sin\psi\cos\theta \end{Bmatrix}. \tag{2.1}$$

For small-angle scattering $\psi \lesssim 5-10°$, so that the magnitude of the component of $\mathbf{q}$ along the incident-beam direction, $q\sin(\psi/2)$, is much smaller than the other two components. The three-dimensional scattering vector is therefore approximated by a two-dimensional one. This approximation, which violates the condition for elastic scattering ($k_0 = k_1$), is valid for not-too-large scattering angles; for instance, for $\psi = 5°$, $|q_x|/q = \sin(\psi/2) \cong 4.4\%$ and the related error in the intensity is less than 1% [185]. Since

$$q = \frac{4\pi}{\lambda}\sin(\psi/2) = \frac{4\pi}{\lambda}\sin\left(\frac{1}{2}\arctan\left[\frac{r_{\mathrm{D}}}{L_{\mathrm{SD}}}\right]\right) \cong k_0 \frac{r_{\mathrm{D}}}{L_{\mathrm{SD}}}, \tag{2.2}$$

where r_{D} is the radial distance on the detector (measured from the beam center), and L_{SD} denotes the sample-to-detector distance, we see that different momentum transfers

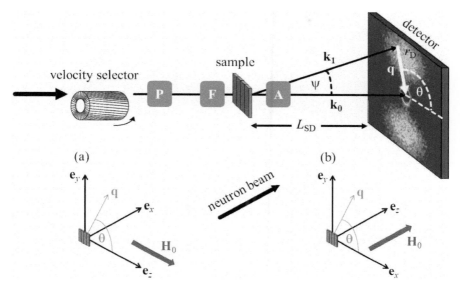

Fig. 2.1: Sketch of the SANS setup and of the two most often employed scattering geometries in magnetic SANS experiments. (a) Applied magnetic field $\mathbf{H}_0$ perpendicular to the incident neutron beam ($\mathbf{k}_0 \perp \mathbf{H}_0$); (b) $\mathbf{k}_0 \parallel \mathbf{H}_0$. The momentum-transfer or scattering vector $\mathbf{q}$ corresponds to the difference between the wave vectors of the incident ($\mathbf{k}_0$) and the scattered ($\mathbf{k}_1$) neutrons, i.e., $\mathbf{q} = \mathbf{k}_0 - \mathbf{k}_1$. Its magnitude for elastic scattering, $q = |\mathbf{q}| = (4\pi/\lambda)\sin(\psi/2)$, depends on the mean wavelength λ of the neutrons and on the scattering angle ψ. For a given λ, sample-to-detector distance L_{SD}, and distance r_{D} from the center of the direct beam to a certain pixel element on the detector, the q-value can be obtained using eqn (2.2). Note that in many graphical representations of two-dimensional SANS data the inverted $\mathbf{q}$-vector, pointing from the center of the beam to a pixel element [as in panels (a) and (b)], is shown. The symbols P, F, and A denote, respectively, the polarizer, spin flipper, and analyzer, which are optional neutron optical devices. Note that a second flipper after the sample has been omitted here. In spin-resolved SANS (POLARIS) using a ^{3}He spin filter, the transmission (polarization) direction of the analyzer can be switched by 180° by means of an rf pulse. SANS is usually implemented as elastic scattering ($k_0 = k_1 = 2\pi/\lambda$), and the component of $\mathbf{q}$ along the incident neutron beam [i.e., q_x in (a) and q_z in (b)] is neglected. The angle θ may be conveniently used in order to describe the angular anisotropy of the recorded scattering pattern on a two-dimensional position-sensitive detector.

(scattering angles) can be accessed by varying L_{SD}, or the neutron wavelength λ. The approximation $q \cong k_0 r_{\mathrm{D}}/L_{\mathrm{SD}}$ in eqn (2.2) is valid for $r_{\mathrm{D}} \ll L_{\mathrm{SD}}$. With conventional SANS instruments it becomes thus possible to cover a q-range of $0.01\,\mathrm{nm}^{-1} \lesssim q \lesssim 5\,\mathrm{nm}^{-1}$, which translates into structure sizes of the order of $1-300\,\mathrm{nm}$.

 The unpolarized neutrons from the source may be polarized by means of a super-mirror transmission polarizer, and the initial neutron polarization P can be reverted

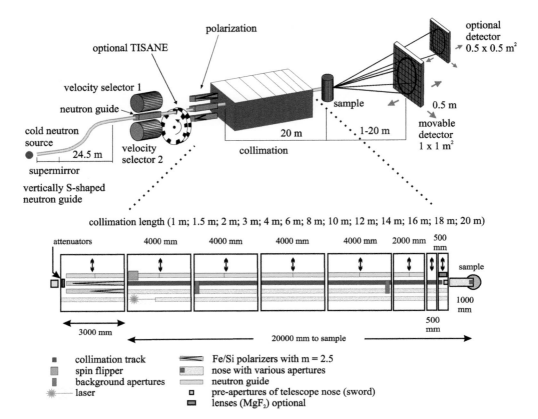

Fig. 2.2: Schematic depiction of the instrument layout of SANS-1 at FRM-II (MLZ) with its major components; see text for further comments. After [182].

by 180° using a radio-frequency (rf) spin flipper of efficiency ϵ mounted before the sample position [186]. We refer to [187] for a description of the operating principle of an rf flipper. In order to discriminate the neutron spin state after interaction with the sample, a ^{3}He spin filter with analyzing power P_A acts as neutron spin analyzer and, correspondingly, is installed behind the sample (sometimes inside the detector housing). The operation principle of a ^{3}He spin polarizer/analyzer is based upon the strongly spin-dependent absorption of neutrons by a nuclear-spin-polarized gas of ^{3}He atoms [127]. Only neutrons with spin component antiparallel to the ^{3}He nuclear spin are absorbed. The transmission direction of the ^{3}He spin filter can be switched by 180° by means of an rf pulse. Magnetic guide fields of the order of 1 mT, typically provided by permanent magnets, serve to maintain the polarization on the path between the polarizer and the ^{3}He filter, i.e., the guide field avoids non-adiabatic transitions of the neutron spin in too-weak magnetic fields of changing direction. Progress in the development of ^{3}He spin filters [188] allows one to perform routinely uniaxial (also called one-dimensional or longitudinal) neutron-polarization analysis on a SANS instrument; for instance, at SANS-1 and KWS-1 at the Heinz Maier-Leibnitz Zentrum, at D22 and D33 at the Institut Laue-Langevin, or at VSANS and NG7 SANS at the NIST Center for Neutron Research. The advantages of ^{3}He spin filters as compared to other polarizing/analyzing devices are (i) that they can be tuned via the gas pressure and optical length to perform at various wavelength bands (from cold to thermal to hot neutrons) and (ii) that they allow for a rather large neutron-energy transfer and scattering angle to be covered. Quan et al. [189] have developed a compact and transportable neutron spin filter for polarized SANS using a pentacene-doped single crystal of naphthalene. This device exhibits a nuclear-spin-lattice relaxation time of $T_1 = 800$ h at 20 mT and 6 K, so that time-dependent corrections are unnecessary and a straightforward spin-leakage correction can be applied to SANS data [190].

The initial neutron polarization P may be estimated from the measurement of the flipping ratio F_R, which is defined as [191–193]:

$$F_R = F_R(P, P_A, \epsilon) = \frac{N_0}{N_1} = \frac{1 + P P_A}{1 - \epsilon P P_A}, \tag{2.3}$$

where N_0 and N_1 denote the neutron count rates, measured without sample, for, respectively, inactive and active spin flipper. Equation (2.3) can be rearranged to yield

$$P = \frac{F_R - 1}{F_R \epsilon P_A + P_A} \geq \frac{F_R - 1}{F_R + 1}, \tag{2.4}$$

where the last inequality is obtained for $\epsilon = P_A = 1$; see the article by Zimmer [192] for further information and for details of how to measure ϵ and P_A.

SANS experiments with a polarized incident beam only and no detection of the polarization of the scattered neutrons provide access to the SANSPOL cross sections $d\Sigma^+/d\Omega$ and $d\Sigma^-/d\Omega$ (see Section 2.7). These (half-polarized) cross sections combine non-spin-flip and spin-flip scattering contributions [compare eqns (1.47) and (1.48)]. The difference between the "spin-up" and "spin-down" SANSPOL cross sections yields information on the polarization-dependent nuclear-magnetic and magnetic-magnetic chiral scattering terms. As demonstrated e.g., in [186, 194] on an Fe_3O_4 glass ceramic, this difference, which is linear in the longitudinal magnetization Fourier component, allows one to highlight weak magnetic contributions relative to strong nuclear scattering (or vice versa) (see Fig. 4.27). Moreover, a further advantageous property of the SANSPOL method is that the difference between the two spin states is independent of the nuclear incoherent scattering (ignoring the very weak nuclear-spin-dependent scattering).

With uniaxial polarization analysis [117] it becomes possible to measure four intensities that connect two neutron-spin states. Here, the externally applied magnetic field at the sample position defines the quantization axis for both the incident and the scattered polarization, whereby the scattered neutron may undergo a spin-reversing event due to the magnetic interaction with the sample. Following [117], the four spin-resolved scattering cross sections are the two non-spin-flip quantities $d\Sigma^{++}/d\Omega$ and $d\Sigma^{--}/d\Omega$ and the two spin-flip cross sections $d\Sigma^{+-}/d\Omega$ and $d\Sigma^{-+}/d\Omega$ [compare eqns (1.42) and (1.43)]. When the rf flipper is off (inactive), we measure, depending on the spin state of the ^{3}He filter, the non-spin-flip or the spin-flip cross section $d\Sigma^{++}/d\Omega$ or $d\Sigma^{+-}/d\Omega$.

Likewise, when the flipper is active, we either measure $d\Sigma^{--}/d\Omega$ or $d\Sigma^{-+}/d\Omega$. The corresponding expressions for the SANS cross sections are denoted as the POLARIS equations (see Section 2.8). The central field of application of the POLARIS method resides in the possibility to fully separate nuclear coherent from magnetic scattering.

We emphasize that the above-described uniaxial polarization-analysis setup, which consists of a supermirror transmission polarizer, an rf spin flipper, and a ^{3}He spin analyzer, represents the configuration which is most often found at SANS instruments. There exist, of course, many other neutron instrumentation devices for polarizing neutron beams and for turning the neutron-spin direction (see, e.g., the textbook by Williams [103] for further details). For more information on polarized neutron scattering and on spherical neutron polarimetry, we refer the reader to the classic papers [104, 109–113, 117, 125, 195–200] and textbooks [76, 86, 103, 201].

2.2 Elementary considerations of resolution

The q-resolution of a pinhole-geometry SANS instrument is mainly related to the wavelength spread of the incident neutrons, the finite collimation of the beam, and the detector resolution (finite pixel size). Taking into account the former two contributions, it is readily verified using $q = \frac{4\pi}{\lambda}\sin(\psi/2)$ [eqn (2.2)] that the root-mean-square (rms) uncertainty in q is given by:

$$\mathrm{rms}(q) = \sqrt{q^2\left(\frac{\Delta\lambda}{\lambda}\right)^2 + \left(k_0^2 - \frac{1}{4}q^2\right)(\Delta\psi)^2}. \qquad (2.5)$$

The angular divergence $\Delta\psi$ of the beam can be experimentally determined from the measured profile of the direct beam [$\mathrm{rms}(q = 0) = k_0\Delta\psi$], or estimated from the relation $\Delta\psi \sim 2r_{\mathrm{s}}/L_{\mathrm{col}}$, where r_{s} is the radius of the source aperture, and L_{col} denotes the collimation length (effective source-to-sample distance). Typical values of $\Delta\psi$ are of the order of 10^{-3}–10^{-2} rad [182]. Equation (2.5) is plotted in Fig. 2.3 demonstrating that wavelength smearing dominates at large q, while angular-divergence effects limit the resolution at small q. For an overview on how to optimize the experimental resolution in a small-angle scattering experiment, see e.g., the article by Mildner and Carpenter [202].

Following Pedersen et al. [203–205], resolution effects can be taken into account by introducing a resolution function $R(\mathbf{q}, \langle\mathbf{q}\rangle)$, where $\langle\mathbf{q}\rangle$ denotes the average scattering vector corresponding to the setting of the instrument. When the SANS instrument is configured to detect neutrons with scattering vector $\langle\mathbf{q}\rangle$, neutrons with scattering vectors $\mathbf{q}$ in a certain range of width σ_{R} around $\langle\mathbf{q}\rangle$ [compare eqn (2.5)] also contribute to the scattering due to the above-mentioned sources of smearing. The resolution function describes the distribution of the scattered neutrons with scattering vector $\mathbf{q}$ contributing to the scattering for the setting $\langle\mathbf{q}\rangle$. The experimental intensity $I(\langle\mathbf{q}\rangle)$ is then related to $R(\mathbf{q}, \langle\mathbf{q}\rangle)$ and to the SANS cross section $d\Sigma/d\Omega$ via (cf. also Dorner and Wildes [206])

$$I(\langle\mathbf{q}\rangle) = \int R(\mathbf{q}, \langle\mathbf{q}\rangle)\frac{d\Sigma}{d\Omega}(\mathbf{q})d\mathbf{q}. \qquad (2.6)$$

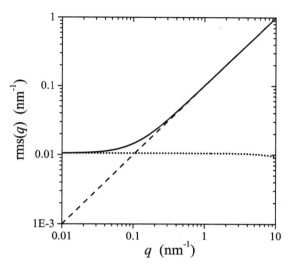

Fig. 2.3: Solid line: plot of rms(q) [eqn (2.5)] (log-log scale). Dashed line: contribution due to wavelength broadening $q\Delta\lambda/\lambda$. Dotted line: contribution due to angular-divergence effects $\sqrt{k_0^2 - q^2/4}\Delta\psi$. The following parameters were used: $\Delta\lambda/\lambda = 0.1$; $\Delta\psi = 0.001\,\mathrm{rad}$; $\lambda = 6\,\mathrm{\AA}$.

Several analytical and numerical investigations (see [203–205] and references therein) utilize Gaussian functions to model $R(\mathbf{q}, \langle\mathbf{q}\rangle)$. For isotropic SANS data, which are azimuthally averaged, the following expression for the resolution function has been suggested [207]:

$$R(q, \langle q\rangle) = \frac{q}{\sigma_{\mathrm{R}}^2} \exp\left(-\frac{q^2 + \langle q\rangle^2}{2\sigma_{\mathrm{R}}^2}\right) I_0\left(\frac{q\langle q\rangle}{\sigma_{\mathrm{R}}^2}\right), \tag{2.7}$$

where

$$\sigma_{\mathrm{R}}^2(\langle q\rangle) = \sigma_{\mathrm{W}}^2 + \sigma_{\mathrm{C}}^2 = \left(\langle q\rangle \frac{\Delta\lambda}{\lambda} \frac{1}{2\sqrt{2\ln 2}}\right)^2 + \left(\frac{k_0 \cos(\psi/2)\Delta\beta}{2\sqrt{2\ln 2}}\right)^2 \tag{2.8}$$

and $I_0(z)$ denotes the modified Bessel function of the first kind and zeroth-order [203, 204]. The parameter σ_{R}, which depends on $\langle q\rangle$, is closely related to the above rms($\langle q\rangle$). The first term (σ_{W}^2) on the right-hand side of eqn (2.8) describes the width of the resolution function due to the wavelength spread, while the second term (σ_{C}^2) is related to the finite collimation. The finite spatial resolution of the detector is not considered. The quantity $\Delta\beta$ depends on the sizes of the source and sample apertures, on the source-to-sample distance, and on the scattering angle [203]. The azimuthally averaged function $I(\langle q\rangle)$ is then given by:

$$I(\langle q\rangle) = \int_0^\infty R(q, \langle q\rangle) \frac{d\Sigma}{d\Omega}(q)dq. \tag{2.9}$$

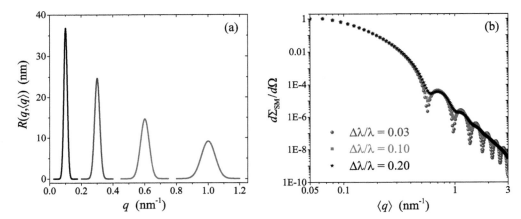

Fig. 2.4: (a) Plot of the resolution function $R(q, \langle q \rangle)$ [eqn (2.7)] at $\langle q \rangle$-values of $0.1\,\mathrm{nm}^{-1}$, $0.3\,\mathrm{nm}^{-1}$, $0.6\,\mathrm{nm}^{-1}$, and $1.0\,\mathrm{nm}^{-1}$. The following parameters were used: $\lambda = 6\,\text{Å}$; $\Delta\lambda/\lambda = 0.1$; $\sigma_\mathrm{C} = 0.01\,\mathrm{nm}^{-1}$. (b) Smeared spin-misalignment SANS cross section $I(\langle q \rangle) = \frac{d\Sigma_\mathrm{SM}}{d\Omega}(\langle q \rangle)$ [eqn (4.14)] at an applied magnetic field of $\mu_0 H_0 = 0.5\,\mathrm{T}$ (log-log scale). For the computation of $\frac{d\Sigma_\mathrm{SM}}{d\Omega}(\langle q \rangle)$ according to eqn (2.9) the form factor of the sphere (with $\xi_\mathrm{H} = \xi_\mathrm{M} = 8\,\mathrm{nm}$) was employed for both S_H and S_M [$S_\mathrm{H} = S_\mathrm{M}$; eqns (3.82)–(3.84)]. Materials parameters for Co were used, $\sigma_\mathrm{C} = 0.01\,\mathrm{nm}^{-1}$, and $\Delta\lambda/\lambda = 0.03; 0.10; 0.20$ (see inset).

One frequently encountered approach to solve eqn (2.9) is to devise a model for $d\Sigma/d\Omega$ and to vary the parameters of the model until the convolution with the resolution function agrees with the measured data. On the other hand, if feasible, the deconvolution may also be carried out by dividing the Fourier transform of the measured data by the Fourier transform of the resolution function. In order to illustrate smearing effects, we display in Fig. 2.4(a) the resolution function $R(q, \langle q \rangle)$ [eqn (2.7)] at selected values of $\langle q \rangle$. It is clearly seen that the broadening increases with increasing $\langle q \rangle$. Figure 2.4(b) depicts the well-known effect of smearing on azimuthally averaged SANS data exhibiting form-factor oscillations. For this particular example, we have chosen the spin-misalignment SANS cross section $d\Sigma_\mathrm{SM}/d\Omega$ [eqn (4.14)]. As can be seen in Fig. 2.4(b), the form-factor oscillations of $d\Sigma_\mathrm{SM}/d\Omega$ are progressively washed out for increasing values of $\Delta\lambda/\lambda$, in particular at high q.

Before moving on we would like to emphasize that the resolution of the instrument, which is always present, needs to be separated from the smearing effect that is related to the particle-size distribution (which is similar in appearance). This can be best seen by considering the nuclear SANS from a dilute assembly of randomly oriented homogeneous particles in a uniform matrix [compare eqn (2.61) in Section 2.4.3]. In this case, the experimental scattering intensity $I(q)$ can be expressed by an integral which is of similar form as eqn (2.9) [208]:

$$I(q) = \int_0^\infty f(R, \langle R \rangle, \sigma_\mathrm{f}) \frac{d\Sigma}{d\Omega}(q, R)\, dR, \qquad (2.10)$$

where

$$\frac{d\Sigma}{d\Omega}(q, R) = \frac{N_{\mathrm{p}}}{V}(\Delta\rho)^2 |F(q, R)|^2 \qquad (2.11)$$

is the theoretical model function for the SANS cross section. In eqns (2.10) and (2.11), N_{p} denotes the number of particles in the scattering volume V, $(\Delta\rho)^2$ is the scattering-length density contrast between particle and matrix (generally assumed to be constant), $F(q, R)$ is the particle form factor [which depends on the particle volume $V_{\mathrm{p}} = V_{\mathrm{p}}(R)$], and the parameters $\langle R \rangle$ and σ_{f} are, respectively, related to the mean particle size and the width of the size distribution $f(R)$ with $\int_0^\infty f(R)dR = 1$. For many problems it is a reasonable assumption to use the form factor of a sphere in eqn (2.11), $F(q, R) = 3V_{\mathrm{s}}j_1(q, R)/(qR)$, where $V_{\mathrm{s}} = \frac{4\pi}{3}R^3$ and $j_1(z)$ denotes the spherical Bessel function of the first order. One can easily convince oneself that for a given two-parameter $f(R)$ (e.g., a Gaussian or log-normal distribution) the scattering curve gets smeared with increasing width σ_{f} of $f(R)$, similar to the resolution-function case. We also refer to Chapter 7.2 in the book by Feigin and Svergun [11] for a discussion and an overview of methods for computing the particle-size distribution from experimental SANS data.

Theoretical descriptions of neutron scattering based on the Born approximation assume infinitely extended plane-wave states for both the incident and the scattered neutron waves [209]. In order to compare the resulting theoretical expressions for the cross sections with experimental data, the non-perfect resolution encountered in any experiment is in the majority of cases only a posteriori taken into account by expressions of the type of eqn (2.9). Gähler et al. [210, 211] have developed a general space-time approach to neutron scattering from many-body systems, which goes beyond the Van Hove formalism [209]. Their wave-optical description takes into account the shaping of the incoming wave function on transmission through the elements of the scattering apparatus (e.g., slits, choppers), and it considers the loss of correlation of the scattered waves after passing through the optical elements in the post-sample flight path.

The above discussion of instrumental q-resolution has to go hand in hand with a consideration of the coherence properties of the neutron beam. As is well known from wave optics, a certain degree of coherence of the illuminating radiation is required in order to observe interference phenomena on the detector [212]. The neutron beam at the sample position can be described in terms of its coherence or correlation volume V_{c}, which is determined by the wavelength distribution of the source (shaped by the velocity selector in conventional SANS experiments) and by the angular divergence (collimation) of the beam, i.e., V_{c} is a property of the neutron beam. Scattering objects in the sample which are separated by distances that are larger than the characteristic dimension of V_{c} are irradiated by neutron waves with different and uncorrelated random phases [78]. Conceptually, one distinguishes two types of coherence with corresponding coherence lengths: the longitudinal coherence length, parallel to the direction of propagation $\mathbf{k}_0$, can be estimated from the wavelength spread $\Delta\lambda$ using the relation

$$l_{\mathrm{coh}}^{\parallel} \cong \frac{\lambda^2}{\Delta\lambda} = \frac{2\pi}{\Delta k_0}, \qquad (2.12)$$

while the transversal coherence length is determined by the angular divergence of the beam and is approximately given by

$$l_{\text{coh}}^{\perp} \cong \frac{L_{\text{col}}}{k_0 r_s} \cong \frac{2}{\Delta q}. \tag{2.13}$$

Here, L_{col} denotes the collimation length or effective source-to-sample distance, in SANS commonly chosen to be equal to the sample-to-detector distance, and r_s is the source-aperture radius [206,213,214]. The expressions (2.12) and (2.13) should be considered as order-of-magnitude relationships [212]. The coherence volume is spanned by the longitudinal and the two transversal coherence lengths. Equation (2.13) implies that $l_{\text{coh}}^{\perp}$ can be increased by increasing L_{col} and by reducing r_s (at the cost of flux). It also illustrates that a neutron beam, which is initially completely incoherent when it emerges e.g., from the moderator of a nuclear reactor (no well-defined phase relationship between neighboring source points due to thermal fluctuations), acquires some coherence by way of downstream traveling. This phenomenon is called creation of coherence via distance [77, 212]. It is also worth emphasizing that even for the idealized situation of a spectrally pure source ($\Delta\lambda/\lambda \to 0$), the coherence volume is limited by the finite spatial extension of the source. For SANS, the coherence volume is typically significantly smaller than the sample size. Assuming typical instrument settings ($\lambda = 1\,\text{nm}$; $\Delta\lambda/\lambda = 0.1$; $L_{\text{col}} = 20\,\text{m}$; $r_s = 1\text{--}2.5\,\text{cm}$), we find $l_{\text{coh}}^{\|} \cong 10\,\text{nm}$ and $l_{\text{coh}}^{\perp} \cong 130\text{--}320\,\text{nm}$, which emphasizes the fact that SANS predominantly probes correlations in the plane normal to the incident beam.

Grigoriev et al. [215] showed that the coherence volume for diffraction in SANS geometry on objects with a two-dimensionally ordered nanostructure and a third non-periodic dimension can significantly expand beyond the value given by the Born approximation: Bragg reflection of coherent neutron waves on a highly ordered hexagonal porous structure of anodic aluminum oxide films, which effectively acts as a two-dimensional neutron grating, leads to an exceptionally elongated coherence length along the beam direction, up to the μm regime. Their result is particularly relevant for studies on colloidal crystals, flux-line lattices [216], or skyrmion structures [217]. On the other hand, the coherence or correlation length of the magnetically ordered state can be conveniently deduced from the peak widths $\delta q^{\star}$ of diffraction peaks via $\xi = 2\pi/\delta q^{\star}$ (including corrections for instrumental resolution).

When the incoming neutrons are scattered by a small angle $\psi \cong q/k_0$ and change direction along $\mathbf{k}_1$, a coherence volume V_c' can be attributed to the outgoing waves. Figure 2.5 depicts the coherence ellipsoids for the incoming and scattered beam in a SANS experiment (see Felber et al. [214] for further reading). The spatial resolution of a SANS instrument is then related to the intersection volume $\overline{V}_c = V_c \cap V_c'$ of the incoming (V_c) and outgoing (V_c') coherence volumes. The characteristic linear dimension of $\overline{V}_c$ (perpendicular to $\mathbf{k}_0$) may be denoted with $\overline{x}_c$, and it can be considered as the spatial resolution of the experiment. An increase of the sample size beyond $\overline{V}_c$ does not yield new information, except for a better counting statistics. This scenario highlights the close link between wavelength band and q-resolution: with increasing scattering-vector magnitude (scattering angle), the beam monochromaticity $\Delta\lambda/\lambda$ has to be enhanced in order to maintain significant overlap (length of $\overline{x}_c$) between V_c and

V'_c [214]. The scattering intensity at momentum-transfer vector $\mathbf{q}$ is obtained by summing up all the products of the particle densities $\rho(\mathbf{r})$ and $\rho(\mathbf{r} + \boldsymbol{\delta})$, where $\boldsymbol{\delta}$ varies within the volume $\overline{V}_c$:

$$I(\mathbf{q}) \sim \int_{\overline{V}_c} C(\boldsymbol{\delta}) \exp\left(-i\mathbf{q} \cdot \boldsymbol{\delta}\right) d^3\delta, \tag{2.14}$$

where

$$C(\boldsymbol{\delta}) \sim \int_{\overline{V}_c} \rho(\mathbf{r})\rho(\mathbf{r} + \boldsymbol{\delta}) d^3r \tag{2.15}$$

is the density-density autocorrelation function within $\overline{V}_c$. The total intensity of the sample, which may be assumed to consist of many coherence volumes, is the average over all the individual coherence volumes $\overline{V}_c$. For magnetic scattering, $\rho(\mathbf{r})$ should be replaced by the component of the magnetization vector field which is perpendicular to the scattering vector [218].

For more detailed studies that address the optimal instrument configuration, instrumental resolution effects, the impact of gravitation, the data-reduction procedure, the general performance of SANS instruments, a survey of background scattering contributions, or the treatment of multiple scattering, see [22, 32, 48, 73, 84, 85, 180, 182–184, 202–205, 219–231] and references therein.

2.3   Influence of inelastic SANS

Typical SANS instrumentation does not allow for energy analysis of the scattered neutrons and the measurable quantity is the energy-integrated macroscopic differential scattering cross section $d\Sigma/d\Omega$ at scattering vector $\mathbf{q}$. Since in this book we are specifically interested in the static magnetic microstructure of magnetic materials, which is probed by elastic scattering, it is necessary to estimate the influence of inelastic scattering to the energy-integrated cross section. Potential sources of inelastic scattering contributions to $d\Sigma/d\Omega$ are spin waves (magnons) and lattice vibrations (phonons). Their influence is estimated in this section. We start out by writing down the conservation laws for momentum and energy transfers in the small-angle region (reciprocal lattice vector: $\mathbf{G} = 0$), which read [76]:

$$\mathbf{q} = \mathbf{k}_0 - \mathbf{k}_1 = \pm\mathbf{W}_{\mathrm{exc}}, \tag{2.16}$$

and

$$\Delta E = \frac{\hbar^2}{2m_{\mathrm{n}}}\left(k_0^2 - k_1^2\right) = \pm\hbar\omega_{\mathrm{sw,ph}}, \tag{2.17}$$

where $\mathbf{q}$ is the momentum-transfer vector, $\mathbf{k}_0$ and $\mathbf{k}_1$ denote, respectively, the wave vectors of the incident and of the scattered neutron, and $\mathbf{W}_{\mathrm{exc}}$ represents the wave vector which characterizes the mode of propagation of the excitation. The $+$ and $-$ signs refer to the creation (neutron energy loss) and annihilation (neutron energy gain) of a

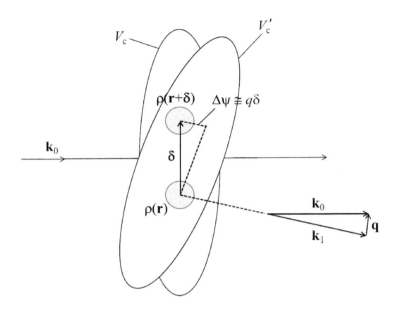

Fig. 2.5: In small-angle neutron scattering the incoming coherence volume V_c has its main extension $l_{coh}^{\perp}$ perpendicular to $\mathbf{k}_0$. This allows one to measure sample correlations within distances $\delta \lesssim l_{coh}^{\perp}$, which may reach values up to μm in SANS instruments. The resolution is limited to the intersection volume $\overline{V}_c = V_c \cap V_c'$ of the ingoing (V_c) and outgoing (V_c') coherence volumes, which shows the relation between monochromaticity and scattering angle. Adding up of all products of scattered waves from $\rho(\mathbf{r})$ and $\rho(\mathbf{r} + \boldsymbol{\delta})$ leads to the Fourier transform of the correlation function $C(\mathbf{r})$ within $\overline{V}_c$. $\Delta\psi \cong q\delta$ denotes the phase difference between waves scattered at $\mathbf{r}$ and at $\mathbf{r} + \boldsymbol{\delta}$. After [214].

single magnon (phonon), m_n is the neutron mass, and $\hbar\omega_{sw,ph}$ represents the dispersion relation for long-wavelength spin waves (sw) or phonons (ph). For ferromagnetic spin waves, we use the simplified expression [232]:

$$\hbar\omega_{sw} \cong \mathrm{D}q_{sw}^2 + g\mu_B\mu_0 H_0 + \Delta, \tag{2.18}$$

where D denotes the spin-wave stiffness constant, q_{sw} is the magnon wave vector magnitude, g is the Landé factor, μ_B is the Bohr magneton, μ_0 is the permeability of free space, H_0 is the applied magnetic field, and Δ depends on the magnetic anisotropy and on the magnetodipolar interaction. Equation (2.18) is plotted in Fig. 2.6. For phonons, the dispersion relation in the long-wavelength limit can be approximated by [232]:

$$\hbar\omega_{ph} \cong \hbar v_s q_{ph}, \tag{2.19}$$

where v_s denotes the velocity of sound, and q_{ph} is the magnitude of the wave vector of the phonon.

In the absence of an applied magnetic field and for $\Delta = 0$, eqns (2.16)–(2.18) predict that the inelastic small-angle scattering due to one-magnon processes is restricted

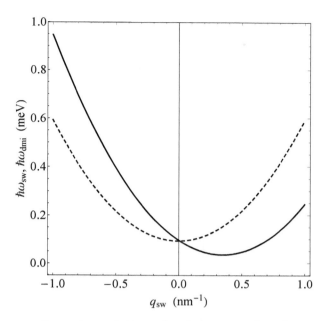

Fig. 2.6: Spin-wave dispersion relations for ferromagnetic spin waves [eqn (2.18)] (dashed line) and in the presence of DMI (Dzyaloshinskii–Moriya interaction) [eqn (2.23)] (solid line). For the calculation, we have used $\Delta = 0$, $g = 2$, $\mu_0 H_0 = 0.8\,\mathrm{T}$, $D = 50\,\mathrm{meV\,\mathring{A}^2}$, $k_s = 0.35\,\mathrm{nm^{-1}}$, and $\mu_0 H_{c2} = 0.5\,\mathrm{T}$. Note that the dispersion relations for the $3d$ band magnets are much steeper than the one shown here.

to scattering angles ψ below the following critical angle (see, e.g., [233, 234] and also example 8.4 with solution in the book by Squires [76]):

$$\psi_{C0} = \arcsin\left(\frac{\hbar^2}{2m_n D}\right) \cong \frac{\hbar^2}{2m_n D}, \tag{2.20}$$

which is independent of the primary neutron wavelength. Note that $2m_n D/\hbar^2 \gg 1$ for most materials (the value is $\sim$111 for Fe, $\sim$203 for Ni, and $\sim$236 for Co at room temperature), which justifies the approximation in eqn (2.20). For $\psi > \psi_{C0}$ the magnetic SANS is elastic. Measurement of ψ_{C0} provides a method for determining the spin-wave stiffness constant D [235–238]. In the presence of a gap in the dispersion relation, $\Delta' = g\mu_B\mu_0 H_0 + \Delta$, the critical angle is further reduced to:

$$\psi_{C0}^{\Delta'} = \arcsin\left(\sqrt{\left(\frac{\hbar^2}{2m_n D}\right)^2 - \frac{\Delta'}{D k_0^2}\left(1 + \frac{\hbar^2}{2m_n D}\right)}\right), \tag{2.21}$$

which now also depends on $k_0 = 2\pi/\lambda$ [239]. For $\Delta' = 0$, eqn (2.21) simplifies to eqn (2.20). The function $\psi_{C0}^{\Delta'}$ is plotted in Fig. 2.7 and shows that $\psi_{C0}^{\Delta'=0} \lesssim 1°$ for the $3d$ band magnets Fe, Ni, and Co. In the absence of an externally applied magnetic field, it is experimentally found that $\Delta' = \Delta \cong 0.05\,\mathrm{meV}$ for the $3d$ band magnets [240], which corresponds to $\Delta'/(g\mu_B) \cong 400\,\mathrm{mT}$.

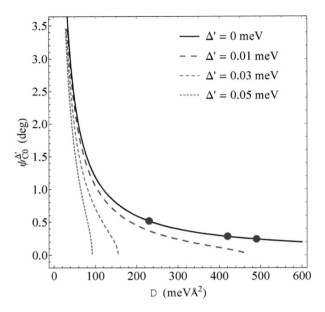

Fig. 2.7: Critical angle for the suppression of one-magnon scattering in the small-angle region, $\psi_{C0}^{\Delta'}$ [eqn (2.21)], as a function of the spin-wave stiffness D and for several values of Δ' (see inset) ($\lambda = 6\,\text{Å}$). For $\Delta' = 0$ [eqn (2.20)], the room-temperature values of ψ_{C0} for the $3d$ band magnets Fe (D = $230\,\text{meVÅ}^2$ [241]), Ni (D = $420\,\text{meVÅ}^2$ [242]), and Co (D = $490\,\text{meVÅ}^2$ [243]) are indicated by the large data points.

Similarly, for phonons, eqns (2.16) and (2.17) together with eqn (2.19) yield the following condition:

$$\left| \pm \frac{v_s}{v_n} + \frac{q_{ph}}{2k_0} \right| \leq 1. \tag{2.22}$$

Assuming that $q_{ph} \ll 2k_0$, eqn (2.22) then suggests that inelastic phonon scattering is not possible at small scattering angles, if the neutron velocity v_n is smaller than v_s. In solids, v_s is typically in the range of several thousand m/s, whereas, for cold neutrons, v_n is of the order of a few hundred m/s (e.g., $v_n = 659\,\text{m/s}$ for neutrons with a wavelength of 6 Å). A more elaborate treatment of inelastic coherent phonon scattering by Lovesey [86] confirms that, for cold neutrons along the forward direction, one-phonon creation processes have a zero cross section, while the coherent cross section for one-phonon annihilation processes is of the order of only 10 millibarns at 300 K.

Based on the simplified isotropic dispersion relations, eqns (2.18) and (2.19), the above results suggest that throughout a considerable part of the parameter space which is probed in typical SANS experiments it is possible to suppress inelastic scattering; in other words, the kinematic requirements of momentum and energy conservation upon absorption or emission of a magnon or phonon cannot generally be satisfied simultaneously for scattering vectors in the small-angle regime, in particular, at large applied fields and/or for strongly anisotropic magnets (large Δ). For a more detailed discussion of inelastic SANS by spin waves using a more advanced dispersion relation than eqn (2.18), we refer to the article by Maleev [234].

For the remainder of this section we address the so-called left-right asymmetry method using polarized neutrons. This technique was developed at the Petersburg Nuclear Physics Institute, in Gatchina, Russia, in the 1980s. It allows one to access the dynamics of spin systems under an applied magnetic field, in particular, in the low-q region [244–256]. Figure 2.8 shows a sketch explaining the method, and Figs. 2.9 and 2.10 illustrate the application of this technique to the determination of the temperature dependence of the spin-wave stiffness constant of the non-centrosymmetric helical magnet MnSi [251]. The helical magnetic order of MnSi below $T_C = 29.5$ K [257] is the result of the competition between the strong isotropic ferromagnetic exchange interaction and the weak antisymmetric Dzyaloshinskii–Moriya interaction (DMI), which has its origin in the lack of inversion symmetry of the B20 crystal structure. The direction of the helical modulation (with a wavelength of $\lambda_h \cong 180$ Å [64]) is pinned along the cubic space diagonals of the crystal structure by weak crystal-field interactions. For ferromagnets in the presence of DMI, the following spin-wave dispersion relation has been predicted by Kataoka [258] (for $H_0 > H_{c2}$):

$$\hbar\omega_{dmi} = D(\mathbf{q}_{sw} - \mathbf{k}_s)^2 + g\mu_B\mu_0(H_0 - H_{c2}), \qquad (2.23)$$

where $\mathbf{k}_s \parallel \mathbf{H}_0$ denotes the wave vector of the helix, and H_{c2} represents the value of the external magnetic field that is required to transform the helix into a ferromagnetic collinear fully polarized state. The magnitude of $\mathbf{k}_s$ depends on the ratio of the DMI constant and the isotropic exchange constant ($k_s = 2\pi/\lambda_h \cong 0.35$ nm^{-1} for MnSi), whereas the sign of the DMI constant determines the direction of $\mathbf{k}_s$ being either parallel or antiparallel with respect to the direction of the external field. The asymmetric dispersion relation eqn (2.23) exhibits three important differences when compared to the symmetric relation eqn (2.18) [251, 258]: (i) the minimum of $\hbar\omega_{dmi}$ is shifted from $\mathbf{q}_{sw} = 0$ to $\mathbf{q}_{sw} = \mathbf{k}_s$; (ii) the sign of the DMI constant determines the preferred direction of propagation of the spin waves; (iii) the spin-wave gap related to the magnetic field is shifted by the value $g\mu_B\mu_0 H_{c2}$ (compare Fig. 2.6).

Following Grigoriev et al. [251–256], the field dependence of the critical angle ψ_C is approximately given by:

$$\psi_C^2(H_0) = \psi_{C0}^2 - \psi_{C0}\frac{g\mu_B\mu_0 H_0}{E_0} + \psi_B^2, \qquad (2.24)$$

where $\psi_{C0} = \hbar^2/(2m_n D)$, E_0 is the energy of the incident neutrons, and ψ_B denotes the Bragg-peak position related to the diffraction on the helix. Equation (2.24) has been derived using the $\mathbf{q}$ and ω-dependent inelastic magnetic chiral contribution to the neutron cross section along with the asymmetric dispersion relation eqn (2.23) [251]. It is straightforward to show that eqn (2.21) can be transformed into eqn (2.24). The inelastic spin-wave scattering of DMI helimagnets is concentrated within a circle of radius ψ_C centered at the Bragg angle ψ_B (compare Fig. 2.8). Hence, ψ_C represents the cutoff angle for this type of scattering and contains the information about the spin-wave stiffness constant D. As the field reaches the value of H_{c2}, the elastic Bragg scattering on the helix disappears completely and only the inelastic diffuse scattering

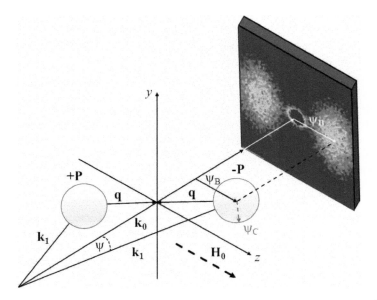

Fig. 2.8: Schematic of inelastic SANS by spin waves from DMI magnets ($\mathbf{k}_0 \perp \mathbf{H}_0$). The distance from the center of the direct beam to the position of the Bragg peak due to the magnetic helix can be related to the scattering angle ψ_B. The inelastic spin-wave scattering of DMI helimagnets is concentrated within a circle of radius ψ_C centered at the Bragg angle ψ_B. If the incident neutrons are polarized along the magnetic field ($-\mathbf{P}$), then the scattering with neutron energy gain is only allowed, while the energy-loss process is forbidden. A flip of the polarization to $+\mathbf{P}$, on the contrary, allows scattering with the magnon creation process and makes the magnon annihilation process forbidden [254].

centered at $\mathbf{q} = \pm\mathbf{k}_\mathrm{s}$ remains (compare Fig. 2.9). This scattering consists of a strong diffuse component in the vicinity of the former Bragg peak and a round spot which is limited by the critical angle ψ_C. The diffuse scattering at $\mathbf{q} = \pm\mathbf{k}_\mathrm{s}$ is maximal at $H_0 \cong H_{c2}$ and strongly suppressed by an increase of the field. The round spot centered at $\mathbf{q} = \pm\mathbf{k}_\mathrm{s}$ can be observed in the field range up to $H_0 \cong 2H_{c2}$. Measurements at low temperature ($\sim 5\,\mathrm{K}$) and high magnetic field ($\sim 5\,\mathrm{T}$) serve as the background signal, since then both the elastic magnetic peak and the spin-wave scattering are fully suppressed [253].

The parameters of eqn (2.24) are found by fitting the azimuthally averaged scattering intensity $I(\psi)$ to a combination of a Lorentzian and sigmoid-type function, according to [compare Fig. 2.10(a)−(d)] [253]:

$$I(\psi) = \frac{I_0}{(\psi - \psi_\mathrm{B})^2 + \kappa_\mathrm{I}^2} \left\{ \frac{1}{2} - \left(\frac{1}{\pi} \arctan\left[\frac{2(\psi - \psi_\mathrm{C})}{\delta_\mathrm{I}} \right] \right) \right\}. \tag{2.25}$$

$I(\psi)$ is here obtained as the difference between two measurements with the neutron spins aligned antiparallel and parallel to the guide field (see Fig. 2.9). The Lorentzian

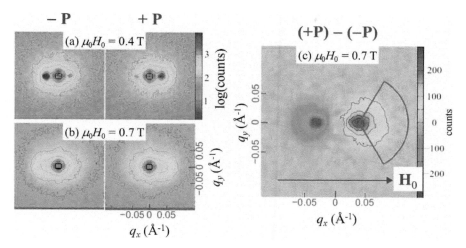

Fig. 2.9: Polarized SANS intensities of single-crystalline MnSi at $T = 15\,\mathrm{K}$ and at an applied magnetic field below H_{c2} (0.4 T) (a) and above H_{c2} (0.7 T) (b). (left column) Incoming polarization **P** antiparallel to the guide field; (right column) **P** parallel to the guide field. (c) Difference between the right and left SANS patterns at $\mu_0 H_0 = 0.7\,\mathrm{T}$. The area enclosed by the solid lines specifies the sector used for computing the azimuthal average. After [251].

function describes the contribution of the diffuse small-angle scattering and its parameter $\kappa_I^2 \cong \frac{\psi_{C0}}{E_0} g\mu_B\mu_0(H_0 - H_{c2})$ reflects the closeness of the system to the critical field H_{c2}. The sigmoid function serves as a step-like function with a cutoff angle ψ_C and a width δ_I, which is related to the spin-wave damping. In Fig. 2.10(e) it is seen that the ψ_C^2 data follow a linear field dependency as predicted by eqn (2.24). The results for $D(T)$ [Fig. 2.10(f)] agree quite well with the values obtained by triple-axis spectroscopy (TAS), which is generally the method of choice for accessing the dynamics of the spin system of single-crystalline magnetic materials. However, TAS—in contrast to the inelastic polarized small-angle method—suffers from a low q-resolution and therefore cannot reveal the chiral anisotropy of the spin-wave spectrum of MnSi in the vicinity of $k_s \cong 0.35\,\mathrm{nm}^{-1}$ [251]. Further advantages of the inelastic SANS technique are that it can be used to characterize the spin-wave stiffness in a wide range of small-sized polycrystalline magnets with an acceptable counting statistics within a reasonable exposure time.

The left-right asymmetry technique has originally been developed and used to study conventional ferromagnets having a centrosymmetric crystal structure without DMI; for instance, it was applied to investigate the spin dynamics of iron in the critical region above T_C or the spin waves in amorphous systems [244–247]. It is important to emphasize that for conventional ferromagnets the externally applied magnetic field $\mathbf{H}_0$ must be inclined at an angle $\varsigma_{(\mathbf{k}_0,\mathbf{H}_0)} \neq 90°$ (or $\varsigma_{(\mathbf{k}_0,\mathbf{H}_0)} \neq 0°$) relative to $\mathbf{k}_0$. This fact is related to the symmetry properties of the polarization-dependent purely inelastic magnetic contribution to the cross section [262], the so-called dynamical chi-

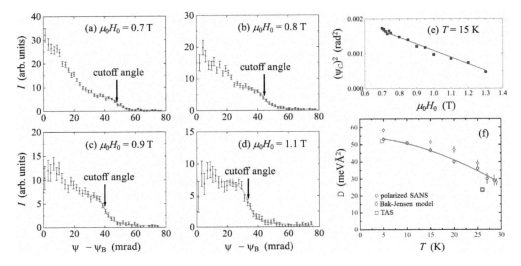

Fig. 2.10: (a)−(d) Azimuthally averaged difference scattering intensity centered at the Bragg-peak position, I versus scattering angle $\psi - \psi_B$, at several applied magnetic fields (see insets). The solid lines represent the results of best fits for determining the cutoff angle ψ_C [eqn (2.25)]. (e) Field dependence of the square of the cutoff angle ψ_C^2 at $T = 15$ K. The solid line is a linear fit to eqn (2.24). (f) Temperature dependence of the spin-wave stiffness $D(T)$: circles—data from polarized SANS; diamonds—estimation from the Bak–Jensen model [259–261]; squares—data from triple-axis spectroscopy (see [251] for details). Solid line is a power-law fit. After [251].

ral function $Z^{+-}(q,\omega) \sim I^+(q,\omega) - I^-(q,\omega)$ [263]. This is an odd function of the energy transfer ω, and, as discussed e.g., in [246], only angles $\varsigma_{(\mathbf{k}_0, \mathbf{H}_0)} \neq 90°$ result in the angular factors for neutron energy loss and gain processes to behave inversely, in this way giving rise to a nonzero integral asymmetry $\int Z^{+-}(q,\omega)d\omega$. However, for DMI helimagnets such as MnSi or Cu$_2$OSeO$_3$, which exhibit an asymmetric magnon spectrum, Grigoriev et al. [251–256] have demonstrated that the case $\varsigma_{(\mathbf{k}_0, \mathbf{H}_0)} = 90°$ ($\mathbf{k}_0 \perp \mathbf{H}_0$; compare Fig. 2.8) yields a non-vanishing asymmetry. As a further reading we recommend the review article by Gukasov [263], who provides a detailed discussion of the left-right asymmetry in polarized neutron scattering. For the remainder of the book, unless otherwise stated, we will treat magnetic SANS within the elastic approximation ($k_0 = k_1$).

2.4   Summary of the basics of nuclear SANS

In order to facilitate connecting the equations for magnetic SANS (to be introduced in Sections 2.5−2.8) to the more familiar results for nuclear SANS, we briefly summarize a few of the most important results of nonmagnetic small-angle scattering. For more details, the reader is referred to the textbooks by Guinier and Fournet [9], Glatter and Kratky [10], Feigin and Svergun [11], Svergun, Koch, Timmins, and May [12], and Gille [13]. We begin the discussion by introducing the well-known expressions for the nuclear coherent and incoherent scattering cross sections (Section 2.4.1). Some selected experiments on nuclear-spin-dependent SANS are highlighted in Section 2.4.2. The basics of nuclear SANS are then presented in Section 2.4.3. We assume in the following that the nuclear atomic scattering lengths b are real-valued quantities (no absorption). Moreover, throughout this book, we consider the nuclei to be fixed in space, i.e., strongly bound to the crystal lattice, so that the relevant atomic scattering lengths are the so-called bound scattering lengths [82]. We remind that the scattering cross section for a free nucleus is for most nuclei slightly smaller than the bound cross section, according to [75, 76]:

$$\sigma_{\text{free}} = \left(\frac{M_{\text{nuc}}}{M_{\text{nuc}} + m_{\text{n}}} \right)^2 \sigma_{\text{bound}}, \qquad (2.26)$$

where M_{nuc} denotes the mass of the nucleus. Equation (2.26) shows that the difference is particularly strong for the case of light hydrogen (factor of four), but becomes negligible for heavier atoms.

2.4.1 Coherent and incoherent nuclear scattering

In terms of the atomic coordinates $\mathbf{r}_k$ in the irradiated sample volume V the macroscopic nuclear elastic differential scattering cross section at momentum-transfer vector $\mathbf{q}$ is given by [76]:

$$\frac{d\Sigma_{\text{nuc}}}{d\Omega}(\mathbf{q}) = \frac{1}{V} \left| \sum_{k=1}^{N} b_k \exp\left(-i\mathbf{q} \cdot \mathbf{r}_k\right) \right|^2 = \frac{1}{V} \sum_{k,l} b_k b_l \exp\left(-i\mathbf{q} \cdot [\mathbf{r}_k - \mathbf{r}_l]\right), \quad (2.27)$$

where the b_k denote the nuclear atomic scattering lengths. It is a particular property of neutron scattering that the nuclear scattering potential varies from one atom to the next for the same chemical element [77]. The associated changes (fluctuations) in the scattering lengths are related (i) to isotope variations and (ii) to variations in the nuclear spin. The scattering lengths are non-systematically distributed on the atomic sites, i.e., there exists no obvious correlation between the values of the scattering lengths and the sites. Figure 2.11 shows the variation of the nuclear scattering lengths with the atomic number. When the system of atoms (particles) is in thermal equilibrium at temperature T, one has to take an ensemble average of eqn (2.27) in order to compare it with experimental data. This ensemble average includes (i) an average over the distribution of positions of particles in the system in all its possible configurations consistent with a given temperature T, and (ii) an average over the distribution of nuclei over various spin and isotopic states [264]. Formally, we take this averaging into account by re-writing eqn (2.27) as:

$$\frac{d\Sigma_{\text{nuc}}}{d\Omega}(\mathbf{q}) = \frac{1}{V} \sum_{k,l} \overline{b_k b_l} \exp\left(-i\mathbf{q} \cdot [\mathbf{r}_k - \mathbf{r}_l]\right), \qquad (2.28)$$

where $\overline{b_k b_l}$ denotes such an average of $b_k b_l$.

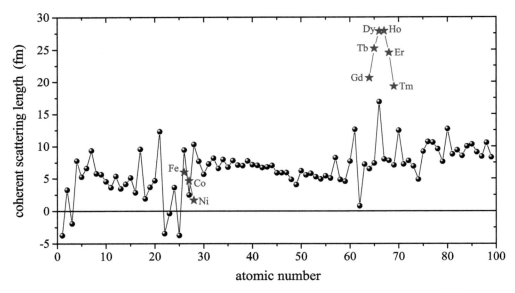

Fig. 2.11: (•) Nuclear coherent scattering length versus atomic number. Values cor‑respond to the naturally occuring isotope composition. For comparison, the magnetic scattering lengths b_m of the magnetic elements from Table 1.2 are also displayed (⋆). Data taken from [102].

In order to further evaluate eqn (2.28), we assume that the sample consists of a single element and that the value of b_k occurs with a relative frequency of f_k, such that [76]

$$\sum_k f_k = 1. \tag{2.29}$$

The variation (from one nucleus to another) of the scattering length is due to the nuclear spin, the presence of isotopes, or both. The isotope distribution function is usually well known for a particular element [102]. The average values of b and b^2 of the specimen, respectively, $\bar{b}$ and $\overline{b^2}$ are then given by:

$$\bar{b} = \sum_k f_k b_k \tag{2.30}$$

and

$$\overline{b^2} = \sum_k f_k b_k^2. \tag{2.31}$$

By assuming that there is no correlation between the values of the scattering length at any pair of different sites k and l and by making use of the relation [76, 103]

$$\overline{b_k b_l} = \bar{b}^2 + \delta_{kl} \left(\overline{b^2} - \bar{b}^2 \right), \tag{2.32}$$

where δ_{kl} is the Kronecker delta function, eqn (2.28) can be re-expressed as

$$\frac{d\Sigma_{\text{nuc}}}{d\Omega}(\mathbf{q}) = \frac{N}{V}\left(\overline{b^2} - \overline{b}^2\right) + \frac{1}{V}\overline{b}^2\sum_{k,l}\exp\left(-i\mathbf{q}\cdot[\mathbf{r}_k - \mathbf{r}_l]\right). \tag{2.33}$$

The term $\overline{b}^2$ in eqn (2.32) collects all the contributions with $k \neq l$, i.e., $(\overline{b_k b_l})_{k\neq l} = \overline{b_k}\,\overline{b_l} = \overline{b}^2$, while the second term $\delta_{kl}(\overline{b^2} - \overline{b}^2)$ represents the $k = l$ self-scattering contributions, i.e., $(\overline{b_k b_l})_{k=l} = \overline{b_k^2} = \overline{b^2}$. The first term on the right-hand side of eqn (2.33) denotes the nuclear incoherent cross section,

$$\frac{d\Sigma_{\text{nuc}}^{\text{inc}}}{d\Omega} = \frac{N}{V}\left(\overline{b^2} - \overline{b}^2\right) = \frac{N}{V}b_{\text{inc}}^2 = \frac{N}{V}\frac{\sigma_{\text{inc}}}{4\pi}, \tag{2.34}$$

whereas the second term is the nuclear coherent contribution,

$$\frac{d\Sigma_{\text{nuc}}^{\text{coh}}}{d\Omega} = \frac{1}{V}\overline{b}^2\sum_{k,l}\exp\left(-i\mathbf{q}\cdot[\mathbf{r}_k - \mathbf{r}_l]\right) = \frac{1}{V}b_{\text{coh}}^2\sum_{k,l}\exp\left(-i\mathbf{q}\cdot[\mathbf{r}_k - \mathbf{r}_l]\right)$$

$$= \frac{1}{V}\frac{\sigma_{\text{coh}}}{4\pi}\sum_{k,l}\exp\left(-i\mathbf{q}\cdot[\mathbf{r}_k - \mathbf{r}_l]\right). \tag{2.35}$$

The incoherent and coherent signals are qualitatively very much different: $d\Sigma_{\text{nuc}}^{\text{inc}}/d\Omega$ adds a q-independent flat background to the elastic SANS signal, whereas $d\Sigma_{\text{nuc}}^{\text{coh}}/d\Omega$ contains the information about the structural correlations of the sample. Equations (2.34) and (2.35) also define the nuclear incoherent σ_{inc} and coherent σ_{coh} scattering cross sections (per atom):

$$\sigma_{\text{inc}} = 4\pi b_{\text{inc}}^2 = 4\pi\left(\overline{b^2} - \overline{b}^2\right) \tag{2.36}$$

and

$$\sigma_{\text{coh}} = 4\pi b_{\text{coh}}^2 = 4\pi\overline{b}^2. \tag{2.37}$$

Values for these quantities are tabulated for most isotopes in [102]. For a system which is composed of a single isotope of zero nuclear spin, $d\Sigma_{\text{nuc}}^{\text{inc}}/d\Omega = 0$ and the nuclear scattering is completely coherent.

For a scattering system consisting of a single isotope with nuclear spin I, the neutron-nucleus interaction gives rise to two compound nuclear states which are characterized by scattering lengths of b_+ and b_-, corresponding, respectively, to spin states of $J_+ = I + \frac{1}{2}$ and $J_- = I - \frac{1}{2}$ [76]. The number of states associated with the compound spin J_+ is $2J_+ + 1 = 2I + 2$, and the number of states corresponding to the spin J_- is $2J_- + 1 = 2I$. This yields a total number of $4I + 2$ equally probable states, provided that the incident neutrons are unpolarized and that the nuclear spins are randomly oriented, resulting in a zero nuclear spin polarization of the sample. This allows one to determine the frequencies f^+ and f^- with which the scattering lengths b_+ and b_- occur as:

$$f^+ = \frac{2I+2}{4I+2} = \frac{I+1}{2I+1} \tag{2.38}$$

and

$$f^- = \frac{2I}{4I+2} = \frac{I}{2I+1}. \tag{2.39}$$

The mean scattering length and the mean of the squared scattering length are then found as:

$$\bar{b} = \frac{I+1}{2I+1}b_+ + \frac{I}{2I+1}b_- = \frac{(I+1)b_+ + Ib_-}{2I+1} \tag{2.40}$$

and

$$\overline{b^2} = \frac{(I+1)b_+^2 + Ib_-^2}{2I+1}. \tag{2.41}$$

Using eqns (2.36) and (2.37), the nuclear coherent and incoherent cross sections are readily computed:

$$\sigma_{\text{coh}} = 4\pi \frac{[(I+1)b_+ + Ib_-]^2}{(2I+1)^2} \tag{2.42}$$

$$\sigma_{\text{inc}} = 4\pi \frac{(b_+ - b_-)^2 \, I(I+1)}{(2I+1)^2}. \tag{2.43}$$

For the proton with $I = 1/2$, one finds $f_+ = 3/4$ for the triplet state and $f_- = 1/4$ for the singlet state. The experimentally determined values of b_+ and b_- (for bound protons) are [101, 265]:

$$b_+ = 10.817 \times 10^{-15}\,\text{m} = 10.817\,\text{fm} \tag{2.44}$$

and

$$b_- = -47.420 \times 10^{-15}\,\text{m} = -47.420\,\text{fm}, \tag{2.45}$$

so that $\bar{b} = b_{\text{coh}} = \frac{3}{4}b_+ + \frac{1}{4}b_- = -3.74\,\text{fm}$, resulting in

$$\sigma_{\text{coh}} = 4\pi\bar{b}^2 = 1.76 \times 10^{-24}\,\text{cm}^2 = 1.76\,\text{barn}. \tag{2.46}$$

Similarly, $\overline{b^2} = \frac{3}{4}b_+^2 + \frac{1}{4}b_-^2 = 6.50\,\text{barn}$, yielding [88]

$$\sigma_{\text{inc}} = 4\pi(\overline{b^2} - \bar{b}^2) = 79.91\,\text{barn}. \tag{2.47}$$

We note that in [102, 265] the value of the incoherent scattering cross section of the proton is reported as $\sigma_{\text{inc}} = 80.27\,\text{barn}$.

For a sample containing several isotopes, the quantities f_+ and f_- need to be multiplied (for each isotope) by the relative abundance of the isotope in order to obtain the relative frequency of the scattering length. It follows that [76]

$$\overline{b} = \sum_{\xi} \frac{c_\xi}{2I_\xi + 1} \left[(I_\xi + 1)b_+^\xi + I_\xi b_-^\xi \right] \qquad (2.48)$$

and

$$\overline{b^2} = \sum_{\xi} \frac{c_\xi}{2I_\xi + 1} \left[(I_\xi + 1)(b_+^\xi)^2 + I_\xi (b_-^\xi)^2 \right], \qquad (2.49)$$

where c_ξ denotes the relative abundance of the ξth isotope, I_ξ is its nuclear spin, and b_+^ξ and b_-^ξ are its scattering lengths.

The magnitude of nuclear incoherent scattering is usually very much smaller than the coherent magnetic SANS that is of interest in the present book; for instance, among the $3d$ band magnets Ni has the largest nuclear incoherent scattering cross section, which is due to the isotope distribution of elemental Ni. Using $\sigma_{inc} = 5.2$ barn [88] and a mass density of $\rho_{m,Ni} = 8.912$ g/cm^3, the macroscopic nuclear incoherent cross section is found at $d\Sigma_{nuc}^{inc}/d\Omega = \frac{N}{V}\frac{\sigma_{inc}}{4\pi} \cong 0.038$ cm^{-1}. For light water, H$_2$O, the nuclear incoherent scattering equals $d\Sigma_{nuc}^{inc}/d\Omega \cong 0.45$ cm^{-1} and is almost exclusively due to the nuclear spin of the proton [73]. At the smallest momentum-transfer vectors these values are negligible as compared to the typical magnetic SANS cross section of a bulk ferromagnet (compare Chapter 4).

2.4.2 Spin-polarized nuclei and polarized neutrons

When the incident neutron beam and the atomic nuclei are polarized (e.g., relative to the same quantization axis), the nuclear incoherent and coherent scattering cross sections become dependent on the incident neutron beam polarization P and on the nuclear spin polarization P_n. In this situation, the $4I+2$ spin states of the nucleus-neutron system are not anymore equally probable. Conceptually, nuclear-spin-dependent scattering is most conveniently discussed by formally introducing the scattering length operator as [266]:

$$\hat{b} = \overline{b} + \frac{1}{2}b_n \mathbf{I} \cdot \boldsymbol{\sigma}_P, \qquad (2.50)$$

where the coherent scattering length $\overline{b}$ and the quantity b_n can be expressed in terms of $b_+ = \overline{b} + \frac{1}{2}b_n I$ and $b_- = \overline{b} - \frac{1}{2}b_n(I + 1)$ as, respectively, eqn (2.40) and

$$b_n = \frac{2(b_+ - b_-)}{2I + 1}. \qquad (2.51)$$

$\boldsymbol{\sigma}_P$ denotes the Pauli spin operator for the neutron. For a sample containing identical nuclei with spin I and ignoring long-range nuclear spin correlations, which may become

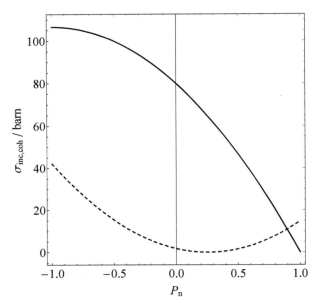

Fig. 2.12: Nuclear-spin-dependent incoherent (solid line) and coherent (dashed line) scattering cross sections of the proton ($I = 1/2$) as a function of the proton spin polarization P_n for $P = 1$ [eqns (2.52) and (2.53)].

relevant at very low temperatures [267], the spin-dependent incoherent and coherent cross sections are according to Glättli and Goldman [266] then given by:

$$\frac{\sigma_{\text{inc}}}{4\pi} = \frac{1}{4}b_n^2 \left(I(I+1) - IPP_n - I^2 P_n^2\right) \tag{2.52}$$

and

$$\frac{\sigma_{\text{coh}}}{4\pi} = \bar{b}^2 + \bar{b}b_n IPP_n + \frac{1}{4}b_n^2 I^2 P_n^2, \tag{2.53}$$

where $-1 \leq P \leq 1$ and $-1 \leq P_n \leq 1$. Equation (2.52) shows that the incoherent scattering vanishes for $PP_n = 1$ and that it takes on its maximum value for $PP_n = -1$. Both cross sections are plotted in Fig. 2.12 for the case of protons ($I = 1/2$; $\bar{b} = -3.74\,\text{fm}$; $b_n = 58.24\,\text{fm}$). The states with $P = 1$ and $P_n = 1$ as well as $P = -1$ and $P_n = -1$ (both resulting in $PP_n = 1$) are eigenstates of the total spin $J_+ = I + 1/2$. Since in this situation all the individual scattering lengths are equal, there is no incoherent scattering. By contrast, the states with $P = 1$ and $P_n = -1$ and $P = -1$ and $P_n = 1$ are mixtures of eigenstates of J_+ and J_- having different scattering lengths, whence the occurrence of incoherent scattering [266].

Under "conventional" experimental conditions, the thermal equilibrium value of P_n, which is given by Boltzmann statistics, is extremely small, so that the strong spin dependence of σ_{inc} and σ_{coh} [eqns (2.52) and (2.53)] cannot be exploited easily in SANS studies on e.g., soft matter via contrast variation. In fact, P_n depends on the ratio of the applied magnetic field $B_0 = \mu_0 H_0$ and the temperature T and is given by the Brillouin function [268]:

$$P_n = \frac{2I+1}{2I} \coth\left(\frac{2I+1}{2I}\frac{\mu_I B_0}{kT}\right) - \frac{1}{2I} \coth\left(\frac{1}{2I}\frac{\mu_I B_0}{kT}\right), \qquad (2.54)$$

which for protons ($I = 1/2$) reduces to

$$P_n = \tanh\left(\frac{\mu_p B_0}{kT}\right), \qquad (2.55)$$

where $\mu_I = \mu_p = 2.793\,\mu_N$ is the magnetic dipole moment of the proton (μ_N: nuclear magneton). Inserting $B_0 = 2.5\,\mathrm{T}$ and $T = 1\,\mathrm{K}$ into eqn (2.55), we see that $P_n \cong 0.26\%$ only, in this way rendering the linear and quadratic terms in eqns (2.52) and (2.53) small as compared to the P_n-independent terms.

However, the method of dynamic nuclear polarization (DNP) [266, 268–270] represents a powerful tool for the alteration of the neutron scattering from nuclei possessing a nonzero nuclear spin (mostly protons). A necessary prerequisite for DNP is the existence of unpaired electrons in the sample, so-called paramagnetic centers, which interact with the nuclei via the magnetodipolar interaction. For a single nuclear magnetic moment μ_I interacting with an electronic magnetic moment μ_S in the presence of an applied magnetic field $\mathbf{B}_0 = \mu_0 \mathbf{H}_0$, the Hamiltonian can be written as:

$$\widehat{H} = -\left(\boldsymbol{\mu}_I + \boldsymbol{\mu}_S\right) \cdot \mathbf{B}_0 + \frac{\mu_0}{4\pi}\left[\frac{\boldsymbol{\mu}_I \cdot \boldsymbol{\mu}_S}{r^3} - 3\frac{(\boldsymbol{\mu}_I \cdot \mathbf{n})(\boldsymbol{\mu}_S \cdot \mathbf{n})}{r^3}\right], \qquad (2.56)$$

where r denotes the distance between $\boldsymbol{\mu}_I$ and $\boldsymbol{\mu}_S$, and $\mathbf{n}$ is a unit vector parallel to this line. On irradiation with microwaves of frequency close to the electron paramagnetic resonance frequency ω_S, the polarization P_e of the electron spin system, which at typical experimental DNP conditions of low temperature and high magnetic field is close to unity ($P_e \to 100\%$), can be transferred to the nearby nuclei, taking advantage of the dipolar electron-nuclear coupling. For samples containing protons with an NMR frequency of $\omega_I \ll \omega_S$, the change of the nuclear polarization induced by DNP, relative to the thermal equilibrium polarization, is typically two orders of magnitude. Since by appropriate choice of the irradiating microwave frequency DNP allows one to polarize the protons either positively or negatively, the ensuing coherent and incoherent scattering cross sections can be varied significantly. In the simplest model for DNP, the so-called solid effect [269, 270], the characteristic DNP frequencies are given by $\omega_S \pm \omega_I$. It is also worth emphasizing that due to the fact that the electron-nucleus dipolar interaction falls off with the third power of the distance between electron and nuclear moments, nuclei close to a paramagnetic center are polarized first, while far away (bulk) nuclei rely on spin diffusion to reach equilibrium in a reasonable time [271].

Nuclear-spin-dependent neutron scattering has already been reported in 1963 by Shull and Ferrier [272] in a single crystal of vanadium, polarized by brute force. Hayter, Jenkin, and White [273] combined DNP with neutron scattering. These authors measured the spin-dependent Bragg diffraction from dynamically polarized protons in a single crystal of lanthanum magnesium nitrate, $La_2Mg_3(NO_3)_{12} \cdot 24H_2O$, doped with 1% of $^{142}Nd^{3+}$. It was already recognized then [273] that combining DNP with neutron scattering offers the unique possibility to label certain spin sites in complex crystals by a suitable paramagnetic center. Later, unpolarized and polarized SANS (e.g., [40, 274–288]) has been employed to probe nuclear-polarization-enhanced scattering in proton-rich organic materials doped with different types of paramagnetic centers.

As an example, we show in Fig. 2.13 time-resolved polarized SANS data obtained on a frozen solution of EHBA-Cr^V molecules in a 98%-deuterated glycerol-water mixture [282]. The EHBA-Cr^V complex [Fig. 2.13(a)] has a diameter of about 1 nm and the electronic spin is surrounded by 20 (close) protons. The SANS experiment (at $T = 1\,K$ and $B = 3.5\,T$) has been carried out such that the nuclear polarization was reversed typically every 10 s, and the acquisition of neutron scattering intensity spectra was triggered synchronously (typically in time intervals of 0.1 s). In order to obtain sufficient statistics, each of these time frames (typically 200) was averaged over several hundred cycles resulting in a typical total counting time of 12 hours. The SANS data [Fig. 2.13(b)] can be described as consisting of a q-dependent coherent signal and of a q-independent incoherent contribution, both however depending on time via $P_n(t)$. In [280], the Cr^V complex was modeled in terms of a spherical core-shell form factor, consisting of a shell of four C_2H_5 residues surrounding the $[CrO_7C_4]^-$ core. Only five free parameters were then sufficient to fit the scattering data of 200 time frames simultaneously in order to deduce the time dependence of the polarization of the protons in the EHBA molecule, i.e., the close protons. The resulting data [Fig. 2.13(c)] can be well described by the sum of two exponentials with time constants of $\tau_1 = 1.1 \pm 0.1\,s$ and $\tau_2 = 5.5 \pm 0.6\,s$, characterizing, respectively, the dynamics of the close and bulk (solvent) proton polarization build-up [282]. Figure 2.13(c) also displays the nuclear polarization of the bulk protons as measured by NMR. Note that the local magnetic field which is created by a paramagnetic moment displaces the Larmor frequency of the close nuclei far enough from that of the bulk nuclei to render them "invisible" by NMR. The difference in the time evolution of the polarization between close and bulk protons reflects the mechanism of DNP: a strong initial gradient develops due to the fast polarization of the protons close to the paramagnetic center, which then spreads to the bulk with a slower rate [280].

2.4.3 Nuclear SANS cross section

Under conventional diffuse SANS conditions, the probing neutron wavelength may be larger than twice the maximum value of the lattice-plane spacing, which defines the so-called Bragg cutoff wavelength. The scattering signal at small momentum transfers does then not carry information about the structure on an atomic scale. It is therefore permissible to replace the discrete sum in eqn (2.27) by an integral over a continuously distributed nuclear scattering-length density $N(\mathbf{r})$ in a coarse-grained description of the actual discrete atomic structure. Generally, it is convenient to work with this continuum approach, since models for SANS are most often also based on a continuum picture, embodying geometric concepts such as spheres or rods, alloy concentration profiles, or—with reference to magnetic scattering—the continuous magnetization profiles $\mathbf{M}(\mathbf{r})$ of micromagnetics.

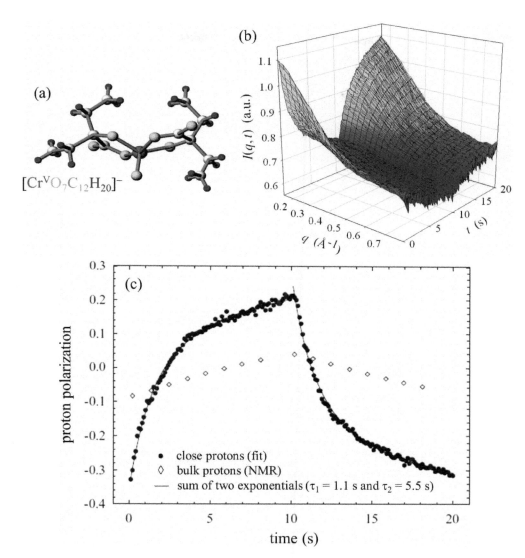

Fig. 2.13: (a) Sketch of the EHBA-Cr^V molecule, where the Cr^V carries the unpaired electron. After [283]. (b) Time-resolved azimuthally averaged SANS cross sections $d\Sigma/d\Omega$ of EHBA-Cr^V in a 98%-deuterated glycerol-water mixture. The concentration of EHBA-Cr^V is 5.0×10^{19} cm^{-3}. During the DNP cycle (positive DNP in the first 10 s and negative DNP in the next 10 s) about 200 SANS cross sections were collected. After [282]. (c) Close proton polarization deduced from the fit of the SANS data shown in (b) to a core-shell form factor model and bulk proton polarization recorded by NMR. After [280].

Let b_α and $\rho_{a,\alpha}(\mathbf{r})$ denote, respectively, the nuclear atomic scattering length and the atomic density of atomic species α, then

$$N(\mathbf{r}) = \sum_\alpha b_\alpha \rho_{a,\alpha}(\mathbf{r}). \tag{2.57}$$

Introducing the Fourier transform $\widetilde{N}(\mathbf{q})$ of $N(\mathbf{r})$ as

$$\widetilde{N}(\mathbf{q}) = \frac{1}{(2\pi)^{3/2}} \int N(\mathbf{r}) \exp\left(-i\mathbf{q} \cdot \mathbf{r}\right) d^3r, \tag{2.58}$$

the nuclear SANS cross section can be formally expressed as

$$\frac{d\Sigma_{\mathrm{nuc}}}{d\Omega}(\mathbf{q}) = \frac{8\pi^3}{V} |\widetilde{N}(\mathbf{q})|^2. \tag{2.59}$$

By employing a microstructural model for $N(\mathbf{r})$ and using standard arguments, eqn (2.59) can be further evaluated; for instance, for the case of a two-phase system consisting of a distribution of N_p particles with uniform scattering-length density $\rho_{\mathrm{nuc}}^\mathrm{p}$ in a homogeneous matrix ($\rho_{\mathrm{nuc}}^\mathrm{m}$), the general expression for the nuclear SANS cross section takes on the following form [22]:

$$\frac{d\Sigma_{\mathrm{nuc}}}{d\Omega}(\mathbf{q}) = \frac{(\Delta\rho)_{\mathrm{nuc}}^2}{V} \left| \sum_{m=1}^{N_\mathrm{p}} F_m(\mathbf{q}) \exp\left(-i\mathbf{q} \cdot \mathbf{r}_m\right) \right|^2, \tag{2.60}$$

where $(\Delta\rho)_{\mathrm{nuc}}^2 = (\rho_{\mathrm{nuc}}^\mathrm{p} - \rho_{\mathrm{nuc}}^\mathrm{m})^2$ is the nuclear scattering-length density contrast, and $F_m(\mathbf{q})$ and $\mathbf{r}_m$ represent, respectively, the form factor and the position vector of the center of mass of particle m. In the dilute and monodisperse limit, eqn (2.60) reduces to:

$$\frac{d\Sigma_{\mathrm{nuc}}}{d\Omega}(\mathbf{q}) = \frac{N_\mathrm{p}}{V} (\Delta\rho)_{\mathrm{nuc}}^2 |F(\mathbf{q})|^2. \tag{2.61}$$

An alternative approach is to relate $d\Sigma_{\mathrm{nuc}}/d\Omega$ to the Fourier transform of the auto-correlation function $C_\mathrm{N}(\mathbf{r})$ of the excess scattering-length density function

$$\delta N(\mathbf{r}) = N(\mathbf{r}) - \langle N \rangle, \tag{2.62}$$

according to

$$\frac{d\Sigma_{\mathrm{nuc}}}{d\Omega}(\mathbf{q}) = \int C_\mathrm{N}(\mathbf{r}) \exp\left(-i\mathbf{q} \cdot \mathbf{r}\right) d^3r, \tag{2.63}$$

where

$$C_\mathrm{N}(\mathbf{r}) = \frac{1}{V} \int \delta N(\mathbf{r}') \delta N(\mathbf{r}' + \mathbf{r}) d^3r'. \tag{2.64}$$

The back-transform of eqn (2.63) is

$$C_\mathrm{N}(\mathbf{r}) = \frac{1}{8\pi^3} \int \frac{d\Sigma_{\mathrm{nuc}}}{d\Omega}(\mathbf{q}) \exp\left(i\mathbf{q} \cdot \mathbf{r}\right) d^3q. \tag{2.65}$$

The average scattering-length density $\langle N \rangle$ may be subtracted in eqn (2.62), since the related scattering is near the origin of reciprocal space, which is not accessible to the experiment. Moreover, we refer the reader to Appendix C for a more detailed discussion of the so-called Fourier–Hankel–Abel cycle, which relates an isotropic autocorrelation function $C_N(r)$ to $\frac{d\Sigma_{\text{nuc}}}{d\Omega}(q)$ via a Fourier transform, $\frac{d\Sigma_{\text{nuc}}}{d\Omega}(q)$ to the projected correlation function $G(r)$ (accessible in a SESANS experiment [289]) via a zero-order Hankel transform, and $G(r)$ to $C_N(r)$ via an Abel transform.

In SANS studies, $d\Sigma_{\text{nuc}}/d\Omega$ is typically known only in a plane of reciprocal space, normal to the incident neutron-beam direction. The following relations are therefore most useful when the scattering can be assumed to be isotropic, so that $d\Sigma_{\text{nuc}}/d\Omega$ depends on $\mathbf{q}$ only through its magnitude q. It is then possible to perform an integration over the entire q-space by extrapolation of the data. Correlation function and scattering cross section are then related via:

$$C_N(r) = \frac{1}{2\pi^2 r} \int_0^\infty \frac{d\Sigma_{\text{nuc}}}{d\Omega}(q) \sin(qr) q \, dq \tag{2.66}$$

and

$$\frac{d\Sigma_{\text{nuc}}}{d\Omega}(q) = \frac{4\pi}{q} \int_0^\infty C_N(r) \sin(qr) r \, dr. \tag{2.67}$$

The definition of the correlation function, eqn (2.64), implies a relation for the mean-square density fluctuation, namely:

$$C_N(0) = \frac{1}{V} \int [\delta N(\mathbf{r}')]^2 d^3 r' = \langle N^2 \rangle - \langle N \rangle^2, \tag{2.68}$$

and an invariant P of the scattering is given by [290]:

$$\text{P} = \langle N^2 \rangle - \langle N \rangle^2 = \frac{1}{2\pi^2} \int_0^\infty \frac{d\Sigma_{\text{nuc}}}{d\Omega}(q) q^2 dq, \tag{2.69}$$

irrespective of the geometry and feature size of the microstructure. A lower bound for P, the so-called Porod invariant, may be obtained by integration of the data in the experimentally accessible range of momentum transfers.

In the analysis of experimental SANS data, the distance distribution function (sometimes also denoted as the pair-distance distribution function)

$$p(r) = r^2 C_N(r) \tag{2.70}$$

is frequently used. The $p(r)$ function corresponds to the distribution of distances between volume elements inside the particle weighted by the excess scattering-length density distribution; see the reviews by Glatter [291] and by Svergun and Koch [39] for detailed discussions of the properties of $p(r)$ and for information on how to compute $p(r)$ by indirect Fourier transformation. Figure 2.14 displays some typical scattering curves $I(q)$ and derived $p(r)$ of selected geometrical bodies with the same maximum

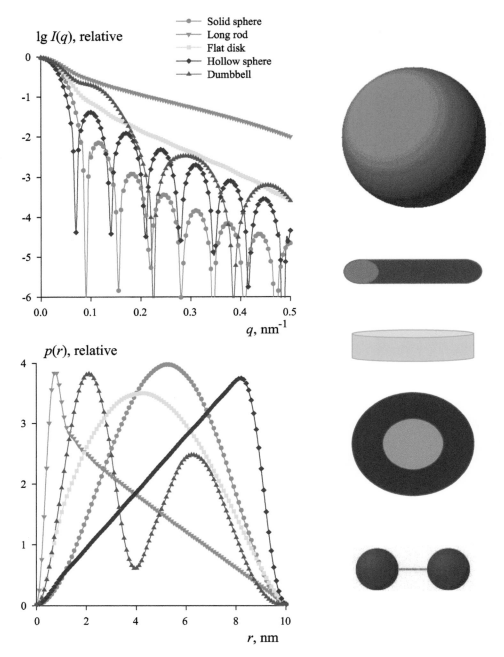

Fig. 2.14: Scattering intensities $I(q)$ and distance distribution functions $p(r)$ of some selected geometrical bodies. After [39].

size. The $I(q)$ and $p(r)$ curves contain the same information, however, the real-space representation is more intuitive and yields sometimes direct information (by visual inspection) about the particle shape. From these $p(r)$ the characteristic dimensions of the particle can be deduced; for instance, the sphere has the characteristic bell-shaped $p(r)$ given by eqn (2.71) [or eqn (1.69)], while the long rod has a skewed $p(r)$ with a clear maximum at small distances corresponding to the radius of the rod's cross section [39]. The depicted examples all exhibit positive $p(r)$ functions. Interparticle interference (dense packing) or particles with an inhomogeneous internal structure may result in negative values of $p(r)$ (e.g., [292, 293]). In this context we also refer to the investigation of atomic magnetic structures via the computation of the magnetic pair-distribution function [294–297]. This formalism provides information about both short-range and long-range magnetic correlations; see Frandsen et al. [294] for the analytical derivation of the magnetic pair-distribution function and for a discussion of various example cases.

In the following, we introduce three examples for correlation functions and the corresponding scattering cross sections, which are of relevance in the discussion of their analogs in magnetic scattering. Note that in the literature on nonmagnetic SANS and small-angle x-ray scattering it is customary to discuss scattering data in terms of $p(r)$, while in the magnetic SANS literature the discussion is commonly based on the magnetic $C(r)$. Due to the r^2-factor, features at medium and large distances are more pronounced in $p(r)$ than in $C(r)$ (compare, e.g., Fig. 6 in [298]). The reason why in magnetic SANS one prefers to display $C(r)$ is related to the fact that for fluctuating magnetic systems the correlation function at the origin differs distinctly from the behavior of particles with a sharp interface [see the discussion related to eqn (6.28)].

- For a uniform sphere of radius R (volume: V_s), we have [299]:

$$C_N(r) = C_s(r) = C_0 \left(1 - \frac{3r}{4R} + \frac{r^3}{16R^3} \right) \quad \text{for} \quad r \leq 2R,$$

$$\frac{d\Sigma_{\mathrm{nuc}}}{d\Omega}(q) = 9C_0 V_s \left(\frac{\sin(qR) - qR\cos(qR)}{(qR)^3} \right)^2. \tag{2.71}$$

Equation (2.71) for $d\Sigma_{\mathrm{nuc}}/d\Omega$ is obtained from eqn (2.61) by setting $N_p = 1, V = V_s, (\Delta\rho)^2_{\mathrm{nuc}} = 1$, and $F(q, R) = 3V_s j_1(q, R)/(qR)$. C_0 is a scaling constant.

- For exponentially correlated density fluctuations (correlation length: κ^{-1}), the scattering cross section has a Lorentzian-squared form [300]:

$$C_N(r) = C_0 \exp(-\kappa r),$$

$$\frac{d\Sigma_{\mathrm{nuc}}}{d\Omega}(q) = \frac{8\pi C_0 \kappa}{(\kappa^2 + q^2)^2}. \tag{2.72}$$

- For a simple example of critical fluctuations (at not-too-small distances r), the Ornstein–Zernike scattering at not-too-large q is a simple Lorentzian [209, 301]:

$$C_N(r) = \frac{C_0}{\kappa r} \exp(-\kappa r),$$

$$\frac{d\Sigma_{\mathrm{nuc}}}{d\Omega}(q) \simeq \frac{4\pi C_0 \kappa^{-1}}{\kappa^2 + q^2}. \tag{2.73}$$

The first two examples above illustrate that the initial slope of the correlation function may often carry information about the structure size (correlation length). Generally, when two uniform phases are separated by sharp interfaces, an inverse correlation length may be obtained by means of the following relation [49, 302, 303]:

$$\frac{d\ln C_N(r)}{dr}\bigg|_{r=0} = -\frac{S}{4V}, \tag{2.74}$$

where S is the total surface or interface area. The surface-to-volume ratio S/V may also be obtained directly from the scattering signal by means of the Debye–Porod law [290, 300, 304]: when there are two phases of uniform scattering-length density and with discontinuous (sharp) interfaces, the scattering in the limit of large q (much larger than the inverse of the characteristic structure scale) obeys (compare Fig. 2.15)

$$\frac{d\Sigma_{\text{nuc}}}{d\Omega}(q) \cong -8\pi(\Delta N)^2 C_N'(0)q^{-4} = 2\pi(\Delta N)^2\frac{S}{V}q^{-4}. \tag{2.75}$$

We refer to the paper by Ciccariello, Goodisman, and Brumberger [305] for a critical discussion of Debye's assumption by which the continuous electron density of a sample is approximated by a discrete-valued one. These authors have provided a generalization of the Debye–Porod law, which relates the value of $C_N'(0)$ to the integral of the discontinuity of the electron density fluctuation along the discontinuity surface.

   The above relation for the high-q limit is supplemented by one for the scattering near the origin of reciprocal space, the so-called Guinier approximation [299]: when the scattering is from a set of non-interfering discrete objects then, in the limit of low $q < 1.3/R_G$ (Fig. 2.15),

$$\frac{d\Sigma_{\text{nuc}}}{d\Omega}(q) \cong \frac{d\Sigma_{\text{nuc}}}{d\Omega}(q=0)\,\exp\left(-\frac{q^2R_G^2}{3}\right). \tag{2.76}$$

For identical scatterers, R_G denotes the individual radius of gyration. In fact, for dilute and monodisperse systems, the Guinier plot ($\ln(d\Sigma_{\text{nuc}}/d\Omega)$ versus q^2) should be a linear function, whose slope yields R_G. As pointed out by Svergun and Koch [39], linearity of the Guinier plot can be considered as a test of the sample homogeneity and deviations indicate attractive or repulsive interparticle interactions leading to interference effects [307]. Furthermore, in the presence of a particle-size distribution $f(R)$, the Guinier radius is related to the ratio of moments of the size distribution. For spherical particles and point collimation, one obtains [11, 308]

$$R_G^2 = \frac{3}{5}\frac{\langle R^8\rangle}{\langle R^6\rangle} = \frac{3}{5}\frac{\int_0^\infty R^8 f(R)dR}{\int_0^\infty R^6 f(R)dR}. \tag{2.77}$$

Thus, experimental R_G-values are strongly weighted towards the largest features in the distribution. For the example of a log-normal distribution function ($\int_0^\infty f(R)dR = 1$) with a median of R_0 and a variance of σ_{LN} [309],

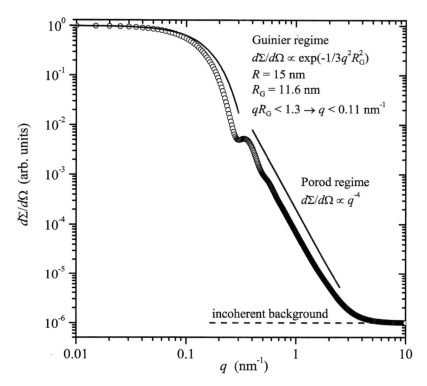

Fig. 2.15: Qualitative variation of the SANS cross section of a spherical particle with a sharp interface ($R = 15\,\text{nm}$; log-log scale). The Guinier and Porod regions are indicated (see text). The sphere form factor has been convoluted with a Gaussian function (to model e.g., the effect of a particle-size distribution) and an incoherent background has been added. Note that the above scattering curve is characteristic for the orientationally averaged SANS cross section of many not-too-strongly shape anisotropic homogeneous particles. For the latter, variations at low q, between the true Guinier and Porod region, may occur [63, 306] (compare also Fig. 2.14).

$$f(R) = \frac{1}{\sqrt{2\pi} R \ln \sigma_{\text{LN}}} \exp\left(-\frac{1}{2}\left[\frac{\ln(R/R_0)}{\ln \sigma_{\text{LN}}}\right]^2\right), \tag{2.78}$$

the radius of gyration equals

$$R_{\text{G}} = \sqrt{\frac{3}{5}} R_0 \exp\left(7 \ln^2 \sigma_{\text{LN}}\right), \tag{2.79}$$

where we have used the relation

$$\int_0^\infty R^n f(R)\, dR = R_0^n \exp\left(\frac{1}{2} n^2 \ln^2 \sigma_{\text{LN}}\right). \tag{2.80}$$

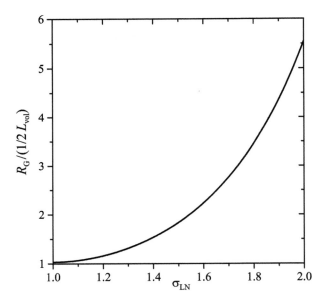

Fig. 2.16: Ratio $R_{\mathrm{G}}/(\frac{1}{2}L_{\mathrm{vol}}) = \sqrt{16/15}\exp\left(\frac{7}{2}\ln^2\sigma_{\mathrm{LN}}\right)$ as a function of the width σ_{LN} of the log-normal size distribution.

The value for R_{G} is often compared to the average particle or crystallite size, which may be obtained by the analysis of wide-angle x-ray diffraction data. As shown by Krill and Birringer [309], for a log-normal grain-size distribution and a spherical crystallite shape, the so-called volume-weighted average column length L_{vol}, which is the outcome of standard x-ray methods for analyzing grain-size-induced Bragg peak broadening, is related to the parameters of the log-normal distribution as follows:

$$\frac{1}{2}L_{\mathrm{vol}} = \frac{3}{4}R_0\exp\left(\frac{7}{2}\ln^2\sigma_{\mathrm{LN}}\right). \tag{2.81}$$

Figure 2.16 displays the ratio

$$\frac{R_{\mathrm{G}}}{\frac{1}{2}L_{\mathrm{vol}}} = \sqrt{\frac{16}{15}}\exp\left(\frac{7}{2}\ln^2\sigma_{\mathrm{LN}}\right) \tag{2.82}$$

as a function of the width σ_{LN} of the log-normal size distribution. The value for σ_{LN} is e.g., determined by the synthesis method; for instance, $\sigma_{\mathrm{LN}} \sim 1.5-1.9$ for inert-gas-condensed materials [309–311], so that $R_{\mathrm{G}}/(\frac{1}{2}L_{\mathrm{vol}}) \sim 1.8-4.4$.

The Porod and Guinier laws have originally been derived for nonmagnetic particle-matrix-type assemblies in the context of the early theoretical developments of the technique of small-angle x-ray scattering [290]. Their application to magnetic materials should be considered with special care. They are certainly applicable to systems consisting of saturated magnetic particles in a nonmagnetic matrix or, likewise, to pores in a saturated matrix. On the other hand, when the smoothly varying magnetization profiles of micromagnetics (see, e.g., Section 6.1) are at the origin of the related magnetic scattering—implying the absence of a sharp interface in the magnetic microstructure—the asymptotic power-law behavior of the cross section differs from the q^{-4}-behavior (see, e.g., Fig. 4.23 and the discussion in Section 4.3.5). This is in agreement with theoretical predictions and experimental observations [312]. Moreover, such a statement does of course not preclude the existence of sharp interfaces in the nuclear grain microstructure of a magnetic material: there may well exist sharp particle-matrix interfaces, but the corresponding spin distribution which decorates these interfaces and which gives rise to the magnetic SANS cross section might be continuous over the defects. Deviations from Porod's law are also well known for nuclear SANS; for instance, for particles having a nonuniform scattering-length density [313] or for those exhibiting smeared or fuzzy interfaces [314–317]; see also Bentley and Cywinski for the "spin emulsion" model [318].

When the objects in a set are randomly arranged in space, the individual scattering intensities are additive. However, even in disordered arrays, deviations from randomness become typically significant when the volume fraction of the objects exceeds a few percent [307,308]. The simplest correlation is a characteristic pair distance $\mathbf{d}$. According to the Debye formula [299], the interparticle interference function is then given by:

$$S(q) = 1 + \frac{\sin(q\mathbf{d})}{q\mathbf{d}}, \tag{2.83}$$

which leads to a maximum in $d\Sigma_{\mathrm{nuc}}/d\Omega$ at

$$q_{\mathrm{max}} = \mathrm{K}^{\star} \frac{2\pi}{\mathbf{d}} \tag{2.84}$$

with $\mathrm{K}^{\star} \cong 1.23$. As discussed by Guinier [299], the order of magnitude of the distance of closest approach of particles, $\mathbf{d}$, can be estimated from the maximum of the principal ring in the SANS pattern according to eqn (2.84) with $\mathrm{K}^{\star} \sim 1.1-1.2$. In Section 2.9 we will again encounter interparticle interference effects when discussing magnetic SANS at saturation.

2.5 Magnetic SANS: general considerations

Since the vast majority of magnetic SANS experiments are carried out by employing the two scattering geometries where the externally applied magnetic field $\mathbf{H}_0$ is either perpendicular [Fig. 2.1(a)] or parallel [Fig. 2.1(b)] to the wave vector $\mathbf{k}_0$ of the incoming neutron beam, we restrict attention to these two specific situations. We adopt a Cartesian laboratory coordinate system with corresponding unit vectors $\mathbf{e}_x$, $\mathbf{e}_y$, and $\mathbf{e}_z$. The field $\mathbf{H}_0$ is assumed to be always parallel to $\mathbf{e}_z$. For the perpendicular scattering geometry ($\mathbf{k}_0 \perp \mathbf{H}_0$), the angle θ on the two-dimensional detector is then measured between $\mathbf{H}_0$ and the momentum-transfer vector (compare Fig. 2.1)

$$\mathbf{q}_{\perp} \cong \left\{ \begin{array}{c} 0 \\ q_y \\ q_z \end{array} \right\} = q \left\{ \begin{array}{c} 0 \\ \sin\theta \\ \cos\theta \end{array} \right\}, \tag{2.85}$$

whereas for $\mathbf{k}_0 \parallel \mathbf{H}_0$, θ is the angle between a specified direction on the detector (e.g., $\mathbf{e}_x$) and

$$\mathbf{q}_\parallel \cong \left\{ \begin{array}{c} q_x \\ q_y \\ 0 \end{array} \right\} = q \left\{ \begin{array}{c} \cos\theta \\ \sin\theta \\ 0 \end{array} \right\}. \tag{2.86}$$

In eqns (2.85) and (2.86) we have neglected the respective $\mathbf{q}$-component along the incident beam, which is permissible within the small-angle regime (compare Section 2.1) [185].

The discrete atomic structure of condensed matter is generally of no relevance for SANS, and the magnetization state of the sample can be represented by a continuous magnetization vector field $\mathbf{M}(\mathbf{r})$, which is defined at each point $\mathbf{r} = \{x, y, z\}$ inside the material. Magnetic SANS is a consequence of nanoscale variations in both the orientation and/or magnitude of $\mathbf{M}(\mathbf{r})$. With $\mathbf{H}_0 \parallel \mathbf{e}_z$, $\widetilde{M}_z$ denotes the longitudinal magnetization Fourier component, whereas $\widetilde{M}_x$ and $\widetilde{M}_y$ are the transversal components, giving rise to spin-misalignment scattering [compare the definition of the three-dimensional Fourier-transform pair of the magnetization, eqns (1.36) and (1.37)]. The nuclear SANS cross section, which is due to nanoscale density and/or compositional fluctuations, is not generally of interest here. It is characterized by $\widetilde{N}(\mathbf{q})$, the Fourier transform of the continuous scattering-length density $N(\mathbf{r})$ (see Section 2.4.3).

For the understanding of magnetic neutron scattering, the Halpern–Johnson vector (sometimes also denoted as the magnetic interaction or magnetic scattering vector)

$$\widetilde{\mathbf{Q}} = \hat{\mathbf{q}} \times \left(\hat{\mathbf{q}} \times \widetilde{\mathbf{M}}(\mathbf{q}) \right) = \hat{\mathbf{q}} \left(\hat{\mathbf{q}} \cdot \widetilde{\mathbf{M}}(\mathbf{q}) \right) - \widetilde{\mathbf{M}}(\mathbf{q}), \tag{2.87}$$

where $\hat{\mathbf{q}}$ is the unit scattering vector, is of utmost importance [109]. It is a manifestation of the dipolar origin of magnetic neutron scattering (see Section 1.4) and it emphasizes the fact that only the components of the magnetization vector $\mathbf{M}$ which are perpendicular to $\mathbf{q}$ are relevant for magnetic neutron scattering (see Fig. 2.17). We note that different symbols for the Halpern–Johnson vector such as $\mathbf{M}_\perp$, $\mathbf{Q}_\perp$, $\mathbf{S}_\perp$, or $\mathbf{q}$, as in the original paper by Halpern and Johnson [109], can be found in the literature. Likewise, in many textbooks (e.g., [76, 86]) $\widetilde{\mathbf{Q}}$ is defined with a minus sign and normalized by the factor $2\mu_B$, which makes it dimensionless. $\widetilde{\mathbf{Q}}$ is a linear vector function of the components of $\widetilde{\mathbf{M}}$. Both $\widetilde{\mathbf{Q}}(\mathbf{q})$ and $\widetilde{\mathbf{M}}(\mathbf{q})$ are in general complex vectors. For $\mathbf{k}_0 \perp \mathbf{H}_0$ and $\mathbf{k}_0 \parallel \mathbf{H}_0$, we obtain, respectively:

$$\widetilde{\mathbf{Q}}_\perp = \left\{ \begin{array}{c} -\widetilde{M}_x \\ -\widetilde{M}_y \cos^2\theta + \widetilde{M}_z \sin\theta\cos\theta \\ \widetilde{M}_y \sin\theta\cos\theta - \widetilde{M}_z \sin^2\theta \end{array} \right\}, \tag{2.88}$$

$$\widetilde{\mathbf{Q}}_\parallel = \left\{ \begin{array}{c} -\widetilde{M}_x \sin^2\theta + \widetilde{M}_y \sin\theta\cos\theta \\ \widetilde{M}_x \sin\theta\cos\theta - \widetilde{M}_y \cos^2\theta \\ -\widetilde{M}_z \end{array} \right\}. \tag{2.89}$$

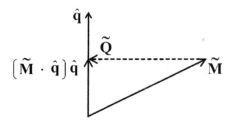

Fig. 2.17: Sketch illustrating the geometrical relationship between $\hat{\mathbf{q}}$, $\widetilde{\mathbf{M}}$, and $\widetilde{\mathbf{Q}}$. Regarding the sign of $\widetilde{\mathbf{Q}}$, we prefer the original notation of Halpern and Johnson [109] [eqn (2.87)], instead of its negative, which is most often used in the literature (e.g., [76, 117]).

By assuming that the direction of the incident neutron polarization is along $\mathbf{e}_z$, i.e., $\mathbf{P} = \mathbf{e}_z$, the elastic non-spin-flip ($++$ and $--$) and spin-flip ($+-$ and $-+$) SANS cross sections are given by eqns (1.42) and (1.43). These are rewritten here for convenience:

$$\frac{d\Sigma^{\pm\pm}}{d\Omega} = \frac{8\pi^3}{V} b_{\mathrm{H}}^2 \left[b_{\mathrm{H}}^{-2}|\widetilde{N}|^2 \pm b_{\mathrm{H}}^{-1}(\widetilde{N}\widetilde{Q}_z^* + \widetilde{N}^*\widetilde{Q}_z) + |\widetilde{Q}_z|^2 \right], \tag{2.90}$$

$$\frac{d\Sigma^{\pm\mp}}{d\Omega} = \frac{8\pi^3}{V} b_{\mathrm{H}}^2 \left[|\widetilde{Q}_x|^2 + |\widetilde{Q}_y|^2 \mp i(\widetilde{Q}_x\widetilde{Q}_y^* - \widetilde{Q}_x^*\widetilde{Q}_y) \right]. \tag{2.91}$$

Several comments are in place [113]: it is seen that the transversal components $\widetilde{Q}_x$ and $\widetilde{Q}_y$ give rise to spin-flip scattering, while the longitudinal component $\widetilde{Q}_z$ results in non-spin-flip scattering. We also note that the nuclear coherent scattering, the nuclear incoherent scattering which is due to isotope disorder, as well as 1/3 of the nuclear-spin incoherent scattering are all non-spin-flip scattering. The remaining 2/3 of the nuclear-spin incoherent scattering reverses the neutron spin, but, since its magnitude is generally small compared to the nuclear coherent scattering and, in particular, relative to the magnetic SANS, we ignore it in the spin-flip channels. Furthermore, if we set $\theta = 0°$ in eqn (2.88), which corresponds to the case that the scattering vector is along the neutron polarization, we see that

$$\widetilde{\mathbf{Q}}_{\perp}^{\theta=0°} = \left\{ \begin{array}{c} -\widetilde{M}_x \\ -\widetilde{M}_y \\ 0 \end{array} \right\}, \tag{2.92}$$

so that nuclear coherent and magnetic scattering are fully separated in the perpendicular scattering geometry. In the case $\mathbf{k}_0 \parallel \mathbf{H}_0$ [eqn (2.89)], spin-flip scattering probes only the transversal magnetization Fourier components $\widetilde{M}_{x,y}$, whereas the longitudinal scattering is entirely contained in the non-spin-flip channel, in contrast to the $\mathbf{k}_0 \perp \mathbf{H}_0$ geometry. This property can be used to reveal the direction of the magnetic anisotropy in single-crystalline spin systems, e.g., an easy plane versus an easy axis anisotropy or the confinement of the propagation vector along certain crystallographic directions in chiral and other exotic magnets [319, 320].

Although the SANS cross sections have already been discussed in Sections 1.4.1 and 1.4.2, we prefer to display them here explicitly in terms of the Cartesian Fourier components $\widetilde{M}_{x,y,z}$ of the magnetization and for the two most relevant (perpendicular and parallel) scattering geometries. For magnetic SANS, the representation of the cross sections in terms of the $\widetilde{M}_{x,y,z}$ is certainly more intuitive, albeit less compact and elegant, than expressing them using the Cartesian components $\widetilde{Q}_{x,y,z}$ of the Halpern–Johnson vector (which depend on the $\widetilde{M}_{x,y,z}$). This is particularly important for relating magnetic SANS data to the outcome of e.g., magnetization measurements and micromagnetic computations.

In the equations for the various SANS cross sections that follow in Sections 2.6–2.8, V denotes the scattering volume, the atomic magnetic scattering length b_m and the constant $b_\mathrm{H} = b_\mathrm{m}\mu_\mathrm{a}^{-1} = 2.91 \times 10^8\,\mathrm{A}^{-1}\mathrm{m}^{-1}$ have been introduced in eqns (1.31) and (1.33), and the magnetic form factor in the small-angle regime was set to unity ($f(\mathbf{q}) \cong 1$) (compare Section 1.4.2). By inserting the solutions for the Halpern–Johnson vector [eqns (2.88) and (2.89)] into the equations for the non-spin-flip and spin-flip SANS cross sections [eqns (2.90) and (2.91)], and by noting the relations between the various cross sections [eqns (1.47)–(1.49)], one can conveniently express the SANS cross sections in terms of the $\widetilde{M}_{x,y,z}$. We remind that the applied-field direction $\mathbf{H}_0$ specifies the z-direction of a Cartesian laboratory coordinate system for both scattering geometries. The polarizer efficiency p represents the percentage of $+$ polarized neutrons after the polarizer has acted on the unpolarized beam [compare eqns (1.40) and (1.41)]. It is related to the polarization P of the beam via $P = 2p - 1$, so that $p = 1/2$ for an unpolarized beam [321]. The efficiency of the spin flipper is denoted with $\epsilon^\pm$ ($\epsilon^+ = 0$ for flipper off and $\epsilon^- = \epsilon \cong 1$ for flipper on), and p_A is the analyzer efficiency. The values of the parameters $p, \epsilon, p_\mathrm{A}$ are usually close to their ideal ones, but nevertheless correction procedures have to be performed in order to account for the imperfect neutron optics.

Nuclear-spin incoherent SANS, which is partly spin-flip scattering and gives rise to a flat q-independent background signal, is neglected in the SANS cross sections. This is motivated by the fact that its magnitude is usually small relative to the here-relevant coherent magnetic SANS of nonhydrogenated samples. We refer to the review by Stuhrmann [40] for a detailed treatment of nuclear-spin-dependent SANS.

2.6   Unpolarized SANS cross sections

2.6.1 $\mathbf{k}_0 \perp \mathbf{H}_0$

$$\frac{d\Sigma}{d\Omega}(\mathbf{q}) = \frac{8\pi^3}{V} b_\mathrm{H}^2 \left(b_\mathrm{H}^{-2}|\widetilde{N}|^2 + |\widetilde{M}_x|^2 + |\widetilde{M}_y|^2 \cos^2\theta + |\widetilde{M}_z|^2 \sin^2\theta \right.$$
$$\left. - (\widetilde{M}_y\widetilde{M}_z^* + \widetilde{M}_y^*\widetilde{M}_z)\sin\theta\cos\theta \right). \qquad (2.93)$$

2.6.2 $\mathbf{k}_0 \parallel \mathbf{H}_0$

$$\frac{d\Sigma}{d\Omega}(\mathbf{q}) = \frac{8\pi^3}{V}b_{\mathrm{H}}^2\left(b_{\mathrm{H}}^{-2}|\widetilde{N}|^2 + |\widetilde{M}_x|^2\sin^2\theta + |\widetilde{M}_y|^2\cos^2\theta + |\widetilde{M}_z|^2\right.$$
$$\left. -(\widetilde{M}_x\widetilde{M}_y^* + \widetilde{M}_x^*\widetilde{M}_y)\sin\theta\cos\theta\right). \qquad (2.94)$$

2.7 SANSPOL cross sections

The half-polarized SANS cross sections [eqns (2.95) and (2.98)] can be obtained directly and corrected for non-ideal neutron polarization provided that the parameters p and $\epsilon = \epsilon^-$ are known from reference measurements.

2.7.1 $\mathbf{k}_0 \perp \mathbf{H}_0$

$$\frac{d\Sigma^{\pm}}{d\Omega}(\mathbf{q}) = \frac{8\pi^3}{V}b_{\mathrm{H}}^2\left(b_{\mathrm{H}}^{-2}|\widetilde{N}|^2 + |\widetilde{M}_x|^2 + |\widetilde{M}_y|^2\cos^2\theta + |\widetilde{M}_z|^2\sin^2\theta\right.$$
$$-(\widetilde{M}_y\widetilde{M}_z^* + \widetilde{M}_y^*\widetilde{M}_z)\sin\theta\cos\theta + W^{\pm}b_{\mathrm{H}}^{-1}(\widetilde{N}\widetilde{M}_z^* + \widetilde{N}^*\widetilde{M}_z)\sin^2\theta$$
$$\left. -W^{\pm}b_{\mathrm{H}}^{-1}(\widetilde{N}\widetilde{M}_y^* + \widetilde{N}^*\widetilde{M}_y)\sin\theta\cos\theta + iW^{\pm}\chi\right), \qquad (2.95)$$

where

$$W^{\pm} = (2p-1)(2\epsilon^{\pm}-1) = P(2\epsilon^{\pm}-1), \qquad (2.96)$$

and the chiral function $\chi(\mathbf{q})$ is given by [compare eqns (2.88) and (2.91)]

$$\chi(\mathbf{q}) = \mathbf{e}_z \cdot (\widetilde{\mathbf{Q}} \times \widetilde{\mathbf{Q}}^*) = \widetilde{Q}_x\widetilde{Q}_y^* - \widetilde{Q}_x^*\widetilde{Q}_y$$
$$= \left(\widetilde{M}_x\widetilde{M}_y^* - \widetilde{M}_x^*\widetilde{M}_y\right)\cos^2\theta - \left(\widetilde{M}_x\widetilde{M}_z^* - \widetilde{M}_x^*\widetilde{M}_z\right)\sin\theta\cos\theta. \qquad (2.97)$$

We note that $\chi(\mathbf{q})$ vanishes at complete magnetic saturation ($M_x = M_y = 0$), or for purely real-valued magnetization Fourier components (irrespective of the value of the field). For the case that the magnetization distribution is an even function of the position, i.e., $\mathbf{M}(\mathbf{r}) = \mathbf{M}(-\mathbf{r})$, the corresponding Fourier transform $\widetilde{\mathbf{M}}(\mathbf{q})$ is also an even and real-valued function, with the consequence that $\chi(\mathbf{q})$ vanishes. On the other hand, if $\mathbf{M}(\mathbf{r})$ is an odd function, then $\widetilde{\mathbf{M}}(\mathbf{q})$ is an odd and imaginary function [322].

2.7.2 $\mathbf{k}_0 \parallel \mathbf{H}_0$

$$\frac{d\Sigma^{\pm}}{d\Omega}(\mathbf{q}) = \frac{8\pi^3}{V}b_{\mathrm{H}}^2\left(b_{\mathrm{H}}^{-2}|\widetilde{N}|^2 + |\widetilde{M}_x|^2\sin^2\theta + |\widetilde{M}_y|^2\cos^2\theta + |\widetilde{M}_z|^2\right.$$
$$\left. -(\widetilde{M}_x\widetilde{M}_y^* + \widetilde{M}_x^*\widetilde{M}_y)\sin\theta\cos\theta + W^{\pm}b_{\mathrm{H}}^{-1}(\widetilde{N}\widetilde{M}_z^* + \widetilde{N}^*\widetilde{M}_z)\right). \qquad (2.98)$$

Note that $\chi(\mathbf{q}) = 0$ for $\mathbf{k}_0 \parallel \mathbf{H}_0$.

2.8 POLARIS cross sections

For the determination of the spin-resolved POLARIS cross sections [eqns (2.99)−(2.102)], it is generally necessary to measure all four partial cross sections in order to correct for spin leakage between the different channels [321]. This requirement may be relaxed by exploiting certain scattering properties of the sample; for instance, if the chiral term and the nuclear-magnetic interference terms are negligible, then it is sufficient to measure only the non-spin-flip and the spin-flip SANS cross section. Such corrections can e.g., be accomplished by means of the *BerSANS* [323, 324], *Pol-Corr* [325, 326], and GRASP [327] software tools. In the following, we will first write down the POLARIS equations for the case of a perfectly working neutron optics ($p = \epsilon = p_A = 1$). The realistic case of a non-perfect neutron optics is considered separately in Section 2.8.3.

2.8.1 $\mathbf{k}_0 \perp \mathbf{H}_0$

$$\frac{d\Sigma^{\pm\pm}}{d\Omega}(\mathbf{q}) = \frac{8\pi^3}{V} b_H^2 \left(b_H^{-2} |\widetilde{N}|^2 + |\widetilde{M}_y|^2 \sin^2\theta \cos^2\theta + |\widetilde{M}_z|^2 \sin^4\theta \right.$$
$$- (\widetilde{M}_y \widetilde{M}_z^* + \widetilde{M}_y^* \widetilde{M}_z) \sin^3\theta \cos\theta$$
$$\left. \mp b_H^{-1} (\widetilde{N}\widetilde{M}_z^* + \widetilde{N}^* \widetilde{M}_z) \sin^2\theta \pm b_H^{-1}(\widetilde{N}\widetilde{M}_y^* + \widetilde{N}^* \widetilde{M}_y) \sin\theta\cos\theta \right). \qquad (2.99)$$

$$\frac{d\Sigma^{\pm\mp}}{d\Omega}(\mathbf{q}) = \frac{8\pi^3}{V} b_H^2 \left(|\widetilde{M}_x|^2 + |\widetilde{M}_y|^2 \cos^4\theta + |\widetilde{M}_z|^2 \sin^2\theta \cos^2\theta \right.$$
$$\left. - (\widetilde{M}_y \widetilde{M}_z^* + \widetilde{M}_y^* \widetilde{M}_z) \sin\theta \cos^3\theta \mp i\chi \right). \qquad (2.100)$$

Although the spin-flip SANS cross section does not depend on the longitudinal Fourier component $\widetilde{Q}_z$ of the Halpern–Johnson vector, eqn (2.100) demonstrates that $d\Sigma^{\pm\mp}/d\Omega$ does depend on the longitudinal Fourier component $\widetilde{M}_z$ of the magnetization. This term gives rise to a pronounced $\sin^2\theta \cos^2\theta$ angular anisotropy of the spin-flip SANS cross section in the saturated state ($M_x = M_y = 0$).

2.8.2 $\mathbf{k}_0 \parallel \mathbf{H}_0$

$$\frac{d\Sigma^{\pm\pm}}{d\Omega}(\mathbf{q}) = \frac{8\pi^3}{V} b_H^2 \left(b_H^{-2} |\widetilde{N}|^2 + |\widetilde{M}_z|^2 \mp b_H^{-1}(\widetilde{N}\widetilde{M}_z^* + \widetilde{N}^* \widetilde{M}_z) \right). \qquad (2.101)$$

$$\frac{d\Sigma^{\pm\mp}}{d\Omega}(\mathbf{q}) = \frac{8\pi^3}{V} b_H^2 \left(|\widetilde{M}_x|^2 \sin^2\theta + |\widetilde{M}_y|^2 \cos^2\theta - (\widetilde{M}_x \widetilde{M}_y^* + \widetilde{M}_x^* \widetilde{M}_y) \sin\theta \cos\theta \right).$$
$$(2.102)$$

Due to the neglect of nuclear-spin-dependent SANS, and since $\chi(\mathbf{q}) = 0$ for $\mathbf{k}_0 \parallel \mathbf{H}_0$, $d\Sigma^{\pm\mp}/d\Omega$ is independent of the incoming polarization for $\mathbf{k}_0 \parallel \mathbf{H}_0$.

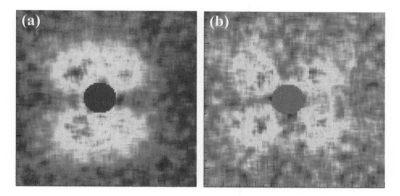

Fig. 2.18: Example for a spin-leakage correction. (a) Uncorrected spin-flip SANS cross section $d\Sigma^{+-}/d\Omega$ of a saturated soft magnetic Fe-based nanocomposite ($\mathbf{k_0} \perp \mathbf{H_0}$). The expected $\sin^2\theta\cos^2\theta$-type angular anisotropy of the spin-flip term at saturation [compare eqn (2.100)] is hidden by a strong $\sin^4\theta$-type contribution due to the non-spin-flip term [compare eqn (2.99)]. (b) Corrected spin-flip data. The broad band at the top and the bottom in (a) is removed by the correction. The range of momentum transfers roughly corresponds to $0.1\,\mathrm{nm}^{-1} < q < 0.7\,\mathrm{nm}^{-1}$. After [190].

2.8.3 POLARIS equations with non-ideal neutron optics

The neutron optics in POLARIS experiments are never perfect and the raw data are always contaminated with unwanted neutron spin states. It is therefore generally necessary to measure all four partial cross sections in order to correct for spin leakage between the different channels. As an example, Fig. 2.18 depicts the result of such a spin-leakage correction for the case of a saturated soft magnetic nanocomposite.

Wildes [321,328] has provided compact expressions for the POLARIS cross sections in the form of matrix equations. Following [321], we denote with

$$\mathbf{\Sigma} = \{\Sigma^{++}, \Sigma^{+-}, \Sigma^{-+}, \Sigma^{--}\} \tag{2.103}$$

the corrected four partial SANS cross sections and with

$$\mathbf{I} = \{I^{++}, I^{+-}, I^{-+}, I^{--}\} \tag{2.104}$$

the measured spin-dependent intensities (corrected for background scattering, transmission, detector efficiency, etc.). An instrument which is configured to measure a certain spin-dependent count rate, e.g., I^{++}, will measure a linear combination of the corrected count rates $\Sigma^{++}, \Sigma^{+-}, \Sigma^{-+}$, and Σ^{--}, each multiplied by the transmissions of the polarizer and the analyzer for the $+$ and $-$ neutrons. The central aim of the spin-leakage correction is to convert between the measured count rates $\mathbf{I}$ and the count rates $\mathbf{\Sigma}$ corrected for the polarization efficiencies. We will first consider the case that the setup has two flippers, one before and one after the sample. The fixed incident neutron beam polarization, provided in SANS usually by a supermirror polarizer, is specified by the polarizer efficiency p, the efficiency of the pre-sample rf spin flipper is ϵ ($\epsilon^+ = 0$ for flipper off and $\epsilon^- = \epsilon \cong 1$ for flipper on), the flipping efficiency of the post-sample rf spin flipper is f_A, and the analyzer efficiency is denoted with p_A. The relation between $\mathbf{I}$ and $\boldsymbol{\Sigma}$ can then be written as [321, 329, 330]:

$$\boldsymbol{\Sigma} = \overline{\overline{p}}_A \overline{\overline{f}}_A \overline{\overline{p}} \overline{\overline{f}}_P \mathbf{I}, \tag{2.105}$$

where

$$\overline{\overline{p}}_A = \frac{1}{1 - 2p_A} \begin{pmatrix} -p_A & (1 - p_A) & 0 & 0 \\ (1 - p_A) & -p_A & 0 & 0 \\ 0 & 0 & -p_A & (1 - p_A) \\ 0 & 0 & (1 - p_A) & -p_A \end{pmatrix}, \tag{2.106}$$

$$\overline{\overline{f}}_A = \frac{1}{f_A} \begin{pmatrix} f_A & 0 & 0 & 0 \\ (f_A - 1) & 1 & 0 & 0 \\ 0 & 0 & f_A & 0 \\ 0 & 0 & (f_A - 1) & 1 \end{pmatrix}, \tag{2.107}$$

$$\overline{\overline{p}} = \frac{1}{1 - 2p} \begin{pmatrix} -p & 0 & (1 - p) & 0 \\ 0 & -p & 0 & (1 - p) \\ (1 - p) & 0 & -p & 0 \\ 0 & (1 - p) & 0 & -p \end{pmatrix}, \tag{2.108}$$

$$\overline{\overline{f}}_P = \frac{1}{\epsilon} \begin{pmatrix} \epsilon & 0 & 0 & 0 \\ 0 & \epsilon & 0 & 0 \\ (\epsilon - 1) & 0 & 1 & 0 \\ 0 & (\epsilon - 1) & 0 & 1 \end{pmatrix}. \tag{2.109}$$

It is readily verified that $\boldsymbol{\Sigma} = \mathbf{I}$ for a perfect neutron optics. All parameters may depend on the neutron wavelength, the scattering angle, or time, but as long as they are known for the same conditions as the measurement, then the eqns (2.105)−(2.109) can be applied to perform the correction. Note that the externally applied magnetic field to the sample may influence the guide field and, hence, the neutron-spin transport properties, leading to a field-dependent change in the beam polarization.

In many magnetic SANS experiments, the analyzer part consists only of a ^{3}He spin-filter cell (without the need for a post-sample spin flipper). Such a device exploits the strongly spin-dependent absorption of neutrons by a nuclear-spin-polarized gas of ^{3}He atoms [127]. Neutrons with spin component antiparallel to the ^{3}He nuclear spin are absorbed, allowing only the wanted spin state to reach the detector. The orientation of the nuclear magnetic moments (and hence the selection of the spin state) can be reversed by 180° with respect to the static holding field using the adiabatic fast-passage technique developed in nuclear magnetic resonance [331]. For this particular setup (polarizer–flipper–sample–^{3}He cell), Krycka et al. [325, 326] explicitly display the modified expressions for the analyzer efficiencies and the spin-leakage correction matrix. In their data-reduction procedure, time-dependent effects (e.g., a decaying

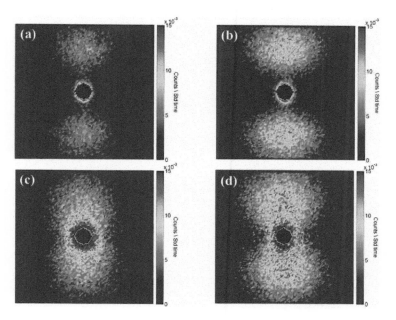

Fig. 2.19: Non-spin-flip SANS cross sections of an Fe–Cr-based nanocrystalline alloy. (a) $d\Sigma^{++}/d\Omega$ and (b) $d\Sigma^{--}/d\Omega$ at a saturating applied magnetic field of 1.31 T ($\mathbf{k}_0 \perp \mathbf{H}_0$) (logarithmic color scale). $\mathbf{H}_0$ is horizontal. (c) $d\Sigma^{++}/d\Omega$ and (d) $d\Sigma^{--}/d\Omega$ at 0.02 T. The range of momentum transfers (from the center to the corner of the detector) roughly corresponds to $0.04\,\text{nm}^{-1} < q < 0.55\,\text{nm}^{-1}$. After [332].

nuclear polarization of the ^{3}He filter) are also taken into account. The T_1-relaxation time of the ^{3}He cell is typically about 200 hours [188].

As discussed in Section 1.4.2, if the system under study does not contain an axial vector, then $d\Sigma/d\Omega$ is independent of the initial neutron polarization $\mathbf{P}$ [116]. This is e.g., the case for paramagnets, superparamagnets, or spin glasses at zero applied magnetic field (random spin orientation), simple antiferromagnets, or ferromagnets with a random domain distribution. For these systems, one has $d\Sigma^{++}/d\Omega = d\Sigma^{--}/d\Omega$ and $d\Sigma^{+-}/d\Omega = d\Sigma^{-+}/d\Omega$ [321]. On the other hand, for many polycrystalline magnetic materials in the presence of a strong applied field, it turns out that $d\Sigma^{++}/d\Omega \neq d\Sigma^{--}/d\Omega$, while $d\Sigma^{+-}/d\Omega = d\Sigma^{-+}/d\Omega$. This is seen in Fig. 2.19, which shows for the case of a soft magnetic two-phase Fe–Cr-based nanocomposite the two non-spin-flip SANS cross sections at saturation and at a lower field. Figure 2.20 displays the two spin-flip SANS cross sections of the same sample. At a saturating field of 1.31 T, when the scattering cross sections are dominated by the longitudinal magnetization Fourier component $\widetilde{M}_z$, the angular anisotropies of $d\Sigma^{++}/d\Omega$ and $d\Sigma^{--}/d\Omega$ [Fig. 2.19(a) and (b)] are of the $\sin^4\theta$-type, whereas $d\Sigma^{+-}/d\Omega$ [Fig. 2.20(a)] exhibits the characteristic $\sin^2\theta\cos^2\theta$ anisotropy [compare eqns (2.99) and (2.100)]. The two-dimensional non-spin-flip and spin-flip data have been corrected for spin leakage, and the azimuthally

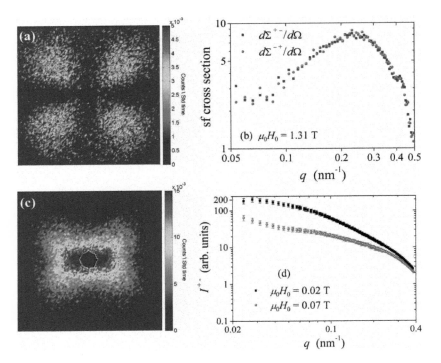

Fig. 2.20: (a) Spin-flip SANS cross section $d\Sigma^{+-}/d\Omega$ of an Fe–Cr-based nanocomposite at a saturating applied magnetic field of 1.31 T ($\mathbf{k}_0 \perp \mathbf{H}_0$) (logarithmic color scale). (b) Azimuthally averaged spin-flip SANS cross sections $d\Sigma^{+-}/d\Omega$ and $d\Sigma^{-+}/d\Omega$ at 1.31 T (log-log scale). (c) $d\Sigma^{+-}/d\Omega$ at 0.07 T. (d) Azimuthally averaged spin-flip data at different applied magnetic fields (see inset) (log-log scale). After [332].

averaged data in Fig. 2.20(b) demonstrate that the scattering in both spin-flip channels is equal, as is expected for this type of polycrystalline alloy. It is also worth mentioning that for applied fields down to about 20 mT the Fe–Cr sample is still in a single-domain state (inferred from magnetization data), so that the depolarization is due to the ultra small-angle spin-flip scattering, and not due to the refraction by a multi-domain structure. Therefore, away from saturation, transversal magnetization Fourier components contribute to both the non-spin-flip and spin-flip scattering and give rise to a different angular anisotropy than in the saturated state. This becomes clearly visible in Fig. 2.20(c), which shows $d\Sigma^{+-}/d\Omega$ at 70 mT. The non-spin-flip scattering [Fig. 2.19(c) and (d)] also changes significantly at a lower field.

2.9 Magnetic SANS at saturation

The completely saturated magnetization state, $\mathbf{M} = \{0, 0, M_z = M_s(\mathbf{r})\}$, is often used as a reference, for instance, when discussing the results of SANS measurements at lower applied magnetic fields. In fact, it turns out that the analysis of the spin-misalignment scattering (see Section 4.4) is best performed when the nuclear and magnetic SANS cross section at a saturating field—also called the residual SANS cross section—has been subtracted. We find it therefore useful to explicitly display in the following the SANS cross-section expressions for saturated magnetic microstructures. It is emphasized that the task of finding $|\widetilde{N}(\mathbf{q})|^2$ and $|\widetilde{M}_s(\mathbf{q})|^2$ is entirely unrelated to micromagnetic theory. It is a purely geometrical problem, which is determined by the size, shape, and arrangement of the microstructural inhomogeneities (see also the discussion in Sections 1.6 and 2.9.4).

2.9.1   Unpolarized SANS cross sections at saturation

$\mathbf{k}_0 \perp \mathbf{H}_0$.

$$\frac{d\Sigma}{d\Omega}(\mathbf{q}) = \frac{8\pi^3}{V}\left(|\widetilde{N}|^2 + b_{\mathrm{H}}^2|\widetilde{M}_s|^2 \sin^2\theta\right). \tag{2.110}$$

$\mathbf{k}_0 \parallel \mathbf{H}_0$.

$$\frac{d\Sigma}{d\Omega}(\mathbf{q}) = \frac{8\pi^3}{V}\left(|\widetilde{N}|^2 + b_{\mathrm{H}}^2|\widetilde{M}_s|^2\right). \tag{2.111}$$

2.9.2   SANSPOL cross sections at saturation

$\mathbf{k}_0 \perp \mathbf{H}_0$.

$$\frac{d\Sigma^{\pm}}{d\Omega}(\mathbf{q}) = \frac{8\pi^3}{V}\left(|\widetilde{N}|^2 + b_{\mathrm{H}}^2|\widetilde{M}_s|^2 \sin^2\theta + W^{\pm}b_{\mathrm{H}}(\widetilde{N}\widetilde{M}_s^* + \widetilde{N}^*\widetilde{M}_s)\sin^2\theta\right). \tag{2.112}$$

$\mathbf{k}_0 \parallel \mathbf{H}_0$.

$$\frac{d\Sigma^{\pm}}{d\Omega}(\mathbf{q}) = \frac{8\pi^3}{V}\left(|\widetilde{N}|^2 + b_{\mathrm{H}}^2|\widetilde{M}_s|^2 + W^{\pm}b_{\mathrm{H}}(\widetilde{N}\widetilde{M}_s^* + \widetilde{N}^*\widetilde{M}_s)\right). \tag{2.113}$$

2.9.3   POLARIS cross sections at saturation

$\mathbf{k}_0 \perp \mathbf{H}_0$.

$$\frac{d\Sigma^{\pm\pm}}{d\Omega}(\mathbf{q}) = \frac{8\pi^3}{V}\left(|\widetilde{N}|^2 + b_{\mathrm{H}}^2|\widetilde{M}_s|^2 \sin^4\theta \mp b_{\mathrm{H}}(\widetilde{N}\widetilde{M}_s^* + \widetilde{N}^*\widetilde{M}_s)\sin^2\theta\right). \tag{2.114}$$

$$\frac{d\Sigma^{\pm\mp}}{d\Omega}(\mathbf{q}) = \frac{8\pi^3}{V}b_{\mathrm{H}}^2|\widetilde{M}_s|^2 \sin^2\theta \cos^2\theta. \tag{2.115}$$

$\mathbf{k}_0 \parallel \mathbf{H}_0$.

$$\frac{d\Sigma^{\pm\pm}}{d\Omega}(\mathbf{q}) = \frac{8\pi^3}{V}\left(|\widetilde{N}|^2 + b_{\mathrm{H}}^2|\widetilde{M}_s|^2 \mp b_{\mathrm{H}}(\widetilde{N}\widetilde{M}_s^* + \widetilde{N}^*\widetilde{M}_s)\right). \tag{2.116}$$

$$\frac{d\Sigma^{\pm\mp}}{d\Omega}(\mathbf{q}) = 0. \tag{2.117}$$

We remind that nuclear-spin-dependent SANS has been ignored in the above equations, hence, $\frac{d\Sigma^{\pm\mp}}{d\Omega} = 0$ for $\mathbf{k}_0 \parallel \mathbf{H}_0$.

Figure 2.21 qualitatively shows the SANS cross sections at saturation for $\mathbf{k}_0 \perp \mathbf{H}_0$ and for different ratios R of nuclear to magnetic scattering,

$$R(q) = \frac{|\widetilde{N}|^2}{b_{\mathrm{H}}^2 |\widetilde{M}_{\mathrm{s}}|^2}, \tag{2.118}$$

assuming for simplicity the sphere form factor for both nuclear ($\widetilde{N}$) and longitudinal magnetic ($\widetilde{M}_{\mathrm{s}}$) scattering amplitudes (dilute limit). Note, however, that the parameter R may in general be q-dependent, for instance, for magnetic nanoparticles with a "dead" surface layer (core-shell-type structure). For statistically isotropic microstructures the case $\mathbf{k}_0 \parallel \mathbf{H}_0$ is of low interest, since the corresponding SANS cross sections are all isotropic (no θ-dependence). While most of the images in Fig. 2.21 have been reported countless times in the literature, we would like to draw the attention of the reader to the cross-shaped angular anisotropy in the non-spin-flip (++) channel depicted in Fig. 2.21(b). This type of scattering pattern has been observed in an Fe-based two-phase nanocomposite [see Fig. 2.22(a)] [333], and later independently also for a Cu–Ni–Fe alloy [287]. Analysis of eqn (2.114) reveals that, for this class of materials, the cross-shaped anisotropy is only observable at saturation in $\frac{d\Sigma^{++}}{d\Omega}$, provided that the ratio of nuclear to magnetic scattering is smaller than unity (roughly $R \sim 0.1$–0.4), as is experimentally observed [see Fig. 2.22(b)]. The q-dependence of R may be seen as an indication for the existence of a core-shell-type structure of the precipitates. The observation of the cross-shaped anisotropy represents one example where POLARIS provides information that is not straightforwardly accessible via conventional unpolarized SANS or SANSPOL techniques.

2.9.4   Relation to particle-matrix approach

The magnetic SANS of bulk magnetic materials (e.g., single-phase elemental ferromagnets, Nd–Fe–B-based permanent magnets, or steels) is to a large extent determined by long-wavelength magnetization fluctuations due to defect-induced spin misalignment. Away from magnetic saturation, all three magnetization components $\mathbf{M} = \{M_x, M_y, M_z\}$ govern the magnetic SANS cross section, whereas in the saturated state, when $\mathbf{M} = \{0, 0, M_z = M_{\mathrm{s}}(\mathbf{r})\}$, the total cross section is determined by the nuclear SANS and by the Fourier transform $\widetilde{M}_{\mathrm{s}}(\mathbf{q})$ of the spatially dependent saturation-magnetization profile $M_{\mathrm{s}}(\mathbf{r})$. It is important to realize that the spin-misalignment SANS cross section depends primarily on the magnetic interactions, but also on the underlying grain microstructure, while the SANS at saturation is entirely determined by the geometry of the microstructure; in other words, $|\widetilde{N}|^2$ and $|\widetilde{M}_{\mathrm{s}}|^2$ depend only on the size, shape, arrangement of and interaction potential between the particles, and on the scattering-length density contrast of the particles relative to the matrix.

Following Schlömann [335], the magnitude square of the Fourier transform of a saturated microstructure consisting of a distribution of $i = 1, \ldots, N_{\mathrm{p}}$ spherical particles with saturation magnetizations $M_{\mathrm{s},i}^{\mathrm{p}}$ and volumes $V_{\mathrm{p},i}$ in a matrix of saturation magnetization $M_{\mathrm{s}}^{\mathrm{m}}$ can be written as (see Appendix D):

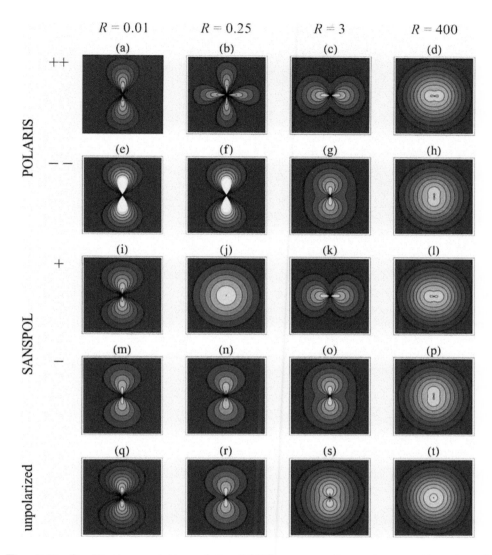

Fig. 2.21: Qualitative variation of the SANS cross sections at magnetic saturation for $\mathbf{k}_0 \perp \mathbf{H}_0$ and for different ratios R of nuclear to magnetic scattering $[R = |\widetilde{N}|^2/(b_H^2|\widetilde{M}_s|^2)]$. (from left to right column) $R = 0.01$, $R = 0.25$, $R = 3$, $R = 400$. (a)$-$(d) $\frac{d\Sigma^{++}}{d\Omega}$; (e)$-$(h) $\frac{d\Sigma^{--}}{d\Omega}$; (i)$-$(l) $\frac{d\Sigma^{+}}{d\Omega}$; (m)$-$(p) $\frac{d\Sigma^{-}}{d\Omega}$; (q)$-$(t) unpolarized $\frac{d\Sigma}{d\Omega}$. For the calculation of the cross sections, we have assumed the sphere form factor (sphere diameter: 16 nm) for both $|\widetilde{N}|^2$ and $|\widetilde{M}_s|^2$. Yellow-green color corresponds to high intensity and blue color to low intensity. Note that the individual images are not drawn on the same color scale. After [59].

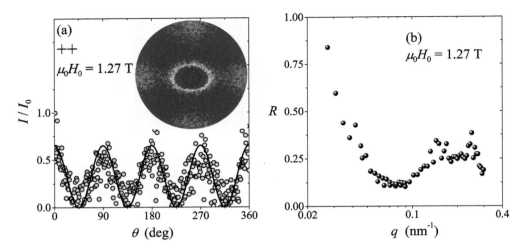

Fig. 2.22: (a) ($\circ$) Normalized non-spin-flip ($++$) SANS intensity I/I_0 as a function of the angle θ at $\mu_0 H_0 = 1.27\,\mathrm{T}$ and at $q = (0.26 \pm 0.01)\,\mathrm{nm}^{-1}$ ($\mathbf{k}_0 \perp \mathbf{H}_0$). Solid line: $I/I_0 \propto \cos^2(2\theta)$. The inset shows $\frac{d\Sigma^{++}}{d\Omega}$ on the two-dimensional detector (logarithmic color scale). Applied-field direction is horizontal. The sample under study is a magnetic nanocomposite from the NANOPERM family of alloys with a nominal composition of $(\mathrm{Fe}_{0.985}\mathrm{Co}_{0.015})_{90}\mathrm{Zr}_7\mathrm{B}_3$ [334]. (b) ($\bullet$) Experimental momentum-transfer dependence (at 1.27 T) of the ratio R of nuclear to magnetic SANS, $R = \frac{d\Sigma_{\mathrm{nuc}}}{d\Omega} / \frac{d\Sigma_M}{d\Omega}$ (log-linear scale). After [333].

$$|\widetilde{M}_{\mathrm{s}}(\mathbf{q})|^2 = \frac{1}{8\pi^3} \sum_{i=1}^{N_{\mathrm{P}}} \sum_{j=1}^{N_{\mathrm{P}}} (M_{\mathrm{s},i}^{\mathrm{P}} - M_{\mathrm{s}}^{\mathrm{m}})(M_{\mathrm{s},j}^{\mathrm{P}} - M_{\mathrm{s}}^{\mathrm{m}}) F_i(\mathbf{q}) F_j^*(\mathbf{q}) \exp\left(-i\mathbf{q} \cdot [\mathbf{r}_i - \mathbf{r}_j]\right),$$

$$(2.119)$$

where

$$F_i(\mathbf{q}) = \int_{V_{\mathrm{p},i}} \exp\left(-i\mathbf{q} \cdot \mathbf{r}\right) d^3 r \qquad (2.120)$$

denotes the particle form factor, and $\mathbf{r}_i$ is the position vector of particle i. For a sphere

$$F_i(qR_i) = 3V_{\mathrm{s},i} \frac{j_1(qR_i)}{qR_i}, \qquad (2.121)$$

where $V_{\mathrm{s},i} = \frac{4\pi}{3} R_i^3$, and $F_i(0) = V_{\mathrm{s},i}$. An analogous expression to eqn (2.119) describes the corresponding nuclear small-angle scattering [compare eqn (2.60)]. Obviously, eqn (2.119) is difficult to evaluate for the general case.

In the monodisperse and dilute limit, $|\widetilde{M}_{\mathrm{s}}(\mathbf{q})|^2$ simplifies to:

$$|\widetilde{M}_{\mathrm{s}}(\mathbf{q})|^2 = \frac{1}{8\pi^3} N_{\mathrm{P}} (\Delta M)^2 |F(\mathbf{q})|^2, \qquad (2.122)$$

where $\Delta M = M_{\mathrm{s}}^{\mathrm{P}} - M_{\mathrm{s}}^{\mathrm{m}}$. Inserting eqn (2.122) into eqns (2.110) and (2.111), the magnetic SANS cross section (e.g., for $\mathbf{k}_0 \perp \mathbf{H}_0$) takes on the familiar form:

$$\frac{d\Sigma_{\mathrm{M}}^{\mathrm{sat}}}{d\Omega}(\mathbf{q}) = \frac{N_{\mathrm{p}}}{V}(\Delta\rho)_{\mathrm{mag}}^2 |F(\mathbf{q})|^2 \sin^2\theta, \tag{2.123}$$

where

$$(\Delta\rho)_{\mathrm{mag}}^2 = b_{\mathrm{H}}^2 (\Delta M)^2 \tag{2.124}$$

denotes the magnetic scattering-length density contrast; for $\mu_0 \Delta M = 1\,\mathrm{T}$, we have $(\Delta\rho)_{\mathrm{mag}} = 2.316 \times 10^{14}\,\mathrm{m}^{-2}$. Equation (2.123) represents the well-known expression—embodying the particle-matrix concept—which is employed in many magnetic SANS investigations, even in situations where the material under study is not fully saturated. As the derivation has shown, eqn (2.123) relies on the special assumption of homogeneously magnetized domains, and for various reasons (see the discussion in Section 5.1.1) it does not describe the magnetic SANS of bulk magnets (unless fully saturated). In fact, the strong field response of $d\Sigma/d\Omega$ by several orders of magnitude that is typically observed in bulk magnets cannot be reconciled with SANS cross sections based on eqn (2.123).

When the particle concentration increases, the interference of the radiation scattered from different particles becomes relevant. This is taken into account by introducing the structure factor $S(\mathbf{q})$, which is generally defined as [74]:

$$S(\mathbf{q}) = \frac{1}{N_{\mathrm{p}}} \sum_{i=1}^{N_{\mathrm{p}}} \sum_{j=1}^{N_{\mathrm{p}}} \exp\left(-i\mathbf{q}\cdot[\mathbf{r}_i - \mathbf{r}_j]\right) = 1 + \frac{1}{N_{\mathrm{p}}} \sum_{\substack{i,j=1 \\ i \neq j}}^{N_{\mathrm{p}}} \exp\left(-i\mathbf{q}\cdot[\mathbf{r}_i - \mathbf{r}_j]\right), \tag{2.125}$$

where in the last expression on the right-hand side of eqn (2.125) the diagonal $i = j$ terms in the double sum have been separated to yield a constant. In diffuse nuclear and magnetic SANS on non-periodic structures, $S(\mathbf{q})$ does not exhibit sharp Bragg peaks, as it is the case for long-range-ordered crystalline and magnetic systems, but shows some sort of short-range ordering depending on the particle density and on the strength of the interparticle interaction. For statistically isotropic systems, the structure factor depends only on the magnitude of the scattering vector, i.e., $S = S(q)$. Moreover, for very low particle concentrations, when correlations in the particle positions can be neglected, $S(q)$ tends to unity. In a continuum picture, the structure factor can be related to the static pair-correlation function $g(r)$ via the following Fourier transform [264]:

$$S(q) = 1 + 4\pi \frac{N_{\mathrm{p}}}{V} \int_0^\infty (g(r) - 1) \frac{\sin qr}{qr} r^2 dr, \tag{2.126}$$

where $g(r)$ describes the arrangement of the particles, and may be seen as a measure for the probability that a pair of particles is a distance r apart. Using liquid-state theory [264], $S(q)$ can be found by solving the Ornstein–Zernike integral equation, which includes the interparticle interaction potential.

Returning to the discussion of magnetic SANS, one can then use eqn (2.125) in eqn (2.119) to obtain an expression for the azimuthally averaged magnetic SANS cross section of a dense array of identical and saturated spherically symmetric particles in a homogeneous saturated or nonmagnetic matrix [336, 337]:

$$
\frac{d\Sigma_{\mathrm{M}}^{\mathrm{sat}}}{d\Omega}(q) = \frac{1}{2}\frac{N_{\mathrm{P}}}{V}(\Delta\rho)_{\mathrm{mag}}^2 F^2(q)\left(1 + \frac{1}{N_{\mathrm{P}}}\sum_{\substack{i,j=1 \\ i \neq j}}^{N_{\mathrm{P}}} \exp\left(-i\mathbf{q}\cdot[\mathbf{r}_i - \mathbf{r}_j]\right)\right)
$$
$$
= P(q)S(q), \tag{2.127}
$$

where the factor of $1/2$ is due to the azimuthal average of the $\sin^2\theta$ factor, $P(q) = \frac{1}{2}\frac{N_{\mathrm{P}}}{V}(\Delta\rho)_{\mathrm{mag}}^2 F^2(q)$ characterizes the single-particle form factor, and $S(q)$ incorporates the effect of wave interferences due the positional correlations between the particles. The so-called Percus–Yevick hard-sphere structure factor is one of the very few known analytical expressions for $S(q)$ [338]. Within this model, two identical spheres with a center-to-center distance of r interact via the following potential:

$$
V_{\mathrm{HS}}(r) = \begin{cases} \infty & \text{if } r < 2R_{\mathrm{HS}} \\ 0 & \text{if } r > 2R_{\mathrm{HS}}, \end{cases} \tag{2.128}
$$

where R_{HS} is the hard-sphere radius. Using the methods of statistical mechanics [338], the final expression for the structure factor then reads (e.g., [35, 339]):

$$
S(q) = \frac{1}{1 + 24\eta_{\mathrm{HS}}G(\mathcal{A})/\mathcal{A}}, \tag{2.129}
$$

where $\mathcal{A} = 2qR_{\mathrm{HS}}$, and η_{HS} denotes the hard-sphere volume fraction. The function $G(\mathcal{A})$ can be written as:

$$
G(\mathcal{A}) = \frac{\alpha_1\left(\sin\mathcal{A} - \mathcal{A}\cos\mathcal{A}\right)}{\mathcal{A}^2} + \frac{\alpha_2\left(2\mathcal{A}\sin\mathcal{A} + (2 - \mathcal{A}^2)\cos\mathcal{A} - 2\right)}{\mathcal{A}^3}
$$
$$
+ \frac{\alpha_3\left(-\mathcal{A}^4\cos\mathcal{A} + 4\left[(3\mathcal{A}^2 - 6)\cos\mathcal{A} + (\mathcal{A}^3 - 6\mathcal{A})\sin\mathcal{A} + 6\right]\right)}{\mathcal{A}^5}, \tag{2.130}
$$

where the coefficients $\alpha_1, \alpha_2, \alpha_3$ are given by:

$$
\alpha_1 = \frac{(1 + 2\eta_{\mathrm{HS}})^2}{(1 - \eta_{\mathrm{HS}})^4} \quad , \quad \alpha_2 = \frac{-6\eta_{\mathrm{HS}}\left(1 + \eta_{\mathrm{HS}}/2\right)^2}{(1 - \eta_{\mathrm{HS}})^4} \quad , \quad \alpha_3 = \frac{\eta_{\mathrm{HS}}\alpha_1}{2}. \tag{2.131}
$$

The normalized cross section eqn (2.127) is plotted in Fig. 2.23 for several values of η_{HS}. From there it is seen that for low-volume fractions the sphere form factor provides a good description, while a pronounced interference peak develops for larger η_{HS}-values.

Equation (2.127) is valid for a collection of monodisperse spherically symmetric particles. For polydisperse systems, it is in general not possible to express the SANS cross section as the product of a form factor $P(q)$ and a structure factor $S(q)$. Pedersen [337] has derived an equation, known as the local monodisperse approximation, which for concentrated systems with hard-sphere interactions includes the effects of polydispersity. In this approach it is assumed that a particle of a certain size is always

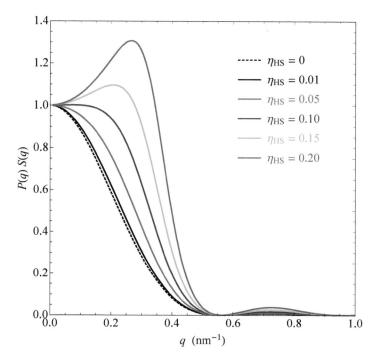

Fig. 2.23: Graphical illustration of $\frac{d\Sigma_{\mathrm{M}}^{\mathrm{sat}}}{d\Omega}(q) = P(q)S(q)$ [eqn (2.127)] using the sphere form factor for $P(q)$ and the hard-sphere structure factor for $S(q)$ [eqns (2.129)–(2.131)] with $R = R_{\mathrm{HS}} = 8\,\mathrm{nm}$. The constant prefactor $\frac{1}{2}\frac{N_{\mathrm{p}}}{V}(\Delta\rho)^2_{\mathrm{mag}}$ in eqn (2.127) has been set to unity, and $\frac{d\Sigma_{\mathrm{M}}^{\mathrm{sat}}}{d\Omega}(q)$ has then been normalized by $\frac{d\Sigma_{\mathrm{M}}^{\mathrm{sat}}}{d\Omega}(q=0) = V_{\mathrm{s}}^2\alpha_1^{-1}$. η_{HS} increases from bottom to top.

surrounded by particles with the same size. Within the local monodisperse approximation the scattering intensity can be written as:

$$\frac{d\Sigma_{\mathrm{M}}^{\mathrm{sat}}}{d\Omega}(q) = \int_0^\infty P(q,R)S(q,R)f(R)dR, \qquad (2.132)$$

where $f(R)$ is the particle-size distribution function. Equation (2.132) suggests that $d\Sigma_{\mathrm{M}}^{\mathrm{sat}}/d\Omega$ corresponds to the incoherent sum of the scattering intensities of monodisperse subsystems weighted with the size distribution. Alternatively, one can regard the intensity as originating from a system in which the size of the particles varies slowly with the position, so that every particle is surrounded by particles of the same size; in other words, there is a strong correlation between the particle position and the particle size [337]. This is in contrast to the so-called decoupling approximation, introduced by Kotlarchyk and Chen [208], which assumes that the particle size is uncorrelated with the particle position. Model calculations by Pedersen [337] show that the local monodisperse approximation works better than the decoupling approximation for systems with larger polydispersities and higher concentrations.

As a final remark we note that analytical expressions for particle form factors have been derived for practically all particle shapes and there exist also a few closed-form results for the structure factor [9–11, 35]. Most of the structure-factor models (e.g., hard-sphere, sticky hard-sphere, or screened Coulomb potential with or without poly-dispersity) have been derived for particles with spherical symmetry interacting through a spherically symmetric potential. We refer to the review article by Pedersen [35] for a detailed discussion of this topic. Likewise, several software packages [340–342] are available which contain an extensive collection of particle form factors (including particle-size distributions) and structure factors. The *SASfit* software package also provides a numerical algorithm to solve the Ornstein–Zernike equations for monodisperse particles for several interaction potentials and closure relations [343].

3

BASICS OF STATIC MICROMAGNETISM

In this chapter, we discuss the basic concepts of static micromagnetism as far as they are relevant for the analysis of magnetic SANS. We start out in Section 3.1 by recalling the main micromagnetic energy contributions. Section 3.2 provides the high-field solution of Brown's balance-of-torques equation for the Fourier components of the magnetization, whereas Section 3.3 establishes the relation between micromagnetics and SANS. We refer the reader to [119, 134, 154, 344] for standard textbooks on micromagnetics. Analytical micromagnetic calculations of the type presented in Section 3.2 have already been carried out by other authors; for instance, for the study of the approach to magnetic saturation in porous ferromagnets [335, 345] and in amorphous alloys [346, 347]. A commonality of the investigations by Schlömann and by Kronmüller and Fähnle is to solve the micromagnetic problem by representing the magnetization in Fourier space. The solutions for the magnetization Fourier components can then easily be adapted to the magnetic SANS problem.

3.1 Magnetic energy contributions

Micromagnetics is a phenomenological continuum theory which has been developed in order to compute the magnetization vector field $\mathbf{M}$ of an arbitrarily shaped ferromagnetic body, provided that the applied magnetic field, the geometry of the ferromagnet, and the magnetic materials parameters are known. The characteristic length scale which is addressed by micromagnetic calculations ranges between a few nanometers and a few hundreds of nanometers [348]. This is a size regime that overlaps with the resolution range of the conventional SANS technique as well as with the methods of ultra and very small-angle neutron scattering (USANS and VSANS). It is therefore straightforward to use micromagnetic theory for computing magnetic SANS cross sections. This approach has been pioneered by Kronmüller, Seeger, and Wilkens [1], who calculated the magnetic SANS due to spin disorder related to the strain fields of dislocations. The magnetic energy contributions that are taken into account in standard micromagnetic computations of the static magnetic microstructure are due to the isotropic and symmetric Heisenberg exchange interaction, the antisymmetric Dzyaloshinskii–Moriya interaction, the magnetocrystalline and magnetoelastic anisotropy, and the external and magnetodipolar field.

Magnetic Small-Angle Neutron Scattering: A Probe for Mesoscale Magnetism Analysis. Andreas Michels, Oxford University Press (2021). © Andreas Michels. DOI: 10.1093/oso/ 9780198855170.003.0003

The magnetization vector field

$$\mathbf{M} = \mathbf{M}(\mathbf{r}) = \left\{ \begin{array}{l} M_x(x,y,z) \\ M_y(x,y,z) \\ M_z(x,y,z) \end{array} \right\} \cong V^{-1} \sum_{i=1}^{N} \boldsymbol{\mu}_{a,i} \tag{3.1}$$

is introduced as the local thermodynamical average over N discrete atomic magnetic moments $\boldsymbol{\mu}_a$ within a small (but still macroscopic) volume V. It is considered to be a smooth and continuous function of the position $\mathbf{r} = \{x, y, z\}$ inside the material. The transition from a discrete atomic to a coarse-grained picture reflects the basic idea of the continuum theory of micromagnetics. Micromagnetic calculations are usually applicable far below the Curie point and subject to the constraint of constant magnetization magnitude within a domain, i.e.,

$$|\mathbf{M}|^2 = M_x^2 + M_y^2 + M_z^2 = M_s^2, \tag{3.2}$$

where M_s denotes the spontaneous or saturation magnetization, which is only a function of temperature, i.e., $M_s = M_s(T)$. Equation (3.2) implies that there are only two independent components of the magnetization, say, M_x and M_y, so that $M_z = (M_s^2 - M_x^2 - M_y^2)^{1/2}$.

Experimental conditions are commonly realized in such a way that the Gibbs free energy G represents the appropriate thermodynamic potential for micromagnetic problems. Therefore, the temperature T and the external magnetic field $\mathbf{B}_0 = \mu_0 \mathbf{H}_0$ may be taken as the independent variables [349]. In terms of energy densities one may write [134, 344]:

$$G = G(T, \mathbf{B}_0, \dots) = \int (U' - TS' - \mathbf{B}_0 \cdot \mathbf{M} + \dots) dV, \tag{3.3}$$

where U' and S' denote, respectively, the internal energy density and the entropy per unit volume. U' includes all the intrinsic magnetic contributions such as the exchange energy, magnetocrystalline anisotropy energy, magnetodipolar energy, and the energy due to the magnetostrictive deformations. The continuum-theoretical expressions for these internal energies can be derived by employing symmetry arguments [350], or by transforming quantum-mechanical energy expressions into a quasi-classical picture, e.g., by treating the spin operator in the Heisenberg Hamiltonian as a classical vector field [351]. According to Brown [344], for constant temperature and applied field, the stable magnetization state is obtained by means of variational calculus: one first has to find an equilibrium state for which the first variation of G vanishes, $\delta G = 0$, and then one has to show that for this particular state the second variation of G is positive, $\delta^2 G > 0$. Both steps are carried out for arbitrary variations of the internal magnetic variables of the system, $M_{x,y,z}$, subject to the constraint of eqn (3.2).

In the following, we consistently use the symbol E (instead of G) for denoting magnetic energies. A recurring issue is whether to group the energies according to physical origin, mathematical form, or to some other category; for instance, for low-symmetry crystal structures, the magnetodipolar interaction gives rise to a local energy contribution which has a mathematical form that is similar to the one of the spin-orbit-interaction-induced magnetocrystalline anisotropy [154]. Likewise, isotropic Heisenberg exchange, which is due to the combined action of the Coulomb interaction and the Pauli principle, contains first-order spatial derivatives of the magnetization, as does the Dzyaloshinskii–Moriya energy, which has its origin in the spin-orbit coupling [352]. Following Brown [344], we prefer to introduce the magnetic energies according to their form, not origin, and we start out with the exchange energy.

3.1.1 Exchange energy

Exchange interaction is a purely quantum-mechanical effect which has its origin in the electrostatic interaction between the electrons, taking their fermionic character into account (Pauli principle). There exist various forms of direct and indirect exchange mechanisms (see [353–358] for further reading). Their fine details are beyond the scope of this book. We start the discussion by writing down the general phenomenological expression for the exchange Hamiltonian between two spins $\mathbf{S}$ at sites i and j, which can be expressed in the following way [358]:

$$\widehat{H}_{\mathrm{ex}} = \sum_{\alpha\beta} J_{\alpha\beta} S_{i,\alpha} S_{j,\beta}, \tag{3.4}$$

where $\alpha, \beta = x, y, z$, and $J_{\alpha\beta}$ denotes the exchange tensor. This expression can be regrouped into three contributions [355, 356],

$$\widehat{H}_{\mathrm{ex}} = -J\mathbf{S}_i \cdot \mathbf{S}_j + \mathbf{D} \cdot (\mathbf{S}_i \times \mathbf{S}_j) + \mathbf{S}_i \cdot \Lambda \cdot \mathbf{S}_j, \tag{3.5}$$

where J is a scalar (exchange integral), $\mathbf{D}$ is the Dzyaloshinskii–Moriya vector, and Λ is the anisotropy tensor. The first term describes the isotropic and symmetric exchange, and the second and third terms are, respectively, the antisymmetric and symmetric parts of the anisotropic exchange. We refer to the review article by Zakharov et al. [358] for details on how the exchange parameters J, $\mathbf{D}$, and Λ are related to the entries of $J_{\alpha\beta}$. The energy of the isotropic exchange coupling [first term in eqn (3.5)] depends only on the relative orientation of the spins $\mathbf{S}_i$ and $\mathbf{S}_j$, and not on their orientation relative to the crystal (bond) axes, in contrast to the other two contributions. The dominant term in eqn (3.5) is generally due to isotropic exchange with J-values ranging between about $0.01-0.1\,\mathrm{eV}$, while the strength Λ of the anisotropic symmetric exchange is typically only about 5% of the isotropic coupling [358]. For $J > 0$ the parallel (ferromagnetic) alignment of the spins is favored, whereas for $J < 0$ the antiparallel (antiferromagnetic) orientation is energetically preferred. The anisotropic symmetric contribution plays an important role for the magnetic anisotropy of a number of low-dimensional spin systems. It will not be further considered in this book. Anisotropic antisymmetric exchange is discussed in Section 3.1.2.

In the limit of a continuous ferromagnetic material, the case that is of importance for the micromagnetic description of magnetic SANS, the isotropic term in the exchange energy can be expressed as [232, 359]:

$$E_{\mathrm{ex}} = \int A \left[(\nabla m_x)^2 + (\nabla m_y)^2 + (\nabla m_z)^2 \right] d^3 r, \tag{3.6}$$

where $A > 0$ denotes the exchange-stiffness constant, $\nabla = \mathbf{e}_x\,\partial/\partial x + \mathbf{e}_y\,\partial/\partial y + \mathbf{e}_z\,\partial/\partial z$ is the gradient (also called del or nabla) operator with $\mathbf{e}_x$, $\mathbf{e}_y$, $\mathbf{e}_z$ being the unit vectors along the Cartesian laboratory axes, and the $m_{x,y,z}(x, y, z)$ are the Cartesian components of the unit vector $\mathbf{m} = \mathbf{M}/M_s$ in the direction of the magnetization. Unless otherwise stated, all the integrals in this chapter extend over the sample volume V, where the local magnetization vector is nonzero. The exchange energy, as it is expressed by eqn (3.6), is a positive-definite quantity which favors the parallel (collinear) alignment of neighboring atomic magnetic moments in the crystal lattice. One should, however, remember that E_{ex} is defined up to a constant, which corresponds to the reference state when all the spins are fully aligned [360]. Any nonuniformity in $\mathbf{M}$ goes along with an energy cost. Values for A are typically taken from experiment and are in the $10\,\mathrm{pJ/m}$ range (see also Table 4.2) [134, 361]. Note that in multiphase magnets the exchange constant is a function of the position inside the material, i.e., $A = A(\mathbf{r})$ [362–364].

Kittel [351] has shown that eqn (3.6) can be derived by a continuum Taylor expansion based on the discrete Heisenberg Hamiltonian $-J\mathbf{S}_i \cdot \mathbf{S}_j$ [first term in eqn (3.5)]. Equation (3.6) follows by treating the $\mathbf{S}_i$ as quasi-classical continuous vectors, taking into account only nearest-neighbor exchange interactions of equal strength J, and by assuming small angles between neighboring spins. Therefore, the applicability of the above continuum expression for the exchange energy is restricted to situations where $\mathbf{M}(\mathbf{r})$ is a slowly varying function of the position $\mathbf{r}$ inside the material. Strongly inhomogeneous magnetization configurations might be realized e.g., at internal interfaces such as grain boundaries, where the materials parameters may take on different values as compared to their single-crystalline bulk values [365]. Moreover, for crystal lattices of low symmetry, the exchange interaction depends on the crystallographic orientation and the tensor character of A should be taken into account [232, 359]; for instance, hexagonal or tetragonal crystal structures would require two exchange constants. In fact, eqn (3.6) has been derived for cubic crystals [344]. Here, $A = JS^2 c^\star/a$, where S is the spin quantum number (measured in units of $\hbar$), a is the lattice constant, and $c^\star = 1, 2, 4$ for simple cubic, body-centered cubic, and face-centered cubic lattices, respectively. However, we would like to mention that the vast majority of micromagnetic computations that are reported in the literature model nonuniform spin states in single-phase materials by a single stiffness parameter; in other words, irrespective of the actual exchange mechanism, eqn (3.6) is expected to phenomenologically describe the stiffness effect of the spin structure to first order. Instead of Taylor expansion, Landau and Lifshitz (see § 43 in [350]) have employed pure symmetry arguments to motivate that eqn (3.6) is the simplest possible continuum expression for the exchange interaction which is invariant with respect to time inversion and rotation of the whole body (see also [232]).

3.1.2 Dzyaloshinskii–Moriya energy

It was Dzyaloshinskii [366] who realized, purely on the grounds of symmetry considerations, that the exchange interaction between spins may contain an antisymmetric contribution. Moriya [352] has explained such antisymmetric spin couplings in his theory of anisotropic superexchange involving the spin-orbit interaction. Anisotropic antisymmetric exchange [second term in eqn (3.5)], also known as the Dzyaloshinskii–Moriya interaction (DMI), is crucial for the understanding of the spin structures of many important "weakly ferromagnetic" antiferromagnets and, in particular, for the recently discovered skyrmions crystals (see [64] for a review). The DMI is due to relativistic spin-orbit coupling, and in low-symmetry crystal structures lacking inversion symmetry it gives rise to antisymmetric magnetic interactions. Particularly well investigated classes of materials where a DMI is operative are ultrathin film nanostructures and non-centrosymmetric B20 transition-metal compounds (e.g., MnSi, $Fe_{1-x}Co_xSi$, or FeGe). In these materials, the DMI plays an important role for the stabilization of various kinds of noncollinear spin structures such as long-wavelength spirals, vortex states, and skyrmion textures (see, e.g., [257, 259–261, 367–398] and references therein).

According to Moriya's theory of anisotropic superexchange interaction [352], the magnitude of the DMI is roughly $\Delta g/g$ times the isotropic superexchange contribution, where g is the Landé factor and Δg is its deviation from the free-electron value. The direction of the DMI vector $\mathbf{D}$ depends on the symmetry of the magnetic exchange path and is for many materials parallel or perpendicular to the line connecting the two magnetic spins $\mathbf{S}_i$ and $\mathbf{S}_j$. In a centrosymmetric crystal-field environment the vector $\mathbf{D}$ vanishes (see, e.g., [352, 368, 399] for a detailed discussion). The DMI energy $E_{\mathrm{dmi}} = \mathbf{D} \cdot (\mathbf{S}_i \times \mathbf{S}_j)$ gives rise to canted spin structures, since it tries to force $\mathbf{S}_i$ and $\mathbf{S}_j$ to be at right angles. This is in contrast to the isotropic exchange term, which favors ferro- or antiferromagnetically coupled spins (at least for nearest-neighbor interactions). The DMI is of particular importance for the weak ferromagnetism found in otherwise antiferromagnetic structures such as α-Fe_2O_3, $MnCO_3$, $CoCO_3$, or CrF_3 [352, 366].

On the phenomenological level of the continuum theory of micromagnetics the DMI energy can be described in terms of so-called Lifshitz invariants $L_{ij}^{(k)}$ [367–369], which are antisymmetric terms that are linear in the first spatial derivatives of the magnetization; in particular [400],

$$E_{\mathrm{dmi}} = \int \sum_{i,j,k} D_{ijk} L_{ij}^{(k)} d^3 r = \int \sum_{i,j,k} D_{ijk} \left(m_i \frac{\partial m_j}{\partial k} - m_j \frac{\partial m_i}{\partial k} \right) d^3 r, \qquad (3.7)$$

where the i, j, k are combinations of the Cartesian coordinates x, y, z, and the parameters D_{ijk} (in units of J/m^2) denote effective DMI constants taking on positive or negative values depending on the material. The $L_{ij}^{(k)} = m_i \frac{\partial m_j}{\partial k} - m_j \frac{\partial m_i}{\partial k}$ in eqn (3.7) reflect the crystallographic symmetry, and for non-centrosymmetric cubic (c) crystal structures (e.g., MnSi and FeGe) with $D_{ijk} = D$ one obtains [368, 401]

$$L_c = L_{zy}^{(x)} + L_{xz}^{(y)} + L_{yx}^{(z)} = \mathbf{m} \cdot (\nabla \times \mathbf{m}). \qquad (3.8)$$

Typical D-values for interfacial DMI are in the range of a few mJ/m^2 (e.g., [388, 394]), whereas bulk DMI values are about an order of magnitude smaller (e.g., [402–404]).

The DMI energy represents a pseudoscalar contribution to the total magnetic energy, and it is of particular relevance for systems without a center of inversion. Pseudoscalar terms change their sign under the inversion operation, and render one type of chirality (corresponding to a negative value of the pseudoscalar) preferred with respect to the other type. Of such terms, constructed from the magnetization vector-field components, the eqn (3.8) is probably the simplest, but there can be more complex expressions of this kind, like eqn (3.7) and others. Chirality selection happens entirely due to the pseudoscalar terms in the magnetic potential energy. The inversion transformation maps the right-handed structures into the left-handed structures. Therefore, when the inversion symmetry is unbroken, the right- and left-handed structures are equivalent.

Antisymmetric interactions are encountered in many fields of physics, e.g., in liquid crystals [405–407], ferroelectric, magnetoelectric, and multiferroic materials [408–410], and in certain classes of magnets [366, 368, 375, 382, 411–413]. Inclusion of eqn (3.8) into the micromagnetic energy functional may result in explicitly complex-valued magnetization Fourier components and in a polarization-dependent purely magnetic and asymmetric contribution to the SANS cross section (see Section 4.3.3) [414]. We note that such an asymmetry has also been reported for the case of spin-wave and ferromagnetic resonance spectroscopy (see, e.g., [258, 401–403, 415–419] and references therein).

Besides the above-discussed intrinsic DMI mechanism, it is known that lattice imperfections in the microstructure of crystalline magnetic materials (including spin glasses) are accompanied by the presence of local DMI couplings due to the breaking of inversion symmetry at defect sites. In 1963, Arrott [420] suggested that the DMI is present in the vicinity of any lattice defect and that it gives rise to inhomogeneous magnetization states: for two magnetic ions which are ferromagnetically coupled by the isotropic exchange interaction, the DMI, when acting on the exchange path, produces an antiferromagnetic (anticollinear) component, while for the two ions being antiferromagnetically aligned by isotropic exchange, the DMI causes a ferromagnetic (collinear) component. In a sense, microstructural defects act as a source of additional local chiral interactions, similar to the above intrinsic DMI in non-centrosymmetric crystal structures. Based on the long-ranged Ruderman–Kittel–Kasuya–Yosida Hamiltonian amended by an additional DMI term that describes the spin-orbit scattering of the conduction electrons by nonmagnetic impurities, Fert and Levy [411–413] have shown that the effect of the DMI can explain the enhancement of the anisotropy fields observed in certain spin glasses and that the magnitude of the DMI term can be quite large, about $10-20\%$ of the unperturbed ground state energy. Therefore, it is important to realize that defect-induced DMI is generally operative in polycrystalline and disordered materials, even in high-symmetry lattices, where the usual intrinsic DMI term vanishes. This point of view has already been adopted by Fedorov et al. [421], who studied the impact of torsional-strain-induced DMI couplings near dislocations on the helix domain populations in polycrystalline Ho metal. Similarly, Grigoriev et al. [422, 423] investigated the field-induced chirality in the helix structures of Dy/Y and Ho/Y multilayer films and provided evidence for interface-induced DMI. Beck and Fähnle [424] combined ab-initio density functional electron theory with a micromagnetic model to study the DMI vectors arising from a fabrication-induced perpendicular strain gradient in a film of bcc Fe. Kitchaev et al. [425] theoretically showed that a DMI may arise in a material with any symmetry when coupled to a strain field. Kim et al. [404] have investigated Gd–Fe–Co amorphous ferrimagnetic alloys and report the observation of a bulk DMI. An asymmetric distribution of elemental content (composition gradient) is supposed to be at the origin of the broken inversion symmetry in the ferrimagnetic layer. Butenko and Rößler [426] have developed a continuum micromagnetic model for dislocation-induced DMI couplings. These authors considered a disk-like film element with a screw dislocation at its center and showed that the associated defect-induced DMI leads to a chirality selection of the vortex state. Their continuum expression for the DMI energy density is the following:

$$\omega_{\text{dmi}}^{\text{dis}} = D \left(\mathbf{M} \times \mathbf{G}\mathbf{M} \right) \cdot \left(\nabla \times \mathbf{G}\mathbf{u} \right), \tag{3.9}$$

where D is a DMI constant for the torsional magnetoelastic couplings, $\mathbf{G} = \partial/\partial x + \partial/\partial y + \partial/\partial z$, and $\mathbf{u}$ is the elastic displacement field around the dislocation. Equation (3.9) may be a new starting point for the investigation of dislocation-induced or, more generally speaking, defect-induced effects on the magnetic SANS cross section [1, 5].

3.1.3 Magnetocrystalline anisotropy

The magnetocrystalline anisotropy energy E_{mc} expresses the dependence of the thermodynamic potential of a ferromagnet on the orientation of the magnetization $\mathbf{M}$ relative to the crystal axes. The origin of magnetocrystalline anisotropy is related to the combined action of the spin-orbit coupling, which connects the spins with the crystal lattice, and the crystal-field interaction. Dipole-dipole interactions may also contribute to E_{mc} [427, 428], however, dipolar anisotropy is generally small and vanishes for ideal cubic and hexagonal lattices [351]. On a phenomenological level

$$E_{\text{mc}} = \int \omega_{\text{mc}} \, d^3 r \tag{3.10}$$

is determined by an expression for the anisotropy energy density ω_{mc} [344]. For a polycrystalline ferromagnet, ω_{mc} may be a function of the position and of the magnetization, i.e.,

$$\omega_{\text{mc}} = \omega_{\text{mc}}(x, y, z, M_x, M_y, M_z). \tag{3.11}$$

However, we remind that due to the micromagnetic constraint [eqn (3.2)], ω_{mc} depends only on two independent components of $\mathbf{M}$. The magnetocrystalline anisotropy energy favors the orientation of $\mathbf{M}$ along certain crystallographic directions, the so-called easy axes or easy planes. Since expressions for E_{mc} should be invariant under time reversal and since $\mathbf{M}$ is an odd function under time reversal, it follows then that E_{mc} must be an even function of the components of the magnetization [350].

In the book by Kronmüller and Fähnle [134] various anisotropy expressions for orthorhombic, tetragonal, cubic, and hexagonal crystal structures can be found. For the case of cubic (c) and uniaxial (u) anisotropy, ω_{mc} can be expanded in powers of the reduced magnetization components $\mathbf{m} = \mathbf{M}/M_{\text{s}} = (m_x, m_y, m_z)$ as [154]:

$$\omega_{\text{mc}}^{\text{c}} = K_{\text{c0}} + K_{\text{c1}} \left(m_x^2 m_y^2 + m_x^2 m_z^2 + m_y^2 m_z^2 \right) + K_{\text{c2}} m_x^2 m_y^2 m_z^2 + \ldots \tag{3.12}$$

and

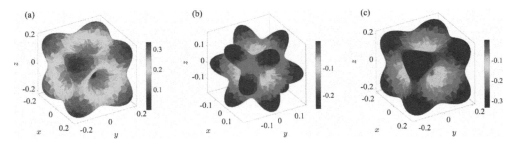

Fig. 3.1: Energy surfaces for cubic magnetocrystalline anisotropy [eqn (3.12) with $K_{c0} = 0$]. (a) Easy axis along $\langle 100 \rangle$ directions with $K_{c1} = 1$ and $K_{c2} = 0$; (b) easy axis along $\langle 110 \rangle$ directions with $K_{c1} = -1$ and $K_{c2} = 5$; (c) easy axis along $\langle 111 \rangle$ directions with $K_{c1} = -1$ and $K_{c2} = 0$ (all K_{ci} in units of J/m^3).

$$\omega_{\mathrm{mc}}^{\mathrm{u}} = K_{\mathrm{u}0} + K_{\mathrm{u}1}(1 - m_z^2) + K_{\mathrm{u}2}(1 - m_z^2)^2 + \ldots$$
$$= K_{\mathrm{u}0} + K_{\mathrm{u}1}\sin^2\gamma + K_{\mathrm{u}2}\sin^4\gamma + \ldots, \tag{3.13}$$

where the respective K_{ci} and $K_{\mathrm{u}i}$ represent the temperature-dependent anisotropy constants. In eqn (3.12) the cubic axes are chosen as coordinate axes, and γ in eqn (3.13) is the angle between **m** and the direction of the uniaxial anisotropy axis (here chosen to be parallel to $\mathbf{e}_z$). The anisotropy constants may be either positive or negative.

Figures 3.1 and 3.2 display these energy densities for selected combinations (magnitude and sign) of the anisotropy constants. The easy directions for the magnetization are found by solving the extrema conditions $\partial\omega_{\mathrm{mc}}/\partial m_i = 0$ and $\partial^2\omega_{\mathrm{mc}}/\partial m_{i,j}^2 > 0$, where $i, j = x, y, z$; see [134] for the phase diagrams of cubic and hexagonal crystals. For the examples of α-Fe (bcc) and Ni (fcc), the respective easy directions at room temperature are the cube axes $\langle 100 \rangle$ and the body diagonals $\langle 111 \rangle$, whereas in hcp Co single crystals, the magnetic moments at room temperature are aligned along the c-axis and the basal plane is a so-called hard plane for the magnetization. Values for the anisotropy constants are usually taken from experiment and range between about 10^2 J/m^3 (for soft magnetic materials) and 10^7 J/m^3 (for hard magnets) [134, 361, 429]; see Table B.1 for K-values of some magnetic materials at room temperature. Note also that, as with the other magnetic materials parameters (A, D, M_{s}), the K's may depend on the position, i.e., $K = K(\mathbf{r})$.

3.1.4 Magnetoelastic anisotropy

When an unstressed ferromagnetic specimen is magnetized by an increasing applied magnetic field, the dimensions of the body change. The process continues until elastic counterforces provide for a state of stable equilibrium. This phenomenon is called magnetostriction, and it demonstrates that there is a connection between the elastic and magnetic properties of magnetic materials [430]. Magnetostriction occurs because the magnetic energy of a ferromagnetic crystal depends on the components of the strain tensor, and it may be energetically favorable for an unstrained lattice to spontaneously deform and lower its energy. The associated interaction energy is called the

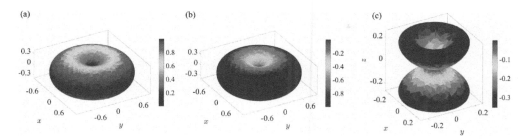

Fig. 3.2: Same as Fig. 3.1, but for uniaxial magnetocrystalline anisotropy [eqn (3.13) with $K_{u0} = 0$]. (a) Easy-axis anisotropy with $K_{u1} = 1$ and $K_{u2} = 0$; (b) easy-plane anisotropy with $K_{u1} = -1$ and $K_{u2} = 0$; (c) easy-cone anisotropy with $K_{u1} = -\sqrt{2}$ and $K_{u2} = \sqrt{2}$.

magnetoelastic energy E_{me}, which can also be defined via a volume integral over some energy density ω_{me} [compare eqn (3.10)],

$$E_{me} = \int \omega_{me} \, d^3 r. \tag{3.14}$$

The general treatment of magnetostrictive effects in deformable magnets is a highly complicated task [134], and we refer the reader to Chapter 8 in Brown's book [344] for a critical discussion of this point.

For polycrystalline and defect-rich single-crystalline magnetic materials it is well known that the mechanical lattice strain which is related to microstructural imperfections couples to the spin distribution and may give rise to nonuniform magnetization patterns [7, 431]; for instance, the long-range stress field of a straight dislocation line (stress: $\sigma \propto 1/r$) produces a characteristic magnetic structure, which can be resolved via the SANS technique [6]; see Fig. 1.7 for qualitative spin distributions around vacancy-type defects and dislocations and Section 4.6 for a summary of the basics of magnetic SANS due to dislocations.

We will close this subsection by stating the relevant expressions for the magneto-elastic coupling energy of cubic crystals and for the case of isotropic magnetostriction. These form the basis for determining the spin distribution in the vicinity of stress-active defects. For cubic crystals, the magnetoelastic coupling energy density (in cubic crystal coordinates) can be expressed as [134]:

$$\omega_{me}^c = -\frac{3}{2}\lambda_{100}\left(\sigma_{xx}m_x^2 + \sigma_{yy}m_y^2 + \sigma_{zz}m_z^2\right)$$
$$-\frac{3}{2}\lambda_{111}\left(\sigma_{xy}m_xm_y + \sigma_{xz}m_xm_z + \sigma_{yz}m_ym_z\right), \tag{3.15}$$

where λ_{100} and λ_{111} (either positive or negative) denote the saturation magnetostriction constants (i.e., the fractional length changes) along the indicated crystallographic directions, and the components of the stress tensor $\boldsymbol{\sigma}$ are due to internal stresses related to microstructural defects or to externally applied stresses. Values for the λ's are between about 10^{-6}–10^{-3} [134]. For isotropic materials (e.g., amorphous metals), eqn (3.15) reduces to ($\lambda_{100} = \lambda_{111} = \lambda_{\text{sat}}$):

$$\omega_{\text{me}}^{\text{iso}} = -\frac{3}{2}\lambda_{\text{sat}}\left(\sigma_{xx}m_x^2 + \sigma_{yy}m_y^2 + \sigma_{zz}m_z^2 + \sigma_{xy}m_xm_y + \sigma_{xz}m_xm_z + \sigma_{yz}m_ym_z\right). \tag{3.16}$$

An expression for the magnetoelastic coupling energy of hexagonal crystals can be found in [432].

3.1.5   Zeeman interaction

An externally applied magnetic field $\mathbf{H}_0$, commonly assumed to be constant in space, imposes a torque on the magnetization vector and tries to rotate the magnetic moments along its direction. The corresponding energy is called the Zeeman energy E_z and is given by:

$$E_z = -\mu_0 \int \mathbf{M} \cdot \mathbf{H}_0 d^3 r, \tag{3.17}$$

where $\mu_0 = 4\pi \times 10^{-7}$ Tm/A is the permeability of free space. For $\mathbf{H}_0 =$ constant, one obtains $E_z = -\mu_0 \mathbf{H}_0 \cdot \mathbf{\Pi}_{\text{tot}}$, where $\mathbf{\Pi}_{\text{tot}}$ is the total magnetic moment of the specimen. It is emphasized that E_z is linear in $\mathbf{M}$, and a local energy contribution (as are E_{ex}, E_{dmi}, and E_{mc}), i.e., the value of E_z at some location point of a magnetic moment inside the sample, depends only on the magnetic moment at that particular point [compare also the discrete version for E_z, eqn (7.1)]. This is in contrast to the magnetodipolar or magnetostatic interaction (see next Section 3.1.6), which is of second order in $\mathbf{M}$ and non-local, i.e., the magnetostatic energy at some point depends on the distribution of the magnetic moments of the whole specimen.

3.1.6   Magnetodipolar interaction

The treatment of the magnetostatic self-interaction energy E_{m} starts with Maxwell's equations [154]. One of them states that the magnetic induction or magnetic flux density $\mathbf{B}$ is divergence-free, i.e.,

$$\nabla \cdot \mathbf{B} = 0, \tag{3.18}$$

where

$$\mathbf{B} = \mu_0 \left(\mathbf{H}_0 + \mathbf{H}_d + \mathbf{M}\right), \tag{3.19}$$

and $\mathbf{H}_d(\mathbf{r})$ denotes the magnetodipolar interaction field that is created by the magnetization $\mathbf{M}$. We emphasize that in the literature $\mathbf{H}_d$ is also sometimes denoted as the magnetostatic field or the magnetostatic stray field. From eqns (3.18) and (3.19) it follows (assuming a uniform applied magnetic field $\mathbf{H}_0$) that

$$\nabla \cdot \mathbf{H}_d = -\nabla \cdot \mathbf{M}, \tag{3.20}$$

and again from Maxwell's equations for the static case and no macroscopic currents

$$\nabla \times \mathbf{H}_d = 0. \tag{3.21}$$

In analogy to the corresponding equations for the electrostatic case, the term $\rho_V = -\nabla \cdot \mathbf{M}$ in eqn (3.20) denotes the magnetic "volume charges". Equation (3.21) implies that

$$\mathbf{H}_d = -\nabla U(\mathbf{r}), \qquad (3.22)$$

which relates $\mathbf{H}_d$ to the magnetostatic potential function $U(\mathbf{r})$. Inserting eqn (3.22) into eqn (3.20) yields:

$$\Delta U_{\text{in}} = \nabla \cdot \mathbf{M} \qquad (3.23)$$

inside the ferromagnetic body ($\mathbf{H}_d = -\nabla U_{\text{in}}$) and

$$\Delta U_{\text{out}} = 0 \qquad (3.24)$$

outside the ferromagnetic body ($\mathbf{H}_d = -\nabla U_{\text{out}}$), where $\mathbf{M} = 0$ and $\mathbf{B} = \mu_0(\mathbf{H}_0 + \mathbf{H}_d)$ ($\Delta = \nabla^2 = \partial^2/\partial x^2 + \partial^2/\partial y^2 + \partial^2/\partial z^2$ is the Laplace operator) [154]. By taking into account the conditions for the fields $\mathbf{B}$ and $\mathbf{H}_d$ on the boundary (S) between two media, namely, that the perpendicular component of $\mathbf{B}$ and the parallel component of $\mathbf{H}_d$ are continuous, one arrives at the following boundary conditions:

$$U_{\text{in}} - U_{\text{out}} = 0 \quad \text{and} \quad \frac{\partial U_{\text{in}}}{\partial \mathbf{n}} - \frac{\partial U_{\text{out}}}{\partial \mathbf{n}} = \mathbf{n} \cdot \mathbf{M} \quad \text{on } S, \qquad (3.25)$$

where $\mathbf{n}$ denotes the unit normal vector on the surface (taken to be positive in the outward direction), and $\partial U/\partial \mathbf{n} = \nabla U \cdot \mathbf{n}$ is the directional derivative of U along $\mathbf{n}$. Moreover, the potential function has to be regular at infinity, i.e., $|rU_{\text{out}}|$ and $|r^2 \nabla U_{\text{out}}|$ remain finite for $r \to \infty$. The differential eqns (3.23) and (3.24) together with the boundary conditions (3.25) then have a unique solution, which can be formally expressed as [154]:

$$U(\mathbf{r}) = \frac{1}{4\pi} \left(-\int_V \frac{\nabla_{\mathbf{r}'} \cdot \mathbf{M}(\mathbf{r}')}{|\mathbf{r} - \mathbf{r}'|} d^3 r' + \int_S \frac{\mathbf{n} \cdot \mathbf{M}(\mathbf{r}')}{|\mathbf{r} - \mathbf{r}'|} d^2 r' \right), \qquad (3.26)$$

so that

$$\mathbf{H}_d(\mathbf{r}) = \frac{1}{4\pi} \left(-\int_V \frac{\nabla_{\mathbf{r}'} \cdot \mathbf{M}(\mathbf{r}') \, (\mathbf{r} - \mathbf{r}')}{|\mathbf{r} - \mathbf{r}'|^3} d^3 r' + \int_S \frac{\mathbf{n} \cdot \mathbf{M}(\mathbf{r}') \, (\mathbf{r} - \mathbf{r}')}{|\mathbf{r} - \mathbf{r}'|^3} d^2 r' \right). \qquad (3.27)$$

The subscript $\mathbf{r}'$ to ∇ means that the nabla operator contains the derivatives with respect to the components of the source point $\mathbf{r}'$. The first integral in eqn (3.26) is over the volume (V) and the second integral over the surface (S) of the ferromagnetic body. The sources of $\mathbf{H}_d$ are due to inhomogeneities in $\mathbf{M}$, either in orientation and/or in magnitude.

Within the Lorentz continuum approximation (see, e.g., p.33 in [344]), the magnetostatic field $\mathbf{H}_d(\mathbf{r})$ is then related to a magnetostatic self-energy [154]

$$E_m = -\frac{1}{2}\mu_0 \int \mathbf{M} \cdot \mathbf{H}_d \, d^3 r = +\frac{1}{2}\mu_0 \int_{\text{all space}} |\mathbf{H}_d|^2 \, d^3 r. \qquad (3.28)$$

Note that the second integral in eqn (3.28) extends over all space, whereas the first one over the sample volume. The magnetostatic energy is a self-energy and without the factor $\frac{1}{2}$ the interaction of each pair of magnetic moments is counted twice. The last expression on the right-hand side of eqn (3.28) embodies the pole-avoidance principle: the magnetostatic self-energy $E_m \geq 0$ tries to avoid any sort of fictitious magnetic volume ($\rho_V = -\nabla \cdot \mathbf{M}$) or surface ($\rho_S = \mathbf{n} \cdot \mathbf{M}$) charges. In contrast to the previously discussed local energy terms [eqns (3.6), (3.7), (3.10), and (3.17)], which involve only a single integration over the sample volume, eqn (3.28) requires integrating twice over the same volume.

Very useful relations can be derived for the special case of a uniformly magnetized ($\nabla \cdot \mathbf{M} = 0$) general ellipsoid. In this situation, the magnetostatic field in the interior of the ellipsoid, $\mathbf{H}_d^{in}$, can be expressed as [154]:

$$\mathbf{H}_d^{in} = -\overline{\overline{N}}_d \mathbf{M},\tag{3.29}$$

where $\overline{\overline{N}}_d$ denotes the demagnetization tensor. If, moreover, the particle is oriented in such a way that its principal axes are aligned along the Cartesian laboratory coordinates, $\overline{\overline{N}}_d$ takes on the simple form:

$$\overline{\overline{N}}_d = \begin{pmatrix} N_{dx} & 0 & 0 \\ 0 & N_{dy} & 0 \\ 0 & 0 & N_{dz} \end{pmatrix}\tag{3.30}$$

with $\mathrm{Tr}(\overline{\overline{N}}_d) = N_{dx} + N_{dy} + N_{dz} = 1$ in the SI system, and $\mathrm{Tr}(\overline{\overline{N}}_d) = 4\pi$ in the cgs system. Note that in the next Section 3.2 the field $\mathbf{H}_d^{in}$, which is exclusively due to the surface charges $\mathbf{n} \cdot \mathbf{M}$, will be denoted as $\mathbf{H}_d^s$. The demagnetizing factors $0 < N_{dx,y,z} < 1$ depend on the ratio of the sample dimensions and are tabulated for the general ellipsoid in [433]. For non-ellipsoidal particle shapes $\mathbf{H}_d^{in}$ is nonuniform, and certain averages of $\mathbf{H}_d^{in}(\mathbf{r})$ have been introduced in the literature, leading to the concepts of the ballistic and magnetometric demagnetizing factors. One should, however, consider these concepts with some caution, since the replacement of a highly inhomogeneous internal field by a constant field is not always a good approximation. Formulas and tables for approximately correcting experimental data on cylinders and rectangular prisms can be found e.g., in [434–436]. The field $\mathbf{H}_d^{in}$ is produced by the magnetization, as is $\mathbf{H}_d^{out}$. In the presence of an externally applied magnetic field $\mathbf{H}_0$, produced e.g., by a permanent magnet or by current-carrying Cu wires, $\mathbf{H}_0$ and $\mathbf{H}_d^{in}$ superpose vectorially.

For the example of a sphere with radius R, homogeneously magnetized along the z-direction ($\mathbf{M} = M_s \mathbf{e}_z$), the magnetostatic potential is given by [154]:

$$U(r, \vartheta) = \frac{M_s}{3} \cos \vartheta \begin{cases} r & \text{if } r \leq R \\ R^3/r^2 & \text{if } r \geq R \end{cases},\tag{3.31}$$

where $r = |\mathbf{r}|$, and ϑ denotes the polar angle [$\vartheta = \angle(\mathbf{r}, \mathbf{e}_z)$]. Straightforward algebra shows that this potential fulfills eqns (3.23) and (3.24) (with $\nabla \cdot \mathbf{M} = 0$) as well as the boundary conditions eqns (3.25). The magnetostatic field which is derived from this potential equals ($z = r \cos \vartheta$):

$$\mathbf{H}_d^{in} = -\nabla U_{in} = -\frac{M_s}{3} \mathbf{e}_z \quad \text{for } r \leq R,\tag{3.32}$$

and [compare eqn (1.58)]

$$
\mathbf{H}_{\mathrm{d}}^{\mathrm{out}} = -\nabla U_{\mathrm{out}} = M_{\mathrm{s}} \left(\frac{R}{r}\right)^3 \left\{ \begin{array}{c} \sin\vartheta\cos\vartheta\cos\zeta \\ \sin\vartheta\cos\vartheta\sin\zeta \\ \cos^2\vartheta - \frac{1}{3} \end{array} \right\} \quad \text{for } r \geq R. \tag{3.33}
$$

We emphasize that $\mathbf{H}_{\mathrm{d}}^{\mathrm{in}}$ and $\mathbf{H}_{\mathrm{d}}^{\mathrm{out}}$ are both expressed in Cartesian coordinates. It can be seen that $\mathbf{H}_{\mathrm{d}}^{\mathrm{in}}$ is homogeneous, everywhere the same inside the sphere, whereas the stray field $\mathbf{H}_{\mathrm{d}}^{\mathrm{out}}$ is long-range and highly inhomogeneous.

Inserting eqn (3.29) into eqn (3.28) yields for the magnetostatic self-energy of a uniformly magnetized ellipsoid with volume V [437]:§

$$
E_{\mathrm{m}} = \frac{1}{2}\mu_0 V \left(N_{\mathrm{d}x}M_x^2 + N_{\mathrm{d}y}M_y^2 + N_{\mathrm{d}z}M_z^2\right). \tag{3.34}
$$

This term is known as the magnetic shape anisotropy energy. For an ellipsoid with two equal axes, it is readily verified that the corresponding demagnetizing factors are the same; for instance, a prolate spheroid (egg-shaped particle elongated along $\mathbf{e}_z$) is characterized by $N_{\mathrm{d}x} = N_{\mathrm{d}y}$ [154], so that

$$
E_{\mathrm{m}} = \frac{1}{2}\mu_0 V \left(N_{\mathrm{d}z} - N_{\mathrm{d}x}\right) M_z^2. \tag{3.35}
$$

In writing down eqn (3.35) we have used the constraint that $M_x^2 + M_y^2 + M_z^2 = M_s^2$ [eqn (3.2)] and neglected the ensuing constant term $\frac{1}{2}\mu_0 V N_{\mathrm{d}x}M_s^2$. Mathematically, eqn (3.35) has the same form as the first-order uniaxial magnetocrystalline anisotropy energy [eqn (3.13)], but, of course, it is of different physical origin. A uniformly magnetized sphere [$N_{\mathrm{d}x} = N_{\mathrm{d}y} = N_{\mathrm{d}z} = \frac{1}{3}$; compare eqn (3.32)] has a constant magnetostatic self energy of $E_{\mathrm{m}} = \frac{1}{6}\mu_0 V_{\mathrm{s}}M_s^2$ and no shape anisotropy. The idealized infinite circular cylinder, assumed to be uniformly magnetized along the long z-direction, has $N_{\mathrm{d}z} = 0$, so that $\mathbf{H}_{\mathrm{d}}^{\mathrm{in}} = 0$ and $E_{\mathrm{m}} = 0$ (no surface). If the same cylinder is uniformly magnetized along the short axis, say, the x-direction, then it follows from $\mathrm{Tr}(\overline{\overline{N}}_{\mathrm{d}}) = 1$ that $N_{\mathrm{d}x} = \frac{1}{2}$, $\mathbf{H}_{\mathrm{d}}^{\mathrm{in}} = -\frac{1}{2}M_{\mathrm{s}}\mathbf{e}_x$, and the magnetostatic energy per unit length along z amounts to $E_{\mathrm{m}} = \frac{1}{4}\mu_0\pi R^2 M_s^2$. For the infinite plate, uniformly magnetized along the surface normal $\parallel \mathbf{e}_z$, one finds $N_{\mathrm{d}z} = 1$, $\mathbf{H}_{\mathrm{d}}^{\mathrm{in}} = -M_{\mathrm{s}}\mathbf{e}_z$, and the magnetostatic energy per unit area is $E_{\mathrm{m}} = \frac{1}{2}\mu_0 t M_s^2$, where t is the thickness of the plate. We refer to the book by Aharoni [154] for further details regarding the basics of magnetostatics.

---

§As discussed in the book by Brown [438], the magnetostatic energy of any arbitrarily shaped uniformly magnetized ferromagnetic body is a quadratic form in the components of $\mathbf{M}$. The diagonalized expression (3.34) represents, in fact, the most general form for E_{m} of a uniformly magnetized particle. It applies to any shape, and not only to ellipsoids; in other words, "an arbitrary single-domain particle in a uniform applied field behaves like a suitably chosen ellipsoid of the same volume" [sic]. This statement, which refers to the magnetostatic energy, is known as the Brown–Morrish theorem [437]. One should, however, remember that the demagnetizing field inside a uniformly magnetized ellipsoid is uniform, which is not the case for a non-ellipsoidal body.

3.2 Balance of torques

In order to find the equilibrium magnetization distribution, the total magnetic energy of a ferromagnet,

$$E_{\text{tot}} = E_{\text{ex}} + E_{\text{dmi}} + E_{\text{mc}} + E_{\text{me}} + E_{\text{z}} + E_{\text{m}}, \tag{3.36}$$

should be considered as a functional of its magnetization state, i.e.,

$$E_{\text{tot}} = \int_V \omega_{\text{tot}} \left[\mathbf{m}(\mathbf{r}), \nabla \mathbf{m}(\mathbf{r}) \right] d^3 r. \tag{3.37}$$

The ferromagnet's total energy density ω_{tot} depends on the magnetization components $m_{x,y,z}$—via the magnetocrystalline anisotropy energy density in a highly nonlinear manner [compare eqns (3.12) and (3.13)]—and on the spatial derivatives of the components of $\mathbf{m}$ [symbolized by the shorthand notation $\nabla \mathbf{m}(\mathbf{r})$ in eqn (3.37)]. We also remind that ω_{tot} depends, by virtue of the non-locality of the magnetostatic energy E_{m} [compare eqns (3.27) and (3.28)], on the magnetization distribution of the whole sample. The state which delivers a (local) minimum to this functional corresponds to an equilibrium magnetization configuration. Therefore, a necessary condition is that at static equilibrium the variation of E_{tot} with respect to the $m_{x,y,z}$ must vanish:

$$\delta E_{\text{tot}} = \delta(E_{\text{ex}} + E_{\text{dmi}} + E_{\text{mc}} + E_{\text{me}} + E_{\text{z}} + E_{\text{m}}) = 0. \tag{3.38}$$

For stability of the equilibrium, the second variation of E_{tot} must be positive [344]. As detailed in the pertinent textbooks [119, 134, 154, 344], variational calculus leads to a set of nonlinear partial differential equations for the bulk along with quite complex boundary conditions for the magnetization at the surface. We refer to Section 5.1.3 for a complete formulation of the micromagnetic equilibrium problem. This is appropriate e.g., for the treatment of the magnetic microstructure and the related magnetic SANS of a finite-sized magnetic nanoparticle, where the boundary conditions at the external surface have to be taken into account. Here, in keeping with our interest in the magnetic microstructure within the bulk of macroscopic magnetic bodies, we restrict attention to the bulk equilibrium conditions. The justification for such an approximation may be taken from an argument put forward by Brown [7, 344]: when the specimen dimensions are large in comparison with the distance over which the effect of a local disturbance becomes negligible, as characterized by the field-dependent exchange length $l_{\text{H}} \sim 1-100\,\text{nm}$ [eqn (3.66)], the sample may be regarded as infinite and boundary conditions can be ignored.

The static equations of micromagnetics (so-called Brown's equations) can be conveniently expressed as a balance-of-torques equation [119, 134, 154, 344],

$$\mathbf{M}(\mathbf{r}) \times \mathbf{H}_{\text{eff}}(\mathbf{r}) = 0. \tag{3.39}$$

Equation (3.39) expresses the fact that at static equilibrium the torque on the magnetization $\mathbf{M}(\mathbf{r})$ due to an effective magnetic field $\mathbf{H}_{\text{eff}}(\mathbf{r})$ vanishes everywhere inside the material. The effective field is obtained as the functional derivative of ω_{tot} with respect to the magnetization, i.e.,

$$\mathbf{H}_{\text{eff}}(\mathbf{r}) = -\frac{1}{\mu_0}\frac{\delta\omega_{\text{tot}}}{\delta\mathbf{M}}$$
$$= \mathbf{H}_0 + \mathbf{H}_{\text{d}}(\mathbf{r}) + \mathbf{H}_{\text{p}}(\mathbf{r}) + \mathbf{H}_{\text{ex}}(\mathbf{r}) + \mathbf{H}_{\text{dmi}}(\mathbf{r}). \tag{3.40}$$

$\mathbf{H}_{\text{eff}}$ is composed of a uniform applied magnetic field $\mathbf{H}_0$, the magnetostatic field $\mathbf{H}_{\text{d}}(\mathbf{r})$, the perturbing magnetic anisotropy field $\mathbf{H}_{\text{p}}(\mathbf{r})$ (sometimes also denoted with the symbol $\mathbf{H}_{\text{K}}$), the exchange field

$$\mathbf{H}_{\text{ex}}(\mathbf{r}) = l_{\text{M}}^2 \nabla^2 \mathbf{M} = l_{\text{M}}^2 \left\{ \nabla^2 M_x, \nabla^2 M_y, \nabla^2 M_z \right\}, \tag{3.41}$$

and of the field $\mathbf{H}_{\text{dmi}}$ which is due to the DMI. In [414], the effect of the DMI on the spin-polarized SANS cross section has been investigated assuming the specific expression [based on eqn (3.8)]

$$\mathbf{H}_{\text{dmi}}(\mathbf{r}) = -l_{\text{D}} \nabla \times \mathbf{M}, \tag{3.42}$$

which is valid for materials with a non-centrosymmetric cubic crystal structure. This particular form for $\mathbf{H}_{\text{dmi}}$ will be used in the following calculation. The micromagnetic length scales

$$l_{\text{M}} = \sqrt{\frac{2A}{\mu_0 M_{\text{s}}^2}} \tag{3.43}$$

and

$$l_{\text{D}} = \frac{2D}{\mu_0 M_{\text{s}}^2} \tag{3.44}$$

are, respectively, related to the magnetostatic interaction and to the DMI. For typical values of the materials parameters A, D, and M_{s}, these length scales are of order of a few nm [see Fig. 3.5(a)] [134].

Approach to saturation: small-misalignment approximation

In the following analysis, we consider a special case, namely, we assume the material to be nearly saturated by a strong applied magnetic field $\mathbf{H}_0 \parallel \mathbf{e}_z$, and we express the magnetization as [414]:

$$\mathbf{M}(\mathbf{r}) = \left\{ M_x(\mathbf{r}), M_y(\mathbf{r}), M_z(\mathbf{r}) = M_{\text{s}}(\mathbf{r}) \right\}, \tag{3.45}$$

where

$$M_x \ll M_{\text{s}} , \ M_y \ll M_{\text{s}} , \ M_z \cong M_{\text{s}}. \tag{3.46}$$

This is known as the small-misalignment approximation. The local saturation magnetization is assumed to be a function of the position $\mathbf{r} = \{x, y, z\}$ inside the material [335, 345, 439, 440]:

$$M_{\text{s}}(\mathbf{r}) = M_{\text{s}}[1 + I_{\text{m}}(\mathbf{r})], \tag{3.47}$$

where I_{m} is an inhomogeneity function, small in magnitude, which describes the local variation of $M_{\text{s}}(\mathbf{r})$ (see Fig. 3.3). The spatial average of I_{m} vanishes, $\langle I_{\text{m}}(\mathbf{r}) \rangle = 0$, so

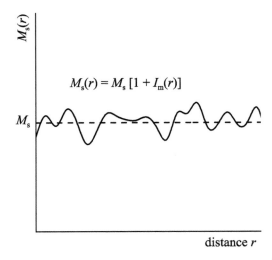

Fig. 3.3: Sketch illustrating the spatial variation of the local saturation magnetization $M_\mathrm{s}(\mathbf{r})$ around the mean value $M_\mathrm{s} = \langle M_\mathrm{s}(\mathbf{r}) \rangle = \frac{1}{V} \int_V M_\mathrm{s}(\mathbf{r}) d^3r$ (dashed horizontal line). After [345].

that $\langle M_\mathrm{s}(\mathbf{r}) \rangle = M_\mathrm{s}$ is the saturation magnetization of the sample, which can be measured with a magnetometer.[¶] By defining the Fourier transform $\widetilde{F}(\mathbf{q})$ of a continuous function $F(\mathbf{r})$ as [compare eqns (1.36) and (1.37)]

$$\widetilde{F}(\mathbf{q}) = \frac{1}{(2\pi)^{3/2}} \int F(\mathbf{r}) \exp\left(-i\mathbf{q}\cdot\mathbf{r}\right) d^3r \tag{3.48}$$

and

$$F(\mathbf{r}) = \frac{1}{(2\pi)^{3/2}} \int \widetilde{F}(\mathbf{q}) \exp\left(i\mathbf{q}\cdot\mathbf{r}\right) d^3q, \tag{3.49}$$

where $i^2 = -1$, and $\mathbf{q} = \{q_x, q_y, q_z\}$ is the wave vector, one can write for the magnetization Fourier component $\widetilde{\mathbf{M}}(\mathbf{q})$ up to the first order in $\widetilde{I}_\mathrm{m}$:

$$\widetilde{\mathbf{M}}(\mathbf{q}) = \left\{ \widetilde{M}_x(\mathbf{q}), \widetilde{M}_y(\mathbf{q}), \widetilde{M}_z(\mathbf{q}) = M_\mathrm{s}[\delta(\mathbf{q}) + \widetilde{I}_\mathrm{m}(\mathbf{q})] \right\}, \tag{3.50}$$

where $\delta(\mathbf{q})$ is Dirac's delta function.

According to eqn (3.27), the magnetostatic field $\mathbf{H}_\mathrm{d}(\mathbf{r})$ can be written as the sum of the surface demagnetizing field $\mathbf{H}_\mathrm{d}^\mathrm{s}(\mathbf{r})$ (denoted as $\mathbf{H}_\mathrm{d}^\mathrm{in}$ in Section 3.1.6) and of the magnetostatic field $\mathbf{H}_\mathrm{d}^\mathrm{b}(\mathbf{r})$ which is related to volume (bulk) charges, i.e.,

$$\mathbf{H}_\mathrm{d}(\mathbf{r}) = \mathbf{H}_\mathrm{d}^\mathrm{s}(\mathbf{r}) + \mathbf{H}_\mathrm{d}^\mathrm{b}(\mathbf{r}). \tag{3.51}$$

In the high-field limit, when the magnetization is close to saturation, and for samples with an ellipsoidal shape with $\mathbf{H}_0$ directed along a principal axis of the ellipsoid, one

---

[¶]Note that the symbol M_s is used here for denoting both the local saturation-magnetization profile $M_\mathrm{s}(\mathbf{r})$ as well as its (constant) mean value M_s, the saturation magnetization of the sample.

may approximate the demagnetizing field due to the surface charges by the uniform field [154]

$$\mathbf{H}_d^s = -N_d M_s \mathbf{e}_z, \tag{3.52}$$

where $0 < N_d < 1$ denotes the corresponding demagnetizing factor [compare eqn (3.29)]. The field $H_d^s = -N_d M_s$ can then be combined with the applied magnetic field H_0 to yield the internal magnetic field

$$H_i = H_0 - N_d M_s, \tag{3.53}$$

i.e., the $\mathbf{q} = 0$ Fourier component of $\mathbf{H}_d$. In Fourier space, at $\mathbf{q} \neq 0$, the magnetostatic relations eqns (3.18)$-$(3.21) suggest the following expression for the Fourier component $\widetilde{\mathbf{H}}_d^b(\mathbf{q})$ of $\mathbf{H}_d^b(\mathbf{r})$ [232]:

$$\widetilde{\mathbf{H}}_d^b(\mathbf{q}) = \left\{ \widetilde{H}_{dx}^b(\mathbf{q}), \widetilde{H}_{dy}^b(\mathbf{q}), \widetilde{H}_{dz}^b(\mathbf{q}) \right\} = -\frac{\mathbf{q}\left[\mathbf{q} \cdot \widetilde{\mathbf{M}}(\mathbf{q})\right]}{q^2}, \tag{3.54}$$

so that the total $\widetilde{\mathbf{H}}_d$ is

$$\widetilde{\mathbf{H}}_d(\mathbf{q}) = H_i\, \delta(\mathbf{q})\, \mathbf{e}_z - \frac{\mathbf{q}\left[\mathbf{q} \cdot \widetilde{\mathbf{M}}(\mathbf{q})\right]}{q^2}. \tag{3.55}$$

In the approach-to-saturation regime, we assume that the anisotropy-energy density $\omega = \omega(\mathbf{r}, \mathbf{M})$ depends only linearly on the components of the magnetization, i.e., $\omega \cong g_0^\star + g_1^\star(\mathbf{r})M_x + g_2^\star(\mathbf{r})M_y$, where the $g_i^\star$ are functions of the position [7, 344]. As a consequence, the resulting anisotropy field

$$\mathbf{H}_p = -\mu_0^{-1}\left\{\partial\omega/\partial M_x, \partial\omega/\partial M_y, \partial\omega/\partial M_z\right\} \tag{3.56}$$

is independent of $\mathbf{M}$ and, therefore, also independent of the applied magnetic field, implying that near saturation $\mathbf{H}_p = \mathbf{H}_p(\mathbf{r})$. As mentioned in Section 3.1, due to the micromagnetic constraint of $|\mathbf{M}| = M_s$, an anisotropy-energy density of the form $\omega_1 = \omega_1(\mathbf{r}, M_x, M_y, M_z)$ may be re-expressed as $\omega_2 = \omega_2(\mathbf{r}, M_x, M_y, (M_s^2 - M_x^2 - M_y^2)^{1/2})$. The anisotropy fields which are computed from ω_1 and ω_2 differ by a vector in the direction of $\mathbf{M}$, which is physically ineffective in producing a torque on the magnetization [344]. In the small-misalignment approximation, when $\mathbf{M}$ is nearly aligned parallel to the external magnetic field $\mathbf{H}_0 \parallel \mathbf{e}_z$, only those components of $\mathbf{H}_p$ which are normal to $\mathbf{H}_0$ give rise to a torque on $\mathbf{M}$ [439]. Therefore, the Fourier transform of $\mathbf{H}_p(\mathbf{r})$ may be approximated by

$$\widetilde{\mathbf{H}}_p(\mathbf{q}) = \left\{ \widetilde{H}_{px}(\mathbf{q}), \widetilde{H}_{py}(\mathbf{q}), 0 \right\}. \tag{3.57}$$

This term is a source of spin disorder, since it increases the magnitude of the transversal Fourier components [compare eqns (3.68) and (3.69)]. The field $\mathbf{H}_p(\mathbf{r})$ contains the information about the sample's microstructure (e.g., crystallite size, inhomogeneous lattice strain, crystallographic texture) [441]. Note that no assumption has been made here about the particular form of the magnetic anisotropy (magnetocrystalline and/or magnetoelastic).

The Fourier representations of the exchange field $\mathbf{H}_{\text{ex}}$ [eqn (3.41)] and of the DMI field $\mathbf{H}_{\text{dmi}}$ [eqn (3.42)] read, respectively:

$$\widetilde{\mathbf{H}}_{\text{ex}}(\mathbf{q}) = -l_{\text{M}}^2 q^2 \left\{ \widetilde{M}_x, \widetilde{M}_y, M_{\text{s}}\widetilde{I}_{\text{m}} \right\} \tag{3.58}$$

and

$$\widetilde{\mathbf{H}}_{\text{dmi}}(\mathbf{q}) = -il_{\text{D}} \left\{ q_y M_{\text{s}}\widetilde{I}_{\text{m}} - q_z \widetilde{M}_y, q_z \widetilde{M}_x - q_x M_{\text{s}}\widetilde{I}_{\text{m}}, q_x \widetilde{M}_y - q_y \widetilde{M}_x \right\}. \tag{3.59}$$

In case of a weak spatial dependency of the parameters A and D, with fluctuations of the order of the saturation-magnetization fluctuation $I_{\text{m}}(\mathbf{r})$, the quantities l_{M} and l_{D} must be understood as spatial averages of the corresponding (now position-dependent) eqns (3.43) and (3.44) [439].

By using eqns (3.55)–(3.59) in the balance-of-torques equation (3.39) and by neglecting terms of higher than linear order in $\widetilde{M}_x$, $\widetilde{M}_y$, and $\widetilde{I}_{\text{m}}$ (including terms such as $\widetilde{M}_x \widetilde{I}_{\text{m}}$ and $\widetilde{H}_{\text{p}x}\widetilde{I}_{\text{m}}$), we obtain, in Fourier space, and for a general orientation of the wave vector $\mathbf{q} = \{q_x, q_y, q_z\}$, the following set of linear equations for $\widetilde{M}_x$ and $\widetilde{M}_y$:

$$\widetilde{M}_x \left(1 + p\frac{q_x^2}{q^2} \right) + \widetilde{M}_y \left(p\frac{q_x q_y}{q^2} - ip\,l_{\text{D}}q_z \right) = O_1, \tag{3.60}$$

$$\widetilde{M}_y \left(1 + p\frac{q_y^2}{q^2} \right) + \widetilde{M}_x \left(p\frac{q_x q_y}{q^2} + ip\,l_{\text{D}}q_z \right) = O_2, \tag{3.61}$$

where

$$O_1 = p \left(\widetilde{H}_{\text{p}x} - M_{\text{s}}\widetilde{I}_{\text{m}} \left[\frac{q_x q_z}{q^2} + il_{\text{D}}q_y \right] \right) \tag{3.62}$$

and

$$O_2 = p \left(\widetilde{H}_{\text{p}y} - M_{\text{s}}\widetilde{I}_{\text{m}} \left[\frac{q_y q_z}{q^2} - il_{\text{D}}q_x \right] \right). \tag{3.63}$$

The dimensionless function (see Fig. 3.4)

$$p(q, H_{\text{i}}) = \frac{M_{\text{s}}}{H_{\text{eff}}(q, H_{\text{i}})} \tag{3.64}$$

depends via the effective magnetic field [not to be confused with $\mathbf{H}_{\text{eff}}(\mathbf{r})$ in the balance-of-torques eqn (3.39)]

$$H_{\text{eff}}(q, H_{\text{i}}) = H_{\text{i}} \left(1 + l_{\text{H}}^2 q^2 \right) \tag{3.65}$$

on the micromagnetic exchange length of the field

$$l_{\text{H}}(H_{\text{i}}) = \sqrt{\frac{2A}{\mu_0 M_{\text{s}} H_{\text{i}}}}. \tag{3.66}$$

The parameter l_{H} takes on values between a few and a few hundred nanometers [see Fig. 3.5(a)], and it characterizes the range over which perturbations in the spin structure decay [442–444]. Likewise, when magnetostrictive interactions are explicitly taken into account, then an additional exchange length of the stress appears [134]:

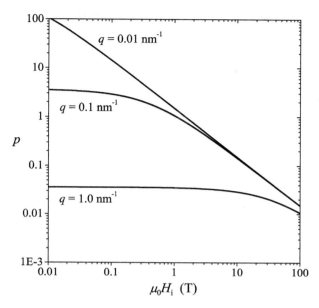

Fig. 3.4: Field dependence of the function $p(q, H_i) = M_s/[H_i + 2Aq^2/(\mu_0 M_s)]$ [eqn (3.64)] at selected values of q (see insets) (log-log scale). Materials parameters: $A = 2.5 \times 10^{-11}$ J/m; $\mu_0 M_s = 1.5$ T. After [59].

$$l_\sigma = \sqrt{\frac{2A}{3\lambda\sigma}}, \tag{3.67}$$

where λ is a magnetostriction constant, and σ is some appropriate average of the internal stress. For typical values of λ and σ, l_σ takes on values of the order of a few tens of nanometers [134]. Such length scales fall within the resolution range of the SANS technique [compare Fig. 3.5(a)].

The effective magnetic field H_{eff} [eqn (3.65)] consists of a contribution due to the internal field H_i and of the exchange field $2Aq^2/(\mu_0 M_s)$. An increase of H_i increases the effective field only at the smallest q-values, whereas H_{eff} at the larger q is always very large ($\sim 10-100$ T) and independent of H_i [see Fig. 3.5(b)]. The latter statement may be seen as a manifestation of the fact that exchange forces tend to dominate on small length scales [154]. Since H_{eff} appears predominantly in the denominators of the final expressions for $\widetilde{M}_x$ and $\widetilde{M}_y$ [eqns (3.68) and (3.69)], its role is to suppress the high-q Fourier components of the magnetization, which correspond to sharp real-space fluctuations. On the other hand, long-range magnetization fluctuations, at small q, are effectively suppressed when H_i is increased.

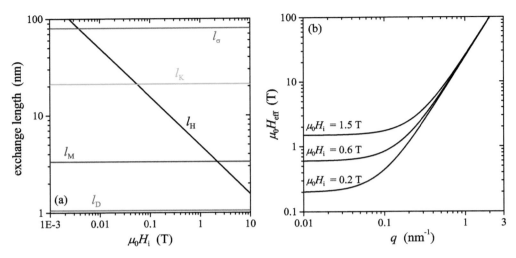

Fig. 3.5: (a) Micromagnetic length scales l_H, l_σ, l_M, l_D, and the wall-width parameter $l_K = \sqrt{A/K_1}$ (log-log scale). These length scales fall into the resolution ranges of the SANS/USANS/VSANS techniques. (b) The effective magnetic field $H_{\text{eff}}(q, H_i)$ (log-log scale). Materials parameters for iron at room temperature were used [134]: $A = 21\,\text{pJ/m}$, $\mu_0 M_s = 2.185\,\text{T}$, $K_1 = 4.8 \times 10^4\,\text{J/m}^3$, and $\lambda_{100} = 22 \times 10^{-6}$; and $D = 2\,\text{mJ/m}^2$ and $\sigma = 10^8\,\text{Pa}$.

The solutions of eqns (3.60) and (3.61) are [414]:

$$\widetilde{M}_x = \frac{p\left(\widetilde{H}_{px}\left[1 + p\frac{q_y^2}{q^2}\right] - \widetilde{M}_z \frac{q_x q_z}{q^2}\left[1 + p\,l_D^2 q^2\right] - \widetilde{H}_{py} p\frac{q_x q_y}{q^2}\right)}{1 + p\frac{q_x^2 + q_y^2}{q^2} - p^2 l_D^2 q_z^2}$$

$$- i\frac{p\left(\widetilde{M}_z(1 + p)l_D q_y - \widetilde{H}_{py} p\,l_D q_z\right)}{1 + p\frac{q_x^2 + q_y^2}{q^2} - p^2 l_D^2 q_z^2}, \tag{3.68}$$

$$\widetilde{M}_y = \frac{p\left(\widetilde{H}_{py}\left[1 + p\frac{q_x^2}{q^2}\right] - \widetilde{M}_z \frac{q_y q_z}{q^2}\left[1 + p\,l_D^2 q^2\right] - \widetilde{H}_{px} p\frac{q_x q_y}{q^2}\right)}{1 + p\frac{q_x^2 + q_y^2}{q^2} - p^2 l_D^2 q_z^2}$$

$$+ i\frac{p\left(\widetilde{M}_z(1 + p)l_D q_x - \widetilde{H}_{px} p\,l_D q_z\right)}{1 + p\frac{q_x^2 + q_y^2}{q^2} - p^2 l_D^2 q_z^2}, \tag{3.69}$$

where $\widetilde{M}_z = M_s \widetilde{I}_m$ at $q \neq 0$ [compare eqn (3.50)]. Note the symmetry of the equations under the exchange of the x- and y-coordinates. By projecting the results for the transversal Fourier components [eqns (3.68) and (3.69)] into the plane of the two-dimensional detector ($q_x = 0$ for $\mathbf{k}_0 \perp \mathbf{H}_0$ and $q_z = 0$ for $\mathbf{k}_0 \parallel \mathbf{H}_0$), it becomes possible to compute the magnetic SANS cross section (see Section 4.2). Furthermore, the above expressions for $\widetilde{M}_x$ and $\widetilde{M}_y$ can also be used for obtaining the autocorrelation function of the spin-misalignment [312] and the approach-to-saturation law [134]; for instance, one can investigate the impact of different types of microstructural defects and of the DMI on the high-field magnetization. Since the magnetic SANS cross section also depends on the longitudinal Fourier component $\widetilde{M}_z$, one needs to develop a microstructural model for this quantity, e.g., in terms of a particle form factor and structure factor [compare eqn (2.119)]. In the following Section 3.3, we will use for simplicity the sphere form factor for $\widetilde{M}_z$ in order to graphically display the Fourier components.

3.3 Connecting micromagnetics and SANS

3.3.1 The general case of a ferromagnet with anisotropy-field and magnetostatic fluctuations

Equations (3.68) and (3.69) describe, in Fourier space, the magnetization response to a spatially varying magnetic anisotropy field $\mathbf{H}_p(\mathbf{r})$ and saturation magnetization $M_s(\mathbf{r})$. In order to compute the magnetic SANS cross section, the Fourier components are evaluated in the plane of the two-dimensional detector and an average over the orientations of the magnetic anisotropy field is performed. Regarding SANS experiments, two scattering geometries are commonly of relevance: the perpendicular scattering geometry, which has the wave vector $\mathbf{k}_0$ of the incoming neutron beam perpendicular to the applied magnetic field $\mathbf{H}_0$, and the parallel geometry, where $\mathbf{k}_0 \parallel \mathbf{H}_0$ (see Fig. 2.1). Note that throughout this book we choose $\mathbf{H}_0 \parallel \mathbf{e}_z$ for both geometries. For $\mathbf{k}_0 \perp \mathbf{H}_0$ ($q_x = 0$), eqns (3.68) and (3.69) reduce to:

$$\widetilde{M}_x = \frac{p\left(\widetilde{H}_{px}\left[1 + p\frac{q_y^2}{q^2}\right] - i\left[\widetilde{M}_z(1+p)l_D q_y - \widetilde{H}_{py}p\,l_D q_z\right]\right)}{1 + p\frac{q_y^2}{q^2} - p^2 l_D^2 q_z^2}, \tag{3.70}$$

$$\widetilde{M}_y = \frac{p\left(\widetilde{H}_{py} - \widetilde{M}_z\frac{q_y q_z}{q^2}\left[1 + p\,l_D^2 q^2\right] - i\widetilde{H}_{px}p\,l_D q_z\right)}{1 + p\frac{q_y^2}{q^2} - p^2 l_D^2 q_z^2}. \tag{3.71}$$

For $\mathbf{k}_0 \parallel \mathbf{H}_0$ ($q_z = 0$), we obtain:

$$\widetilde{M}_x = \frac{p\left(\widetilde{H}_{px}\left[1 + p\frac{q_y^2}{q^2}\right] - \widetilde{H}_{py}p\frac{q_x q_y}{q^2} - i\widetilde{M}_z(1+p)l_D q_y\right)}{1 + p}, \tag{3.72}$$

$$\widetilde{M}_y = \frac{p\left(\widetilde{H}_{py}\left[1 + p\frac{q_x^2}{q^2}\right] - \widetilde{H}_{px}p\frac{q_x q_y}{q^2} + i\widetilde{M}_z(1+p)l_D q_x\right)}{1 + p}. \tag{3.73}$$

Several comments are in place:

- We note that both Fourier components $\widetilde{M}_x(\mathbf{q})$ and $\widetilde{M}_y(\mathbf{q})$ are explicitly complex functions, which is due to the DMI ($\propto l_D$).

- At $\mathbf{q} \neq 0$, $\widetilde{M}_{x,y}$ depend on the longitudinal magnetization Fourier component [compare eqn (3.50)]

$$\widetilde{M}_z(\mathbf{q}) = M_s \widetilde{I}_m(\mathbf{q}). \tag{3.74}$$

  This term is related to spatial variations in the saturation magnetization; for instance, since $\widetilde{M}_z \propto \Delta M$ [335], this term models inhomogeneities in the magnetic microstructure that are due to jumps ΔM in the magnetization at internal interfaces (e.g., at particle-matrix interphases in a nanocomposite).

- The Fourier transform of the magnetic anisotropy field $\widetilde{\mathbf{H}}_p$ and $\widetilde{M}_z$ both increase the amplitudes of the transversal Fourier components and are, thus, the sources of spin disorder in the system. This is best seen by inspecting the averaged squared functions, eqns (3.76)−(3.81).

- Terms in $\widetilde{M}_x$ and $\widetilde{M}_y$ such as q_x^2/q^2, q_y^2/q^2, or $q_x q_y/q^2$ are due to the long-range and anisotropic magnetodipolar interaction [compare to the expression (3.55) for $\widetilde{\mathbf{H}}_d$]. These contributions, as well as terms related to the DMI, already give rise to an angular anisotropy of the transversal magnetization Fourier components (see Figs. 3.6 and 3.7). This θ-dependence is in addition to the anisotropy that is related to the trigonometric functions in the cross section, which appear in the magnetic scattering vector $\widetilde{\mathbf{Q}}$ due to the dipolar neutron-magnetic interaction [eqns (2.87)−(2.89)].

- The denominator of $\widetilde{M}_x$ and $\widetilde{M}_y$ has for $\mathbf{k}_0 \perp \mathbf{H}_0$ a singularity for $1 + pq_y^2/q^2 = p^2 l_D^2 q_z^2$. This may become noticeable at small fields and for small q along the horizontal field direction ($\theta = 0°$), where $q_y = 0$. However, we remind that the present theory is valid in the approach-to-saturation regime, when the sample consists of a single magnetic domain and one considers small deviations of magnetic moments due to spatially varying $\mathbf{H}_p$, M_s, and due to the DMI relative to the applied-field direction. For large q or large H_i, the effective magnetic field H_{eff} [eqn (3.65)] takes on large values, so that $p = M_s/H_{\text{eff}} \ll 1$ (compare Fig. 3.4) and the term $p^2 l_D^2 q_z^2$ is much smaller than $1 + pq_y^2/q^2$.

- Note the symmetry of the equations in the parallel scattering geometry, which is absent in the perpendicular case. This is due to the fact that for $\mathbf{k}_0 \parallel \mathbf{H}_0$ the two transversal magnetization components lie in the detector plane, whereas for $\mathbf{k}_0 \perp \mathbf{H}_0$ only one transversal component lies in the detector plane and the other one is pointing along the incident-beam direction.

Since the magnetic SANS cross section depends on the magnetization Fourier components (see Sections 2.6−2.8), it is necessary to compute appropriate averages of functions such as $|\widetilde{M}_x|^2$, $|\widetilde{M}_y|^2$, and the cross terms $CT_{\mathbf{k}_0 \perp \mathbf{H}_0} = -(\widetilde{M}_y \widetilde{M}_z^* + \widetilde{M}_y^* \widetilde{M}_z)$ and $CT_{\mathbf{k}_0 \parallel \mathbf{H}_0} = -(\widetilde{M}_x \widetilde{M}_y^* + \widetilde{M}_x^* \widetilde{M}_y)$. For this purpose, we assume that $\widetilde{M}_z = M_s \widetilde{I}_m$ is real-valued ($\widetilde{M}_z = \widetilde{M}_z^*$) and isotropic [$\widetilde{M}_z = \widetilde{M}_z(q)$], and that the Fourier component $\widetilde{\mathbf{H}}_p(\mathbf{q})$ of the magnetic anisotropy field $\mathbf{H}_p(\mathbf{r})$ is isotropically (randomly) distributed in the plane perpendicular to $\mathbf{H}_0$, i.e.,

$$\widetilde{\mathbf{H}}_p(\mathbf{q}) = \left\{ \widetilde{H}_{px}, \widetilde{H}_{py}, 0 \right\} = \left\{ \widetilde{H}_p \cos\beta, \widetilde{H}_p \sin\beta, 0 \right\}, \tag{3.75}$$

where the angle β specifies the orientation of $\widetilde{\mathbf{H}}_{\mathrm{p}}$.[||] This allows us to average $|\widetilde{M}_x(q, \theta, H_i, \beta)|^2$, and the other combinations of Fourier components, over the angle β, according to

$$(2\pi)^{-1} \int_0^{2\pi} (\ldots)\, d\beta.$$

The results for the projected magnetization Fourier components for the perpendicular scattering geometry ($\mathbf{k}_0 \perp \mathbf{H}_0$) are:

$$|\widetilde{M}_x|^2 = \frac{p^2}{2} \frac{\widetilde{H}_{\mathrm{p}}^2 \left(\left[1 + p\sin^2\theta\right]^2 + p^2 l_{\mathrm{D}}^2 q^2 \cos^2\theta \right) + 2\widetilde{M}_z^2 (1+p)^2 l_{\mathrm{D}}^2 q^2 \sin^2\theta}{\left(1 + p\sin^2\theta - p^2 l_{\mathrm{D}}^2 q^2 \cos^2\theta\right)^2}, \quad (3.76)$$

$$|\widetilde{M}_y|^2 = \frac{p^2}{2} \frac{\widetilde{H}_{\mathrm{p}}^2 \left(1 + p^2 l_{\mathrm{D}}^2 q^2 \cos^2\theta\right) + 2\widetilde{M}_z^2 \left(1 + p\, l_{\mathrm{D}}^2 q^2\right)^2 \sin^2\theta \cos^2\theta}{\left(1 + p\sin^2\theta - p^2 l_{\mathrm{D}}^2 q^2 \cos^2\theta\right)^2}, \quad (3.77)$$

$$CT_{\mathbf{k}_0 \perp \mathbf{H}_0} = -(\widetilde{M}_y \widetilde{M}_z^* + \widetilde{M}_y^* \widetilde{M}_z) = \frac{2\widetilde{M}_z^2 p \left(1 + p\, l_{\mathrm{D}}^2 q^2\right) \sin\theta\cos\theta}{1 + p\sin^2\theta - p^2 l_{\mathrm{D}}^2 q^2 \cos^2\theta}. \quad (3.78)$$

The results for the parallel scattering geometry ($\mathbf{k}_0 \parallel \mathbf{H}_0$) are:

$$|\widetilde{M}_x|^2 = \frac{p^2}{2} \frac{\widetilde{H}_{\mathrm{p}}^2 \left(1 + p(2+p)\sin^2\theta\right) + 2\widetilde{M}_z^2 (1+p)^2 l_{\mathrm{D}}^2 q^2 \sin^2\theta}{(1+p)^2}, \quad (3.79)$$

$$|\widetilde{M}_y|^2 = \frac{p^2}{2} \frac{\widetilde{H}_{\mathrm{p}}^2 \left(1 + p(2+p)\cos^2\theta\right) + 2\widetilde{M}_z^2 (1+p)^2 l_{\mathrm{D}}^2 q^2 \cos^2\theta}{(1+p)^2}, \quad (3.80)$$

$$CT_{\mathbf{k}_0 \parallel \mathbf{H}_0} = -(\widetilde{M}_x \widetilde{M}_y^* + \widetilde{M}_x^* \widetilde{M}_y) = p^2 \frac{\left(\widetilde{H}_{\mathrm{p}}^2 p(2+p) + 2\widetilde{M}_z^2 (1+p)^2 l_{\mathrm{D}}^2 q^2\right) \sin\theta\cos\theta}{(1+p)^2}. \quad (3.81)$$

One can graphically display these Fourier components by specifying the materials parameters (A, M_{s}, D), the value of the internal magnetic field H_{i}, and by choosing a model function for $\widetilde{H}_{\mathrm{p}}^2(q)$ and $\widetilde{M}_z^2(q)$, e.g., a particle form factor with or without a distribution of sizes, or a Lorentzian or Gaussian function. Equations (3.76)−(3.81) then depend only on two independent parameters, the magnitude q and the orientation θ of the (scattering) wave vector $\mathbf{q}$, with $q = (q_y^2 + q_z^2)^{1/2}$ and $\theta = \arctan(q_y/q_z)$ for $\mathbf{k}_0 \perp \mathbf{H}_0$, and $q = (q_x^2 + q_y^2)^{1/2}$ and $\theta = \arctan(q_y/q_x)$ for $\mathbf{k}_0 \parallel \mathbf{H}_0$ (compare Fig. 2.1).

[||]In [441], a texture in the orientation distribution of the magnetic anisotropy field $\mathbf{H}_{\mathrm{p}}(\mathbf{r})$ was considered. It was concluded that its effect on the magnetic SANS cross section is small, in particular at large effective magnetic fields H_{eff}. Therefore, we will in this book only consider the isotropic case.

In the following (for plotting the Fourier components), we employ the sphere form factor for both $\widetilde{H}_{\mathrm{p}}^2$ and $\widetilde{M}_z^2$, i.e. [312],

$$\widetilde{H}_{\mathrm{p}}^2(q) = \frac{H_{\mathrm{p}}^2}{8\pi^3} P(q), \tag{3.82}$$

$$\widetilde{M}_z^2(q) = \frac{(\Delta M)^2}{8\pi^3} P(q), \tag{3.83}$$

$$P(q) = F^2(q) = 9V_{\mathrm{s}}^2 \frac{j_1^2(qR)}{(qR)^2}, \tag{3.84}$$

where $V_{\mathrm{s}} = \frac{4\pi}{3} R^3$, and $j_1(z)$ denotes the spherical Bessel function of the first order. We emphasize, however, that the characteristic structure sizes of $\widetilde{H}_{\mathrm{p}}^2$ and $\widetilde{M}_z^2$ need not to be identical. These are related, respectively, to the spatial extent of regions with uniform magnetic anisotropy field (characterized by the correlation length ξ_{H}) and saturation magnetization (ξ_{M}) (see also Section 4.2). Unless otherwise stated and when the sphere form factor is used, we set $\xi_{\mathrm{H}} = \xi_{\mathrm{M}} = R$. As mentioned already above, corresponding distribution functions characterizing $\widetilde{H}_{\mathrm{p}}^2$ and $\widetilde{M}_z^2$ can also be included.

Figures 3.6 and 3.7 visualize, respectively, for $\mathbf{k}_0 \perp \mathbf{H}_0$ and $\mathbf{k}_0 \parallel \mathbf{H}_0$ the angular anisotropies of the Fourier components of the magnetization projected onto the two-dimensional detector [eqns (3.76)−(3.81)]. The emphasis in these contour plots is not on the magnitude of the Fourier components, which decreases with increasing field, but rather on the change of their angular anisotropy with field. In contrast to Figs. 2 and 3 in [414], the DMI has not been taken into account in these images, i.e., $l_{\mathrm{D}} = 0$ has been set. It is seen that in the parallel scattering geometry, all Fourier components are highly anisotropic (θ-dependent), while $|\widetilde{M}_x|^2$ for $\mathbf{k}_0 \perp \mathbf{H}_0$ is isotropic. The displayed anisotropies are clearly a consequence of the magnetodipolar interaction, i.e., they are due to terms such as $p \sin^2 \theta$ or $p \cos^2 \theta$. Note that both cross terms change sign at the borders between quadrants [Fig. 3.6(g)−(i) and Fig. 3.7(g)−(i)]. On multiplication with the additional factor $\sin \theta \cos \theta$, the respective positive-definite contribution to the unpolarized SANS cross section is obtained [compare eqns (2.93) and (2.94) and Fig. 7.9]. One should also keep in mind that introducing the DMI or changing the ratio between anisotropy and magnetostatic scattering ($H_{\mathrm{p}}/\Delta M = 1$ in Figs. 3.6 and 3.7) can result in different angular anisotropies. The analytical results for the Fourier components shown here agree qualitatively with the outcome of numerical micromagnetic simulations (compare Figs. 7.8 and 7.10). In Section 4.2 in the next chapter, we will see how the averaged Fourier components, eqns (3.76)−(3.81), can be combined to obtain the magnetic SANS cross section [compare, e.g., eqn (4.5)].

3.3.2 The case of a single-phase ferromagnet

A special class of systems are single-phase ferromagnets, which are characterized by a homogeneous saturation-magnetization profile, i.e., $M_{\mathrm{s}} = \text{constant}$. Neglecting the DMI ($l_{\mathrm{D}} = 0$) and terms related to $M_{\mathrm{s}}(\mathbf{r})$, we obtain the following simplified expressions for the Fourier components:

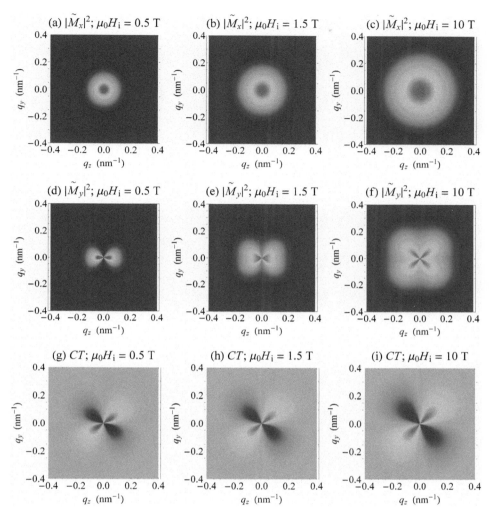

Fig. 3.6: Contour plots of the Fourier components of the magnetization [eqns (3.76)−(3.78)] at selected applied magnetic fields (see insets) ($\mathbf{k}_0 \perp \mathbf{H}_0$; $l_\mathrm{D} = 0$). For both $\widetilde{H}_\mathrm{p}^2(qR)$ and $\widetilde{M}_z^2(qR)$ the sphere form factor was chosen [eqns (3.82)−(3.84) with $H_\mathrm{p} = \Delta M = 1$ and $\xi_\mathrm{H} = \xi_\mathrm{M} = R = 8\,\mathrm{nm}$]. $\mathbf{H}_0 \parallel \mathbf{e}_z$ is horizontal in the plane. In (a)−(f) red-yellow color corresponds to high positive values and blue color to low positive values of the Fourier components, while in (g)−(i) red-yellow color corresponds to positive and blue color to negative values of the cross term (CT). After [414].

For $\mathbf{k}_0 \perp \mathbf{H}_0$,

$$|\widetilde{M}_x|^2 = \frac{p^2}{2} \widetilde{H}_\mathrm{p}^2, \qquad (3.85)$$

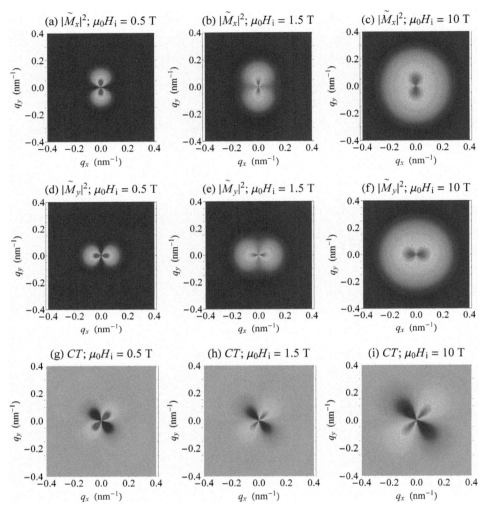

Fig. 3.7: Same as Fig. 3.6, but for $\mathbf{k}_0 \parallel \mathbf{H}_0$ [eqns (3.79)–(3.81)]. $\mathbf{H}_0 \parallel \mathbf{e}_z$ is normal to the plane. After [414].

$$|\widetilde{M}_y|^2 = \frac{p^2}{2} \frac{\widetilde{H}_p^2}{\left(1 + p\sin^2\theta\right)^2},\tag{3.86}$$

$$CT_{\mathbf{k}_0 \perp \mathbf{H}_0} = -(\widetilde{M}_y\widetilde{M}_z^* + \widetilde{M}_y^*\widetilde{M}_z) = 0.\tag{3.87}$$

For $\mathbf{k}_0 \parallel \mathbf{H}_0$,

$$|\widetilde{M}_x|^2 = \frac{p^2}{2} \frac{\widetilde{H}_p^2\left(1 + p(2 + p)\sin^2\theta\right)}{(1 + p)^2},\tag{3.88}$$

$$|\widetilde{M}_y|^2 = \frac{p^2}{2} \frac{\widetilde{H}_{\mathrm{p}}^2 \left(1 + p(2+p)\cos^2\theta\right)}{(1+p)^2}, \tag{3.89}$$

$$CT_{\mathbf{k}_0\|\mathbf{H}_0} = -(\widetilde{M}_x\widetilde{M}_y^* + \widetilde{M}_x^*\widetilde{M}_y) = p^2 \frac{\widetilde{H}_{\mathrm{p}}^2 p(2+p)\sin\theta\cos\theta}{(1+p)^2}. \tag{3.90}$$

Here, we see that $\widetilde{M}_y\widetilde{M}_z$ correlations do not show up in the transversal geometry. Accordingly, as might be anticipated, clover-leaf-type angular anisotropies with maxima roughly along the diagonals of the detector [e.g., [121, 445–449], see also Figs. 4.7(d) and 4.9] cannot be reproduced by unpolarized SANS cross sections based on the single-phase eqns (3.85) and (3.86).

4

MAGNETIC SANS OF BULK FERROMAGNETS

In the previous Chapter 3, the Fourier components of the magnetization $\widetilde{M}_{x,y,z}(\mathbf{q})$ were computed in the small-misalignment approximation using the continuum theory of micromagnetics. As we will see in Section 4.2, in this way one can obtain closed-form expressions for any desired magnetic SANS cross section (scattering geometry) as a function of momentum-transfer vector, applied magnetic field, magnetic interaction parameters [exchange, Dzyaloshinskii–Moriya interaction (DMI), anisotropy, magnetostatics], and microstructural quantities such as particle size, shape, arrangement, and crystallographic texture. The averaged projected Fourier components, eqns (3.76)−(3.81), need only to be multiplied by the corresponding trigonometric functions (given by the scattering geometry) and summed up in order to obtain the magnetic SANS cross section. Section 4.3 highlights the basic features of magnetic SANS on bulk magnetic materials, i.e., the angular anisotropy, role of the dipolar interaction and the DMI, magnetic Guinier law, asymptotic power-law behavior, and the polarization dependence of the magnetic SANS cross section. In Section 4.4, we will discuss selected experimental data on a soft magnetic iron-based nanocomposite, on a hard magnetic Nd–Fe–B-based nanocomposite, and on nanocrystalline Co. The remarkable SANS response of nanocrystalline rare-earth magnets in the paramagnetic temperature regime is reviewed in Section 4.5, and the early magnetic SANS studies on dislocations in cold-worked metals are the subject of Section 4.6. We begin this chapter with a brief historical overview of the developments regarding the theoretical aspects of magnetic SANS.

4.1 Historical overview of magnetic SANS theory

As discussed in the preface to this book, theoretical research in magnetic SANS commenced in 1963 with the seminal paper by Kronmüller, Seeger, and Wilkens [1], who computed the magnetic small-angle scattering due to dislocation-induced spin misalignment. Their work, published well before the first dedicated SANS instruments were developed, was in the early 1970s extended by Schmatz et al. [4, 5], who provided, in reciprocal space, an analytical expression for the dilatation and stress tensor due to dislocations in anisotropic cubic crystals. Magnetic SANS research on dislocations essentially ceased in the 1980s with the paper by Göltz et al. [6] summarizing the state-of-the-art (see Section 4.6 for further details). Maleev [234] investigated the inelastic SANS due to magnons, Maleev and Ruban [139, 141] analyzed the critical

Magnetic Small-Angle Neutron Scattering: A Probe for Mesoscale Magnetism Analysis. Andreas Michels,
Oxford University Press (2021). © Andreas Michels. DOI: 10.1093/oso/ 9780198855170.003.0004

depolarization of neutrons by ferromagnets in the context of SANS, and Maleev and Toperverg [450] studied the multiple small-angle scattering by static inhomogeneities. Regarding magnetic SANS on nanoparticles, Ernst, Schelten, and Schmatz [451, 452] performed SANS experiments on nanosized Co precipitates in Cu single crystals and realized (already in 1971) that the experimental data cannot be reconciled with the conventional particle-matrix approach for the magnetic SANS cross section, which assumes homogeneously magnetized domains (see also the discussion in Section 5.1.1). Intraparticle spin misalignment was identified as the source of discrepancy and micromagnetic model calculations assuming different domain configurations suggested a possible solution of the problem. Schärpf et al. [453] studied the diffraction and refraction of neutrons on domains and domain walls beyond the Born approximation by solving the Schrödinger equation. By contrast, Lermer and Steyerl [218] investigated the transmission and scattering of low-energy neutrons on polycrystalline Fe, Ni, Co and analyzed their results within the Born approximation. The early magnetic SANS work and the developments in SANS instrumentation are authoritatively summarized in [2, 17]. In 1995, Del Moral and Cullen [454] provided an extended theory for the magnetic correlations in random anisotropy magnets in the quasi-saturated regime. Besides corroborating the origins of the well-known Lorentzian (L) and Lorentzian-squared (L^2) terms in the SANS cross section, they obtained new cross sections of $L^{1/2}$ and $L^{3/2}$ types. At the end of the 1990s, Weissmüller et al. [441, 455, 456] continued with theoretical and experimental magnetic SANS research on polycrystalline ferromagnets with a nanometer crystallize size (so-called nanocrystalline ferromagnets). These authors considered magnetic SANS due to nanoscale variations in the orientation and magnitude of the magnetic anisotropy field, as appropriate for random-anisotropy-type nanocrystalline ferromagnets (see [49] for a review). In 2005, Löffler et al. [457] provided a magnetic SANS theory of nanostructured metals based on the Stoner–Wohlfarth model, which is able to explain anisotropic scattering profiles with an intensity enhancement for scattering vectors parallel to the field direction. Following the experimental results by Cywinski et al. [458], Pynn et al. [459], and by Bellouard, Mirebeau, and Hennion [460, 461], magnetic SANS due to single-domain particles exhibiting superparamagnetism has been addressed by Kohlbrecher, Wiedenmann, and Heinemann [462–466] in the early 2000s, who derived field-dependent cross-section expressions in terms of the Langevin function. In 2013, Honecker and Michels [444] extended and completed the micromagnetic SANS theory of Weissmüller by taking into account spatial variations in the saturation magnetization as a second perturbing effect on the magnetic microstructure (in addition to the spatially dependent anisotropy fields). In [414], the DMI was added to the Hamiltonian, which gives rise to a polarization dependence of the spin-flip SANS cross section via the chiral function. This then provided quite general expressions for the magnetic SANS cross section of bulk ferromagnets as a function of the magnetic interactions (isotropic exchange, DMI, magnetic anisotropy, magnetostatics), the anisotropy-field microstructure, and of the magnitude and orientation of the scattering vector; in particular, it became possible to describe clover-leaf-type angular anisotropies with maxima roughly along the detector diagonals, which were otherwise not contained in the previous treatments. For isolated nanodots, Metlov and Michels [467] in 2016, and similarly Mirebeau et al. [468] in 2018, provided a magnetic SANS theory by using the analytical expressions for the magnetization textures of thin submicron-sized cylinders in vortex state. In 2020, Honecker, Barquín, and Bender [469] developed a theoretical expression for the magnetic structure factor of interacting single-domain particles. These authors explicitly considered the case of polarized neutrons and the dependency of the spin-spin correlation function on the applied magnetic field. Starting with the study of Ogrin et al. [470] from 2006, numerical simulations of magnetic SANS based on solving Brown's nonlinear equations of micromagnetics have greatly contributed to the fundamental understanding of magnetic SANS from bulk nanocomposites as well as from isolated nanoparticles and inverse opal-like structures (see Chapter 7 for a detailed account) [58,162,163,470–485].

4.2 Closed-form results for the spin-misalignment scattering

4.2.1 Unpolarized neutrons ($k_0 \perp H_0$)

For the perpendicular scattering geometry, the unpolarized elastic SANS cross section $d\Sigma/d\Omega$ at momentum-transfer vector $\mathbf{q}$ [eqn (2.93)] can be formally written as:

$$\frac{d\Sigma}{d\Omega}(\mathbf{q}) = \frac{d\Sigma_{\mathrm{res}}}{d\Omega}(\mathbf{q}) + \frac{d\Sigma_{\mathrm{SM}}}{d\Omega}(\mathbf{q}), \tag{4.1}$$

where

$$\frac{d\Sigma_{\mathrm{res}}}{d\Omega}(\mathbf{q}) = \frac{8\pi^3}{V}\left(|\widetilde{N}|^2 + b_{\mathrm{H}}^2|\widetilde{M}_{\mathrm{s}}|^2\sin^2\theta\right) \tag{4.2}$$

represents the nuclear and magnetic residual SANS cross section, which is measured at complete magnetic saturation (infinite field, all magnetic moments parallel to $\mathbf{H}_0$), and the remaining term

$$\frac{d\Sigma_{\mathrm{SM}}}{d\Omega}(\mathbf{q}) = \frac{8\pi^3}{V}b_{\mathrm{H}}^2\left(|\widetilde{M}_x|^2 + |\widetilde{M}_y|^2\cos^2\theta + \left(|\widetilde{M}_z|^2 - |\widetilde{M}_{\mathrm{s}}|^2\right)\sin^2\theta\right.$$
$$\left. - (\widetilde{M}_y\widetilde{M}_z^* + \widetilde{M}_y^*\widetilde{M}_z)\sin\theta\cos\theta\right), \tag{4.3}$$

denotes the spin-misalignment SANS cross section, which vanishes at saturation when the real-space magnetization is given by $\mathbf{M} = \{0, 0, M_z = M_{\mathrm{s}}(\mathbf{r})\}$. An analogous decomposition can of course be done for the $d\Sigma/d\Omega$ in the parallel geometry (see Section 4.2.2). If we now restrict the considerations to the approach-to-saturation regime, where

$$\widetilde{M}_z \cong \widetilde{M}_{\mathrm{s}}, \tag{4.4}$$

and ignore the DMI ($l_{\mathrm{D}} = 0$), then $d\Sigma_{\mathrm{SM}}/d\Omega$ can be expressed in closed-form by making use of the micromagnetic results for the magnetization Fourier components of Section 3.3. By inserting the (into the detector plane $q_x = 0$) projected Fourier components [eqns (3.76)–(3.78)] into eqn (4.3), the final result for $d\Sigma_{\mathrm{SM}}/d\Omega$ is [444]:

$$\frac{d\Sigma_{\mathrm{SM}}}{d\Omega}(\mathbf{q}) = \frac{8\pi^3}{V}b_{\mathrm{H}}^2\left(|\widetilde{M}_x|^2 + |\widetilde{M}_y|^2\cos^2\theta - (\widetilde{M}_y\widetilde{M}_z^* + \widetilde{M}_y^*\widetilde{M}_z)\sin\theta\cos\theta\right)$$
$$= S_{\mathrm{H}}(\mathbf{q})R_{\mathrm{H}}(q, \theta, H_{\mathrm{i}}) + S_{\mathrm{M}}(\mathbf{q})R_{\mathrm{M}}(q, \theta, H_{\mathrm{i}}). \tag{4.5}$$

The magnetic small-angle scattering due to transversal spin components, with related Fourier amplitudes $\widetilde{M}_x(\mathbf{q})$ and $\widetilde{M}_y(\mathbf{q})$, is contained in $d\Sigma_{\mathrm{SM}}/d\Omega$, which decomposes into a contribution $S_H R_H$ due to perturbing magnetic anisotropy fields and a part $S_M R_M$ related to magnetostatic fields. The micromagnetic SANS theory considers a uniform exchange interaction and a random distribution of magnetic easy axes, but takes explicitly into account spatial variations in the saturation magnetization [via the function S_M, see eqn (4.7)]. We emphasize that it is $d\Sigma_{SM}/d\Omega$ which depends on the magnetic interactions (exchange, DMI, anisotropy, magnetostatics), while $d\Sigma_{res}/d\Omega$ is determined by the geometry of the underlying grain microstructure (e.g., the particle shape or the particle-size distribution). Therefore, in order to access the magnetic interactions, it is advantageous to subtract the residual scattering cross section (measured at a saturating field) from the total $d\Sigma/d\Omega$ at a lower field, provided that this is possible in the given experimental situation. If in a SANS experiment the approach-to-saturation regime can be reached for a particular magnetic material, then the residual SANS can be obtained by an analysis of field-dependent data via the extrapolation to infinite field (see Section 4.4). In a sense, for a bulk ferromagnet, the scattering at saturation resembles the topographical background in Kerr-microscopy experiments, which needs to be subtracted in order to access the magnetic domain structure of the sample [486].

Throughout Section 4.2 we will make the assumption that $\widetilde{M}_z \cong \widetilde{M}_s$ in the approach-to-saturation regime. Since from the neutron scattering point of view it is difficult to experimentally determine the function $|\widetilde{M}_z|^2$, we employ here micromagnetic simulations for studying the applied-field dependence of this term. Figure 4.1 shows the results of numerical micromagnetic computations for the field variation of $|\widetilde{M}_z(q)|^2$. The system under study is a two-phase soft magnetic nanocomposite consisting of $\sim$10 nm-sized single-domain iron particles embedded in an amorphous magnetic matrix (compare Fig. 7.3). As is seen, the field dependence of $|\widetilde{M}_z|^2$ is very weak, mainly restricted to $q \lesssim 0.4$ nm^{-1}, in this way supporting the assumption that within the small-misalignment approximation the longitudinal Fourier component may be well represented by the one of the saturated state, i.e., $|\widetilde{M}_z|^2(q) \cong |\widetilde{M}_s|^2(q)$. The central peak at about 0.78 nm^{-1} is related to the dense packing of the iron nanoparticles (volume fraction: 40%) that are in a single-domain state, whereas the shoulder at about 1.26 nm^{-1} is an artifact related to the 5 nm mesh-element size chosen to discretize the matrix phase. It is also worth emphasizing that $|\widetilde{M}_z|^2$ increases with increasing field, in contrast to the transversal Fourier components $|\widetilde{M}_x|^2$ and $|\widetilde{M}_y|^2$, which decrease with increasing field (compare, e.g., Fig. 7.11).

Although we have neglected the DMI in eqn (4.5), there is no principal difficulty in including it. However, since the DMI is a weak interaction, we believe that the effect of the DMI on the unpolarized SANS cross section is masked by the many other contributions to $d\Sigma_{SM}/d\Omega$. As we will see in Section 4.3.3, polarized neutron scattering (half-polarized and uniaxial polarization analysis) provides a means to access the impact of the DMI via the measurement of the chiral function $\chi(\mathbf{q})$ [eqn (2.97)]. The function $\chi(\mathbf{q})$ characterizes the defect-induced symmetry breaking and can be obtained e.g., from the difference measurement of the spin-flip SANS cross sections.

The anisotropy-field scattering function

$$S_H(\mathbf{q}) = \frac{8\pi^3}{V} b_H^2 |\widetilde{\mathbf{H}}_p(\mathbf{q})|^2 \tag{4.6}$$

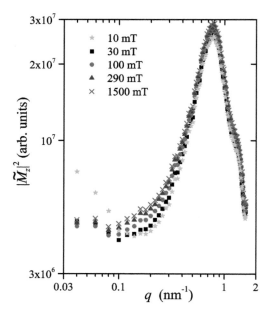

Fig. 4.1: Applied-field dependence of the longitudinal Fourier component $|\widetilde{M}_z|^2(q)$ obtained from micromagnetic simulations of a two-phase iron-based nanocomposite (log-log scale); see Section 7.4.1 for further details. Data courtesy of Sergey Erokhin, General Numerics Research Lab, Jena, Germany.

depends on the Fourier transform $\widetilde{\mathbf{H}}_{\mathrm{p}}(\mathbf{q})$ of the magnetic anisotropy field, whereas the scattering function of the longitudinal magnetization

$$S_{\mathrm{M}}(\mathbf{q}) = \frac{8\pi^3}{V} b_{\mathrm{H}}^2 |\widetilde{M}_z(\mathbf{q})|^2 \tag{4.7}$$

characterizes the spatial variations of the saturation magnetization (see Sections 4.4.1 and 4.3.4); for instance, $S_{\mathrm{M}}(\mathbf{q})$ provides information on the mean magnitude $\Delta M \propto \widetilde{M}_z$ of the magnetization jump at internal (e.g., particle-matrix) interfaces. Quite frequently, S_{H} and S_{M} are described by particle form factors (very often the one of the sphere) or special scattering profiles (e.g., squared Lorentzians), and it is therefore convenient to introduce the corresponding structure sizes (correlation lengths) ξ_{H} and ξ_{M}. A simple example where $\xi_{\mathrm{H}} = \xi_{\mathrm{M}}$ is a collection of homogeneous and defect-free magnetic nanoparticles in a nonmagnetic matrix. If, on the other hand, atomically sharp grain boundaries are introduced into such particles, then the direction of the magnetic anisotropy field changes due to the changing set of crystallographic directions, but the value of M_{s} may remain the same, so that $\xi_{\mathrm{H}} < \xi_{\mathrm{M}}$.

The dimensionless micromagnetic response functions for unpolarized neutrons can be expressed as:

$$R_{\mathrm{H}}(q, \theta, H_{\mathrm{i}}) = \frac{p^2}{2} \left(1 + \frac{\cos^2\theta}{\left(1 + p\sin^2\theta\right)^2} \right) \tag{4.8}$$

and

$$R_{\mathrm{M}}(q,\theta,H_{\mathrm{i}}) = \frac{p^2 \sin^2\theta \cos^4\theta}{\left(1 + p\sin^2\theta\right)^2} + \frac{2p\sin^2\theta \cos^2\theta}{1 + p\sin^2\theta}, \tag{4.9}$$

where $p(q, H_{\mathrm{i}})$ has been defined via eqn (3.64). When the functions $|\widetilde{N}|^2$, $|\widetilde{M_z}|^2 \cong |\widetilde{M_{\mathrm{s}}}|^2$, and $|\widetilde{\mathbf{H}_{\mathrm{p}}}|^2$ depend only on the magnitude q of the scattering vector, for instance, for statistically isotropic magnetic materials, one can perform an azimuthal average of eqn (4.1). The resulting expressions for the response functions then read:

$$R_{\mathrm{H}}(q,H_{\mathrm{i}}) = (2\pi)^{-1} \int\limits_0^{2\pi} R_{\mathrm{H}}(q,\theta,H_{\mathrm{i})}d\theta = \frac{p^2}{4}\left(2 + \frac{1}{\sqrt{1+p}}\right) \tag{4.10}$$

and

$$R_{\mathrm{M}}(q,H_{\mathrm{i}}) = (2\pi)^{-1} \int\limits_0^{2\pi} R_{\mathrm{M}}(q,\theta,H_{\mathrm{i})}d\theta = \frac{\sqrt{1+p}-1}{2} \approx \frac{p}{4} \text{ for } p \ll 1, \tag{4.11}$$

so that the azimuthally averaged total nuclear and magnetic SANS cross section can be written as:

$$\frac{d\Sigma}{d\Omega}(q) = \frac{d\Sigma_{\mathrm{res}}}{d\Omega}(q) + \frac{d\Sigma_{\mathrm{SM}}}{d\Omega}(q), \tag{4.12}$$

where

$$\frac{d\Sigma_{\mathrm{res}}}{d\Omega}(q) = \frac{8\pi^3}{V}\left(|\widetilde{N}|^2 + \frac{1}{2}b_{\mathrm{H}}^2|\widetilde{M_{\mathrm{s}}}|^2\right) \tag{4.13}$$

and

$$\frac{d\Sigma_{\mathrm{SM}}}{d\Omega}(q) = S_{\mathrm{H}}(q)R_{\mathrm{H}}(q,H_{\mathrm{i}}) + S_{\mathrm{M}}(q)R_{\mathrm{M}}(q,H_{\mathrm{i}}). \tag{4.14}$$

From eqn (4.13) it is seen that one may combine the magnetic part of the azimuthally averaged residual scattering cross section, $\frac{1}{2}\frac{8\pi^3}{V}b_{\mathrm{H}}^2|\widetilde{M_{\mathrm{s}}}|^2 = \frac{1}{2}S_{\mathrm{M}}$, with the contribution $S_{\mathrm{M}}R_{\mathrm{M}}$ of $d\Sigma_{\mathrm{SM}}/d\Omega$ to yield a term $S_{\mathrm{M}}\left(R_{\mathrm{M}} + \frac{1}{2}\right)$, so that the total unpolarized $d\Sigma/d\Omega$ equals:

$$\frac{d\Sigma}{d\Omega} = \frac{8\pi^3}{V}|\widetilde{N}|^2 + S_{\mathrm{H}}R_{\mathrm{H}} + S_{\mathrm{M}}\left(R_{\mathrm{M}} + \frac{1}{2}\right). \tag{4.15}$$

By using the defining expressions for $p(q,H_{\mathrm{i}})$, i.e., the effective magnetic field $H_{\mathrm{eff}}(q,H_{\mathrm{i}})$ and the exchange length $l_{\mathrm{H}}(H_{\mathrm{i}})$ [eqns (3.64)–(3.66)], it is easily seen that the response functions vary asymptotically as (see Fig. 4.2):

$$R_{\mathrm{H}}(q,H_{\mathrm{i}}) \propto q^{-4} \quad \text{and} \quad R_{\mathrm{M}}(q,H_{\mathrm{i}}) \propto q^{-2}. \tag{4.16}$$

At $q = 0$, the values of R_{H} and R_{M} are determined by $p_0 = p(q = 0, H_{\mathrm{i}}) = M_{\mathrm{s}}/H_{\mathrm{i}}$. The behavior of S_{H} and S_{M} depends on the microstructural model chosen. Employing a simple phenomenological sphere form factor model [cf. eqns (3.82)–(3.84)], it is immediately seen that (except for prefactors) the well-known approximations established for nuclear SANS and small-angle x-ray scattering (SAXS) do apply (see Section 2.4.3);

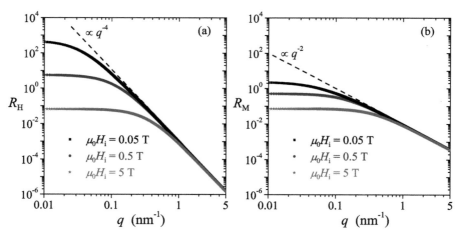

Fig. 4.2: The dimensionless micromagnetic response functions $R_H(q, H_i)$ (a) and $R_M(q, H_i)$ (b) [eqns (4.10) and (4.11)] at selected magnetic fields (increasing from top to bottom, see insets) (log-log scale). Materials parameters: $A = 2.5 \times 10^{-11}$ J/m; $\mu_0 M_s = 1.5$ T. After [312].

for instance, both functions then vary asymptotically as $S_{H,M} \propto q^{-4}$ and exhibit a Guinier behavior at low q (see Section 4.3.4 for further details). As a consequence, the spin-misalignment scattering cross section does generally not exhibit the Porod q^{-4}-behavior, but varies asymptotically more steeply (see Section 4.3.5).

Figure 4.3 displays both response functions R_H [eqn (4.10)] and R_M [eqn (4.11)] as a function of the dimensionless parameter p (shown in Fig. 3.4). Assuming that S_H and S_M in eqn (4.14) are of comparable magnitude, it is seen that at large applied fields or large momentum transfers (when $p \ll 1$), $d\Sigma_{SM}/d\Omega$ is dominated by the term $S_M R_M$, whereas at small fields and small momentum transfers (when $p \gg 1$), $d\Sigma_{SM}/d\Omega$ is governed by the $S_H R_H$ contribution. The field-induced transition from anisotropy-field dominated to magnetostatic scattering, which is accompanied by a change in the angular anisotropy of $d\Sigma_{SM}/d\Omega$, is depicted in the upper row of Fig. 4.7.

Finally, we state the result for a homogeneous single-phase ferromagnet, where $\widetilde{M}_z$-fluctuations become negligible. Using eqns (3.85) and (3.86), $d\Sigma_{SM}/d\Omega$ evaluates to [441]

$$\frac{d\Sigma_{SM}}{d\Omega}(\mathbf{q}) = \frac{8\pi^3}{V} b_H^2 \left(|\widetilde{M}_x|^2 + |\widetilde{M}_y|^2 \cos^2 \theta \right)$$
$$= S_H(\mathbf{q}) R_H(q, \theta, H_i). \tag{4.17}$$

S_H and R_H are, respectively, given by eqns (4.6) and (4.8), and the azimuthally averaged expression for R_H equals eqn (4.10). The single-phase material SANS cross section eqn (4.17) is graphically displayed in the lower row of Fig. 4.7 and compared to the more general expression eqn (4.5), which takes into account both spatial variations in the magnetic anisotropy field and in the saturation magnetization.

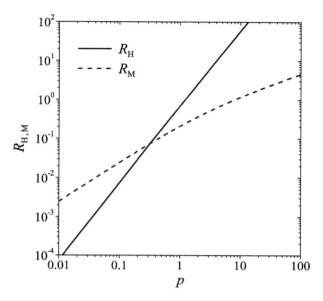

Fig. 4.3: Micromagnetic response functions R_H [eqn (4.10)] and R_M [eqn (4.11)] versus $p(q, H_i)$ (log-log scale). After [59].

4.2.2 Unpolarized neutrons ($\mathbf{k}_0 \parallel \mathbf{H}_0$)

For the parallel scattering geometry, the total unpolarized $d\Sigma/d\Omega$ [eqn (2.94)] can be decomposed analogously into the scattering at saturation and into a term which takes into account the magnetic SANS away from saturation:

$$\frac{d\Sigma}{d\Omega}(\mathbf{q}) = \frac{d\Sigma_{\text{res}}}{d\Omega}(\mathbf{q}) + \frac{d\Sigma_{\text{SM}}}{d\Omega}(\mathbf{q}), \tag{4.18}$$

$$\frac{d\Sigma_{\text{res}}}{d\Omega}(\mathbf{q}) = \frac{8\pi^3}{V}\left(|\widetilde{N}|^2 + b_H^2|\widetilde{M}_s|^2\right), \tag{4.19}$$

$$\frac{d\Sigma_{\text{SM}}}{d\Omega}(\mathbf{q}) = \frac{8\pi^3}{V}b_H^2\left(|\widetilde{M}_x|^2\sin^2\theta + |\widetilde{M}_y|^2\cos^2\theta + \left(|\widetilde{M}_z|^2 - |\widetilde{M}_s|^2\right)\right.$$
$$\left. -(\widetilde{M}_x\widetilde{M}_y^* + \widetilde{M}_x^*\widetilde{M}_y)\sin\theta\cos\theta\right). \tag{4.20}$$

In the small-misalignment approximation, where $\widetilde{M}_z \cong \widetilde{M}_s$ [eqn (4.4)], the azimuthally averaged $d\Sigma_{\text{SM}}/d\Omega$ can be written in closed form as ($l_D = 0$) [444]:

$$\frac{d\Sigma_{\text{SM}}}{d\Omega}(q) = \frac{8\pi^3}{V}b_H^2\left(|\widetilde{M}_x|^2\sin^2\theta + |\widetilde{M}_y|^2\cos^2\theta - (\widetilde{M}_x\widetilde{M}_y^* + \widetilde{M}_x^*\widetilde{M}_y)\sin\theta\cos\theta\right)$$
$$= S_H(q)R_H(q, H_i), \tag{4.21}$$

where the response function R_H simplifies to

$$R_H(q, H_i) = \frac{p^2}{2},$$

(4.22)

and varies asymptotically as $R_H \propto q^{-4}$. Equation (4.21) is obtained by using the expressions for the (into the detector plane $q_z = 0$) projected Fourier components [eqns (3.79)–(3.81)]. S_H is given by eqn (4.6), and we note that in this geometry $d\Sigma_{SM}/d\Omega$ does not depend on the $\widetilde{M}_z$-fluctuations and equals the expression for the single-phase material case [456]; in other words, inhomogeneities in the saturation magnetization are for $\mathbf{k}_0 \parallel \mathbf{H}_0$ only contained in $d\Sigma_{res}/d\Omega$, and not in $d\Sigma_{SM}/d\Omega$. We also re-emphasize that although the individual Fourier contributions to $d\Sigma_{SM}/d\Omega$ are highly anisotropic (compare Fig. 3.7), their sum in eqn (4.21) is isotropic (θ-independent) for statistically isotropic ferromagnets (compare Figs. 4.31 and 7.6). This is not the case for the perpendicular scattering geometry.

Figure 4.4 shows the field-dependent spin-misalignment SANS cross section for $\mathbf{k}_0 \parallel \mathbf{H}_0$ using the materials parameters of Co. For the scattering function $S_H(q)$ of the magnetic anisotropy field, we have assumed a Lorentzian-squared dependency with a correlation length of $\xi_H = 8$ nm. It can be clearly seen that $d\Sigma_{SM}/d\Omega = S_H R_H$ exhibits an extremely large field response by several orders of magnitude. The corresponding results for $d\Sigma_{SM}/d\Omega = S_H R_H + S_M R_M$ in the perpendicular geometry are qualitatively similar. While the correlation length $\xi_H = 8$ nm in S_H describes the characteristic field-independent size of volumina where the magnetic anisotropy field is constant, we observe that with increasing field the point with the largest curvature in $d\Sigma_{SM}/d\Omega$ evolves towards larger q-values (see inset in Fig. 4.4). This is the signature of the second field-dependent correlation length $l_H \propto H_i^{-1/2}$, which is contained in both response functions R_H and R_M [compare eqns (3.64)–(3.66)]. In Section 4.3.4, we will see that under certain assumptions the magnetic Guinier law can be obtained, which allows one to determine these two (structural and magnetic) correlations lengths.

The described scenario illustrates the general convolution relationship between the nuclear grain microstructure and micromagnetic response functions. In Fourier space, and in the high-field limit, $d\Sigma_{SM}/d\Omega$ is the sum of products of functions which contain information about the microstructure (S_H and S_M) and functions which provide information about the field-dependent spin perturbations related to anisotropy-field fluctuations (R_H) and variations in the saturation magnetization (R_M). Consistent with this assessment, we show in Chapter 6 that experimentally determined correlation lengths l_C consist of a field-independent part, which is related to the size of the dominant microstructural defect causing the spin perturbation (some weighted average of ξ_H and ξ_M), and of a field-dependent contribution, which is a measure for the characteristic decay length of the perpendicular magnetization around the defect. We also emphasize that the field dependence of $d\Sigma_{SM}/d\Omega$ is restricted to small and intermediate values of the momentum-transfer vector (compare Fig. 4.4), and we refer to Section 4.4 for an estimate of the related characteristic q-value above which $d\Sigma_{SM}/d\Omega$ is practically field independent.

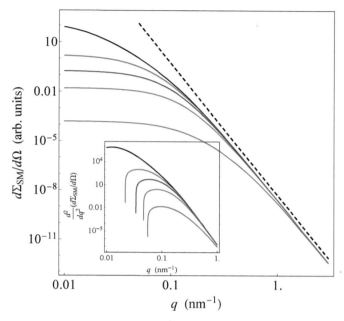

Fig. 4.4: Field dependence of the theoretical $d\Sigma_{\mathrm{SM}}/d\Omega$ for the parallel scattering geometry [eqn (4.21)] (log-log scale). The internal magnetic field H_{i} increases from top to bottom (in T): 0.01; 0.1; 0.3; 1.0; 10. $S_{\mathrm{H}}(q) = H_{\mathrm{p}}^2/(1 + \xi_{\mathrm{H}}^2 q^2)^2$ with $\xi_{\mathrm{H}} = 8\,\mathrm{nm}$ $(8\pi^3 V^{-1} b_{\mathrm{H}}^2 = 1)$. Materials parameters for Co were used: $A = 2.8 \times 10^{-11}\,\mathrm{J/m}$; $\mu_0 M_{\mathrm{s}} = 1.79\,\mathrm{T}$. Dashed line: $d\Sigma_{\mathrm{SM}}/d\Omega \propto q^{-8}$. Inset: second q-derivative of $d\Sigma_{\mathrm{SM}}/d\Omega$ as a function of q (fields increase from top to bottom as in the main figure).

4.2.3 Polarized neutrons (POLARIS) ($\mathbf{k}_0 \perp \mathbf{H}_0$)

The two half-polarized SANSPOL cross sections $d\Sigma^{\pm}/d\Omega$ [eqns (2.95) and (2.98)] differ from the corresponding unpolarized SANS cross sections [eqns (2.93) and (2.94)] by the two nuclear-magnetic interference terms $\propto \widetilde{N}\widetilde{M}_z$ and $\propto \widetilde{N}\widetilde{M}_y$, and by the chiral contribution $\chi(\mathbf{q})$, which will be treated separately in Section 4.3.3. For a statistically isotropic magnetic material, the volume average of the transversal magnetization component M_y vanishes, $\langle M_y \rangle = \frac{1}{V} \int_V M_y(\mathbf{r}) d^3 r = 0$, so that nuclear-magnetic interference terms $\propto \widetilde{N}\widetilde{M}_y$ do generally not contribute to $d\Sigma^{\pm}/d\Omega$. Therefore, in the absence of $\chi(\mathbf{q})$, the measurement of the SANSPOL "spin-up" and "spin-down" cross sections does not provide significantly more information regarding $d\Sigma_{\mathrm{SM}}/d\Omega$ than can already be learned by the measurement of the unpolarized SANS cross section alone (see also the discussion in Section 4.3.6).

In the following, we concentrate on the micromagnetic description of the spin-flip SANS cross section, which does not contain the nuclear coherent scattering signal. Moreover, we neglect the chiral term $\chi(\mathbf{q})$ (see Section 4.3.3) and nuclear-spin incoherent SANS, so that the spin-flip SANS is independent of the incoming neutron polarization, i.e., $d\Sigma^{+-}/d\Omega = d\Sigma^{-+}/d\Omega$. The latter situation is realized for many scattering systems; for instance, for all disordered systems such as gases and liquids, statistically isotropic polycrystalline ferromagnets, paramagnets, or single crystals with a center of symmetry [76]. Figure 2.20(b) demonstrates the absence of a polarization dependence of the spin-flip SANS cross section for the example of an Fe–Cr-based polycrystalline nanocomposite (see also the discussion in Section 2.8.3).

At saturation, the residual spin-flip SANS cross section for $\mathbf{k}_0 \perp \mathbf{H}_0$ equals [compare eqn (2.100) and Fig. 2.20(a)]:

$$\frac{d\Sigma_{\text{res}}^{+-}}{d\Omega}(\mathbf{q}) = \frac{8\pi^3}{V} b_{\text{H}}^2 |\widetilde{M}_{\text{s}}|^2 \sin^2\theta \cos^2\theta, \tag{4.23}$$

which, when azimuthally averaged $(\frac{1}{2\pi}\int_0^{2\pi}(\ldots)d\theta)$ assuming that $|\widetilde{M}_{\text{s}}|^2 = |\widetilde{M}_{\text{s}}|^2(q)$, reduces to

$$\frac{d\Sigma_{\text{res}}^{+-}}{d\Omega}(q) = \frac{\pi^3}{V} b_{\text{H}}^2 |\widetilde{M}_{\text{s}}|^2. \tag{4.24}$$

The remaining spin-misalignment part of the spin-flip cross section reads $(\widetilde{M}_z \cong \widetilde{M}_{\text{s}})$:

$$\frac{d\Sigma_{\text{SM}}^{+-}}{d\Omega}(\mathbf{q}) = \frac{8\pi^3}{V} b_{\text{H}}^2 \left(|\widetilde{M}_x|^2 + |\widetilde{M}_y|^2 \cos^4\theta - (\widetilde{M}_y \widetilde{M}_z^* + \widetilde{M}_y^* \widetilde{M}_z) \sin\theta \cos^3\theta \right). \tag{4.25}$$

Inserting the expressions for the Fourier components [eqns (3.76)–(3.78)] and ignoring the DMI yields [444]:

$$\frac{d\Sigma_{\text{SM}}^{+-}}{d\Omega}(\mathbf{q}) = S_{\text{H}}(\mathbf{q}) R_{\text{H}}^{+-}(q,\theta,H_{\text{i}}) + S_{\text{M}}(\mathbf{q}) R_{\text{M}}^{+-}(q,\theta,H_{\text{i}}), \tag{4.26}$$

where $S_{\text{H}}(\mathbf{q}) = 8\pi^3 V^{-1} b_{\text{H}}^2 \widetilde{H}_{\text{p}}^2(\mathbf{q})$ and $S_{\text{M}}(\mathbf{q}) = 8\pi^3 V^{-1} b_{\text{H}}^2 \widetilde{M}_z^2(\mathbf{q})$ remain unchanged, but the two response functions take on a different form as compared to the unpolarized case:

$$R_{\text{H}}^{+-}(q,\theta,H_{\text{i}}) = \frac{p^2}{2}\left(1 + \frac{\cos^4\theta}{(1+p\sin^2\theta)^2}\right), \tag{4.27}$$

$$R_{\text{M}}^{+-}(q,\theta,H_{\text{i}}) = \frac{p^2 \sin^2\theta \cos^6\theta}{(1+p\sin^2\theta)^2} + \frac{2p\sin^2\theta \cos^4\theta}{1+p\sin^2\theta}. \tag{4.28}$$

The azimuthal averages of the response functions are obtained as:

$$R_{\text{H}}^{+-}(q,H_{\text{i}}) = \frac{2 + 2p^2 - (2-p)\sqrt{1+p}}{4}, \tag{4.29}$$

$$R_{\text{M}}^{+-}(q,H_{\text{i}}) = \frac{8(\sqrt{1+p}-1) - p[16 - 12\sqrt{1+p} + p(9 - 4\sqrt{1+p})]}{8p^2}, \tag{4.30}$$

and the total magnetic-field-dependent spin-flip SANS cross section is expressed as:

$$\frac{d\Sigma^{+-}}{d\Omega}(q,H_{\text{i}}) = \frac{d\Sigma_{\text{res}}^{+-}}{d\Omega}(q) + \frac{d\Sigma_{\text{SM}}^{+-}}{d\Omega}(q,H_{\text{i}})$$

$$= \frac{d\Sigma_{\text{res}}^{+-}}{d\Omega}(q) + S_{\text{H}}(q) R_{\text{H}}^{+-}(q,H_{\text{i}}) + S_{\text{M}}(q) R_{\text{M}}^{+-}(q,H_{\text{i}}). \tag{4.31}$$

Figure 4.34 features the application of eqn (4.31) to the analysis of the field-dependent spin-flip SANS cross section of nanocrystalline Co.

4.2.4 Polarized neutrons (POLARIS) ($k_0 \parallel H_0$)

The first thing to be noted is that for the parallel scattering geometry the chiral function vanishes, i.e., $\chi(\mathbf{q}) = 0$. Secondly, neglecting nuclear-spin incoherent scattering (as is done throughout this book), it is seen that the two spin-flip SANS cross sections $d\Sigma^{+-}/d\Omega = d\Sigma^{-+}/d\Omega$ [eqn (2.102)] are, in the approach-to-saturation regime, identical to the corresponding spin-misalignment SANS cross section $d\Sigma_{SM}/d\Omega$ for unpolarized neutrons [eqn (4.21)]. The latter is obtained by subtracting from eqn (2.94) the nuclear and magnetic scattering at saturation. With respect to experiment, this then suggests that the indicated subtraction procedure of unpolarized data may approximately yield the spin-flip SANS cross section, provided, of course, that the nuclear SANS is field independent. The advantages of such an approach are obvious: complicated and time-consuming polarization-analysis experiments and related spin-leakage corrections may be avoided (see the discussion in Section 6.5.1). The two non-spin-flip SANS cross sections $d\Sigma^{\pm\pm}/d\Omega$ [eqn (2.101)] depend on the nuclear coherent, on the longitudinal magnetic, and on the nuclear-magnetic interference terms. Near saturation, no micromagnetic computation is required for this case.

4.2.5 Third-order magnetic SANS effect

The micromagnetic SANS theory which is presented in Sections 3.2 and 3.3 describes heterogeneous multiphase magnets with inhomogeneous saturation magnetization and magnetic anisotropy. It is based on a first-order [in the inhomogeneities amplitude $I_m(\mathbf{r})$] expansion of the magnetization around the saturated state [414, 444], which has its origin in the theory of the approach to magnetic saturation [335, 345]. The inhomogeneity function $I_m(\mathbf{r}) \ll 1$ describes the local variation of $M_s(\mathbf{r})$ around its mean value (compare eqn (3.47) and Fig. 3.3). We emphasize that in addition to spatial inhomogeneities in the saturation magnetization there can be similar inhomogeneities present in the exchange stiffness and in the anisotropy constants. In order to investigate its range of validity, the first-order magnetic SANS theory [414, 444] has been extended to higher orders in [439]. The experimental and numerical (micromagnetic) verification can be found in [440]. Here, we only sketch the main ideas and we quote the most important relations. A more detailed exposition is beyond the scope of this book.

In general, for a weakly inhomogeneous and anisotropic magnet, there will be a small deviation from uniformity, which—using the magnitude of $I_m(\mathbf{r})$ as a small parameter—can be represented via a Taylor-series expansion. In Fourier space, this can be written as [439]:

$$\widetilde{\mathbf{M}} = \{0, 0, M_s\}\delta(\mathbf{q}) + \widetilde{\mathbf{M}}^{(1)} + \widetilde{\mathbf{M}}^{(2)} + \dots, \tag{4.32}$$

where the first term on the right-hand side of eqn (4.32) corresponds to the saturated state, and $\widetilde{\mathbf{M}}^{(i)}$ contains the terms of the order i in $\widetilde{I}_m(\mathbf{q})$, which is the Fourier transform of $I_m(\mathbf{r})$. We note that since the magnetization Fourier components start with the first order in $\widetilde{I}_m$, the lowest-order terms in the magnetic SANS cross section will be of the second order [compare eqn (3.50)].

Higher-order contributions to physical properties are usually very small, and if lower-order effects are present at the same time they may be completely masked by them. On the other hand, analysis of the higher-order effects may allow one to extract independently additional information about the system, which the lower-order effects do not provide. Therefore, it is desirable to establish the experimental conditions when the lower-order effects are canceled out and only the higher-order terms contribute, thus, enabling their analysis. As shown in [439], for $\mathbf{k}_0 \perp \mathbf{H}_0$ and unpolarized neutrons, the following combination of SANS cross-section values,

$$\Delta\Sigma_{\rm SM} = \left.\frac{d\Sigma_{\rm SM}}{d\Omega}\right|_{\theta=0} - 2\left.\frac{d\Sigma_{\rm SM}}{d\Omega}\right|_{\theta=\pi/2}, \tag{4.33}$$

is exactly zero in second order. This can be easily verified by inserting eqns (4.5)−(4.9) into eqn (4.33). The cancellation of the second-order terms in $\Delta\Sigma_{\rm SM}$ (first order in the magnetization) is a universal property of the SANS cross section, which is independent of the specific model used. In the next significant order, which is the third one, $\Delta\Sigma_{\rm SM}$ is nonzero and takes on an especially simple form:**

$$\Delta\Sigma_{\rm SM} = 32\pi^3 V b_{\rm H}^2 \left.\langle F_{\rm Z}\widetilde{I}_{\rm m}\rangle\right|_{\substack{q_x=0 \\ q_z=0}}, \tag{4.34}$$

where $q = q_y$, the angular brackets denote a triple (configurational, directional, and anisotropy direction) average, and the function $F_{\rm Z}$ is given by the following sum of convolutions (denoted by $\otimes$) [439, 440]

$$F_{\rm Z} = \frac{\widetilde{M}_x^{(1)} \otimes \widetilde{M}_x^{(1)} + \widetilde{M}_y^{(1)} \otimes \widetilde{M}_y^{(1)}}{2}. \tag{4.35}$$

The $\widetilde{M}_{x,y}^{(1)}$ correspond to the first-order solutions, eqns (3.68) and (3.69) with $l_{\rm D} = 0$. Each of the $\widetilde{M}_x^{(1)}$ and $\widetilde{M}_y^{(1)}$ is proportional to $\widetilde{I}_{\rm m}$ in the first order, so that their product multiplied by $\widetilde{I}_{\rm m}$ is of the third order in $\widetilde{I}_{\rm m}$ [compare eqns (4.34) and (4.35)]. For spherical Gaussian defects, eqn (4.34) can be evaluated to yield [439]:

$$\Delta\Sigma_{\rm SM} = 32\pi^3 b_{\rm H}^2 M_{\rm s}^2 \rho_{\rm d} v_{\rm d}^2 \langle a_{\rm n}^3\rangle \left[\kappa_{\rm A}^2 g_{\rm A}(\mathbf{m}, h_{\rm i}, \lambda_{\rm s}) + g_{\rm MS}(\mathbf{m}, h_{\rm i}, \lambda_{\rm s})\right], \tag{4.36}$$

where $\rho_{\rm d}$ is the defect density, $v_{\rm d}$ is the volume of a single inhomogeneity with size s and random amplitude $a_{\rm n}$, and $\kappa_{\rm A}$ is a small number (of the order of unity) which describes the anisotropy-field inhomogeneity. The dimensionless functions $g_{\rm A}$ and $g_{\rm MS}$ are, respectively, related to spatial fluctuations of the magnetic anisotropy (A) and of the saturation magnetization (MS). They depend on the reduced scattering vector $\mathbf{m} = qs$, the reduced magnetic field $h_{\rm i} = H_{\rm i}/M_{\rm s}$, and on the dimensionless parameter $\lambda_{\rm s} = l_{\rm M}/s$, where $l_{\rm M} \sim 3{-}10\,{\rm nm}$ [134] is the magnetostatic exchange length [eqn (3.43)].

The full expressions for $g_{\rm A}$ and $g_{\rm MS}$ are too complex to be presented here (see the Mathematica file in the Supplemental Material to [439]). The functions $g_{\rm A}$ and $g_{\rm MS}$

**Note that in [439], and unlike in the present book [compare eqns (3.48) and (3.49)], the definition of the forward Fourier transform involves a prefactor of $1/V$, so that the Fourier transform has the same dimension as the transformed quantity. Therefore, in eqns (4.34) and (4.35), $I_{\rm m}$ and $\widetilde{I}_{\rm m}$ are dimensionless, and $\mathbf{M}$ and $\widetilde{\mathbf{M}}$ are in A/m, so that $\Delta\Sigma_{\rm SM}$ has the dimension of 1/length.

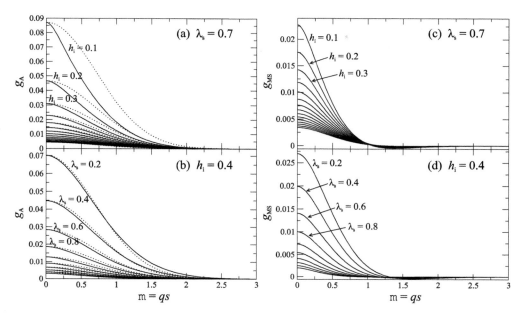

Fig. 4.5: [(a) and (b)] The functions $g_A(m, h_i, \lambda_s)$ (solid lines) and their approximation by decaying exponentials (dotted lines) for different values of h_i at fixed λ_s (a) and for different values of λ_s at fixed h_i (b). [(c) and (d)] The functions $g_{MS}(m, h_i, \lambda_s)$. The curves correspond to the same values of parameters as those in (a) and (b). After [439].

are plotted as solid lines in Fig. 4.5. At large fields, g_A can be approximated by a Gaussian function [dotted lines in Fig. 4.5(a) and (b)] [439],

$$g_A(h_i \gg 1) = \frac{\exp\left(-\frac{3}{4}m^2\right)}{30\sqrt{2}h_i^2}. \tag{4.37}$$

The dependence of the third-order perpendicular magnetic difference SANS cross section, eqn (4.36), on m for the considered Gaussian defect model is mostly a featureless decaying exponential. Only for small values of h_i does this dependence become sharper at small values of m. In the case of a very small amplitude of the anisotropy inhomogeneities κ_A, such that the cross section is dominated by the function g_{MS}, it is possible to have negative values of $\Delta\Sigma_{SM}$ for $m \cong 1.5$ [compare Fig. 4.5(c) and (d)]. The third-order magnetic SANS effect is certainly a weak contribution, which supports the validity of the second-order SANS theory (first order in **M**).

Figure 4.6 features experimental data on the two-phase alloy NANOPERM as well as the results of micromagnetic simulations on nanoporous iron [440]. Both data sets in Fig. 4.6(a) and (d) exhibit a monotonous decay of $\Delta\Sigma_{SM}(q)$ and seem to support the existence of the third-order magnetic SANS effect. In fact, the nonzero values of $\Delta\Sigma_{SM}$ prove the existence of the effect, since it cannot be described by the second-order SANS theory. In [440], a defect size of $s \cong 19.5\,\mathrm{nm}$ has been estimated by fitting the field-dependent $\Delta\Sigma_{SM}$ data to the theory. This value agrees very well with the

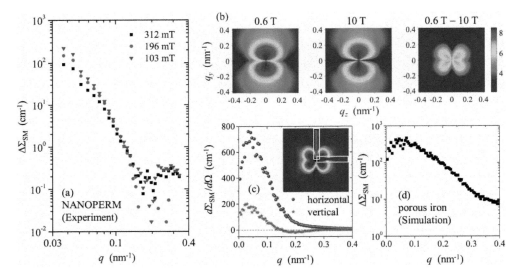

Fig. 4.6: (a) Experimental third-order magnetic SANS cross sections $\Delta\Sigma_{SM}(q)$ of NANOPERM [$(Fe_{0.985}Co_{0.015})_{90}Zr_7B_3$] at selected applied magnetic fields (see inset) (log-log scale). The $\Delta\Sigma_{SM}$ have been computed according to eqn (4.33) ($\pm 10°$ sector averages). (b)–(d) Results of micromagnetic simulations on nanoporous iron. (b) Illustration of the subtraction procedure between the magnetic SANS cross sections at 0.6 T and at 10.0 T to obtain $d\Sigma_{SM}/d\Omega$ (logarithmic color scale). (c) Azimuthally averaged $d\Sigma_{SM}/d\Omega$ data along the horizontal and vertical directions (see inset). (d) $\Delta\Sigma_{SM}(q)$ computed according to eqn (4.33) (log-linear scale). After [440].

particle size of $15 \pm 2\,nm$ of the NANOPERM alloy determined by the analysis of x-ray diffraction data.

4.3 Basic features of magnetic SANS

4.3.1 Angular anisotropy

Although the averages of the magnetization Fourier components for $\mathbf{k}_0 \parallel \mathbf{H}_0$ are highly anisotropic (see Fig. 3.7), the ensuing magnetic SANS cross section, which is obtained by summing up these Fourier components weighted by the trigonometric functions according to eqn (4.21), is isotropic (θ-independent) at all fields and momentum transfers. Therefore, we discuss in the following only the unpolarized and spin-polarized SANS cross sections for the perpendicular scattering geometry.

The spin-misalignment SANS cross section $d\Sigma_{SM}/d\Omega$ is shown in Fig. 4.7 for $\mathbf{k}_0 \perp \mathbf{H}_0$ and for $l_D = 0$. The upper row displays $d\Sigma_{SM}/d\Omega = S_H R_H + S_M R_M$ [eqn (4.5)], which considers both anisotropy-field and saturation-magnetization fluctuations, while the lower row shows $d\Sigma_{SM}/d\Omega = S_H R_H$ [eqn (4.17)], which only takes anisotropy-field variations into account and is appropriate for single-phase ferromagnets with negligible $\widetilde{M}_z$-fluctuations. Again, the emphasis in these graphs is on the change of the angular anisotropy with field, and not on the magnitude of $d\Sigma_{SM}/d\Omega$. At the largest fields, $d\Sigma_{SM}/d\Omega$ for the two-phase case [Fig. 4.7(c) and (d)] exhibits maxima roughly along the diagonals of the detector—the so-called clover-leaf anisotropy—which was observed in the Fe-based two-phase alloy NANOPERM (see Fig. 4.9) [446]. This is in agreement with the observation that at large applied fields or large momentum transfers (when $p \ll 1$), $d\Sigma_{SM}/d\Omega$ is dominated by the term $S_M R_M$ (compare Fig. 4.3). The position of the maxima in $d\Sigma_{SM}/d\Omega$ depends on q and H_i [58]. The clover-leaf pattern is absent in the $d\Sigma_{SM}/d\Omega$ for the single-phase case [Fig. 4.7(g) and (h)]. Reducing the field [Fig. 4.7(a), (b), (e), (f)], one observes in both $d\Sigma_{SM}/d\Omega$ an elongation of the spin-misalignment scattering along the field direction, with a "flying-saucer-type" pattern taking over at small q and H_i. The sharp spike in Fig. 4.7(a) and (e) is due to the magnetostatic interaction, more specifically, it is related to the term $p \sin^2 \theta$ in the denominator of eqn (4.8). This angular anisotropy was first predicted by Weissmüller et al. [456] and experimentally observed [487] on a Nd–Fe–B-based permanent magnet (see Fig. 4.8). The pole-avoidance principle, eqn (3.28), in combination with the expression for the $\mathbf{q} \neq 0$ Fourier component of the magnetostatic field, eqn (3.54), suggests that such Fourier modes of the magnetization are energetically preferred when they have $\widetilde{\mathbf{M}} \perp \mathbf{q}$. In the $\mathbf{k}_0 \perp \mathbf{H}_0$ scattering geometry, the condition $\widetilde{M}_x \mathbf{e}_x \perp \mathbf{q}$ is fulfilled for any orientation of $\mathbf{q}$ in the detector plane, giving rise to the circular part of the scattering pattern, while $\widetilde{M}_y \mathbf{e}_y$ is normal to $\mathbf{q}$ only for the special angles of $\theta = 0°$ and $\theta = 180°$, in this way giving rise to the sharp spike along the horizontal direction. It is of interest to find out the real-space spin structure behind the spike anisotropy.

For the case of a soft magnetic two-phase nanocomposite, Fig. 4.9 provides a qualitative comparison between experiment, the analytical SANS theory, and numerical micromagnetic simulations for the field dependence of $d\Sigma_{SM}/d\Omega$ [488]. The figure demonstrates that the experimental anisotropy (θ-dependence) of $d\Sigma_{SM}/d\Omega$ (upper row in Fig. 4.9) can be well reproduced by the theory [eqn (4.5)]. At the largest field of 163 mT, one observes the clover-leaf-shaped angular anisotropy with maxima in $d\Sigma_{SM}/d\Omega$ approximately along the diagonals of the detector. This feature is related to the magnetostatic term $S_M R_M$ in $d\Sigma_{SM}/d\Omega$ [compare eqn (4.9)]. Reducing the field to 45 mT results in the emergence of a scattering pattern that is more of a $\cos^2 \theta$-type, with maxima along the horizontal field direction, as described by the term $S_H R_H$ in $d\Sigma_{SM}/d\Omega$ [compare eqn (4.8)]. The observed transition in the experimental data—from anisotropy-field-dominated scattering at low field to magnetostatically dominated SANS at large field—is qualitatively reproduced by the analytical micromagnetic theory (middle row) and by the results of full-scale three-dimensional micromagnetic simulations for $d\Sigma_{SM}/d\Omega$ (lower row). For further details on the micromagnetic simulation methodology, we refer the reader to [58, 474–476] and to Chapter 7.

The preceding discussion of $d\Sigma_{SM}/d\Omega$ is based on the approximation underlying eqn (4.4). When the largest available field in an experiment is not sufficient to saturate the sample, then the subtraction procedure (low field minus high field) may result in a significant contribution $\propto (\Delta \widetilde{M}_z^2) \sin^2 \theta$. As a consequence, the overall angular anisotropy of the difference scattering pattern may then be of the $\sin^2 \theta$-type (elongated perpendicular to the field), although spin-misalignment contributions are present, but are masked by the longitudinal scattering [489].

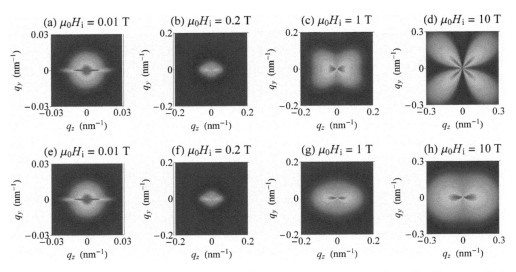

Fig. 4.7: (a)−(d) Contour plots of $d\Sigma_{SM}/d\Omega = S_H R_H + S_M R_M$ [eqn (4.5)] at applied magnetic fields as indicated ($\mathbf{k}_0 \perp \mathbf{H}_0$; $\mathbf{H}_0$ is horizontal; $l_D = 0$). (e)−(h) Field dependence of $d\Sigma_{SM}/d\Omega = S_H R_H$ for a homogeneous single-phase ferromagnet [eqn (4.17)]. Field values and materials parameters in (e)−(h) are the same as in (a)−(d). Red-yellow color corresponds to high and blue color to low values of $d\Sigma_{SM}/d\Omega$. Note the changing q-scale. After [444].

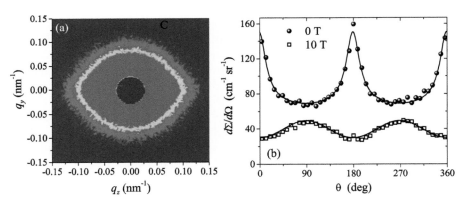

Fig. 4.8: Spike anisotropy observed on a sintered Nd–Fe–B-based permanent magnet. Shown is the total unpolarized $d\Sigma/d\Omega$ ($\mathbf{k}_0 \perp \mathbf{H}_0$). (a) Two-dimensional $d\Sigma/d\Omega$ in the remanent state. (b) $d\Sigma/d\Omega$ versus azimuthal angle θ at $q = 0.10 \pm 0.02\,\mathrm{nm}^{-1}$ and for applied fields of $0\,\mathrm{T}$ and $10\,\mathrm{T}$ (see inset). After [487].

4.3.2 Effect of magnetodipolar interaction

The micromagnetic analysis demonstrates that the anisotropic and long-range magnetodipolar interaction plays a decisive role for the understanding of magnetic SANS. In fact, it is this interaction which is responsible for the angular anisotropy, i.e.,

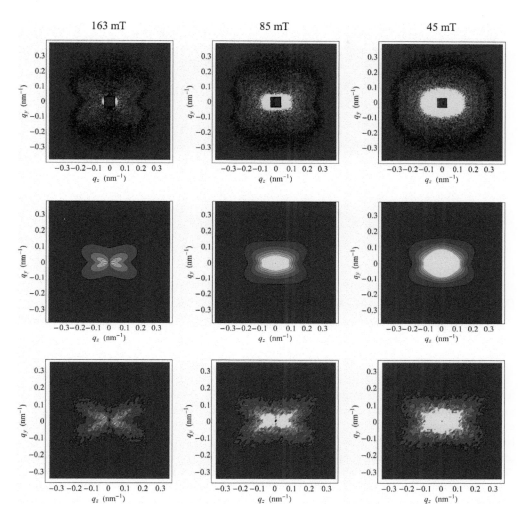

Fig. 4.9: Qualitative comparison between experiment, analytical theory, and numerical micromagnetic simulations for the case of the soft magnetic two-phase alloy $Fe_{89}Zr_7B_3Cu$ (NANOPERM) [445]. Upper row: experimental spin-misalignment SANS cross sections $d\Sigma_{SM}/d\Omega$ in the plane of the two-dimensional detector at selected applied magnetic fields (see insets). The $d\Sigma_{SM}/d\Omega$ were obtained by subtracting the scattering at a saturating field of 1994 mT. $\mathbf{H}_0$ is horizontal. Middle row: prediction by the analytical micromagnetic theory for $d\Sigma_{SM}/d\Omega$ [eqn (4.5)] at the same field values as above. For $\widetilde{H}_p^2(qR)$ and $\widetilde{M}_z^2(qR)$, we used the form factor of the sphere with a radius of $R = 6$ nm [eqns (3.82)−(3.84)]. Lower row: results of full-scale three-dimensional micromagnetic simulations for $d\Sigma_{SM}/d\Omega$ [58, 474, 475]. Yellow color corresponds to high values and blue color to low values of $d\Sigma_{SM}/d\Omega$. After [488].

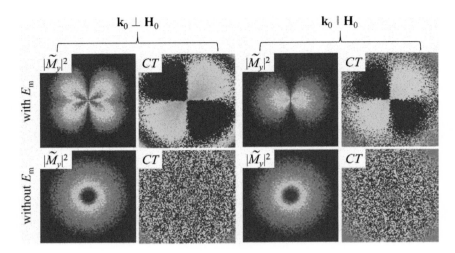

Fig. 4.10: Influence of the dipolar interaction energy E_m on the Fourier components of the magnetization. Shown are $|\widetilde{M}_y|^2$ and both cross terms $CT = -(\widetilde{M_y}\widetilde{M}_z^* + \widetilde{M}_y^*\widetilde{M}_z)$ ($\mathbf{k}_0 \perp \mathbf{H}_0$) and $CT = -(\widetilde{M_x}\widetilde{M}_y^* + \widetilde{M}_x^*\widetilde{M}_y)$ ($\mathbf{k}_0 \parallel \mathbf{H}_0$). The Fourier components were computed from a real-space magnetic microstructure with a normalized magnetization of 99.0%. Applied magnetic fields of 290 mT (with dipolar interaction) and 7 mT (without dipolar interaction) were required in order to achieve this magnetization value. Pixels in the corners of the images have $q \cong 0.9\,\mathrm{nm}^{-1}$. After [58].

for the θ-dependence of the magnetization Fourier components (see Figs. 3.6 and 3.7) and, hence, of $d\Sigma_{\mathrm{SM}}/d\Omega$. The impact of the dipolar interaction energy on $d\Sigma_{\mathrm{SM}}/d\Omega$ can be conveniently studied in micromagnetic computations, since it is straightforward to "switch on" and "off" this energy term. Figure 4.10 shows for both scattering geometries the results of micromagnetic simulations for $|\widetilde{M}_y|^2$ and for both cross terms, obtained with and without the dipolar interaction [58]. When the dipolar interaction is ignored in the micromagnetic computations, all Fourier components are isotropic at all q and H_i investigated. This clearly shows that without the magnetostatic energy the experimentally observed clover-leaf anisotropy pattern, which exhibits maxima along the diagonals of the detector (see, e.g., upper left image in Fig. 4.9), cannot be explained. Similarly, the spike pattern displayed in Fig. 4.8 has its origin in the dipole-dipole interaction.

To be more quantitative, we compare in the following for $\mathbf{k}_0 \perp \mathbf{H}_0$ the azimuthally averaged transversal Fourier component

$$\overline{|\widetilde{M}_y|^2}(q) = (2\pi)^{-1} \int\limits_0^{2\pi} |\widetilde{M}_y|^2(q,\theta)d\theta \qquad (4.38)$$

with and without the dipolar interaction. Using eqn (3.77) with $l_{\mathrm{D}} = 0$ to describe a material with both $\widetilde{H}_{\mathrm{p}}$ and $\widetilde{M}_z$-fluctuations, and performing the indicated integration in eqn (4.38), we find:

$$\overline{|\widetilde{M}_y|^2}(q, H_i) = \frac{\widetilde{H}_p^2 p^2 (2+p) + 2\widetilde{M}_z^2 (1+p) (2+p-2\sqrt{1+p})}{4(1+p)^{3/2}} \quad (4.39)$$

with dipolar interaction, and

$$\overline{|\widetilde{M}_y|^2}(q, H_i) = \frac{p^2}{2} \widetilde{H}_p^2 \quad (4.40)$$

without this term. The functions $\widetilde{H}_p^2$ and $\widetilde{M}_z^2$ are generally depending on the magnitude of q. Figure 4.11 displays the ratio R_{dip} of eqns (4.39) and (4.40),

$$R_{\mathrm{dip}}(p, \iota) = \frac{p^2 (2+p) + 2\iota^{-1} (1+p) (2+p-2\sqrt{1+p})}{2p^2 (1+p)^{3/2}}, \quad (4.41)$$

as a function of p for several values of the parameter $\iota = \widetilde{H}_p^2 / \widetilde{M}_z^2$, which is here assumed to be q-independent. We remind that the function $p(q, H_i)$ is smaller than unity at large q or at large internal magnetic field H_i, and that p is larger than unity at small q and small H_i (compare Fig. 3.4). Therefore, depending on the q-range and on the value of H_i, as probed by the parameter p, we see that R_{dip} may exhibit values that are larger or smaller than unity. Obviously, the relative contribution of anisotropy-field-related ($\widetilde{H}_p^2$) and magnetostatic ($\widetilde{M}_z^2$) perturbations plays an important role for the dipolar ratio. When $\iota \gg 1$, then $\overline{|\widetilde{M}_y|^2}$ is larger over most of the (q, H_i)-space when the dipolar term is ignored. This can be seen e.g., from eqn (3.86), where the term $p \sin^2 \theta$ in the denominator of $|\widetilde{M}_y|^2$ is due to the magnetostatic interaction; to be more precise, it is due to the inclusion of the $\mathbf{q} \neq 0$ Fourier component of the magnetostatic field $\widetilde{\mathbf{H}}_d^b(\mathbf{q})$ [eqn (3.54)]. Neglecting this term increases the amplitude of $|\widetilde{M}_y|^2$. On the other hand, for $\iota \lesssim 1$, we see that increasing M_s-fluctuations result in an R_{dip} which is larger than unity, most notably at small p. For large real-space length scales (small q) and small applied fields (large p), the parameter R_{dip} is generally smaller than unity. The observations in Figs. 4.10 and 4.11 suggest that for a correct description of magnetic SANS, the magnetostatic interaction has to be taken into account. This is further highlighted by micromagnetic simulations of magnetic SANS, which are extensively discussed in Chapter 7.

When spatial variations in the saturation magnetization can be neglected, e.g., for a single-phase material, then it becomes possible to derive a closed-form expression for the mean-square magnetostatic stray field $\langle |\mathbf{H}_d^b|^2 \rangle$ [compare eqn (3.51)] due to nonzero magnetic volume charges $\nabla \cdot \mathbf{M}$ [490]. Based on the expressions for the magnetization Fourier components in the high-field limit (see Section 3.3.2), $\langle |\mathbf{H}_d^b|^2 \rangle$ can be expressed as an integral over the anisotropy-field scattering function $S_H(q)$:

$$\langle |\mathbf{H}_d^b|^2 \rangle = \frac{1}{V} \int |\mathbf{H}_d^b(\mathbf{r})|^2 \, d^3r \quad (4.42)$$

$$= \frac{1}{8\pi^2 b_H^2} \int_0^\infty S_H(q) q^2 p^2 \left(\frac{(1+2p)\ln(\sqrt{p}+\sqrt{1+p}) - \sqrt{p+p^2}}{(p+p^2)^{3/2}} \right) dq.$$

Using experimental data for $S_H(q)$ (see, e.g., Figs. 4.22 and 4.30 or Fig. 2 in [490]), eqn (4.42) can be numerically evaluated. Figure 4.12 depicts the results for the field

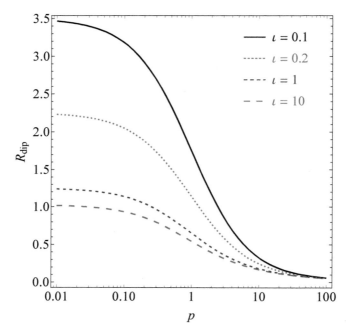

Fig. 4.11: Dipolar ratio R_{dip} [eqn (4.41)] as a function of $p(q, H_i)$ for several values of $\iota = \widetilde{H}_p^2 / \widetilde{M}_z^2$ (see inset) (log-linear scale) ($\mathbf{k}_0 \perp \mathbf{H}_0$) ($l_D = 0$).

dependence of $\langle |\mathbf{H}_d^b|^2 \rangle$ for nanocrystalline Ni and Co, and for a cold-worked microcrystalline Ni specimen. For all samples investigated, the values of $\langle |\mathbf{H}_d^b|^2 \rangle$ decrease when the applied field is increased. This can be explained by the fact that the contribution of the volume charges $\nabla \cdot \mathbf{M}$ to the total magnetostatic field is expected to decrease for an increasing degree of uniformity of the magnetic microstructure. The magnitude of $\langle |\mathbf{H}_d^b|^2 \rangle$ is largest in the nanocrystalline samples, taking on values in the range between $10{-}50\,\mathrm{mT}$, and smallest in microcrystalline Ni, for which it lies between $5{-}20\,\mathrm{mT}$. Therefore, one may conclude that the reduction of relevant microstructural length scales down to the nanometer range (here: the average grain size) is accompanied by a concomitant enhancement of the strength of the effect of magnetic volume charges. We refer to [490] for a discussion and comparison of the temperature variation of $\langle |\mathbf{H}_d^b|^2 \rangle$.

4.3.3 Effect of Dzyaloshinskii–Moriya interaction

For bulk magnetic materials containing many lattice imperfections (e.g., nanocrystalline or mechanically deformed magnets), the relativistic Dzyaloshinskii–Moriya interaction (DMI) may result in nonuniform spin textures due to the lack of inversion symmetry in the near vicinity of the defect cores (e.g., at interface regions). This idea was already put forward by Arrott in 1963 [420]. Within the framework of the continuum theory of micromagnetics, the impact of the DMI on the elastic magnetic SANS cross section has been explored [414, 491]. It gives rise to a polarization-dependent

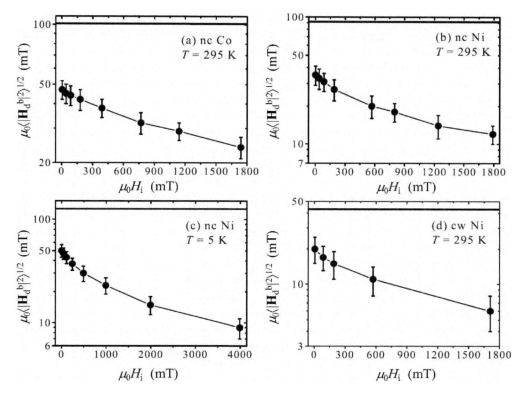

Fig. 4.12: Lower bounds for the root-mean-square magnetostatic stray field $\langle|\mathbf{H}_d^b|^2\rangle^{1/2}$ as a function of the internal magnetic field H_i. $\langle|\mathbf{H}_d^b|^2\rangle$ was determined by numerical integration of eqn (4.42) using experimental data for $S_H(q)$. (a) Nanocrystalline (nc) Co at 295 K; (b) nc Ni at 295 K; (c) nc Ni at 5 K; (d) coarse-grained and cold-worked (cw) polycrystalline Ni at 295 K. Lines are a guide to the eyes. The horizontal line in each subfigure indicates the respective value of the root-mean-square anisotropy field $\langle|\mathbf{H}_p|^2\rangle^{1/2}$, computed using eqn (4.63). After [490].

asymmetric term in the SANS cross section and the analysis of this feature provides a means to determine the DMI constant. The effect of interfacial DMI in multilayer systems on the reflection of polarized neutrons has been studied in [492].

As Fig. 4.13 demonstrates, the DMI has also an effect on the unpolarized SANS cross section. Shown is the (over 2π) azimuthally averaged $d\Sigma_{SM}/d\Omega$ at $\mu_0 H_i = 0.8$ T with and without the DMI term. Since the DMI introduces nonuniformity into the spin structure, the spin-misalignment scattering cross section is larger when this term is included. However, since there are many other magnetic interactions (besides the DMI) contributing to the unpolarized SANS cross section, it might be difficult to unambiguously identify the effect of the DMI on unpolarized SANS data, in particular, in view of the plethora of possible angular anisotropies (compare Figs. 4.7−4.9).

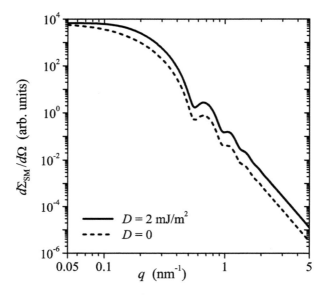

Fig. 4.13: Azimuthally averaged $d\Sigma_{\mathrm{SM}}/d\Omega$ at $\mu_0 H_{\mathrm{i}} = 0.8\,\mathrm{T}$ with and without the DMI term (see inset) (log–log scale) ($\mathbf{k}_0 \perp \mathbf{H}_0$). Both $d\Sigma_{\mathrm{SM}}/d\Omega$ have been convoluted with the same Gaussian distribution function for both $\widetilde{H}_{\mathrm{p}}^2$ and $\widetilde{M}_z^2$ (for which the sphere form factor was assumed). After [414].

Polarized neutrons provide a means to access the effect of the DMI. The chiral function [eqn (2.97)], which characterizes the defect-induced symmetry breaking, can be obtained from (i) the difference measurement of the spin-flip SANS cross sections [eqns (2.100)], according to ($\mathbf{k}_0 \perp \mathbf{H}_0$)

$$\frac{d\Sigma^{+-}}{d\Omega} - \frac{d\Sigma^{-+}}{d\Omega} = \frac{8\pi^3}{V} b_{\mathrm{H}}^2 \left[-2i\chi(\mathbf{q}) \right], \tag{4.43}$$

or, more easily from the experimental point of view, (ii) from the difference $\Delta\Sigma$ between flipper-on ($-$) and flipper-off ($+$) SANS cross sections [eqns (2.95)]:

$$\Delta\Sigma = \frac{d\Sigma^-}{d\Omega} - \frac{d\Sigma^+}{d\Omega} = \frac{8\pi^3}{V} b_{\mathrm{H}}^2 P\epsilon \left[2b_{\mathrm{H}}^{-1}(\widetilde{N}\widetilde{M}_z^* + \widetilde{N}^*\widetilde{M}_z)\sin^2\theta + 2i\chi \right], \tag{4.44}$$

where polarization-dependent scattering contributions $\propto \widetilde{N}\widetilde{M}_y \sin\theta \cos\theta$ have been neglected. Using the expressions for the Fourier components $\widetilde{M}_x$ and $\widetilde{M}_y$ [eqns (3.70) and (3.71)], one obtains for the chiral function [eqn (2.97)] [414]:

$$2i\chi(\mathbf{q}) = -\frac{2\widetilde{H}_{\mathrm{p}}^2 p^3 \left(2 + p\sin^2\theta\right) l_{\mathrm{D}} q \cos^3\theta + 4\widetilde{M}_z^2 p(1+p)^2 l_{\mathrm{D}} q \sin^2\theta \cos\theta}{\left(1 + p\sin^2\theta - p^2 l_{\mathrm{D}}^2 q^2 \cos^2\theta\right)^2}. \tag{4.45}$$

Equation (4.45) has been derived using the expression for the intrinsic DMI energy of cubic systems [compare eqn (3.42)]. In the absence of a more appropriate expression for

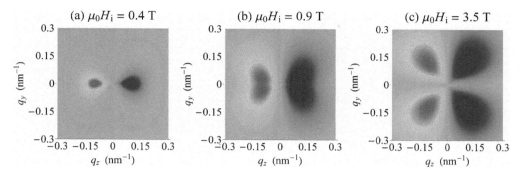

Fig. 4.14: Contour plots of the chiral function $2i\chi(\mathbf{q})$ [eqn (4.45)] at selected applied magnetic fields (see insets) ($\mathbf{k}_0 \perp \mathbf{H}_0$) [414]. For the graphical representation, the sphere form factor for both $\widetilde{H}_p^2(q\xi_H)$ and for $\widetilde{M}_z^2(q\xi_M)$ [eqns (3.82)–(3.84)] was used with $R = \xi_H = \xi_M = 8\,\text{nm}$ and $H_p/\Delta M = 1$. Red color corresponds to positive and blue color to negative values of $2i\chi(\mathbf{q})$. After [414].

defect-related DMI, this seems to be suitable in order to capture the overall effect, i.e., to reproduce the experimental asymmetry. As discussed in Section 3.1.2, the expression for dislocation-induced DMI derived by Butenko and Rößler [eqn (3.9)] may be used as a starting point for the study of the effect of defect-related DMI on the magnetic SANS cross section [426]. Further theoretical work in this direction is required.

Equation (4.45) is graphically displayed in Fig. 4.14. At small fields, two extrema parallel and antiparallel to the field axis are observed [Fig. 4.14(a)], whereas at larger fields additional maxima and minima appear approximately along the detector diagonals [Fig. 4.14(b) and (c)]. This change in anisotropy is due to the different field dependencies of the terms proportional to $\widetilde{H}_p^2$ and $\widetilde{M}_z^2$ in the enumerator of eqn (4.45). Note that $2i\chi(\mathbf{q})$ describes an asymmetry arising in the elastic SANS cross section due to the effect of the DMI on the static magnetic microstructure. It is very much different than the so-called left-right asymmetry [247, 251, 263], which is related to inelastic features in the spin system. We refer to [251, 493] for studies which address the inelastic and critical scattering related to the DMI term. Moreover, the DMI also shows up as an asymmetry in spin-wave and ferromagnetic resonance spectroscopy measurements (e.g., [258, 401–403, 415–419]).

Evaluating eqn (4.45) along the horizontal direction ($\theta = 0°$ and $\theta = 180°$) and assuming that $\widetilde{H}_p^2$ depends only on the magnitude of $\mathbf{q}$, we have (independent of $\widetilde{M}_z^2$)

$$2i\chi(q, H_i) = \mp \frac{4\widetilde{H}_p^2 p^3 l_D q}{\left(1 - p^2 l_D^2 q^2\right)^2}, \tag{4.46}$$

which can be used to analyze experimental data. Note that $\chi(\mathbf{q}) = 0$ for $\theta = 90°$, which allows its clear separation from the nuclear-magnetic interference term $\propto \widetilde{N}\widetilde{M}_z \sin^2\theta$ (provided that both $\widetilde{N}$ and $\widetilde{M}_z$ are isotropic). In [491], the anisotropy-field Fourier component was described by a squared Lorentzian:

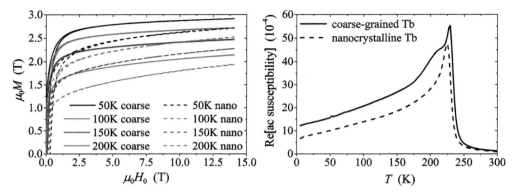

Fig. 4.15: (a) Magnetization curves of nanocrystalline ($D_{gs} = 40$ nm) and coarse-grained ($D_{gs} > 100$ nm) Tb at several temperatures (see inset). (b) Temperature dependence of the ac magnetic susceptibility of nanocrystalline and coarse-grained Tb. After [491].

$$\widetilde{H}_{\mathrm{p}}^2(q\xi_{\mathrm{H}}) = \frac{\langle H_{\mathrm{p}}^2 \rangle}{(1 + \xi_{\mathrm{H}}^2 q^2)^2}, \tag{4.47}$$

where $\langle H_{\mathrm{p}}^2 \rangle$ is the mean-square anisotropy field, and ξ_{H} is the correlation length of the anisotropy.

The following results on the defect-induced DMI were mainly obtained on the heavy rare-earth metal terbium, which crystallizes in the hexagonal close-packed (hcp) structure. In zero applied field, Tb exhibits an antiferromagnetic helical phase between the Néel temperature ($T_{\mathrm{N}} = 230$ K) and the Curie point ($T_{\mathrm{C}} = 220$ K) [120]. Below T_{C}, the spins are ferromagnetically aligned within the basal planes of the hcp lattice. Figure 4.15 demonstrates the effect of small grain size (large defect density) on the magnetization of polycrystalline Tb. At 14 T and 50 K, the magnetization of 40 nm-sized Tb is reduced by about 6.7% relative to the coarse-grained specimen. This is a well-known effect, and for nanocrystalline Gd it could be shown that the magnetization reduction scales with the inverse grain size, i.e., with the volume fraction of grain boundaries (compare Fig. 6.21) [449]. Similarly, the ac susceptibility of the nanocrystalline Tb sample is reduced over a large part of the temperature range, and the region of the ferro-to-paramagnetic phase transition at around 220−230 K is smeared out.

Figure 4.16 depicts the results of half-polarized SANS measurements on nanocrystalline Tb at $T = 100$ K (deep within the ferromagnetic state) and at an applied magnetic field of $\mu_0 H_0 = 5$ T. Both spin-resolved data sets $d\Sigma^-/d\Omega$ [Fig. 4.16(a)] and $d\Sigma^+/d\Omega$ [Fig. 4.16(b)] are characterized by a maximum of the scattering intensity along the horizontal applied-field direction. This is the signature of spin-misalignment scattering due to the presence of transversal magnetization components [cf. the term $|\widetilde{M}_y|^2 \cos^2\theta$ in eqns (2.95)]. The difference between the two SANSPOL cross sections, $\Delta\Sigma = d\Sigma^-/d\Omega - d\Sigma^+/d\Omega$ [Fig. 4.16(c)], clearly exhibits an asymmetric contribution related to the chiral function. The asymmetry is most pronounced along the horizontal direction ($\theta = 0°$), which by comparison to the theoretical prediction [eqn (4.45)]

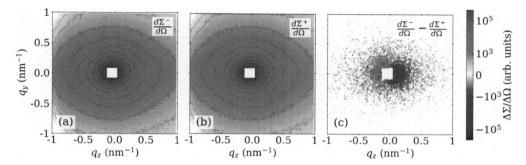

Fig. 4.16: DMI asymmetry in nanocrystalline Tb. Shown are polarized SANS results of nanocrystalline Tb at $100\,\mathrm{K}$ and $5\,\mathrm{T}$ ($\mathbf{H}_0 \parallel \mathbf{e}_z$). (a) Two-dimensional flipper-on SANS cross section $d\Sigma^-/d\Omega$; (b) flipper-off SANS cross section $d\Sigma^+/d\Omega$; (c) $\Delta\Sigma = d\Sigma^-/d\Omega - d\Sigma^+/d\Omega$. After [491].

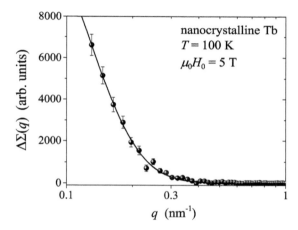

Fig. 4.17: Determination of the defect-induced DMI strength. ($\bullet$) Azimuthally averaged $\Delta\Sigma(q)$ of the data shown in Fig. 4.16(c) (log-linear scale). For the data analysis, the arithmetic mean of both horizontal $\pm 10°$ sector-averaged $\Delta\Sigma$ branches were employed (taking into account their different signs). Solid line: fit to eqn (4.46) with the $+$ sign. After [491].

can be attributed to the anisotropy-field term $\propto -\widetilde{H}_{\mathrm{p}}^2 \cos^3\theta$. Difference data taken at smaller momentum transfers additionally show the "usual" symmetric $\sin^2\theta$-type anisotropy (with maxima at $\theta = 90°$ and $\theta = 270°$), which is due to the polarization-dependent nuclear-magnetic interference term in the SANS cross section [491].

By taking an angular average of $\Delta\Sigma$ along the horizontal direction and by carrying out a weighted nonlinear least-squares fit of the resulting data to eqn (4.46) (solid line in Fig. 4.17), the following parameters are obtained: $D = 0.45 \pm 0.07\,\mathrm{mJ/m^2}$; $A = 8.2 \pm 2.0 \times 10^{-11}\,\mathrm{J/m}$; $\xi_{\mathrm{H}} = 7.6 \pm 1.0\,\mathrm{nm}$. The value for D is comparable to bulk DMI values (e.g., [402–404]), the effective A-value is slightly increased but still within the

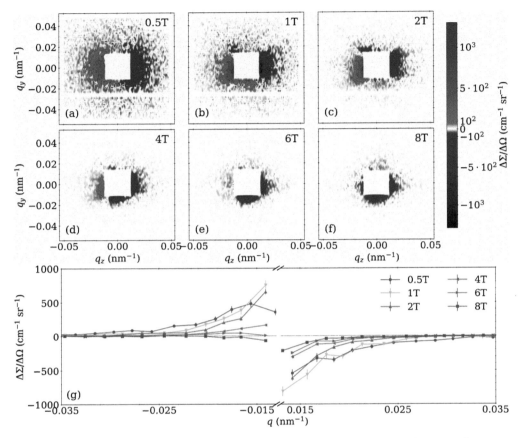

Fig. 4.18: Field dependence of DMI asymmetry in mechanically deformed Co. Displayed are polarized SANS results of cold-rolled Co at $T = 300\,\mathrm{K}$ and at a series of applied magnetic fields (see insets). $\mathbf{H}_0 \parallel \mathbf{e}_z$ and parallel to the rolling (texture) direction. (a)−(f) Two-dimensional flipper-on minus flipper-off data $\Delta\Sigma(\mathbf{q})$. (g) Azimuthally averaged $\Delta\Sigma(q)$ ($\pm 10°$ horizontal sector averages). After [491].

range of experimental data [134, 303], while the correlation length ξ_H of the anisotropy field is smaller than the average crystallite size of the Tb sample ($D_\mathrm{gs} = 40\,\mathrm{nm}$ [491]). The latter finding indicates that there is a significant spin disorder within the grains. This is in agreement with the results of an unpolarized SANS study [303], which has found that up to applied fields of several Tesla the magnetization remains "locked in" to the basal planes of the hcp crystal lattice of each individual crystallite, but that the in-plane orientation of the spins is highly nonuniform within each grain.

As a further example, magnetic-field-dependent $\Delta\Sigma$ data of a mechanically deformed Co sample are displayed in Fig. 4.18 with the magnetic field applied parallel to the rolling direction (texture axis). Inspection of the Co magnetization curves (see Supplemental Material in [491]) clearly shows that all of the displayed field values fall into the approach-to-saturation regime, where the predictions of the micromagnetic SANS theory are valid [414]. The results confirm, in agreement with the theory, that the asymmetry decreases with increasing field and eventually vanishes for fields in excess of ~ 8 T. Note that $\chi \to 0$ for $H_i \to \infty$ [compare eqn (2.97)]. As shown in the Supplemental Material of [491], polarized SANS data on nanocrystalline Ho (grain size: 33 ± 3 nm) at 1.8 K and 5 T also reveal the scattering asymmetry. At 1.8 K, Ho exhibits a nearly conical ferromagnetic helical structure with an antiferromagnetic moment component in the basal plane and a ferromagnetic component along the *c*-axis [120]. The left-right asymmetry has also been detected in the commercial soft magnetic material Vitroperm ($Fe_{73}Si_{16}B_7Nb_3Cu_1$) [494]. This two-phase nanocrystalline alloy possesses a macroscopic texture axis, which is induced by magnetic-field annealing, and the origin of the effect is presumably related to the broken symmetry at the particle-matrix interfaces. Altogether, these results underline the generic character of the DMI asymmetry for defect-rich polycrystalline magnetic materials.

In the following, we will discuss the possible mesoscale real-space spin structure which lies at the heart of the SANS data shown in Fig. 4.16. Mirebeau et al. [468] have also provided a SANS study, albeit using unpolarized neutrons, of a disordered magnetic material, the re-entrant spin glass $Ni_{0.81}Mn_{0.19}$. Supported by the results of Monte Carlo simulations of a two-dimensional lattice structure with competing ferromagnetic and antiferromagnetic exchange interactions, these authors have suggested the existence of vortex-like chiral spin structures (see Fig. 5.10). While vortices are ubiquitous in magnetism and can very well be at the origin of the underlying mesoscopic spin structure, the data in Fig. 4.16 directly show that there is a general symmetry breaking in the samples (besides the one induced by the external field) resulting in a dominant overall chirality of the spin texture. In the case of the Co sample, the reason for such a symmetry breaking is obviously related to the mechanical deformation (similar to the work by Fedorov et al. [421]). Regarding the non-deformed samples (Tb and Ho), the vorticity in the magnetization texture appears only because the samples are intrinsically inhomogeneous. The shape of the chiral structures is not an equilibrium property of the uniform magnet, but directly reflects the shape of the magnetic inhomogeneities. This is demonstrated in Fig. 4.19, which displays the numerically computed real-space spin structure around defects in the presence of DMI (see [491] for details). An asymmetry in the transversal spin configuration clearly becomes evident, which is a consequence of the DMI: setting $l_D \propto D = 0$ produces a symmetric magnetization pattern with a maximum transversal spin deviation located at the centers of the defects (due to the perturbing effect of the magnetic anisotropy field). By contrast, for $l_D \neq 0$, it is seen that the chirality in the magnetization texture manifests itself at or near the boundaries of the inhomogeneities (boundaries of the shaded areas in Fig. 4.19), precisely where one would expect the antisymmetric DMI to appear. Far from the boundaries, the magnetization is parallel to the external field and is not influenced by the DMI. Thus, the value of the DMI constant obtained from the SANS experiment should mainly reflect this emergent DMI strength.

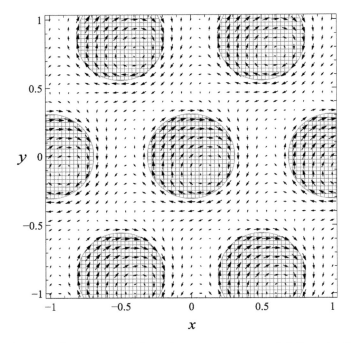

Fig. 4.19: Real-space illustration of the spin structure around defects in the presence of DMI. Shown is the transversal magnetization distribution around a periodic columnar array of circular defects, obtained from the solution of the micromagnetic problem in [414]. The defects (shaded areas) are regions in space where the local saturation magnetization is increased and where the magnetic anisotropy field is nonzero. They form a hexagonal lattice in the $x-y$ plane and are uniform in the z-direction ($\mathbf{H}_0 \parallel \mathbf{e}_z$). The exchange length $l_M = 0.2$ and the DMI length $l_D = 1$ are in units of the x-lattice period. The internal field is $H_i = M_s$, and the anisotropy field is $\widetilde{H}_{px}(\mathbf{q}) = 0.2\,\widetilde{I}_m(\mathbf{q})$ and $\widetilde{H}_{py}(\mathbf{q}) = 0.2\,\widetilde{I}_m(\mathbf{q})$. The defect strength is absorbed as a factor into the arrow length. After [491].

4.3.4 Magnetic Guinier law

In order to derive the Guinier law for magnetic SANS [495], analogous to the corresponding expression for nuclear particle scattering [eqn (2.76)], we look in the following for the low-q behavior of the spin-misalignment SANS cross section $d\Sigma_{SM}/d\Omega$, in particular for the parallel scattering geometry ($\mathbf{k}_0 \parallel \mathbf{H}_0$). The sample volume which is probed by the neutrons typically contains many defects (e.g., crystallites separated by grain boundaries), each one having a different orientation and/or magnitude of the magnetic anisotropy field. To obtain an expression for $S_H \propto |\widetilde{\mathbf{H}}_p|^2$, we make the assumption that the total anisotropy field of the sample is the sum of the anisotropy fields of the individual defects i, i.e., [456],

$$\mathbf{H}_{\mathrm{p}}(\mathbf{r}) = \sum_{i=1}^{N} \mathbf{H}_{\mathrm{p},i}(\mathbf{r}). \tag{4.48}$$

This decomposition also applies to the Fourier transform $\widetilde{\mathbf{H}}_{\mathrm{p}}(\mathbf{q})$ of $\mathbf{H}_{\mathrm{p}}(\mathbf{r})$, i.e.,

$$\widetilde{\mathbf{H}}_{\mathrm{p}}(\mathbf{q}) = \sum_{i=1}^{N} \widetilde{\mathbf{H}}_{\mathrm{p},i}(\mathbf{q}), \tag{4.49}$$

so that

$$\widetilde{\mathbf{H}}_{\mathrm{p}}^{2}(\mathbf{q}) = \sum_{i=1}^{N} \widetilde{\mathbf{H}}_{\mathrm{p},i}^{2}(\mathbf{q}) + \sum_{i \neq j}^{N} \widetilde{\mathbf{H}}_{\mathrm{p},i}(\mathbf{q}) \cdot \widetilde{\mathbf{H}}_{\mathrm{p},j}(\mathbf{q}), \tag{4.50}$$

where we have assumed that the $\widetilde{\mathbf{H}}_{\mathrm{p},i}$ are real-valued quantities. If the orientations of the $\widetilde{\mathbf{H}}_{\mathrm{p},i}$ of the individual defects are statistically uncorrelated (random anisotropy), then terms $\widetilde{\mathbf{H}}_{\mathrm{p},i} \cdot \widetilde{\mathbf{H}}_{\mathrm{p},j}$ with $i \neq j$ take on both signs with equal probability. Consequently, the sum over these terms vanishes, and

$$\widetilde{\mathbf{H}}_{\mathrm{p}}^{2} = \sum_{i=1}^{N} \widetilde{\mathbf{H}}_{\mathrm{p},i}^{2}. \tag{4.51}$$

Equation (4.51) suggests that $\widetilde{\mathbf{H}}_{\mathrm{p}}^{2}$, and hence $S_{\mathrm{H}} \propto \widetilde{\mathbf{H}}_{\mathrm{p}}^{2}$, can be computed for an arbitrary arrangement of defects once the solution for the single-defect case $\widetilde{\mathbf{H}}_{\mathrm{p},i}(\mathbf{q})$ is known [456]. This can e.g., be accomplished for an idealized nanocrystalline ferromagnet, where the crystallites have random crystallographic orientation and where the anisotropy field arises exclusively from the magnetocrystalline anisotropy of the grains. Because each grain is a single crystal, the anisotropy field in the grain is a constant vector, i.e., $\mathbf{H}_{\mathrm{p},i} \neq \mathbf{H}_{\mathrm{p},i}(\mathbf{r})$, and the anisotropy field Fourier amplitude is obtained by the following form-factor integral:

$$\widetilde{\mathbf{H}}_{\mathrm{p},i}(\mathbf{q}) = \frac{1}{(2\pi)^{3/2}} \int_{V_{\mathrm{p},i}} \mathbf{H}_{\mathrm{p},i} \exp\left(-i\mathbf{q} \cdot \mathbf{r}\right) d^{3}r = \frac{\mathbf{H}_{\mathrm{p},i}}{(2\pi)^{3/2}} \int_{V_{\mathrm{p},i}} \exp\left(-i\mathbf{q} \cdot \mathbf{r}\right) d^{3}r, \tag{4.52}$$

where the integral extends over the volume of grain i. For the example of a spherical grain shape ($\xi_{\mathrm{H},i} = R_i$ and $V_{\mathrm{p},i} = V_{\mathrm{s},i} = \frac{4\pi}{3} R_i^3$), we obtain the well-known result that

$$\widetilde{\mathbf{H}}_{\mathrm{p},i}(\mathbf{q}) = \widetilde{\mathbf{H}}_{\mathrm{p},i}(qR_i) = \frac{\mathbf{H}_{\mathrm{p},i}}{(2\pi)^{3/2}} 3V_{\mathrm{s},i} \frac{j_1(qR_i)}{qR_i}, \tag{4.53}$$

where $j_1(z)$ denotes the spherical Bessel function of the first order. Note that S_{H} is independent of the applied magnetic field in the approach-to-saturation regime (compare the discussion in Section 3.2). The square of eqn (4.53) is identical, except for the prefactor, to the nuclear SANS cross section of an array of non-interfering spherical particles, and general asymptotic results at small and large q are therefore immediately transferable; in particular, the Guinier approximation relates $S_H(q) \propto \widetilde{H}_p^2(q)$ at small scattering vectors to the radius of gyration R_{GH} of the magnetic anisotropy field, according to (compare Section 2.4.3):

$$S_H(q) \cong S_H(q=0) \exp\left(-\frac{q^2 R_{GH}^2}{3}\right). \tag{4.54}$$

By inspecting the expression for $\widetilde{M}_z^2(q)$ [compare, e.g., eqns (2.122) and (D.13)], it is straightforward to conclude that for a dilute assembly of uniform second-phase particles a Guinier expression is also valid for the scattering function $S_M(q) \propto \widetilde{M}_z^2(q)$ of the longitudinal magnetization, i.e.,

$$S_M(q) \cong S_M(q=0) \exp\left(-\frac{q^2 R_{GM}^2}{3}\right). \tag{4.55}$$

Analysis of the experimental S_H and S_M data in terms of the Guinier approximation yields the radius of gyration of, respectively, the magnetic anisotropy field and the magnetostatic interaction. Similar to nuclear SANS and SAXS, where R_G is a measure for the particle size, R_{GH} and R_{GM} deduced, respectively, from S_H and S_M may be seen as a measure for the size of regions over which the magnetic anisotropy field $\mathbf{H}_p(\mathbf{r})$ and the saturation magnetization $M_s(\mathbf{r})$ are homogeneous. In this context, we refer to the study by Burke [496], who investigated the influence of magnetic shape anisotropy on the Guinier law of fine ferromagnetic single-domain particles.

Note that information on the nuclear grain microstructure is here obtained by analysis of $S_H(q)$ and/or $S_M(q)$. However, since within the micromagnetic SANS theory of Section 4.2 the spin-misalignment scattering cross section, e.g., for $\mathbf{k}_0 \parallel \mathbf{H}_0$, is given by $d\Sigma_{SM}/d\Omega = S_H(q)R_H(q, H_i)$, it is not permissible to derive information on the nuclear microstructure by analyzing $d\Sigma_{SM}/d\Omega$ directly in terms of the Guinier or Porod approximations, and neglecting the magnetic response functions. This is, of course, not surprising, since both the Guinier and Porod law have been derived in the context of nonmagnetic particle scattering and are not a priori transferable to systems with nonuniform electron or magnetization density [290]. In fact, as we will see below, the Guinier radius that is obtained from the analysis of $d\Sigma_{SM}/d\Omega$ depends on the applied field H_i and on the magnetic interactions. Moreover, due to the q-dependence of the response functions R_H and R_M [compare eqns (4.16) and Fig. 4.2] $d\Sigma_{SM}/d\Omega$ does generally not exhibit the classical Porod q^{-4}-behavior (see Section 4.3.5).

In an attempt to derive a magnetic Guinier expression for $d\Sigma_{SM}/d\Omega$, it was shown in [456, 495] that for the parallel scattering geometry ($\mathbf{k}_0 \parallel \mathbf{H}_0$), Taylor expansion of $R_H = p^2/2$ [eqn (4.22)] around $q = 0$ yields:

$$R_H(q, H_i) \cong \frac{p_0^2}{2}\left(1 - 2l_H^2 q^2\right) \cong \frac{p_0^2}{2}\exp\left(-\frac{6q^2 l_H^2}{3}\right), \tag{4.56}$$

where $l_H(H_i)$ is the exchange length of the field [eqn (3.66)], and $p_0 = M_s/H_{eff}(q = 0) = M_s/H_i$. Inserting eqns (4.54) and (4.56) into $d\Sigma_{SM}/d\Omega = S_H R_H$, it follows that [495]:

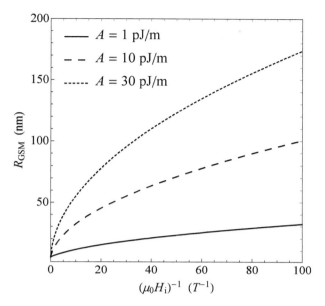

Fig. 4.20: Magnetic Guinier radius R_{GSM} [eqn (4.58)] as a function of H_i^{-1} for different A-values (see inset) ($R_{\mathrm{GH}} = 5\,\mathrm{nm}$; $\mu_0 M_s = 1.5\,\mathrm{T}$).

$$\frac{d\Sigma_{\mathrm{SM}}}{d\Omega}(q) \cong \frac{d\Sigma_{\mathrm{SM}}}{d\Omega}(q = 0)\exp\left(-\frac{q^2 R_{\mathrm{GSM}}^2}{3}\right), \qquad (4.57)$$

where

$$R_{\mathrm{GSM}}^2(H_i) = R_{\mathrm{GH}}^2 + 6l_{\mathrm{H}}^2(H_i) = R_{\mathrm{GH}}^2 + \frac{12A}{\mu_0 M_s H_i} \qquad (4.58)$$

represents the field-dependent magnetic Guinier radius of the spin-misalignment SANS, and the forward-scattering cross section obeys

$$\frac{d\Sigma_{\mathrm{SM}}}{d\Omega}(q = 0) \propto p_0^2 = \left(\frac{M_s}{H_i}\right)^2. \qquad (4.59)$$

Equations (4.57) and (4.58) provide a means to determine the exchange-stiffness constant A from field-dependent SANS measurements (see Fig. 4.20). The observation that R_{GSM} depends on R_{GH}, which is field independent, and on the micromagnetic exchange length l_{H}, can be seen as a manifestation of the fact that the magnetic microstructure in real space (for which R_{GSM} is representative) corresponds to the convolution of the nuclear grain microstructure (R_{GH}) with field-dependent micromagnetic response functions (l_{H}). Figure 4.21 features the magnetic Guinier analysis for the case of nanocrystalline Co [495].

Up to now, we have only discussed the magnetic Guinier approximation for the parallel scattering geometry ($\mathbf{k}_0 \parallel \mathbf{H}_0$), where 2π-averaged magnetic SANS data can be used for the analysis in terms of eqns (4.57)–(4.59). In the perpendicular geometry

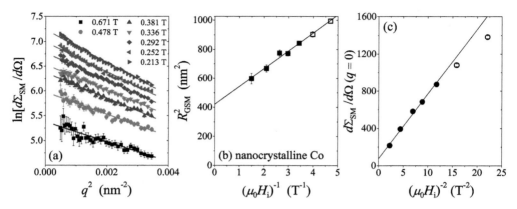

Fig. 4.21: Magnetic Guinier analysis on nanocrystalline Co. (a) Guinier plot $\ln[d\Sigma_{\mathrm{SM}}/d\Omega]$ versus q^2 and fits (solid lines) to eqn (4.57) at selected values of the internal magnetic field (see inset). (b) Plot of R_{GSM}^2 versus H_{i}^{-1} and fit (solid line) to eqn (4.58). In the fitting routine, R_{GH} and A were treated as adjustable parameters. (c) Field dependence of $\frac{d\Sigma_{\mathrm{SM}}}{d\Omega}(q=0)$. Solid line: $\frac{d\Sigma_{\mathrm{SM}}}{d\Omega}(0) \propto H_{\mathrm{i}}^{-2}$. In (b) and (c), the last two data points (open symbols), corresponding to internal fields of 0.213 T and 0.252 T, have been excluded from the fit analysis. After [495].

($\mathbf{k}_0 \perp \mathbf{H}_0$), an additional scattering term $S_{\mathrm{M}} R_{\mathrm{M}}$, related to magnetostatic fluctuations, appears in $d\Sigma_{\mathrm{SM}}/d\Omega$ [compare eqn (4.5)], which complicates the discussion. Two comments are then in place. (i) Since $S_{\mathrm{M}} \propto \widetilde{M}_z^2(\mathbf{q})$, the $S_{\mathrm{M}} R_{\mathrm{M}}$ contribution to $d\Sigma_{\mathrm{SM}}/d\Omega$ can be neglected for single-phase ferromagnets, where M_{s}-fluctuations are weak. A similar expression to eqn (4.58) is then found for large applied fields ($p_0 \ll 1$) [456]. (ii) Inspection of the expression for the magnetostatic response function R_{M} in the perpendicular geometry [eqn (4.9)] shows that this function vanishes by taking an average of the two-dimensional $d\Sigma_{\mathrm{SM}}/d\Omega$ along $\theta = 0°$ (or $\theta = 180°$). On the other hand, the corresponding $R_{\mathrm{H}}(\theta = 0°) = p^2$ [eqn (4.8)] is almost equal (besides a factor of $1/2$) to $R_{\mathrm{H}}(\theta = 0°) = p^2/2$ in the parallel geometry. This then implies that eqn (4.57) can also be employed to analyze ($\theta = 0°$) sector-averaged data in the $\mathbf{k}_0 \perp \mathbf{H}_0$ geometry.

In an idealized single-phase nanocrystalline ferromagnet with purely magnetocrystalline anisotropy and negligible M_{s}-fluctuations, the radii of gyration of the crystallites and of the magnetic anisotropy field will coincide. Figure 4.22(a) depicts $S_{\mathrm{H}}(q)$ of nanocrystalline Ni at 5 K and at 295 K, and Fig. 4.22(b) shows the corresponding Guinier plots ($\ln S_{\mathrm{H}}$ versus q^2) of the low-q data [441]. The temperature variation of S_{H} can be qualitatively understood by the notion that the magnetic anisotropy generally increases in strength with decreasing temperature. The limited experimental q-range of $0.02\,\mathrm{nm}^{-1} \lesssim q \lesssim 0.25\,\mathrm{nm}^{-1}$ suggests that important contributions to the anisotropy-field microstructure are related to structures smaller than roughly $2\pi/q_{\mathrm{max}} \cong 25\,\mathrm{nm}$ and larger than $2\pi/q_{\mathrm{min}} \cong 300\,\mathrm{nm}$. The Guinier plots in Fig. 4.22(b) are approximately linear, and straight-line fits yield $R_{\mathrm{GH}} = 20\,\mathrm{nm}$ at 5 K and $R_{\mathrm{GH}} = 22\,\mathrm{nm}$ at 295 K. For idealized spherical grains of diameter $D_{\mathrm{gs}} = 2R$, the radius of gyration is $R_{\mathrm{G}}^2 = \frac{3}{5}R^2$; in other words, $R_{\mathrm{G}} = 19\,\mathrm{nm}$ would be inferred from the experimental grain size of

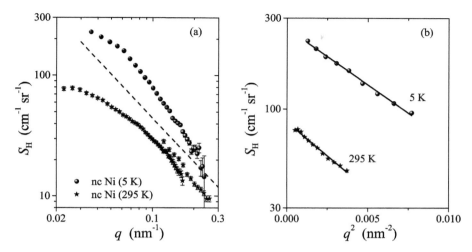

Fig. 4.22: (a) Anisotropy-field scattering functions $S_H(q)$ of nanocrystalline (nc) Ni at $T = 5\,\mathrm{K}$ and at $T = 295\,\mathrm{K}$ (log-log scale). Dashed line: $S_H(q) \propto q^{-1.2}$. (b) Guinier (log-linear) plots of the $S_H(q)$. Solid lines: fit to eqn (4.54). After [497].

the nanocrystalline Ni sample, $D_{gs} = 49\,\mathrm{nm}$, determined by the analysis of wide-angle x-ray diffraction data [441]. The apparent agreement between the radii of gyration of the anisotropy field and of the crystallites suggests that the anisotropy is coherently aligned in one direction within each grain. This would be expected if magnetocrystalline anisotropy dominated the net anisotropy field. In view of the possible roles of magnetostriction and interfacial anisotropy [441], this is a nontrivial observation. Moreover, recall that when there is a set of objects with a distribution of sizes, as in the studied polycrystalline Ni sample, then the experimental R_G is heavily weighted towards the largest objects [cf. eqn (2.77) and Fig. 2.16]. Thus, additional features in $\mathbf{H}_p$, varying on a smaller scale, are not precluded by the result for R_{GH} (see [441] for further details). For the determination of the scaling relation between the radii of gyration of the crystallites and of the magnetic anisotropy field, knowledge on the particle-size distribution is required.

4.3.5 Asymptotic power-law exponent

In nuclear SANS and SAXS, the asymptotic power-law exponent m of the cross section, $d\Sigma_{nuc}/d\Omega \propto q^{-m}$, is frequently discussed in the literature, since the value of m allows one to draw important conclusions on the underlying microstructure. The term asymptotic means that $qL \gg 1$, where L is some characteristic dimension of the scatterer. Examples that were already addressed in Section 2.4.3 include the Porod q^{-4}-law for particles with sharp interfaces (in the orientation-averaged limit), the Lorentzian-squared q^{-4}-dependency for exponentially correlated fluctuations, and the simple Lorentzian for critical Ornstein–Zernike q^{-2}-fluctuations. Power-law scattering plays also an important role for disordered, porous, and fractal systems (e.g., [23,28,65,498]). Additionally, we mention here SANS on dislocation structures in mechanically deformed metals and complex alloys, which exhibits a variety of power-law exponents

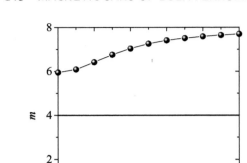
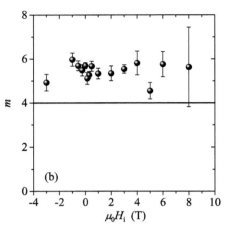

Fig. 4.23: (a) Theoretical variation of the power-law exponent m in $d\Sigma_{SM}/d\Omega = S_H R_H + R_H R_M = K/q^m$ with the ratio $H_p/\Delta M$ ($\mu_0 H_i = 1.0\,\mathrm{T}$). Line is a guide to the eyes. For both $\widetilde{H}_p^2(qR)$ and $\widetilde{M}_z^2(qR)$, we used the sphere form factor. After [488]. (b) Field dependence of the power-law exponent m determined from the experimental $d\Sigma_{SM}/d\Omega$ of a Nd–Fe–B-based nanocomposite. Solid horizontal lines in (a) and (b): $m = 4$. After [526].

(see, e.g., [1, 2, 4–6, 8, 19, 499–523] and references therein). Long, Levine, and Thomson [524, 525] analyzed the nuclear small-angle scattering from dislocation structures. These authors showed that single dislocations result in an asymptotic q^{-2}-dependency, dislocation dipoles give rise to a q^{-3}-law, while the scattering from dislocation walls may be more complicated and strongly depends upon the dislocation configurations and the sample/beam geometry: in the small-q regime, a $q^{-4}(1 - v^2 q^2)$-behavior is predicted for the dislocation walls, which for the case of finite-thickness walls with infinitely sharp boundaries (corresponding to $v = 0$) reduces to $d\Sigma_{nuc}/d\Omega \propto q^{-4}$.

With respect to magnetic SANS, we emphasize that the situation is more complex, since the exponent m is field-dependent and its value depends on the underlying magnetic interactions; for instance, for nanostructures with sharp interfaces, both $S_H \propto \widetilde{H}_p^2(q)$ and $S_M \propto \widetilde{M}_z^2(q)$ vary asymptotically as q^{-4} [compare eqns (3.82)–(3.84)]. Noting furthermore that the field-dependent response functions R_H and R_M vary asymptotically as $R_H \propto q^{-4}$ and $R_M \propto q^{-2}$ [compare eqns (4.16) and Fig. 4.2], we see that in this situation $d\Sigma_{SM}/d\Omega = S_H R_H + S_M R_M$ exhibits power-law exponents between $m \cong 6$–8. Therefore, depending on the relative magnitude of both contributions to $d\Sigma_{SM}/d\Omega$, one observes for this particular case different asymptotic power-law exponents of $d\Sigma_{SM}/d\Omega$ and, hence, of the total $d\Sigma/d\Omega$. This is illustrated in Fig. 4.23(a), where m in $d\Sigma_{SM}/d\Omega = K/q^m$ is plotted at $\mu_0 H_i = 1.0\,\mathrm{T}$ as a function of the ratio $H_p/\Delta M$ using the sphere form factor for both $\widetilde{H}_p^2(q)$ and $\widetilde{M}_z^2(q)$. Such steep decays of $d\Sigma_{SM}/d\Omega$ are not in accordance with simple exponentially decaying magnetization correlations, which would give rise to $m = 4$ [compare eqn (2.72)].

Other models for the anisotropy-field microstructure may result in different exponents; in particular, the $\widetilde{H}_p^2(q)$ that are related to the long-range stress fields of microstructural defects will give rise to asymptotic power laws that are different from the (sharp-interface) Porod exponent of $m = 4$. An open question is therefore the actual behavior of $\widetilde{H}_p^2(q)$ [and of $\widetilde{M}_z^2(q)$] for real materials. The problem is the lack of knowledge of the characteristic structure sizes of $\widetilde{H}_p^2$ and $\widetilde{M}_z^2$ for polycrystalline ferromagnets. As mentioned already in Section 3.3, the correlation lengths of $\widetilde{H}_p^2$ and $\widetilde{M}_z^2$ are related, respectively, to the spatial extent of regions with uniform magnetic anisotropy field (ξ_H) and saturation magnetization (ξ_M). These characteristic length scales are not necessarily related to some characteristic structural length scale such as the average crystallite or particle size. In Section 4.4, we will provide a method which allows one to determine the functions $\widetilde{H}_p^2$ and $\widetilde{M}_z^2$ model-independently from the analysis of field-dependent SANS data. At any rate, there is ample theoretical and experimental evidence that the spin-misalignment scattering from nanocrystalline magnets is characterized by power-law exponents that are larger than $m = 4$ [441,527]; see, e.g., Fig. 4.23(b), which shows the experimental field dependence of m for a Nd–Fe–B-based nanocomposite [526].

The above-discussed results for $d\Sigma_{SM}/d\Omega$ agree qualitatively with the magnetic SANS due to dislocations: the long-range strain field of the dislocations results in transversal fluctuations of the magnetization by means of magnetoelastic coupling. These fluctuations represent a contrast for magnetic SANS and can be computed by means of micromagnetic theory for samples close to saturation. As shown in [1,5], for $q \gg l_H^{-1}$, the orientationally averaged magnetic scattering due to a dislocation network varies asymptotically according to $d\Sigma_{SM}/d\Omega \propto q^{-7}$ (see Section 4.6 for further details).

Since the continuum theory of micromagnetics is also applicable to amorphous ferromagnets [134], the predictions of the micromagnetic SANS theory are expected to be transferable to these systems as well. However, an important difference between conventional (nano)crystalline ferromagnets and amorphous magnets relates to the characteristic length scales over which perturbations in the magnetic anisotropy field and in the saturation magnetization manifest: while in nanocrystalline systems $\mathbf{H}_p$ and M_s may vary on a scale of the grain size, say, $\xi_H \sim \xi_M \sim 10$ nm, in amorphous magnets this length scale is defined by the interatomic distance due to atomic-site anisotropy and randomness. As a consequence, the functions $S_H(q)$ and $S_M(q)$ are almost constant (i.e., q-independent) over a large part of the experimentally accessible range of momentum transfers (0.01 nm$^{-1} \lesssim q \lesssim 1.0$ nm^{-1}), so that the asymptotic power-law dependence of the spin-misalignment scattering cross section $d\Sigma_{SM}/d\Omega = S_H(q)R_H(q, H_i) + S_M(q)R_M(q, H_i)$ is determined by the asymptotic behavior of both response functions [compare eqns (4.16) and Fig. 4.2]; in other words, for amorphous magnets, $d\Sigma_{SM}/d\Omega \propto q^{-m}$ with $2 \leq m \leq 4$ is expected. The results in Fig. 4.24 for the field-dependent exponent m, obtained from the total $d\Sigma/d\Omega$ of a rare-earth-containing bulk metallic glass [528], seem to confirm this prediction; see also the review by Boucher and Chieux [30] for a discussion of power-law exponents found for rare-earth-based amorphous alloys. Table 4.1 lists some of the expected power-law exponents for magnetic SANS.

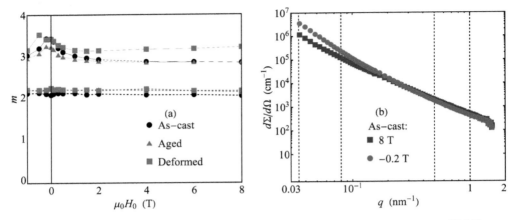

Fig. 4.24: (a) Field dependence of the power-law exponent m in $d\Sigma/d\Omega = K/q^m$ for the total $d\Sigma/d\Omega$ of the as-cast, aged, and deformed bulk metallic glass $(Nd_{60}Fe_{30}Al_{10})_{92}Ni_8$. The dashed vertical lines in (b) indicate the respective low-q and high-q ranges over which the fit was carried out. Solid (dashed) connecting lines in (a) correspond to the low-q (high-q) region. After [528].

Table 4.1 Asymptotic power-law exponents m for magnetic SANS.

Origin of scattering/underlying microstructure	m
Exponentially correlated magnetization fluctuations	4
Magnetic critical scattering ($T > T_C$)	2
Dislocation structures (orientationally averaged limit)	7
Nanocrystalline random anisotropy magnets	4–8
Amorphous metals	2–4

4.3.6   Polarization dependence of magnetic SANS

In this section, we discuss the dependence of the magnetic SANS cross section on the polarization of the incident neutron beam. Polarization-dependent chiral scattering terms are not considered here (see Section 4.3.3). For nanostructured bulk ferromagnets, the spin-misalignment SANS cross section $d\Sigma_{SM}/d\Omega$ generally represents the dominant contribution to the total unpolarized SANS cross section, at least for applied fields not-too-close to saturation. This is evidenced by numerous experimental SANS data (see, e.g., [49, 59]) which exhibit an extremely large field dependence; for instance, the SANS signal of nanocrystalline Co [Fig. 6.16(a)] can vary by up to three orders of magnitude at the smallest q between zero field (or coercivity) and a large field close to saturation (compare also Figs. 4.29 and 6.13) [441]. Since $d\Sigma_{SM}/d\Omega$

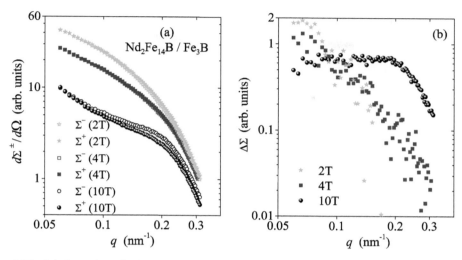

Fig. 4.25: (a) 2π-azimuthally averaged SANSPOL cross sections $d\Sigma^+/d\Omega$ (closed symbols) and $d\Sigma^-/d\Omega$ (open symbols) of a Nd–Fe–B-based nanocomposite as a function of momentum transfer q at selected applied magnetic fields (see inset) ($\mathbf{k}_0 \perp \mathbf{H}_0$) (log-log scale). (b) Corresponding difference SANS cross sections $\Delta\Sigma = d\Sigma^-/d\Omega - d\Sigma^+/d\Omega$ (log-log scale). After [59].

is related to terms $|\widetilde{M}_x|^2$, $|\widetilde{M}_y|^2$, $-(\widetilde{M}_y\widetilde{M}_z^* + \widetilde{M}_y^*\widetilde{M}_z)$, or $-(\widetilde{M}_x\widetilde{M}_y^* + \widetilde{M}_x^*\widetilde{M}_y)$, and since these contributions are independent of the polarization of the incident neutron beam, it is straightforward to conclude that the measurement of the half-polarized SANSPOL spin-up and spin-down cross sections does not provide significantly more information regarding $d\Sigma_{SM}/d\Omega$ than can already be learned by the measurement of the unpolarized cross section alone. This circumstance is illustrated in the following.

The SANSPOL cross sections $d\Sigma^\pm/d\Omega$ [eqns (2.95) and (2.98)] contain terms which do depend on the polarization of the incident neutron beam; in particular, the difference between data taken with the neutron-spin flipper on and off, $\Delta\Sigma = d\Sigma^-/d\Omega - d\Sigma^+/d\Omega$, depends for $\mathbf{k}_0 \perp \mathbf{H}_0$ on terms $\widetilde{N}\widetilde{M}_z \sin^2\theta$ and $\widetilde{N}\widetilde{M}_y \sin\theta\cos\theta$ (ignoring the polarization-dependent chiral term). Figure 4.25 shows the results of magnetic-field-dependent SANSPOL measurements on a Nd–Fe–B-based nanocomposite. Figure 4.25(a) depicts the spin-up and spin-down SANS cross sections at selected field values, while Fig. 4.25(b) shows the corresponding difference $\Delta\Sigma$ of the SANS cross sections between the two spin states. It is seen that the $d\Sigma^\pm/d\Omega$ at a given field are only very weakly dependent on the incoming neutron polarization. A small difference can be detected at an applied field of $10\,\mathrm{T}$, where the spin-misalignment scattering is already largely suppressed. Given that the total unpolarized $d\Sigma/d\Omega$ and, hence, the $d\Sigma_{SM}/d\Omega$ of this sample vary strongly as a function of the external field (compare Fig. 6.13), the observation in Fig. 4.25(b) that the $\Delta\Sigma$ are small in magnitude compared to the unpolarized data and exhibit only a very weak field variation suggests that for this isotropic nanocomposite ferromagnet the interference between nuclear and transverse spin-misalignment scattering amplitudes $\widetilde{N}\widetilde{M}_y \sin\theta\cos\theta$ is of

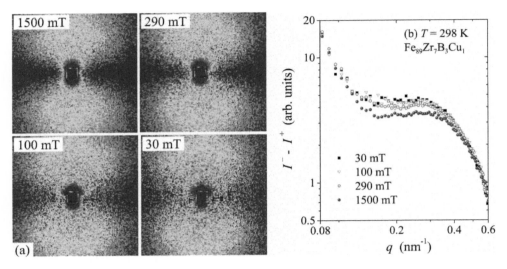

Fig. 4.26: (a) Difference $I^- - I^+$ between flipper-on and flipper-off SANS intensities of NANOPERM ($Fe_{89}Zr_7B_3Cu_1$) as a function of the applied magnetic field (see insets) ($\mathbf{k}_0 \perp \mathbf{H}_0$). $\mathbf{H}_0$ is horizontal. Pixels in the corners of the detector have momentum transfer of $q \cong 0.64\,\mathrm{nm}^{-1}$. (b) Corresponding azimuthally averaged difference between flipper-on and flipper-off data (log-log scale). After [446].

minor relevance. Therefore, the very weak polarization dependence of $d\Sigma^\pm/d\Omega$ can be attributed to terms $\widetilde{N}\widetilde{M_z}\sin^2\theta$.

As a further example, Fig. 4.26 depicts for a two-phase Fe-based nanocomposite the difference between data taken with the neutron-spin flipper on and off at several applied magnetic fields [446]. The angular anisotropy of the scattering pattern [Fig. 4.26(a)] is for all fields shown clearly of the $\sin^2\theta$-type. The corresponding azimuthally averaged data [Fig. 4.26(b)] are small in magnitude compared to the unpolarized data (see Figs. 1 and 4 in [446]) and only very weakly dependent on the applied field, in agreement with the previous conclusions. We also note that scattering contributions related to the term $\widetilde{N}\widetilde{M_y}\sin\theta\cos\theta$ do not become noticeable. Similar results were also obtained on nanocrystalline Co [441].

The observations in Figs. 4.25 and 4.26 strongly suggest that for the study of spin-misalignment scattering, which is related to the transversal spin components, it is sufficient to measure the field dependence of the unpolarized SANS cross section. Moreover, as discussed in Section 6.5.1, the subtraction of the nuclear and magnetic scattering at a saturating field may under certain conditions yield a purely magnetic SANS cross section, which is reminiscent of the spin-flip SANS and which therefore renders time-consuming and low-intensity polarization-analysis experiments obsolete. It is however important to emphasize that, while the usage of the SANSPOL method may not always be beneficial when studying nanocrystalline bulk ferromagnets, where the polarization-independent spin-misalignment scattering is dominant, half-polarized experiments have their merits when investigating nearly homogeneously magnetized single-domain particles such as ferrofluids, particularly when the magnetic signal is weak compared to the nuclear one. The combination of SANSPOL and POLARIS is advisable in order to elucidate the inhomogeneous spin structure of nanoparticles.

Figure 4.27 illustrates very impressively that polarized neutrons are extremely powerful for the detection of weak magnetic scattering contributions on top of strong nuclear scattering signals (or vice versa) in systems consisting of single-domain magnetic nanoparticles [186,194]. The SANSPOL data were taken on superparamagnetic Fe_3O_4 nanoparticles which are embedded in a glass matrix. At an applied field of 1 T, the magnetic moments are aligned along the field and the unpolarized SANS cross section is given by $d\Sigma/d\Omega \propto |\widetilde{N}|^2 + b_{\mathrm{H}}^2 |\widetilde{M}_z|^2 \sin^2\theta$. From the observation that the unpolarized SANS signal $\propto (I^- + I^+)$ is only very slightly anisotropic, one can conclude that the isotropic nuclear SANS $\propto |\widetilde{N}|^2$ dominates over the anisotropic magnetic contribution $\propto |\widetilde{M}_z|^2 \sin^2\theta$. However, the difference between the spin-up and spin-down SANS cross sections $(I^- - I^+)$ yields the nuclear-magnetic interference term $\propto \widetilde{N}\widetilde{M}_z \sin^2\theta$, which is linear in $\widetilde{M}_z$ and which brings the anisotropic magnetic scattering to daylight. Another advantage of the SANSPOL difference method is that there are no contaminations by nuclear incoherent scattering (neglecting the very weak nuclear-spin incoherent SANS) and by spin-misalignment scattering (which does not depend on the polarization of the beam). The measurement of $I^- - I^+$ allows one to determine the longitudinal magnetization Fourier component $\widetilde{M}_z$ for weak magnetic scattering. The results of such kind of experiments can be compared to macroscopic magnetization data, as has been demonstrated by Disch et al. [529] for iron-oxide nanoparticles.

4.4   Selected experimental results

4.4.1   Fe-based soft magnetic nanocomposites

In the previous section, we have discussed several fundamental characteristics of magnetic SANS, and we have illustrated these by also displaying experimental data. In the following, we are going to demonstrate that the micromagnetic SANS theory can be directly fitted to azimuthally averaged experimental data.

The final result for the azimuthally averaged total unpolarized SANS cross section $d\Sigma/d\Omega$ can be written in compact form as [$\mathbf{k}_0 \perp \mathbf{H}_0$; compare eqns (4.12)−(4.14)]:

$$\frac{d\Sigma}{d\Omega}(q, H_{\mathrm{i}}) = \frac{d\Sigma_{\mathrm{res}}}{d\Omega}(q) + S_{\mathrm{H}}(q)R_{\mathrm{H}}(q, H_{\mathrm{i}}) + S_{\mathrm{M}}(q)R_{\mathrm{M}}(q, H_{\mathrm{i}}). \tag{4.60}$$

This expression for $d\Sigma/d\Omega$ is linear in both R_{H} and R_{M}, with a priori unknown functions $d\Sigma_{\mathrm{res}}/d\Omega$, S_{H}, and S_{M}. For given values of the materials parameters A and M_{s}, the numerical values of the response functions R_{H} and R_{M} are known at each value of q and H_{i}. Therefore, by plotting at a particular $q = q^\dagger$ the values of $d\Sigma/d\Omega$ measured at several H_{i} versus $R_{\mathrm{H}}(q^\dagger, H_{\mathrm{i}}, A)$ and $R_{\mathrm{M}}(q^\dagger, H_{\mathrm{i}}, A)$, one can obtain the values of $d\Sigma_{\mathrm{res}}/d\Omega$, S_{H}, and S_{M} at $q = q^\dagger$ by a (weighted) linear least-squares fit. The data-analysis procedure is illustrated in Fig. 4.28 for the case of a soft magnetic two-phase iron-based nanocomposite. Treating the exchange-stiffness constant A in the expression for $p(q, H_{\mathrm{i}})$ as an adjustable parameter, allows one to obtain information on this quantity by means of χ^2 minimization [compare eqn (4.69)]. We note that in

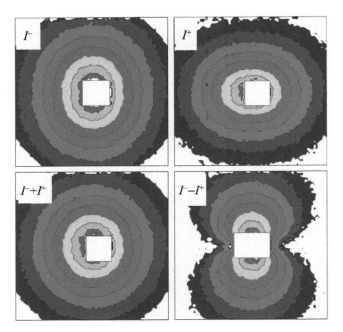

Fig. 4.27: SANSPOL patterns of an Fe_3O_4 glass ceramic for neutron spins antiparallel (I^-) and parallel (I^+) to the horizontal applied magnetic field. The sum $(I^- + I^+)$ corresponds to the pattern of unpolarized neutrons. The difference $(I^- - I^+)$ yields the nuclear-magnetic interference term $\propto \widetilde{N}\widetilde{M_z}\sin^2\theta$. After [194].

order to obtain a best-fit value for A from experimental field-dependent SANS data, it is not necessary that the data are available in absolute units. The described fitting routine does not represent a "continuous" fit in the conventional sense, rather the theoretical cross section at discrete q and at several H_i is computed. For comparison to experiment, these simulated data points may then be connected by straight lines (see Fig. 4.29).

The theoretical SANS cross section $d\Sigma/d\Omega$ [eqn (4.60)] depends on the internal magnetic field H_i via the effective magnetic field H_{eff}, which is contained in the expressions for R_H and R_M. The effective field $H_{eff} = H_i + 2Aq^2/(\mu_0 M_s)$ [eqn (3.65)] becomes field independent when the exchange-field-related term $2Aq^2/(\mu_0 M_s)$ dominates over H_i [compare Fig. 3.5(b)]. The condition $2Aq^2/(\mu_0 M_s) \gtrsim H_i$ then suggests that for scattering vectors larger than about

$$q_C \cong \sqrt{\frac{\mu_0 M_s H_{max}}{2A}}, \tag{4.61}$$

where H_{max} denotes the largest applied magnetic field, H_{eff} and, hence, $d\Sigma/d\Omega$ are practically independent of H_i [compare, e.g., Figs. 4.29, 4.34(a), and 6.16(a)]. Typical values for q_C may range between $\sim 0.2-0.6\,\mathrm{nm}^{-1}$. This circumstance hampers the reliable separation of spin-misalignment $(S_H R_H + S_M R_M)$ and residual $(d\Sigma_{res}/d\Omega)$

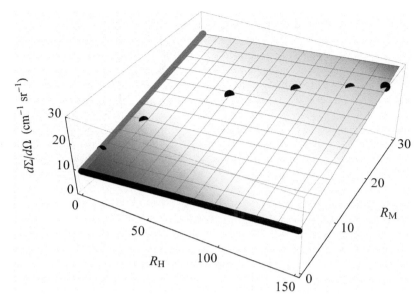

Fig. 4.28: Illustration of the neutron-data analysis procedure according to eqn (4.60). The total $d\Sigma/d\Omega$ ($\bullet$) at $q^\dagger = 0.114\,\mathrm{nm}^{-1}$ and for $A = 4.7\,\mathrm{pJ/m}$ is plotted versus the numerical values of the response functions R_H and R_M at experimental field values (in mT) of 1270, 312, 103, 61, 42, 33. The plane represents a fit to eqn (4.60). The intercept of the plane with the $d\Sigma/d\Omega$-axis provides the residual SANS cross section $d\Sigma_\mathrm{res}/d\Omega$, while S_H and S_M are obtained from the slopes of the plane (slopes of the thick black and red lines). When the fitting procedure is carried out for successive q-values, then the functional dependencies $\frac{d\Sigma_\mathrm{res}}{d\Omega}(q)$, $S_\mathrm{H}(q)$, and $S_\mathrm{M}(q)$ are obtained. After [488].

scattering contributions for $q \gtrsim q_\mathrm{C}$; in other words, a fit of experimental data to eqn (4.60) should be limited to $q \lesssim q_\mathrm{C}$.

As an example, the azimuthally averaged field-dependent SANS cross sections of two soft magnetic nanocomposites from the NANOPERM family of alloys along with the fits to the micromagnetic theory [eqn (4.60)] are displayed in Fig. 4.29. NANO-PERM consists of a dispersion of iron nanoparticles in an amorphous magnetic matrix. It is seen that for both samples the (q, H_i)-dependence of $d\Sigma/d\Omega$ can be excellently described by the micromagnetic prediction. As expected, both residual SANS cross sections $d\Sigma_\mathrm{res}/d\Omega$ ($\circ$) are smaller than the respective total $d\Sigma/d\Omega$, supporting the notion of dominant spin-misalignment scattering in these type of materials. From the fit of the entire (q, H_i) data set to eqn (4.60), one obtains values for the volume-averaged exchange-stiffness constants [compare insets in Fig. 4.29(a) and (b)]. We obtain $A = 3.1 \pm 0.1\,\mathrm{pJ/m}$ for the Co-free alloy and $A = 4.7 \pm 0.9\,\mathrm{pJ/m}$ for the zero-magnetostriction NANOPERM sample. Table 4.2 compiles values for exchange-stiffness constants obtained by this method on a number of magnetic materials. The A-value is related to the spin-wave stiffness constant D via [530]:

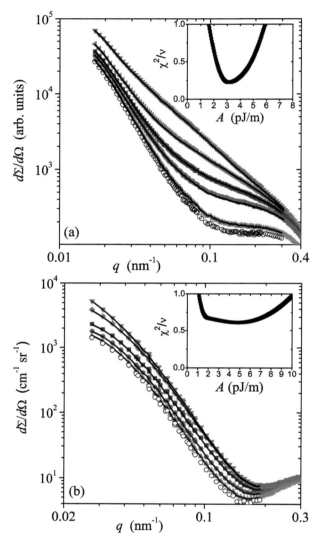

Fig. 4.29: Azimuthally averaged total unpolarized $d\Sigma/d\Omega$ of (a) $Fe_{89}Zr_7B_3Cu$ and (b) $(Fe_{0.985}Co_{0.015})_{90}Zr_7B_3$ at selected applied magnetic fields (log-log scale) ($\mathbf{k}_0 \perp \mathbf{H}_0$). The addition of a small amount of Co results in a vanishing magnetostriction [334]. Field values (in mT) from bottom to top: (a) 1994, 321, 163, 85, 45; (b) 1270, 312, 103, 61, 33. Solid lines in (a) and (b): fit to the micromagnetic theory [eqn (4.60)]. The solid lines connect the computed $d\Sigma/d\Omega$ at each value of q and H_i. (○) Residual scattering cross sections $d\Sigma_{res}/d\Omega$. The insets depict the respective (reduced) weighted mean-square deviation between experiment and fit, χ^2/ν, as a function of the exchange constant A (ν = number of degrees of freedom of the fit). After [488].

Table 4.2 Exchange-stiffness constants A and spin-wave stiffness constants D of various magnetic materials. Ni and Co: bulk nanocrystalline materials prepared by means of electrodeposition. $Nd_2Fe_{14}B/Fe_3B$: two-phase melt-spun nanocomposite (isotropic). $Nd_2Fe_{14}B$: isotropic sintered magnet. $Fe_{89}Zr_7B_3Cu$ and $(Fe_{0.985}Co_{0.015})_{90}Zr_7B_3$: nanocrystalline two-phase alloys prepared by the melt-spinning technique (isotropic). The room-temperature D-value for $Nd_2Fe_{14}B$ has been estimated based on inelastic neutron-scattering data on a single crystal (isotropic average) [531], while the ones for NANOPERM were computed by means of eqn (4.62) using $\mu_0 M_s = 1.26\,T$ ($Fe_{89}Zr_7B_3Cu$), $\mu_0 M_s = 1.64\,T$ [$(Fe_{0.985}Co_{0.015})_{90}Zr_7B_3$], and $g = 2.21$.

Material	$A\ (10^{-12}\ \mathrm{J/m})$	$D\ (\mathrm{meV\mathring{A}^2})$	$T\ (\mathrm{K})$	References
Ni	9.2 ± 0.2	450 ± 10	5	[441, 532]
Ni	7.6 ± 0.3	370 ± 20	295	[441, 532]
Co	28 ± 1	500 ± 20	295	[441, 532]
$Nd_2Fe_{14}B/Fe_3B$	$10-13$	—	300	[533]
$Nd_2Fe_{14}B$	5.9 ± 0.1	220 ± 10	295	[487, 531]
$Fe_{89}Zr_7B_3Cu$	3.1 ± 0.1	80 ± 5	295	[488]
$(Fe_{0.985}Co_{0.015})_{90}Zr_7B_3$	4.7 ± 0.9	90 ± 20	295	[488]

$$D(T) = \frac{2A(T)g\mu_B}{M_s(T)}, \tag{4.62}$$

where g is the Landé factor. Values for D computed by means of eqn (4.62) are also listed in Table 4.2.

In addition to the exchange-stiffness constant, analysis of field-dependent SANS data in terms of eqn (4.60) provides the magnitude squares of the Fourier components of the magnetic anisotropy field $S_H \propto |\widetilde{H}_p(q)|^2$ and of the longitudinal magnetization $S_M \propto |\widetilde{M}_z(q)|^2 \propto (\Delta M)^2$. The obtained results for these functions are shown in Fig. 4.30. It is immediately seen that over most of the displayed q-range $|\widetilde{M}_z|^2$ is orders of magnitude larger than $|\widetilde{H}_p|^2$, suggesting that jumps ΔM in the magnetization at internal interfaces is the dominant source of spin disorder in these alloys. Numerical integration of $S_H(q)$ and $S_M(q)$ over the whole $\mathbf{q}$-space, i.e.,

$$\left\langle |\mathbf{H}_p|^2 \right\rangle = \frac{1}{2\pi^2 b_H^2} \int_0^\infty S_H(q)q^2 dq \tag{4.63}$$

$$\left\langle |M_z|^2 \right\rangle = \frac{1}{2\pi^2 b_H^2} \int_0^\infty S_M(q)q^2 dq \tag{4.64}$$

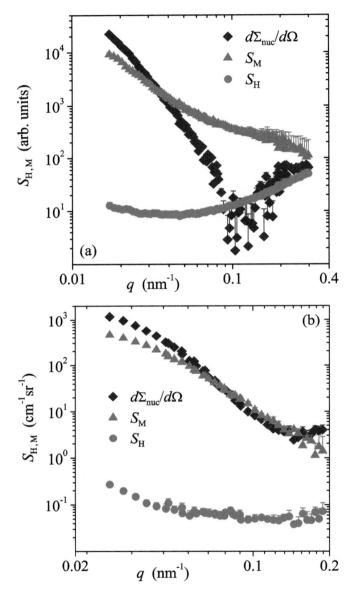

Fig. 4.30: Best-fit results for the scattering function of the magnetic anisotropy field $S_{\mathrm{H}} \propto |\widetilde{H}_{\mathrm{p}}(q)|^2$ and for the scattering function of the longitudinal magnetization $S_{\mathrm{M}} \propto |\widetilde{M}_z(q)|^2$. (a) $Fe_{89}Zr_7B_3Cu$; (b) $(Fe_{0.985}Co_{0.015})_{90}Zr_7B_3$ (log-log scale). $d\Sigma_{\mathrm{nuc}}/d\Omega = (8\pi^3/V)|\widetilde{N}|^2$ denotes the nuclear SANS, which was obtained by subtracting the respective $|\widetilde{M}_z|^2$ scattering from the residual SANS cross section $d\Sigma_{\mathrm{res}}/d\Omega$ [compare eqn (4.13)]. After [488].

yields, respectively, the mean-square anisotropy field $\langle|\mathbf{H}_p|^2\rangle$ and the mean-square longitudinal magnetization fluctuation $\langle|M_z|^2\rangle$. These quantities are, respectively, defined as:

$$\left\langle|\mathbf{H}_p|^2\right\rangle = \frac{1}{V}\int|\mathbf{H}_p(\mathbf{r})|^2 d^3r \qquad (4.65)$$

and

$$\left\langle|M_z|^2\right\rangle = \frac{1}{V}\int|M_z(\mathbf{r})|^2 d^3r. \qquad (4.66)$$

Equations (4.63) and (4.64) can be derived by taking into account the definitions for S_H and S_M [eqns (4.6) and (4.7)] and by making use of the Parseval theorem of Fourier theory [534]: if the volume average of the magnitude square of a vector field $\mathbf{W}(\mathbf{r})$ is defined as

$$\langle|\mathbf{W}|^2\rangle = \frac{1}{V}\int_V|\mathbf{W}(\mathbf{r})|^2 d^3r, \qquad (4.67)$$

then this can be re-expressed as an integral over $\mathbf{q}$-space, according to

$$\langle|\mathbf{W}|^2\rangle = \frac{1}{V}\int_{\mathbf{q}}|\widetilde{\mathbf{W}}(\mathbf{q})|^2 d^3q, \qquad (4.68)$$

where $\widetilde{\mathbf{W}}(\mathbf{q})$ represents the Fourier transform of $\mathbf{W}(\mathbf{r})$. We emphasize that, since experimental data for S_H and S_M are only available within a finite range of momentum transfers (between q_{min} and q_{max}), and since both integrands $S_H q^2$ and $S_M q^2$ do not show signs of convergence, one can only obtain lower bounds for these quantities: for the $(Fe_{0.985}Co_{0.015})_{90}Zr_7B_3$ sample (for which $d\Sigma/d\Omega$ is available in absolute units), we obtain $\mu_0\langle|\mathbf{H}_p|^2\rangle^{1/2} \cong 10\,\text{mT}$ and $\mu_0\langle|M_z|^2\rangle^{1/2} \cong 50\,\text{mT}$. This finding qualitatively supports the notion of dominant spin-misalignment scattering due to magnetostatic fluctuations. Knowledge of $S_M \propto |\widetilde{M}_z|^2$ and of the residual SANS cross section $d\Sigma_{res}/d\Omega$ [eqn (4.2)] allows one to obtain the nuclear scattering $|\widetilde{N}|^2$ (see Fig. 4.30), without using sector-averaging procedures in unpolarized scattering or polarization analysis [332]. Equations (4.63) and (4.64) may be complemented by eqn (4.42), which provides a means to estimate the average magnetostatic field due to magnetic volume charges from experimental $S_H(q)$ data. Finally, we emphasize that field-dependent SANS has also been used for identifying the domain orientation in magnetic-field-annealed textured alloys of this type [535–537].

4.4.2 Nd–Fe–B-based hard magnetic nanocomposites

Figure 4.31 depicts the two-dimensional unpolarized SANS intensity distribution of a Nd–Fe–B-based nanocomposite at selected applied magnetic fields and for $\mathbf{k}_0 \parallel \mathbf{H}_0$. The Nd–Fe–B nanocrystalline sample was prepared by the melt-spinning technique and, as expected for a statistically isotropic microstructure, the total $d\Sigma/d\Omega$ is isotropic in the parallel scattering geometry, although the individual contributions to $d\Sigma/d\Omega$ are highly anisotropic [compare eqn (2.94)]. The scattering becomes more concentrated along the forward direction with increasing field. The corresponding azimuthally averaged $d\Sigma/d\Omega$ are shown in Fig. 4.32 for both scattering geometries $\mathbf{k}_0 \perp \mathbf{H}_0$ and

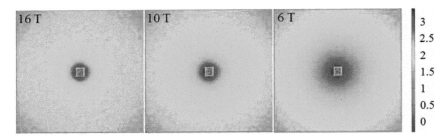

Fig. 4.31: Experimental unpolarized SANS intensity distribution of a Nd–Fe–B-based nanocomposite at selected applied magnetic fields (see insets) ($\mathbf{k}_0 \parallel \mathbf{H}_0$) (logarithmic color scale). $\mathbf{H}_0$ is normal to the detector plane. Pixels in the corners of the detector correspond to $q \cong 0.30 \, \mathrm{nm}^{-1}$. After [59].

$\mathbf{k}_0 \parallel \mathbf{H}_0$ [533]. We note that the magnetization state of the specimen at the indicated field values falls well into the approach-to-saturation regime [compare inset in Fig. 6.13(a)]. It is seen that the prediction by the micromagnetic theory [eqns (4.12) and (4.18)] provides an excellent description of the (q, H_0)-dependence of $d\Sigma/d\Omega$ (solid lines in Fig. 4.32). Further independent support for the dominance of spin-misalignment scattering comes from the observation in Fig. 4.32 that both residual SANS cross sections $d\Sigma_{\mathrm{res}}/d\Omega$ (determined from the micromagnetic fit) are significantly smaller than the total experimental $d\Sigma/d\Omega$ at all q and H_0 (compare also Fig. 4.29). We remind the reader that $d\Sigma_{\mathrm{res}}/d\Omega$ represents the extrapolated nuclear and magnetic $d\Sigma/d\Omega$ at infinite field.

As with the NANOPERM sample (Fig. 4.29), from such global fits one can determine a volume-averaged value for the exchange-stiffness constant A. In the present case, due to deviations of the shape of the Nd–Fe–B ribbon sample from the ideal ellipsoidal shape, we have also treated the value of the (effective) surface demagnetizing field $H_{\mathrm{d}}^{\mathrm{s}}$ in the expression for the internal magnetic field H_{i} [compare eqns (3.52) and (3.53)] as an adjustable parameter. We note that for each scattering geometry the entire (q, H_0) data set was fitted by a single set of values for A and $H_{\mathrm{d}}^{\mathrm{s}}$ (one straight-line fit at each q). For this purpose, we have defined the quantity

$$\chi^2(A, H_{\mathrm{d}}^{\mathrm{s}}) = \sum_{m,n} \frac{1}{\sigma_{m,n}^2} \left(\frac{d\Sigma_{m,n}^{\mathrm{exp}}}{d\Omega} - \frac{d\Sigma_{m,n}^{\mathrm{sim}}}{d\Omega} \right)^2, \tag{4.69}$$

where the indices m and n count, respectively, the scattering vectors and field values, $\sigma_{m,n}$ is the uncertainty in the experimental SANS cross section $\frac{d\Sigma_{m,n}^{\mathrm{exp}}}{d\Omega} = \frac{d\Sigma^{\mathrm{exp}}}{d\Omega}(q_m, H_{0,n})$, and $\frac{d\Sigma_{m,n}^{\mathrm{sim}}}{d\Omega} = \frac{d\Sigma^{\mathrm{sim}}}{d\Omega}(q_m, H_{\mathrm{i},n})$ denotes the micromagnetic fit value. The values of A and $H_{\mathrm{d}}^{\mathrm{s}}$ which minimize $\chi^2(A, H_{\mathrm{d}}^{\mathrm{s}})$ are identified with the experimental best-fit values.

Figure 4.33 displays gray-scale-coded plots of $\chi^2(A, H_{\mathrm{d}}^{\mathrm{s}})$ for $\mathbf{k}_0 \perp \mathbf{H}_0$ [Fig. 4.33(a)] and for $\mathbf{k}_0 \parallel \mathbf{H}_0$ [Fig. 4.33(b)]. The minima in χ^2 are found for $A = 10.2 \pm 2.9 \, \mathrm{pJ/m}$ and $\mu_0 H_{\mathrm{d}}^{\mathrm{s}} = 0.00 \pm 0.07 \, \mathrm{T}$ ($\mathbf{k}_0 \perp \mathbf{H}_0$) and for $A = 13.1 \pm 3.2 \, \mathrm{pJ/m}$ and $\mu_0 H_{\mathrm{d}}^{\mathrm{s}} = 1.08 \pm 0.10 \, \mathrm{T}$ ($\mathbf{k}_0 \parallel \mathbf{H}_0$). Within error bars, both A-values coincide and are in very

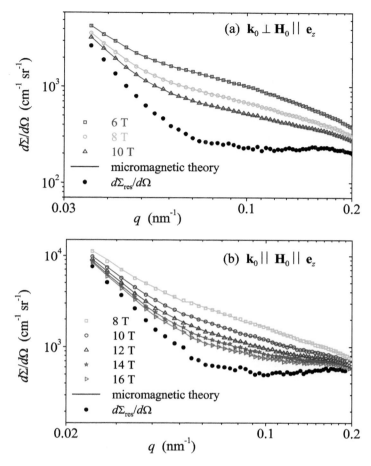

Fig. 4.32: Azimuthally averaged total unpolarized SANS cross section $d\Sigma/d\Omega$ versus momentum transfer q of a $Nd_2Fe_{14}B/Fe_3B$ nanocomposite (log-log scale) ($T = 300\,\mathrm{K}$). (a) $\mathbf{k}_0 \perp \mathbf{H}_0$; (b) $\mathbf{k}_0 \parallel \mathbf{H}_0$. (○) Experimental $d\Sigma/d\Omega$ at several applied magnetic fields (see insets). Solid lines: prediction by the micromagnetic theory. The solid lines connect the computed $d\Sigma/d\Omega$ at each value of q and H_0. (●) Residual SANS cross sections $d\Sigma_{\mathrm{res}}/d\Omega$. After [533].

good agreement with data reported in the literature [538]. The two values for the demagnetizing field are quite different, reflecting the two scattering geometries, which have the applied magnetic field in-plane ($\mathbf{k}_0 \perp \mathbf{H}_0$) and out-of-plane ($\mathbf{k}_0 \parallel \mathbf{H}_0$) of the ribbon sample. The solid lines in Fig. 4.32(a) and (b) have been computed using a value of $A = 12.5\,\mathrm{pJ/m}$ and the corresponding estimated demagnetizing fields.

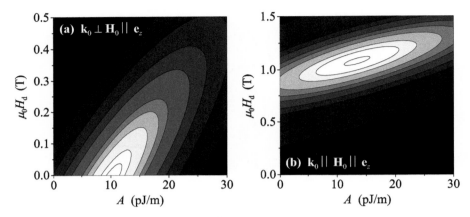

Fig. 4.33: Gray-scale-coded plot of $\chi^2(A, H_d^s)$ [eqn (4.69)] for (a) $\mathbf{k}_0 \perp \mathbf{H}_0$ and (b) $\mathbf{k}_0 \parallel \mathbf{H}_0$. Lighter colors correspond to smaller values of χ^2. For the computation of χ^2, we have used, respectively, the experimental data shown in Fig. 4.32(a) and (b). After [533].

4.4.3 Spin-flip SANS of nanocrystalline cobalt

Figure 4.34 depicts the results of a micromagnetic neutron-data analysis of the field-dependent spin-flip SANS cross section $d\Sigma^{+-}/d\Omega$ of nanocrystalline Co with an average crystallite size of $D_{gs} = 9.5 \pm 3.0 \, \text{nm}$. Global fitting of the $d\Sigma^{+-}/d\Omega$ data [solid lines in Fig. 4.34(a) using eqn (4.31)] provides a room-temperature value of $A = 2.6 \pm 0.1 \times 10^{-11} \, \text{Jm}^{-1}$ for the volume-averaged exchange-stiffness constant [compare inset in Fig. 4.34(b)]. This value agrees with the result of an earlier analysis [539] (neglecting the $S_M R_M$ term) and with literature data on single crystals obtained by means of inelastic neutron scattering [243]. The q-variations of the anisotropy-field scattering function S_H and of the scattering function of the longitudinal magnetization S_M [Fig. 4.34(b)] demonstrate that $S_M < S_H$ over most of the q-range. This finding is in line with the expectation that for a homogeneous single-phase magnet, fluctuations in the magnetic anisotropy field dominate in strength over M_s-fluctuations.

4.5 SANS on systems with random paramagnetic susceptibility

In the previous sections, we have discussed the magnetic SANS of bulk magnetic materials in the ferromagnetic state, far below the Curie point T_C, where a description in terms of the continuum theory of micromagnetics becomes possible. There, we have seen that the magnetic SANS cross section changes strongly (decreases) in response to an increasing applied magnetic field. The dominant contribution to the SANS signal is elastic and its origin resides in the static magnetic microstructure. However, when approaching T_C, one of the central assumptions underlying micromagnetic theory, in particular, the constant length of the magnetization vector $|\mathbf{M}| = M_s$, becomes less and less valid and critical fluctuations take over. At T_C, the magnetic neutron scattering cross section exhibits a pronounced peak due to critical magnetic scattering [540, 541].

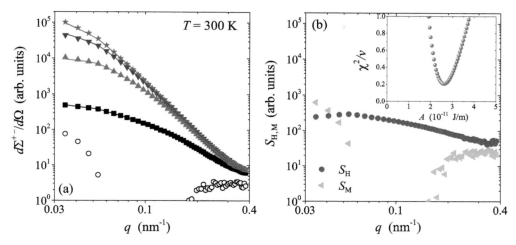

Fig. 4.34: (a) Azimuthally averaged spin-flip SANS cross section $d\Sigma^{+-}/d\Omega$ of nanocrystalline Co (average grain size: $D_{\mathrm{gs}} = 9.5 \pm 3.0\,\mathrm{nm}$) as a function of momentum transfer q at selected applied magnetic fields ($\mathbf{k}_0 \perp \mathbf{H}_0$) (log-log scale) [539]. Field values (in mT) from bottom to top: 1239, 181, 53, 24. The solid lines represent a fit of the data to the micromagnetic theory [eqn (4.31)]. ($\circ$) Residual scattering cross section $d\Sigma_{\mathrm{res}}^{+-}/d\Omega = \frac{\pi^3}{V} b_{\mathrm{H}}^2 |\widetilde{M}_s|^2$. (b) Scattering functions of the anisotropy field S_{H} and of the longitudinal magnetization S_{M} (log-log scale). The inset depicts the (reduced) weighted mean-square deviation between experiment and fit, χ^2/ν, as a function of the exchange-stiffness constant A (ν = number of degrees of freedom of the fit). After [59].

The spin-pair correlation function for critical magnetic scattering is of the Ornstein–Zernike type [eqn (2.73)], giving rise to a Lorentzian q-dependence of the cross section in the static approximation (see, e.g., [542] and Section 5.2.3).

In the following, we will discuss the paramagnetic SANS from polycrystalline rare-earth metals with a nanometer-sized average crystallite size D_{gs}. Such samples can be prepared in bulk form by means of the inert-gas-condensation technique [543–545]. For nanocrystalline Tb with grain sizes between 25–50 nm, Balaji et al. [546] have observed a counterintuitive increase of the scattering intensity by about one order of magnitude with increasing applied field (from 0–5 T) at low momentum transfers q, whereas the scattering was found to decrease at high q-values. The latter was ascribed to the suppression of local paramagnetic spin fluctuations. Consequently, a rather unusual crossover of the scattering curves at different fields was observed, similar to what can be seen in Fig. 4.35(a) for the case of nanocrystalline Gd. A quantitative explanation for the increase of the scattering cross section at low q, similar to what can be seen in Fig. 4.35 for nanocrystalline Gd and Ho, is based on the well-known anisotropy of the paramagnetic susceptibility tensor of Tb [120]. A measure for the anisotropy in the paramagnetic state is the difference in the values of the Curie temperatures along different crystallographic directions (see Table 4.3). For the nanocrystalline material, the random orientation of the crystallographic axes of the individual grains gives rise

Table 4.3   Paramagnetic Curie temperatures Θ_p of some heavy rare-earth metals measured parallel ($\parallel$) and perpendicular ($\perp$) to the hexagonal c-axis [120].

	$\Theta_p^{\parallel}$ (K)	$\Theta_p^{\perp}$ (K)
Gd	317	317
Tb	195	239
Dy	121	169
Ho	73.0	88.0
Er	61.7	32.5
Tm	41.0	−17.0

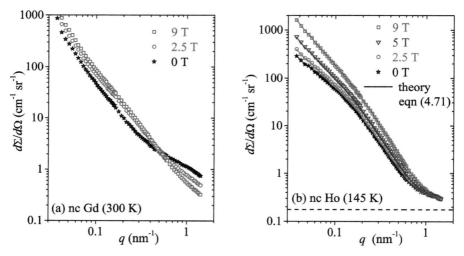

Fig. 4.35: Paramagnetic SANS on nanocrystalline (nc) ^{160}Gd and Ho. Shown are the azimuthally averaged unpolarized SANS intensities of (a) nanocrystalline ^{160}Gd at $T = 300$ K and (b) nanocrystalline Ho at $T = 145$ K for applied-field values as indicated (log-log scale) ($\mathbf{k_0} \perp \mathbf{H_0}$). Average grain sizes: $D_{gs} = 21 \pm 6$ nm (Gd) and $D_{gs} = 51 \pm 7$ nm (Ho). Solid lines in (b) are theoretical predictions by eqn (4.71). Dashed horizontal line in (b) is paramagnetic SANS at zero field, $d\Sigma/d\Omega \cong 0.175$ cm^{-1}sr^{-1} [eqn (4.70)]. After [547].

to a highly nonuniform magnetic field response on the nanoscale due to the anisotropy of the susceptibility tensor; in other words, in these samples the scattering contrast between neighboring grains (and thus the total scattering cross section) is strongly increased with the field, in contrast to the usual suppression of magnetic nanoscale disorder in ferromagnets, which is associated with a decrease of the SANS signal (see, e.g., the previous Section 4.4).

Figure 4.35 shows the results for the field-dependent paramagnetic SANS of nanocrystalline Gd ($4f^7$ for Gd^{3+}; L = 0, S = J = 7/2), as a representative of a low-anisotropy material, and of Ho ($4f^{10}$ for Ho^{3+}; L = 6, S = 2, J = 8) with a strong magnetic anisotropy (compare also Table 4.3) [547]. The order of magnitude of the magnetic scattering signal is very much larger than what may be expected for an idealized paramagnet. For a system consisting of N non-interacting spins (with an angular momentum number of J) in the scattering volume V, the unpolarized macroscopic differential scattering cross section at zero applied field reads [76, 548]:

$$\frac{d\Sigma}{d\Omega} = \frac{N}{V}\frac{2}{3}\left(\frac{1}{2}\gamma_n r_e\right)^2 g^2 J(J+1)|f(\mathbf{q})|^2, \tag{4.70}$$

where $\gamma_n = 1.913$, $r_e = 2.818 \times 10^{-15}$ m is the classical radius of the electron, g is the Landé factor, and the atomic magnetic form factor $f(\mathbf{q}) \cong 1$ in the small-angle region. For Ho ($N/V = 3.21 \times 10^{28}$ m^{-3}; $g = 5/4$; J = 8), the magnitude of this scattering cross section evaluates to $d\Sigma/d\Omega \cong 0.175$ cm^{-1}sr^{-1} [dashed horizontal line in Fig. 4.35(b)].

By exploiting the linear field response of paramagnets, closed-form expressions for $d\Sigma/d\Omega$ were obtained by Balaji et al. [546]; in particular, at small momentum transfers, the azimuthally averaged total nuclear and magnetic $d\Sigma/d\Omega$ in the paramagnetic state can be written as ($\mathbf{k}_0 \perp \mathbf{H}_0$)

$$\frac{d\Sigma}{d\Omega}(q, H_i) = \frac{d\Sigma_{\text{nuc}}}{d\Omega}(q) + S_{\text{mag}}(q)H_i^2. \tag{4.71}$$

The field-independent magnetic scattering function S_{mag} is related to the size and shape of the grains as well as to the main-axis entries of the susceptibility tensor. Since $d\Sigma/d\Omega$ is linear in H_i^2, we see that when the experimental $d\Sigma/d\Omega$ are plotted at each discrete q versus the values of H_i^2, a weighted straight-line fit provides the values of $d\Sigma_{\text{nuc}}/d\Omega$ and S_{mag} at the particular q (see Fig. 4.36). The values of the theoretical cross sections computed in this way are connected by solid lines in Fig. 4.35(b). For Ho, an excellent agreement between the data and eqn (4.71) is found over the whole (q, H_i)-range, whereas for Gd the agreement is quantitative only up to fields of $\sim$0.6 T [547]. Regarding the angular anisotropy on a two-dimensional detector, the theory predicts that the paramagnetic SANS cross section follows a $\sin^2\theta$-behavior, i.e., enhanced scattering in the direction normal to the field.

Integration of the S_{mag} data, according to

$$Q_p = \frac{45\mu_0^2}{13\pi^2 b_H^2} \int_0^\infty S_{\text{mag}}(q)q^2 dq = (\chi_{11} - \chi_{33})^2, \tag{4.72}$$

yields an invariant Q_p of the azimuthally averaged paramagnetic scattering cross section [546]. Note that S_{mag} has a dimension of cm^{-1}sr^{-1}Tesla^{-2} and H_i in eqn (4.71) is in Tesla. Q_p is related to a deviatoric component, $(\chi_{11} - \chi_{33})^2$, of the susceptibility tensor, where χ_{11} and χ_{33} denote, respectively, the susceptibility along the a-axis and c-axis of the hcp lattice. When the integration in eqn (4.72) is carried out we obtain a lower bound for Q_p due to the limited range of experimental scattering vectors

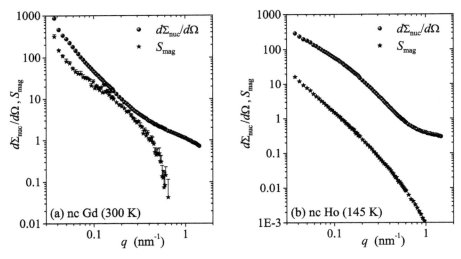

Fig. 4.36: $d\Sigma_{\mathrm{nuc}}/d\Omega$ and S_{mag} of nanocrystalline (nc) ^{160}Gd at $T = 300\,\mathrm{K}$ (a) and of nanocrystalline Ho at $T = 145\,\mathrm{K}$ (b) as obtained from the fit of the data shown in Fig. 4.35 to eqn (4.71) (log-log scale). After [547].

($q_{\min} \leq q \leq q_{\max}$). The resulting value for Ho, $\sqrt{Q_{\mathrm{p}}} \cong 0.060$, compares reasonably well with the value of $|\chi_{11} - \chi_{33}| \cong 0.046$ estimated from single-crystal magnetization measurements [549]. Note that the theory behind eqn (4.71) neglects terms of local demagnetizing field and magnetoelastic coupling, which may result in a reduced value of Q_{p} [546]. Furthermore, the results for nanocrystalline Ho do not provide evidence for grain-boundary-induced effects, as can be seen e.g., from the fact that S_{mag} takes on very low values for $q \gtrsim 1\,\mathrm{nm}^{-1}$ [see Fig. 4.36(b)]. While this does not necessarily imply that grain boundaries have no relevance for the paramagnetic properties of nanocrystalline Ho, we note that the comparatively large crystallite size of 51 nm of the Ho sample is associated with a volume fraction of grain boundaries of only about 8% [449].

The results for S_{mag} of nanocrystalline Gd [Fig. 4.36(a)] indicate that the random paramagnetic susceptibility of the individual grains leads to a significant correlated nanoscale spin disorder at 300 K and moderate fields in accordance with the model proposed in [546]. From eqn (4.72) a value of $\sqrt{Q_{\mathrm{p}}} \cong 0.27$ can be derived. Note that for $q \gtrsim 0.5\,\mathrm{nm}^{-1}$, S_{mag} could not be evaluated from the data due to critical scattering. Since $\sqrt{Q_{\mathrm{p}}}$ compares directly with the difference $|\chi_{11} - \chi_{33}|$ and since Gd is known to exhibit by far the lowest magnetocrystalline anisotropy among the heavy rare-earth metals [550], one may expect to obtain a much lower $\sqrt{Q_{\mathrm{p}}}$-value for Gd than for nanocrystalline Ho. However, the experimental temperature value of 300 K differs only by about 15 K from the Curie temperature of nanocrystalline Gd with $D_{\mathrm{gs}} = 21\,\mathrm{nm}$ [551], whereas for the Ho data there is a difference of more than 55 K to $\Theta_{\mathrm{p}}^{\perp}$, leading to comparatively low susceptibility values [$\chi \propto (T - \Theta_{\mathrm{p}})^{-1}$] and, thus, to a small value of $|\chi_{11} - \chi_{33}|$.

The observation of a finite magnetic anisotropy of the susceptibility tensor of Gd in the paramagnetic state is surprising in view of the fact that the orbital angular momentum of the Gd^{3+}-ion vanishes. This finding indeed addresses fundamental aspects regarding the origin of magnetic anisotropy of Gd metal, which has been investigated in the ferromagnetic temperature regime by means of first-principles density functional theory [428]. The authors of [428] have suggested a mechanism for the magnetic anisotropy energy involving (besides dipolar anisotropy) the spin-orbit interaction of the conduction electrons. In this respect, it would be of interest to investigate in detail the origin of the observed magnetic anisotropy of nanocrystalline Gd in the paramagnetic temperature regime. The magnetic SANS of nanocrystalline Tb and Gd in the ferromagnetic state is discussed in Section 6.5.

4.6 Magnetic SANS on dislocations

The spatial variation of the density which is related to the long-range strain fields around dislocations (or defects in general) gives rise to a nuclear small-angle scattering signal (compare Figs. 4.37 and 4.38). This was recognized in the 1950s when the first SAXS experiments on deformed metals were conducted (e.g., [505, 552, 553]). However, the early SAXS data were dominated by double-Bragg scattering and only the usage of long-wavelength cold neutrons from research reactors aided to overcome this difficulty, so that the "true" small-angle scattering due to the dislocations could be accessed [503, 506]. A nice side effect to using cold neutrons in this regard is the possibility to exploit the neutron spin degree of freedom for the study of dislocations in ferromagnets. The strain field of the defect couples to the magnetization distribution by means of magnetoelastic coupling and gives rise to deviations from the homogeneously magnetized state (compare Fig. 1.7). Kronmüller, Seeger, and Wilkens [1] delivered the first theoretical treatment of magnetic SANS in terms of the continuum theory of micromagnetics. Their results show that magnetic SANS on dislocations [2, 4–6, 8, 510–515] is much larger than the corresponding nuclear SANS: for Ni and an applied magnetic field of 0.13 T, the SANS due to the magnetic fluctuations is roughly a factor of 20 larger than the nuclear scattering on the density fluctuations [1, 5].

Based on the review article by Schmatz [5], which represents an authoritative and detailed exposition of nuclear and magnetic SANS on dislocation structures, we will in the following briefly sketch the main steps to compute the magnetic SANS cross section due to dislocations. Since most of the investigations published so far employed unpolarized neutrons, we will consider only this situation. For an unpolarized incident neutron beam, the total macroscopic nuclear and magnetic SANS cross section of a mono-atomic ensemble of N atoms in the scattering volume V can be written as [5]:

$$\frac{d\Sigma}{d\Omega}(\mathbf{q}) = \frac{d\Sigma_{nuc}}{d\Omega}(\mathbf{q}) + \frac{d\Sigma_{M}}{d\Omega}(\mathbf{q}) = \frac{1}{V}\left(\overline{b}^2|\widetilde{N}|^2 + b_m^2|\widetilde{\mathbf{Q}}|^2\right), \qquad (4.73)$$

where $\overline{b}$ and b_m are the nuclear and magnetic scattering lengths, respectively, and

$$\widetilde{N}(\mathbf{q}) = \frac{1}{\Omega_V}\int_V N(\mathbf{r})\exp\left(-i\mathbf{q}\cdot\mathbf{r}\right)d^3r = \frac{1}{\Omega_V}\int_V \frac{\rho_a(\mathbf{r})-\overline{\rho}}{\overline{\rho}}\exp\left(-i\mathbf{q}\cdot\mathbf{r}\right)d^3r \qquad (4.74)$$

denotes the Fourier transform of the relative density difference $N(\mathbf{r}) = (\rho_a(\mathbf{r})-\overline{\rho})/\overline{\rho}$ (dilatation), with $\Omega_V = V/N$ the average atomic volume. In the above equations, we

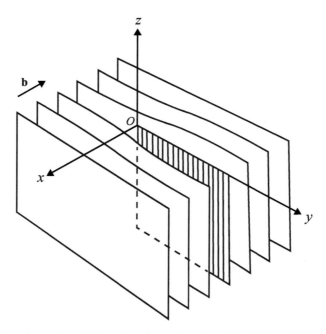

Fig. 4.37: Sketch of an edge dislocation in a simple cubic lattice. The dislocation core lies along the z-axis. The extra half-plane of atoms is indicated by shading. The x-axis is parallel to the Burgers vector **b**, which is perpendicular to the dislocation line. It is clear that there is an increase of density for points with y positive and a decrease for y negative. Nonmagnetic small-angle scattering on dislocations is a consequence of spatial variations in the density of electrons (x-rays) or of nuclei (neutrons) extending over several atomic dimensions. After [504].

have partly adopted the notation of Schmatz, where the Fourier transforms $\widetilde{N}(\mathbf{q})$ and $\widetilde{\mathbf{Q}}(\mathbf{q})$ are dimensionless quantities. Note also that the nuclear scattering-length density function $N(\mathbf{r})$ should not be confused with the number N of nuclei in the sample. The scattering amplitude eqn (4.74) has been evaluated by Seeger and Kröner [507] for an elastic isotropic medium within the limits of linear elasticity theory. For a more recent treatment of nuclear dislocation scattering, including the effect of dislocation walls, see [524, 525]. Since the nuclear scattering due to dislocations can be eliminated by taking the difference between the SANS cross sections at two different magnetic fields (similar to the discussion related to the subtraction of the residual SANS in Section 4.2), we concentrate in the following on the magnetic SANS, which is also treated using linear elasticity theory.

Seeger and Kronmüller [555, 556] have performed analytical calculations of the influence of internal stresses due to dislocations on the approach-to-saturation regime of the magnetization curve of cold-worked ferromagnets. These results form the basis for the evaluation of the corresponding magnetic SANS cross section in [1]. Assuming small spin misalignment, i.e., $m_x \ll 1$, $m_y \ll 1$, and $m_z \cong 1$, the Fourier transform

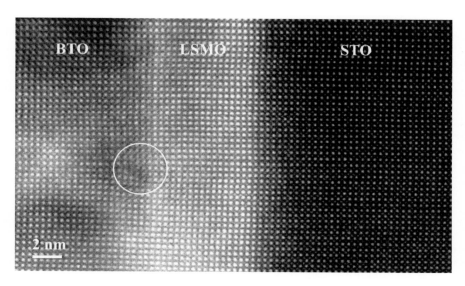

Fig. 4.38: High-resolution scanning transmission electron microscopy image of a $BaTiO_3$(BTO)/$La_{1-x}Sr_xMnO_3$(LSMO)/$SrTiO_3$(STO) epitaxial heterostructure. The white circle marks a dislocation, which has formed at a BTO-LSMO interface. Image courtesy of Saleh Gorji, Di Wang, and Christian Kübel, Karlsruhe Institute of Technology, Karlsruhe, Germany. After [554].

$\widetilde{\mathbf{m}}(\mathbf{q}) = \{\widetilde{m}_x, \widetilde{m}_y, \widetilde{m}_z\}$ of the unit magnetization vector $\mathbf{m}(\mathbf{r}) = \{m_x, m_y, m_z\}$ can be expressed in terms of the Fourier transform of the strain tensor components $\widetilde{\eta}_{ij}$ [see eqn (4.81)]. The direction of the applied magnetic field (saturation direction) coincides here with the z-direction. Taking the magnetoelastic coupling energy, the exchange energy, and the magnetic stray-field energy into account, the Fourier transforms of the magnetization components can be written as [5, 134, 555]:

$$\widetilde{m}_x(q_x, q_y, q_z) = -\frac{\widetilde{g}_x}{q^2 + l_H^{-2}} + l_M^{-2}\frac{q_x}{q^2 + l_H^{-2}}\frac{\widetilde{g}_xq_x + \widetilde{g}_yq_y}{q^4 + (l_H^{-2} + l_M^{-2})q^2 - l_M^{-2}q_z^2}, \quad (4.75)$$

$$\widetilde{m}_y(q_x, q_y, q_z) = -\frac{\widetilde{g}_y}{q^2 + l_H^{-2}} + l_M^{-2}\frac{q_y}{q^2 + l_H^{-2}}\frac{\widetilde{g}_xq_x + \widetilde{g}_yq_y}{q^4 + (l_H^{-2} + l_M^{-2})q^2 - l_M^{-2}q_z^2}, \quad (4.76)$$

$$\widetilde{m}_z(q_x, q_y, q_z) = N\delta(\mathbf{q}), \quad (4.77)$$

where the micromagnetic exchange lengths l_H and l_M are given by eqns (3.66) and (3.43). Since the (single-phase) material is assumed to be nearly saturated along the z-direction, the corresponding Fourier component $\widetilde{m}_z$ reduces to a delta function at the origin of reciprocal space. The second terms on the right-hand sides of the symmetric eqns (4.75) and (4.76) are due to the magnetostatic stray-field energy. Their magnitude relative to the respective first term is 20−50% [5], which emphasizes the importance of the dipolar interaction (compare Section 4.3.2). The functions $\widetilde{g}_{x,y}(\mathbf{q})$ are the Fourier transforms of the partial derivatives of the magnetoelastic coupling energy density ω_{me} with respect to the $m_{x,y}$ [5, 555]:

$$\tilde{g}_{x,y}(\mathbf{q}) = \frac{1}{\Omega_{\mathrm{V}}} \int_V g_{x,y}(\mathbf{r}) \exp\left(-i\mathbf{q}\cdot\mathbf{r}\right) d^3r \tag{4.78}$$

with

$$g_{x,y} = -\frac{1}{2A} \frac{\partial\omega_{\mathrm{me}}}{\partial m_{x,y}}\bigg|_{m_x=m_y=0,m_z=1} \tag{4.79}$$

in the high-field limit (A: exchange-stiffness constant). The magnetoelastic coupling energy depends on the components σ_{ij} of the stress tensor and on the $m_{x,y,z}$. For cubic crystals, ω_{me} is given (in cubic crystal coordinates) by eqn (3.15), which is rewritten here for convenience:

$$\begin{aligned}\omega_{\mathrm{me}}^{\mathrm{c}} = & -\frac{3}{2}\lambda_{100}\left(\sigma_{xx}m_x^2 + \sigma_{yy}m_y^2 + \sigma_{zz}m_z^2\right) \\ & -\frac{3}{2}\lambda_{111}\left(\sigma_{xy}m_xm_y + \sigma_{xz}m_xm_z + \sigma_{yz}m_ym_z\right),\end{aligned} \tag{4.80}$$

where λ_{100} and λ_{111} denote the saturation magnetostriction constants along the indicated crystallographic directions. Combining eqns (4.78)−(4.80) with eqns (4.75) and (4.76) one immediately realizes that the magnetization Fourier components $\tilde{m}_{x,y}$ depend linearly on the Fourier transforms of the components $\tilde{\sigma}_{ij}$ of the stress tensor. These are in turn related to the Fourier-transformed components of the displacement field $\tilde{s}_{ij}$ and of the strain field $\tilde{\eta}_{ij}$ according to [5, 6]:

$$\tilde{\sigma}_{ij}(\mathbf{q}) = -iC_{ijkl}q_l\tilde{s}_k(\mathbf{q}) = C_{ijkl}\tilde{\eta}_{kl}(\mathbf{q}), \tag{4.81}$$

where C_{ijkl} is the elasticity tensor. In the early theory [1], the Fourier transform of the stress tensor for an elastic isotropic medium was calculated from the real-space stress tensor of a dislocation in the axis-system of a cylinder [5]. Later, based on an analytic expression for the elastic Green function of anisotropic cubic crystals [557] and on the representation of the displacement field by a line integral [558], Schmatz, Dederichs, and Scheuer [4] derived analytic expressions for the dilatation and the stress tensor in reciprocal space. In this way, one can obtain analytical closed-form expressions for the $\tilde{m}_{x,y}$, which can then be used to compute the magnetic SANS cross section $d\Sigma_{\mathrm{M}}/d\Omega$. It is also emphasized that the magnetocrystalline anisotropy and spatial variations in the saturation magnetization have been neglected in the above formulas. The algebra to obtain $d\Sigma_{\mathrm{M}}/d\Omega$ (and $d\Sigma_{\mathrm{nuc}}/d\Omega$) for a given scattering geometry, dislocation type (edge or screw dislocation), and for a particular arrangement of dislocations (single dislocation versus dislocation network) is quite complicated and beyond the scope of this book. Before showing some experimental data and closing this section, we state the asymptotic power-law dependencies which are expected for the orientationally averaged nuclear and magnetic SANS cross section on dislocations [5, 507, 508]:

$$\frac{d\Sigma_{\mathrm{nuc}}}{d\Omega} \propto q^{-3}, \tag{4.82}$$

$$\frac{d\Sigma_{\mathrm{M}}}{d\Omega} \propto q^{-3}\left(q^2 + l_{\mathrm{H}}^{-2}\right)^{-2}, \tag{4.83}$$

where $q\overline{L} \gg 1$ is assumed with $\overline{L}$ the average straight dislocation length. Equation (4.83) shows that $d\Sigma_{\mathrm{M}}/d\Omega \propto q^{-7}$ for $q \gg l_{\mathrm{H}}^{-1}$ and that the magnetic scattering

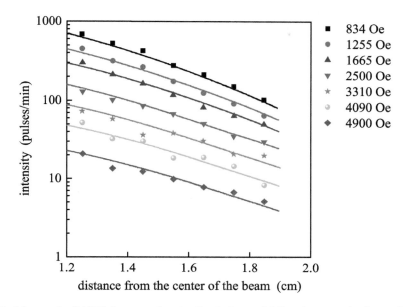

Fig. 4.39: Magnetic SANS from a plastically deformed Ni polycrystal after subtraction of the scattering at 0.603 T. Solid lines are calculated with reasonable dislocation parameters. The distance from the center of the beam, used as abscissa, is a measure for the length of the scattering vector. After [5, 510].

vanishes at infinite field (since $l_H^{-2} \propto H_0$), exactly as the spin-misalignment scattering does (compare Section 4.2).

As an example for magnetic dislocation scattering, Fig. 4.39 displays the field-dependent purely magnetic SANS signal of plastically deformed polycrystalline Ni [510]. The nuclear and magnetic contribution at a large field of 0.603 T has been subtracted from the data taken at lower fields. Moreover, neutron data taken on an un-deformed and annealed polycrystalline reference sample showed practically no field dependence of the cross section. Based on (i) the magnitude of the magnetic SANS cross section relative to the scattering from the un-deformed sample, (ii) the good agreement between the data and the fit to the asymptotic theory prediction (solid lines in Fig. 4.39), and (iii) the fact that reasonable values for the dislocation densities and the average distance between dislocations were obtained, Schmatz [5] concluded that real magnetic dislocation scattering is seen in Fig. 4.39. As mentioned in Section 4.1, magnetic SANS research on dislocations petered out in the 1980s with the paper by Göltz et al. [6] on the dislocation arrangements in plastically deformed Fe single crystals summarizing the state-of-the-art.

5

MAGNETIC SANS OF NANOPARTICLES AND COMPLEX SYSTEMS

In Chapter 4, we have seen that for bulk ferromagnets a theoretical framework of magnetic SANS has been developed and experimentally tested [49, 59, 312, 414, 439, 444, 467, 488]. By contrast, there exists the open problem of calculating the magnetic SANS cross section of *nonuniformly magnetized nanoparticles that are rigidly embedded in a nonmagnetic matrix*. This is the prototypical sample microstructure in many magnetic SANS experiments, and it is this subfield of magnetic SANS where we expect—with the aid of numerical micromagnetic computations—the largest progress to be made in the coming years. In this chapter, we discuss the current state-of-the-art. Moreover, we have devoted several sections to so-called "complex systems", by which we denote material classes such as ferrofluids, magnetic steels, spin glasses and amorphous magnets, or magnetic oxides and alloys. These magnetic materials, as well as the nanoparticle systems, are characterized by distributions of particle sizes and shapes and/or by distributions of magnetic-interaction strengths, i.e., spatial nanometer or atomic-scale variations of the exchange interaction, magnetic anisotropy, and saturation magnetization. For ferrofluids, even when the individual particles are in a single-domain state, the isotropic short-range steric repulsion of particles, the anisotropic and long-range magnetodipolar interaction, the van der Waals interaction, and complex hydrodynamic interactions might be of additional relevance. Overall, the complexity involved is for most problems very difficult to grasp into an analytical treatment of the related magnetic SANS cross section, so that the analysis of experimental data largely relies on phenomenological models. In fact, much of the discussion in this chapter does not directly connect to the micromagnetic SANS results introduced in Chapter 4, but is related to the macrospin-based particle-matrix approach in terms of form and structure factors, adapted from nonmagnetic SANS and small-angle x-ray scattering (SAXS).

This chapter is structured as follows. In Section 5.1.1, we scrutinize the assumptions and implications underlying the conventional particle-matrix-based model used in SANS studies on magnetic nanoparticle assemblies. Section 5.1.2 then displays the results for the cross sections of superparamagnetic single-domain particles. Section 5.1.3 provides the full set of Brown's static equations of micromagnetics, which are the starting point for any computation of the spin structure and ensuing magnetic SANS cross section. Section 5.1.4 summarizes the SANS by magnetic vortices in thin submicron-sized soft ferromagnetic cylinders. Sections 5.2.1, 5.2.2, and 5.2.3 briefly review SANS

Magnetic Small-Angle Neutron Scattering: A Probe for Mesoscale Magnetism Analysis. Andreas Michels, Oxford University Press (2021). © Andreas Michels. DOI: 10.1093/oso/ 9780198855170.003.0005

on, respectively, ferrofluids,[††] magnetic steels, and spin glasses and amorphous magnets. Section 5.2.4 provides a selection of further SANS applications, mainly intended to provide a guide to the literature.

5.1 Nanoparticles

5.1.1 Description by the conventional particle-matrix approach

In order to illuminate the problem, let us examine the expression which is commonly employed for magnetic SANS data analyses on two-phase magnetic nanoparticle-nonmagnetic matrix type microstructures (see also the discussion in [49]). For such systems, the macroscopic nuclear and magnetic elastic differential SANS cross section $d\Sigma/d\Omega$ (for unpolarized neutrons) at scattering vector $\mathbf{q}$ is often expressed as (e.g., [466]):

$$\frac{d\Sigma}{d\Omega}(\mathbf{q}) = \frac{N_p}{V} |F(\mathbf{q})|^2 \left[(\Delta\rho)_{\text{nuc}}^2 + (\Delta\rho)_{\text{mag}}^2 \sin^2\alpha \right], \tag{5.1}$$

where N_p is the number of particles in the scattering volume V, the form factor of an individual particle with volume V_p is given by $F(\mathbf{q}) = \int_{V_p} \exp(-i\mathbf{q}\cdot\mathbf{r})d^3r$, the constants $(\Delta\rho)_{\text{nuc}}^2$ and $(\Delta\rho)_{\text{mag}}^2 = b_{\text{H}}^2(M_s^p - M_s^m)^2$ denote, respectively, the nuclear and magnetic scattering-length density contrasts between the particle (M_s^p) and the matrix ($M_s^m = 0$), and α is the angle between $\mathbf{q}$ and the local magnetization vector $\mathbf{M}$ of the particle. Equation (5.1) describes e.g., small-angle scattering from a dilute system of homogeneous and identical magnetic particles which are embedded in a homogeneous and nonmagnetic matrix. Additionally, it is assumed that the nuclear and magnetic form factors of the particles are identical. The introduction of a particle-size distribution function and of interparticle interference into eqn (5.1) is straightforward, but as with the incorporation of spatially dependent scattering-length densities needed e.g., to describe core-shell particles, diffusion zones, or magnetically "dead" surface layers [315–317], such a procedure affects only the q-dependence of the scattering rather than its variation with the applied magnetic field $\mathbf{H}_0$.

The dependence of eqn (5.1) on $\mathbf{H}_0$ is contained in the $\sin^2\alpha$ term, which can be written as:

$$\sin^2\alpha = 1 - (\hat{\mathbf{q}}\cdot\mathbf{m})^2. \tag{5.2}$$

This factor takes account of the dipolar nature of the neutron-magnetic interaction [109]. For a single-domain particle, the expectation value of $\sin^2\alpha$ depends on the orientation of the unit magnetization vector $\mathbf{m} = \mathbf{M}/M_s$ relative to the two-dimensional unit scattering vector $\hat{\mathbf{q}}$. For a random spin orientation, one may express the unit magnetization vector as $\mathbf{m} = \{\sin\beta_1\cos\beta_2, \sin\beta_1\sin\beta_2, \cos\beta_1\}$, where β_1

[††]The above-sketched complexity of interactions in ferrofluids and the fact that the ferrofluid particles are movable within the carrier liquid render their magnetic SANS response much more complicated than that of magnetic nanoparticles that are rigidly embedded in some solid matrix. That is why we prefer to include a separate subsection on magnetic SANS of ferrofluids into Section 5.2 on complex systems, and not in Section 5.1 on nanoparticles.

and β_2 denote the polar and azimuthal angles. Then, with reference to the perpendicular ($\mathbf{k}_0 \perp \mathbf{H}_0$) and parallel ($\mathbf{k}_0 \parallel \mathbf{H}_0$) scattering geometries, where, respectively, $\hat{\mathbf{q}} = \{0, \sin\theta, \cos\theta\}$ and $\hat{\mathbf{q}} = \{\cos\theta, \sin\theta, 0\}$ [compare Fig. 2.1 and eqns (2.85) and (2.86)], it is readily verified that the expectation value

$$\langle \sin^2 \alpha \rangle = \frac{1}{2\pi} \int_0^{2\pi} \left(\frac{1}{4\pi} \int_0^{4\pi} [1 - (\hat{\mathbf{q}} \cdot \mathbf{m})^2] \sin\beta_1 d\beta_1 d\beta_2 \right) d\theta \tag{5.3}$$

evaluates to $\langle \sin^2 \alpha \rangle = 2/3$ for both geometries. This has already been experimentally demonstrated on a Fe_3O_4 powder sample by Shull, Wollan, and Koehler [195]. Equation (5.3) represents an average over the orientation (β_1, β_2) of the nanoparticle and over the azimuthal angle θ on the detector. If the particle's magnetic moment is oriented along the applied field, corresponding to $\mathbf{m} = \{0, 0, 1\}$ for both $\mathbf{k}_0 \perp \mathbf{H}_0$ and $\mathbf{k}_0 \parallel \mathbf{H}_0$, one obtains $\langle \sin^2 \alpha \rangle = 1/2$ for $\mathbf{k}_0 \perp \mathbf{H}_0$ and $\langle \sin^2 \alpha \rangle = 1$ for $\mathbf{k}_0 \parallel \mathbf{H}_0$. Since the $\langle \sin^2 \alpha \rangle$ factor is per definition the only magnetic-field-dependent term in the azimuthally averaged version of eqn (5.1), it is seen that the magnetic contribution to $d\Sigma/d\Omega$ increases by a factor of $4/3$ between saturation and random spin orientation in the perpendicular geometry, whereas it decreases by a factor of $2/3$ for the parallel case; see [559] for a study of the effect of uniaxial magnetic anisotropy on the field dependence of $\langle \sin^2 \alpha \rangle$.

Equation (5.1) provides information on the magnitude of the particle's magnetization [via $(\Delta\rho)^2_{mag}$] and on the orientation of the average particle magnetization (via $\sin^2 \alpha$). The possible continuous variation of the local orientation of the magnetization vector inside the particle is entirely neglected; in other words, eqn (5.1) ignores the magnetic scattering due to intraparticle spin disorder. Furthermore, eqn (5.1) is not adapted to the magnetic microstructure of a nanoparticle assembly: besides providing information on the nuclear microstructure of a material (e.g., particle size and shape, particle-size distribution), the only information regarding the magnetic microstructure which is contained in eqn (5.1) refers to the magnitude and average orientation of the magnetization. There is no obvious relation to the characteristic magnetic parameters and interactions such as the exchange constant, magnetic anisotropy, or the magnetostatic stray field. Additionally, SANS cross sections based on eqn (5.1) do not comprise characteristic magnetic length scales, which may vary as a function of the applied field. In Sections 3.2 and 3.3, we have seen that the continuum theory of micromagnetics predicts that the response of the magnetization to spatially inhomogeneous anisotropy and magnetostatic fields is such that fluctuations are reduced in magnitude and wavelength as the applied field is increased [via the exchange length $l_H \propto H_i^{-1/2}$, eqn (3.66)]. Therefore, when the SANS cross section is measured at different fields, one might expect that the progressive decrease in a certain characteristic magnetic length scale (with increasing field) manifests itself as a shift of a characteristic feature in the SANS curve (e.g., the point with the largest curvature) to larger momentum transfers (see inset in Fig. 4.4). Equation (5.1) cannot reproduce such behavior, since its q-dependence resides in a term which is linked to the particle microstructure, and not to magnetism.

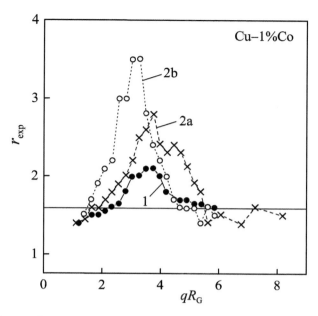

Fig. 5.1: Experimental ratio r_{exp} of the scattered intensity at zero field and at $H_0 = 2.5\,\mathrm{kOe}$ with $\mathbf{q} \parallel \mathbf{H_0}$ plotted versus qR_G, where R_G is the experimentally determined mean radius of gyration. Curve (1) corresponds to a 160-h-aged sample with $\mathbf{q} \parallel [111]$; curves (2a, 2b) to a 300-h-aged sample with $\mathbf{q} \parallel [111]$ and $\mathbf{q} \parallel [100]$, respectively. Horizontal line: $r = 1.59$ [eqn (5.6)]. After [452].

The limitations of eqn (5.1) have already become evident in the early SANS study by Ernst, Schelten, and Schmatz [451, 452]. These authors investigated the transition from single- to multi-domain configurations of Co precipitates in Cu and analyzed the following ratio $r(\mathbf{q})$ of SANS cross sections ($\mathbf{k_0} \perp \mathbf{H_0}$) [452]:

$$r(\mathbf{q}) = \frac{\frac{d\Sigma}{d\Omega}(\mathbf{q})\big|_{H_0=0}}{\frac{d\Sigma}{d\Omega}(\mathbf{q})\big|_{\mathbf{q}\parallel\mathbf{H_0}\to\infty}} = \frac{\left[\frac{d\Sigma_{\mathrm{nuc}}}{d\Omega}(\mathbf{q}) + \frac{d\Sigma_{\mathrm{M}}}{d\Omega}(\mathbf{q})\right]\big|_{H_0=0}}{\frac{d\Sigma_{\mathrm{nuc}}}{d\Omega}(\mathbf{q})\big|_{\mathbf{q}\parallel\mathbf{H_0}}}, \tag{5.4}$$

which is very similar to the so-called A-ratio [eqn (5.51)] used in the SANS data analysis of magnetic steels (see Section 5.2.2). The total unpolarized SANS cross section $d\Sigma/d\Omega$ at zero applied magnetic field equals the sum of nuclear and magnetic contributions, while the cross section at a saturating field $\mathbf{H_0}$ applied parallel to the scattering vector $\mathbf{q}$ yields (for $\mathbf{k_0} \perp \mathbf{H_0}$) the purely nuclear SANS cross section $d\Sigma_{\mathrm{nuc}}/d\Omega$ [compare eqn (2.110)]. We emphasize that the interpretation of $r(\mathbf{q})$ is highly nontrivial [560], since it depends on a number of both structural and magnetic parameters; for instance, on the particle volume fraction, at high packing densities also on the shape and size distribution of the particles, and not the least on the internal spin structure of the nanoparticles, which depends e.g., on the particle size and the applied field, but also on the strength of the magnetodipolar interaction between the particles.

Consider the special case of a dilute assembly of randomly oriented homogeneous single-domain particles: if for $H_0 = 0$ the magnetizations of the particles are randomly oriented, then the unpolarized two-dimensional $d\Sigma/d\Omega$ is isotropic, whereas it exhibits the well-known $\sin^2 \theta$ angular anisotropy for the saturated case, $\mathbf{k}_0 \perp \mathbf{H}_0$, and for a not-too-strong nuclear signal. For this particular situation, the ratio r depends only on the magnitude q of the scattering vector,

$$r(q) = \frac{\left[\frac{d\Sigma_{\text{nuc}}}{d\Omega}(q) + \frac{d\Sigma_{\text{M}}}{d\Omega}(q) \right]}{\frac{d\Sigma_{\text{nuc}}}{d\Omega}(q)} = 1 + \frac{\frac{d\Sigma_{\text{M}}}{d\Omega}(q)}{\frac{d\Sigma_{\text{nuc}}}{d\Omega}(q)}, \qquad (5.5)$$

where the isotropic zero-field SANS cross section may be averaged over 2π. By contrast, for a globally anisotropic microstructure, e.g., for oriented shape-anisotropic particles [452, 561, 562] or for a system exhibiting a large remanence, the r-ratio may depend on the orientation of $\mathbf{q}$. Moreover, if the dilute ensemble of randomly oriented single-domain particles exhibits identical chemical and magnetic particle sizes, then eqn (5.5) further simplifies to the q-independent value:

$$r = 1 + \frac{2}{3} \frac{(\Delta\rho)^2_{\text{mag}}}{(\Delta\rho)^2_{\text{nuc}}}, \qquad (5.6)$$

where the factor of 2/3 results from the orientational average of the $\sin^2 \theta$ factor of the magnetic SANS cross section in the remanent state. Under the above assumptions, deviations from this constant value may indicate the presence of intraparticle spin disorder. Based on the evaluation of $r(\mathbf{q})$ (see Fig. 5.1), Ernst, Schelten, and Schmatz revealed that their particles are shape anisotropic (r depends on the direction of $\mathbf{q}$) and that the transition to a multi-domain state occurs for a Co particle radius of about 20 ± 3 nm. Interestingly, for the analysis of $r(\mathbf{q})$, numerical micromagnetic computations assuming different internal spin structures were performed, however, no details on these calculations are reported in [452]. Supported by large-scale micromagnetic simulations, Bersweiler et al. [560] also employed the intensity-ratio plot [Fig. 5.1 and eqn (5.4)] to find an indication for particle-size-dependent internal spin disorder in Mn–Zn-ferrite nanoparticles.

Besides [452], there exist numerous experimental SANS investigations on ensembles of magnetic nanoparticles which report a nonuniformly magnetized, canted, or core-shell-type internal spin structure (e.g., [460, 529, 560, 563–585]). As an example, Fig. 5.2 displays the results of a detailed polarized SANS study on non-interacting cubic and spherical iron-oxide nanoparticles by Disch et al. [529]. Sketches of the assumed spatial magnetization profiles inside the inorganic particle cores are depicted. Based on the comparison of the experimental magnetic form factor to different models for the nanoparticles' magnetic scattering-length density (uniform sphere, core-shell structure, linear decrease of magnetization), it was concluded that for both nanospheres and nanocubes the magnetic nanoparticle volume is smaller than the nuclear volume, most likely due to a thin surface shell (thickness: ∼0.3−0.5 nm) of gradually decreasing magnetization [Fig. 5.2(c)]. In this context, we also refer to the study by Zákutná et al. [585] who reported surface spin disorder in non-interacting ferrite nanoparticles with a surface-layer thickness of ∼0.3 nm in saturation, increasing

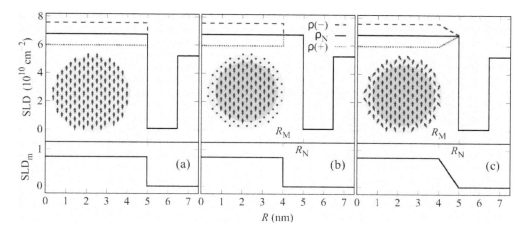

Fig. 5.2: (a)–(c) Contrast profiles for different magnetization-distribution models of iron-oxide nanoparticles (spheres and cubes). Top: scattering-length density (SLD) profiles are given for the two incident neutron polarizations, $\rho(+)$ and $\rho(-)$, with ρ_N being the purely nuclear contrast profile. R_N and R_M denote the radii of, respectively, the nonmagnetic nuclear and the magnetic particle core. Insets depict sketches of the spin structures: (a) uniform magnetization; (b) magnetic core with a nonmagnetic shell; (c) linear decrease in magnetization density towards the surface. Bottom: magnetic scattering-length density profiles SLD_m. After [529].

up to ∼0.7 nm with decreasing applied magnetic field. The quantitative analysis by Disch et al. [529] reveals moreover that even the magnitude of the atomic magnetic moments in the particle core is significantly lower than expected. Therefore, the low magnetization observed macroscopically in nanoparticles may result partially from spin canting at the surface, but to a much larger extent from the reduced magnetization inside the nanoparticle core, e.g., due to the presence of vacancies or antiphase domain boundaries [177].

Figure 5.3 displays the SANS cross sections of several magnetic nanoparticle assemblies—a Cu–Ni–Fe alloy in Fig. 5.3(a) and Co nanowires in Fig. 5.3(b). It is seen that the respective SANS signal at small and intermediate q does change stronger than the prediction for homogeneously magnetized domains, embodied by eqn (5.1). Günther et al. [586] have suggested the presence of significant intraparticle spin disorder when analyzing SANS experiments on a Co nanorod array: the strong field dependence of the SANS signal that these authors have observed [a factor of about 5 between the largest field of 2 T and the coercive field of −0.05 T, see Fig. 5.3(b)] could not be explained by eqn (5.1). Polarized SANS experiments on ferromagnetic nanowires by Napolskii et al. [587] and by Maurer et al. [588,589] additionally indicate that the nanowires' stray fields have to be taken into account in the magnetic form-factor derivation. Significant dipolar interwire interactions in high-density arrays of segmented Fe–Ga/Cu nanowires were also reported by Grutter et al. [590] employing SANS polarization analysis.

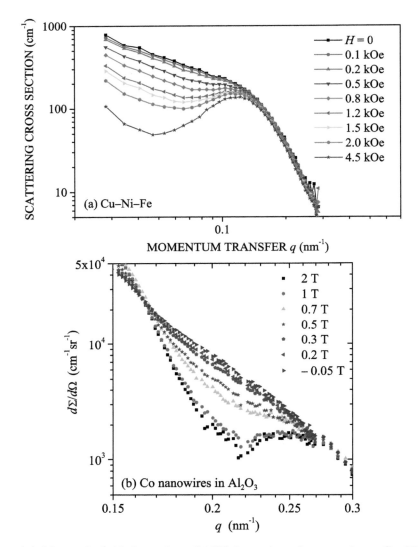

Fig. 5.3: (a) Magnetic-field-dependent SANS intensity of a two-phase Cu–Ni–Fe alloy ($\mathbf{k}_0 \perp \mathbf{H}_0$) (log-log scale). The sample consists of nanometer-sized ferromagnetic clusters which are embedded in a nonmagnetic crystalline bulk matrix. After [43]. (b) Field dependence of the unpolarized SANS cross section $d\Sigma/d\Omega$ of Co nanowires (diameter: 27 ± 3 nm), which are embedded in an Al_2O_3 matrix ($\mathbf{k}_0 \perp \mathbf{H}_0$) (log-log scale). The long rod axes are aligned parallel to the incident neutron beam. After [59].

The previous discussion clearly suggests that the macrospin model which is embodied in eqn (5.1) is oversimplified, since it assumes a homogeneous or stepwise homogeneous nanoparticle magnetization and, thus, ignores the complex spin structures that may appear at the nanoscale. Intraparticle spin disorder may be due to the interplay between different magnetic interactions such as surface anisotropy and dipolar interaction, or to the presence of crystal defects and antiphase boundaries (see, e.g., [162,163,168–179] and references therein). Therefore, magnetic SANS approaches based on eqn (5.1) might be erroneous when the spatial dependence of the magnitude and direction of the magnetization $\mathbf{M}(\mathbf{r})$ is not taken into account. Moreover, it is obvious that nothing can be directly learned from eqn (5.1) on the internal magnetodipolar interaction, the magnetic anisotropy, or on the exchange interaction, simply because the corresponding energy terms are left out in the analysis of the problem.

In summary, there is ample theoretical and experimental evidence that the spin structure of nanoparticles may be nonuniform. Therefore, instead of solving the geometrical (form factor) and statistical-mechanics (structure factor) problems which are inherent to macrospin-based models for the magnetic SANS cross section, it appears to be straightforward to employ the continuum theory of micromagnetics for calculating the nanoparticles' spin structure. The corresponding Fourier image will then naturally provide the desired magnetic SANS cross section. As we will see in Section 5.1.3, this procedure is very challenging, since the governing differential equations are highly nonlinear and boundary conditions for the magnetization at the nanoparticle's surface have to be taken into account. Before proceeding to the discussion of SANS due to inhomogeneous magnetization states, we examine in the following the magnetic SANS cross section of single-domain particles exhibiting superparamagnetism.

5.1.2 SANS of superparamagnetic particles

This topic has first been addressed by Cywinski et al. [458], Pynn et al. [459], Bellouard, Mirebeau, and Hennion [460, 461], and by Kohlbrecher, Wiedenmann, and Heinemann [462–466]. We consider here only the case of a dilute set of superparamagnetic spherically symmetric particles; see the works by Hennion and Mirebeau [461] and by Wagner and Kohlbrecher [43] for unpolarized and spin-polarized cross-section expressions which take interparticle interference into account. For superparamagnetic particles, besides the possible average over the particle-size distribution $f(R)$, an additional averaging procedure over the orientation-distribution function of the individual magnetic moments relative to the field direction becomes relevant. Before we analyze the SANS cross section, we will first briefly discuss the two main relaxation mechanisms of superparamagnetic nanoparticles.

The total magnetic energy of a single-domain particle in the presence of an applied field can be expressed as:

$$E = E_{\mathrm{z}} + E_{\mathrm{uni}} = -\mu_0 V_{\mathrm{p}} \mathbf{H}_0 \cdot \mathbf{M} + K_{\mathrm{u}} V_{\mathrm{p}} \sin^2 \gamma, \qquad (5.7)$$

where $M_{\mathrm{s}} = |\mathbf{M}|$ is the saturation magnetization of the particle with volume V_{p}. Since, in general, magnetic materials are not isotropic but exhibit some kind of anisotropy, we have added in eqn (5.7) a contribution E_{uni} due to an effective magnetic anisotropy. For simplicity, we have assumed a first-order uniaxial anisotropy of strength $K_{\mathrm{u}} > 0$, which may have its origin in magnetocrystalline and/or magnetic shape anisotropy. The magnetic behavior of such a single-domain particle, as probed by a certain experimental technique, depends decisively on the ratio of the anisotropy-energy barrier $K_{\mathrm{u}} V_{\mathrm{p}}$ and kT. Within the Néel–Brown theory of superparamagnetism, the magnetic relaxation time τ_N of a particle exhibiting uniaxial magnetic anisotropy is given by [154, 155, 591, 592]:

$$\tau_N = \tau_0 \exp\left(\frac{K_u V_p}{kT}\right), \tag{5.8}$$

where $\tau_0 \cong 10^{-12} - 10^{-9}$ s is a characteristic microscopic attempt time, which may depend on temperature and applied magnetic field. Note the exponential dependence of τ_N on the particle volume, which implies that (at constant T) tiny changes in the size of a particle may have dramatic consequences for the relaxation time. At zero applied magnetic field, eqn (5.8) describes the average time it takes for the magnetic moment to jump between the two equivalent energy minima of a particle exhibiting uniaxial magnetic anisotropy. If τ_N becomes very large compared to the characteristic timescale t_{exp} of the experimental technique (measuring time), say $\tau_N \gtrsim 100\,t_{exp}$, then one observes stable, i.e., time-independent ferromagnetic properties. The magnetization vector is then pointing along either one of the two energy minimum directions at $\gamma = 0°$ or $\gamma = 180°$. Note that $t_{exp} \sim 10^{-5}–10^{-1}$ s for ac magnetic susceptibility, $t_{exp} \sim 10^{-10}–10^{-5}$ s for muon-spin spectroscopy, $t_{exp} \sim 10^{-10}–10^{-7}$ s for Mössbauer spectrocopy, and $t_{exp} \sim 10^{-12}–10^{-10}$ s for inelastic neutron scattering [155, 593]. The above criterion defines the superparamagnetic blocking temperature

$$T_B \cong \frac{K_u V_p}{k} \frac{1}{\ln\left(\frac{100\,t_{exp}}{\tau_0}\right)}. \tag{5.9}$$

In the "blocked" regime ($T < T_B$), the hysteresis curve of such a particle is described by the well-known Stoner–Wohlfarth model [164].[‡‡] If on the other hand $\tau_N \ll t_{exp}$, then the system behaves as a superparamagnet. During t_{exp} the thermal energy induces many transitions between the local energy minima, and the magnetization curve is described by the Langevin function [eqn (5.16)]. Finally, when $\tau_N \cong t_{exp}$, time-dependent effects become important, since the magnetic properties change during the measurement.

In a typical SANS experiment using a cold neutron beam, the interaction time of the neutrons with the coherence volume is about $10^{-12}–10^{-10}$ s; for instance, for 6 Å neutrons with a velocity of $v_n = 659$ m/s it takes about 1.5×10^{-11} s for the neutrons to cross a particle of 10 nm diameter. Given the strong dependence of the relaxation time [eqn (5.8)] on the particle-size distribution and on the temperature, the question regarding the applicability of the static approximation, i.e., whether or not the neutrons see a static spin structure, should be considered on a case by case basis (see also Section 5.2.3) [594, 595]; see also the reviews [591, 592] for a discussion of the spin dynamics of interacting particle systems, spin glasses, and superspin glasses.

Up to now, we have implicitly assumed that the particle is rigidly embedded in some solid nonmagnetic matrix. When the particles are dispersed in a fluid of viscosity

---

[‡‡]With respect to magnetic SANS, we refer to the work by Löffler et al. [457], who developed a SANS theory for nanocrystalline ferromagnets based on the Stoner–Wohlfarth model.

η_{fl}, an additional relaxation mechanism, the so-called Brownian rotational-diffusion mechanism, comes into play. Brownian relaxation is characterized by a switching time

$$\tau_{\mathrm{B}} = \frac{3\eta_{\mathrm{fl}}V_{\mathrm{H}}}{kT}, \tag{5.10}$$

where V_{H} denotes the hydrodynamic volume of a single particle, which is larger than its magnetic volume. In general, both mechanisms—Néel and Brown relaxation—take place in a ferrofluid, resulting in an effective relaxation time of [596, 597]

$$\tau_{\mathrm{eff}} = \frac{\tau_{\mathrm{N}}\tau_{\mathrm{B}}}{\tau_{\mathrm{N}} + \tau_{\mathrm{B}}}. \tag{5.11}$$

For a given particle size, the shorter of the two time constants τ_{N} and τ_{B} determines the behavior of the effective relaxation time.

For a single-domain particle, one can assume an orientation distribution based on the competition between the thermal and Zeeman energies. If $d\Omega = \sin\vartheta d\vartheta d\zeta$ denotes the element of solid angle in a polar system where the z-direction is given by the external field $\mathbf{H}_0$, then the orientational averaging process is described by the Boltzmann distribution (neglecting magnetic anisotropy [559])

$$p(\vartheta) = \exp\left(-\frac{E}{kT}\right) = \exp\left(\frac{\mu_0 H_0 M_{\mathrm{s}} V_{\mathrm{p}} \cos\vartheta}{kT}\right). \tag{5.12}$$

In case of moderately polydisperse systems, the double average $\int\int f(R)p(\vartheta)dRd\Omega$ may be divided into two separate parts [464]. For a unit magnetization vector $\mathbf{m} = \{m_x, m_y, m_z\} = \{\sin\vartheta\cos\zeta, \sin\vartheta\sin\zeta, \cos\vartheta\}$, the directional averages of the Cartesian components can be computed according to ($i = x, y, z$):

$$\langle m_i\rangle = \frac{\int_0^\pi\int_0^{2\pi} m_i\exp\left(\beta_{\mathrm{L}}\cos\vartheta\right)\sin\vartheta d\zeta d\vartheta}{\int_0^\pi\int_0^{2\pi}\exp\left(\beta_{\mathrm{L}}\cos\vartheta\right)\sin\vartheta d\zeta d\vartheta} \tag{5.13}$$

and

$$\langle m_i m_j\rangle = \frac{\int_0^\pi\int_0^{2\pi} m_i m_j\exp\left(\beta_{\mathrm{L}}\cos\vartheta\right)\sin\vartheta d\zeta d\vartheta}{\int_0^\pi\int_0^{2\pi}\exp\left(\beta_{\mathrm{L}}\cos\vartheta\right)\sin\vartheta d\zeta d\vartheta} \tag{5.14}$$

for the second-order components. The expressions entering the SANS cross sections [eqns (2.93)−(2.102)] then become (see Fig. 5.4):

$$\langle m_x\rangle = \langle m_y\rangle = 0 \text{ and } \langle m_z\rangle = L(\beta_{\mathrm{L}}),$$
$$\langle m_x m_y\rangle = \langle m_x m_z\rangle = \langle m_y m_z\rangle = 0,$$
$$\langle m_x^2\rangle = \langle m_y^2\rangle = \frac{L(\beta_{\mathrm{L}})}{\beta_{\mathrm{L}}} \text{ and } \langle m_z^2\rangle = \left(1 - 2\frac{L(\beta_{\mathrm{L}})}{\beta_{\mathrm{L}}}\right), \tag{5.15}$$

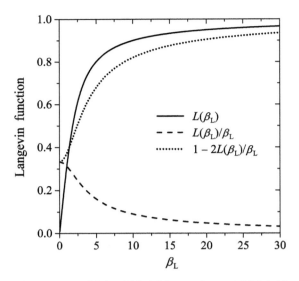

Fig. 5.4: The functions $L(\beta_L)$, $L(\beta_L)/\beta_L$, and $1 - 2L(\beta_L)/\beta_L$ (see inset).

where

$$L(\beta_L) = \coth(\beta_L) - 1/\beta_L \qquad (5.16)$$

denotes the Langevin function with argument $\beta_L = \mu_0 H_0 M_s V_p/(kT)$.

In many studies on single-domain nanoparticles, the form factors take on the following form [36]:

$$F_{\mathrm{nuc}}^{\mathrm{mag}}(\mathbf{q}) = \int_{V_p} \Delta\rho_{\mathrm{nuc}}^{\mathrm{mag}}(\mathbf{r}) \exp\left(-i\mathbf{q}\cdot\mathbf{r}\right) d^3r, \qquad (5.17)$$

where $\Delta\rho_{\mathrm{nuc}}^{\mathrm{mag}}$ are piece-wise constant functions, and the shape of the sphere or a core-shell structure is assumed (compare, e.g., Fig. 5.2). For the remainder of this section we assume that both F_{nuc}^2 and F_{mag}^2 are only functions of q. For the SANSPOL case and the transversal scattering geometry ($\mathbf{k}_0 \perp \mathbf{H}_0$), $d\Sigma^\pm/d\Omega$ of a dilute set of superparamagnetic particles then evaluates to (for details see, e.g., [462–466]):

$$\frac{d\Sigma^\pm}{d\Omega} = \frac{1}{V} \int\limits_0^\infty \left(F_{\mathrm{nuc}}^2 + 2F_{\mathrm{mag}}^2 \frac{L(\beta_L)}{\beta_L} + F_{\mathrm{mag}}^2 \left[1 - 3\frac{L(\beta_L)}{\beta_L}\right] \sin^2\theta \right.$$
$$\left. + W^\pm F_{\mathrm{nuc}} F_{\mathrm{mag}} L(\beta_L) \sin^2\theta \right) f(R) dR, \qquad (5.18)$$

where $W^\pm = (2p-1)(2\epsilon^\pm - 1)$ [eqn (2.96)]. Equation (5.18) shows that, under non-saturating conditions, part of the magnetic SANS is isotropic [the second term on the right-hand side of eqn (5.18)]. Magnetic field (H_0) or temperature (T) variation changes β_L and thereby the fraction of isotropic and anisotropic scattering [464, 465].

Note that $L(\beta_L) \to 1$ and $L(\beta_L)/\beta_L \to 0$ for $\beta_L \to \infty$ (compare Fig. 5.4), so that at saturation:

$$\frac{d\Sigma^{\pm}}{d\Omega} = \frac{1}{V} \int\limits_0^{\infty} \left(F_{\text{nuc}}^2 + F_{\text{mag}}^2 \sin^2\theta + W^{\pm} F_{\text{nuc}} F_{\text{mag}} \sin^2\theta \right) f(R) dR. \tag{5.19}$$

In zero field $(L(\beta_L)/\beta_L \to 1/3$ for $\beta_L \to 0)$ and for unpolarized neutrons $(p = 1/2)$, eqn (5.18) reduces to the following isotropic expression:

$$\frac{d\Sigma}{d\Omega} = \frac{1}{V} \int\limits_0^{\infty} \left(F_{\text{nuc}}^2 + \frac{2}{3} F_{\text{mag}}^2 \right) f(R) dR. \tag{5.20}$$

The difference between flipper-on $(\epsilon^- = \epsilon \cong 1)$ and flipper-off $(\epsilon^+ = 0)$ data results in

$$\frac{d\Sigma^-}{d\Omega} - \frac{d\Sigma^+}{d\Omega} = \frac{2\epsilon(2p-1)}{V} \int\limits_0^{\infty} L(H_0, R) F_{\text{nuc}}(q, R) F_{\text{mag}}(q, R) \sin^2\theta f(R) dR. \tag{5.21}$$

The cross-term intensity [eqn (5.21)] is proportional to $L(\beta_L)$ and therefore to the magnetic moment of the particles. This allows a straightforward testing of the Langevin behavior by magnetic field variation experiments; see Fig. 5.17 in Section 5.2.1 for the application of the Langevin formalism using eqn (5.21) to a Co-based ferrofluid. For the parallel scattering geometry $(\mathbf{k}_0 \parallel \mathbf{H}_0)$, one finds the following SANSPOL cross section (dilute case) [598]:

$$\frac{d\Sigma^{\pm}}{d\Omega} = \frac{1}{V} \int\limits_0^{\infty} \left(F_{\text{nuc}}^2 + F_{\text{mag}}^2 \left[1 - \frac{L(\beta_L)}{\beta_L} \right] + W^{\pm} F_{\text{nuc}} F_{\text{mag}} L(\beta_L) \right) f(R) dR, \tag{5.22}$$

which reduces to

$$\frac{d\Sigma^{\pm}}{d\Omega} = \frac{1}{V} \int\limits_0^{\infty} \left(F_{\text{nuc}}^2 + F_{\text{mag}}^2 + W^{\pm} F_{\text{nuc}} F_{\text{mag}} \right) f(R) dR \tag{5.23}$$

at saturation, and to

$$\frac{d\Sigma}{d\Omega} = \frac{1}{V} \int\limits_0^{\infty} \left(F_{\text{nuc}}^2 + \frac{2}{3} F_{\text{mag}}^2 \right) f(R) dR \tag{5.24}$$

at zero applied field and for $p = 1/2$. Equation (5.24) for $\mathbf{k}_0 \parallel \mathbf{H}_0$ is equal to the corresponding expression for the perpendicular scattering geometry [eqn (5.20)], as it should be for an ideal superparamagnet, where the macroscopic mean magnetization vanishes at zero field since the magnetic moments of the particles point in random directions. Note, however, that this is not true for bulk ferromagnets, where the zero-field magnetic SANS cross sections in the two geometries are generally different. Finally, for $\mathbf{k}_0 \parallel \mathbf{H}_0$, the difference between flipper-on and flipper-off data can be described by

$$\frac{d\Sigma^-}{d\Omega} - \frac{d\Sigma^+}{d\Omega} = \frac{2\epsilon(2p-1)}{V} \int_0^{\infty} L(H_0, R) F_{\mathrm{nuc}}(q, R) F_{\mathrm{mag}}(q, R) f(R) dR. \qquad (5.25)$$

Ridier et al. [598] have employed the superparamagnetic SANS approach to reveal the crossover from a single-particle to a collective-state behavior depending on the particle size in dense assemblies of nanoparticles. The observation of ferromagnetically correlated clusters of 4.8 nm particles, with a temperature-dependent magnetic correlation length, is opposed to the superparamagnetic behavior of larger 8.6 nm particles and was attributed to the much larger surface anisotropy in the smaller nanoparticles.

The previous sections were largely concerned with the discussion of the magnetic SANS from uniformly magnetized nanoparticles, and its limitations. In the following, we specify the complete micromagnetic boundary-value problem. This section could also have been placed into Chapter 3, however, we prefer to have the discussion here, since it highlights the complexity of the problem which must be solved in order to obtain the equilibrium magnetization structure and, hence, the magnetic SANS cross section. In view of the fact that for bulk ferromagnets a micromagnetic theory framework exists (see Chapter 4), this is of particular relevance for future SANS research on magnetic nanoparticles.

5.1.3   The micromagnetic problem

Neglecting the magnetoelastic and Dzyaloshinskii–Moriya interaction (DMI), the following set of nonlinear partial differential equations specify the micromagnetic equilibrium problem completely (see Section 4.1 in [344]):

$$\mathbf{M} \times \left(\mathbf{H}_0 + \mathbf{H}_d + \mathbf{H}_p + l_{\mathrm{M}}^2 \Delta \mathbf{M}\right) = 0 \qquad \text{in } V, \qquad (5.26)$$

$$\mathbf{M} \times \left(l_{\mathrm{M}}^2 \frac{\partial \mathbf{M}}{\partial \mathbf{n}} + \frac{1}{\mu_0} \frac{\partial \omega_{\mathrm{s}}}{\partial \mathbf{M}}\right) = 0 \qquad \text{on } S, \qquad (5.27)$$

$$\mathbf{H}_d = -\nabla U_{\mathrm{in}} \qquad \text{in } V, \qquad (5.28)$$

$$\mathbf{H}_d = -\nabla U_{\mathrm{out}} \qquad \text{outside } V, \qquad (5.29)$$

where

$$\Delta U_{\mathrm{in}} = \nabla \cdot \mathbf{M}, \qquad (5.30)$$

$$\Delta U_{\mathrm{out}} = 0, \qquad (5.31)$$

and on S

$$U_{\mathrm{in}} - U_{\mathrm{out}} = 0, \qquad (5.32)$$

$$\frac{\partial U_{\mathrm{in}}}{\partial \mathbf{n}} - \frac{\partial U_{\mathrm{out}}}{\partial \mathbf{n}} = \mathbf{n} \cdot \mathbf{M}. \qquad (5.33)$$

The fields in the above equations and the magnetostatic boundary-value problem have been introduced in Sections 3.1.6 and 3.2. The quantity $l_{\mathrm{M}} = \sqrt{2A/(\mu_0 M_{\mathrm{s}}^2)}$

[eqn (3.43)] is the magnetostatic exchange length ($l_M \sim 3-10\,$nm for many magnetic materials [134]), $\mathbf{n}$ denotes the unit normal vector on the surface (taken to be positive in the outward direction), $\nabla = \{\partial/\partial x, \partial/\partial y, \partial/\partial z\}$ is the gradient operator, $\Delta = \nabla^2$ the Laplace operator, and the symbols V and S refer, respectively, to the volume and the surface of the nanoparticle. Moreover, the potential function U has to be regular at infinity, i.e., $|rU_{\mathrm{out}}|$ and $|r^2\nabla U_{\mathrm{out}}|$ remain finite for $r \to \infty$. The regularity condition ensures that far from the region of non-vanishing (localized) magnetization the magnetostatic scalar potential corresponds to the one of a dipole with a strength that is given by the total magnetic moment of the magnetization distribution [154, 161]. The boundary conditions on S [eqn (5.27)] may also contain a contribution due to a phenomenological surface anisotropy energy term, which in first-order approximation can be written as [154]:

$$E_{\mathrm{s}} = \int \omega_{\mathrm{s}}\,dS = \frac{1}{2}K_{\mathrm{s}}\int (\mathbf{n}\cdot\mathbf{m})^2\,dS, \qquad (5.34)$$

where ω_{s} denotes the surface anisotropy energy density, K_{s} is the surface anisotropy constant (in units of $\mathrm{J/m^2}$), $\mathbf{m} = \mathbf{M}/M_{\mathrm{s}}$, and $\partial\omega_{\mathrm{s}}/\partial\mathbf{m} = K_{\mathrm{s}}(\mathbf{n}\cdot\mathbf{m})\mathbf{n}$. Depending on the sign of K_{s}, this energy term prefers the alignment of the surface spins either parallel or perpendicular to $\mathbf{n}$. When E_{s} is ignored, then the boundary conditions on S can be written as:

$$\frac{\partial\mathbf{M}}{\partial\mathbf{n}} = 0. \qquad (5.35)$$

This so-called exchange boundary condition follows by combining eqn (5.27) with the identity $\mathbf{M}\cdot\partial\mathbf{M}/\partial\mathbf{n} = 0$, which is a consequence of $\mathbf{M}^2 = M_{\mathrm{s}}^2$ [154, 599]. It should also be remembered that due to the constraint of constant magnetization magnitude there are only two independent differential equations and the third one is a linear combination of the other two. The effective field inside the volume of the magnet, $\mathbf{H}_{\mathrm{eff}} = \mathbf{H}_{\mathrm{eff}}^{\mathrm{V}} = \mathbf{H}_0 + \mathbf{H}_{\mathrm{d}} + \mathbf{H}_{\mathrm{p}} + l_M^2\Delta\mathbf{M}$ [eqn (5.26)], has the unit of A/m, whereas its surface analog, $\mathbf{H}_{\mathrm{eff}}^{\mathrm{S}} = l_M^2\frac{\partial\mathbf{M}}{\partial\mathbf{n}} + \mu_0^{-1}\frac{\partial\omega_{\mathrm{s}}}{\partial\mathbf{M}}$ [eqn (5.27)], comes in the unit of A [344].

Analytical calculations of the magnetic SANS cross section of isolated magnetic nanoparticles based on Brown's static equations of micromagnetics [eqns (5.26)−(5.33)] are mathematically extremely challenging. Up to now, only Metlov and Michels [467] and Mirebeau et al. [468] provided closed-form results for the magnetic SANS of magnetic vortices in thin submicron-sized circular cylinders (dots). The main results of their study are briefly summarized in the next section. We also refer to Section 7.4.2 for some numerical micromagnetic simulation results of the magnetic SANS from cylindrical nanoparticles.

5.1.4 SANS by magnetic vortices in thin submicron-sized soft ferromagnetic cylinders

As shown by Metlov [600], the magnetization textures of thin submicron-sized ferromagnetic cylinders (nanodots) can be approximately expressed via functions of complex variable (see [601–608] for further information). In [467], use of this formalism has been made in order to compute the ensuing magnetic SANS cross section analytically. The circular cylinder with radius R and thickness L_{cyl} is assumed to be thin

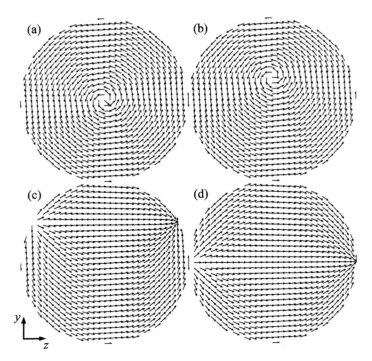

Fig. 5.5: Equilibrium and transient magnetization states in soft magnetic nano-discs [607,608]. (a) Centered magnetic vortex; (b) displaced magnetic vortex; (c) "C"-like state; (d) "leaf" state. After [467].

enough so that the magnetization vector $\mathbf{M}$ is independent of the Cartesian coordinate x along the cylinder's axis. Thus, the components of $\mathbf{M}$ depend on the coordinates in the dot's plane, y and z, as well as on other adjustable parameters [467]. Figure 5.5 depicts some spin structures as they are commonly encountered in submicron-sized soft magnetic cylindrical dots, while Fig. 5.6 displays an experimental magnetization curve, measured on a weakly interacting array of individual permalloy magnetic cylinders [609]. The magnetization distributions in Fig. 5.5 correspond to a local extremum of the exchange energy, which is the most important energy term in submicron-sized magnets, and of the magnetostatic energy related to the magnetic charges on the side faces of the cylinder.

The magnetization curve can be sketched using straight lines only: two parallel-inclined ones and two horizontal ones. The two inclined lines correspond to the magnetic vortex displacement [shown in Fig. 5.5(b)] and the horizontal ones to the dot in the state of magnetic saturation [such as in Fig. 5.5(d)]. The dotted vertical lines on the sketch mark the transitions between these two states, such as that from a displaced vortex to the quasi-uniform state [602]. It is around these transitions that the straight-line sketch of the hysteresis loop in Fig. 5.6 departs most from reality. Nevertheless, as one can see, the discrepancy is not very large, so that one can conclude that for the most part during the in-plane hysteresis loop, the magnetization in the dot assumes

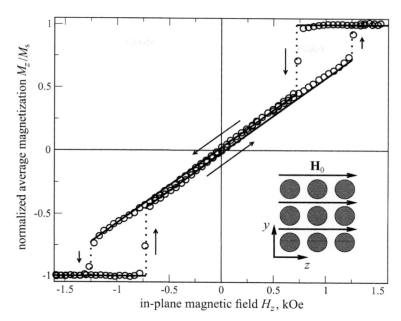

Fig. 5.6: Typical in-plane hysteresis loop of an array of weakly interacting submicron-sized permalloy cylinders (data taken from [609]). The inset depicts schematically the nanodot array, the coordinate-system axes, and the direction of the externally applied magnetic field $\mathbf{H}_0$. After [467].

either the displaced vortex state or the quasi-uniform state. Both configurations can be analytically described [467].

In Section 4.2, we have seen that the definition of the spin-misalignment SANS cross section entails the subtraction of the SANS cross section at magnetic saturation, the so-called residual SANS. In the present case it becomes evident that, in order to unmask the signature of the magnetic vortex state and its displacement, it is advantageous to consider the total magnetic SANS cross section, i.e., to add back the saturated term into the cross section. This procedure should allow one to extract the finer details of the magnetic vortex displacement process, e.g., extracting the position and shape of the magnetic vortex core by fitting the resulting SANS images to the cross-section expressions. Due to the special role of the saturation scattering for nanoparticles, we will in the following always compare cross sections with and without the saturation term.

The linearity of the major hysteresis-loop branches in the vortex state suggests that a linear approximation in the vortex-core displacement is sufficient to model the low-field part of the hysteresis loop. If one neglects the vortex core, which has a size of $\sim 5-20$ nm in many different ferromagnetic materials (e.g., [610]), then the second-order expansion of the perpendicular spin-misalignment SANS cross section, in the vortex state, can be algebraically expressed via Bessel and Struve functions ($\mathbf{k}_0 \perp \mathbf{H}_0$) [467]:

$$\frac{\partial \sigma_{\mathrm{SM}}^{\perp}}{\partial \Omega} = \frac{\pi^2 \left(J_1 H_0 - J_0 H_1\right)^2}{4\mathrm{k}^2} - \frac{J_1^2 \sin^2 \theta}{\mathrm{k}^2}$$
$$+ \mathrm{b}^2 \frac{\left([2\mathrm{k} + \pi(1 - \mathrm{k}^2)H_0]J_1 - [2\mathrm{k}^2 + \pi(1 - \mathrm{k}^2)H_1]J_0\right)^2 \sin^2 \theta}{4\mathrm{k}^4}, \quad (5.36)$$

where $d\Sigma_{\mathrm{SM}}^{\perp}/d\Omega = 4b_{\mathrm{H}}^2 V M_{\mathrm{s}}^2 \partial \sigma_{\mathrm{SM}}^{\perp}/\partial \Omega$, $J_n = J_n(\mathrm{k})$ and $H_n = H_n(\mathrm{k})$ denote, respectively, the Bessel and Struve functions with argument $\mathrm{k} = qR$, $\mathbf{q} = \{0, q_y, q_z\} = q\{0, \sin\theta, \cos\theta\}$, and $V = \pi R^2 L_{\mathrm{cyl}}$ denotes the volume of the thin circular disc. The value of the dimensionless vortex-center displacement parameter b in eqn (5.36) is proportional to the externally applied magnetic field $H_0 = H_z$. The proportionality coefficient can be derived from the relation $M_z/M_{\mathrm{s}} = \frac{2}{3}\mathrm{b}$, which is valid for $\mathrm{b} \ll 1$. We remind the reader of the following. (i) In the perpendicular scattering geometry the incident neutrons travel along the x-axis and the vortex, displaced by the magnetic field, acquires a nonzero z-component of the average magnetization. (ii) The spin-misalignment SANS cross section is defined as [compare eqn (4.3)]:

$$\frac{d\Sigma_{\mathrm{SM}}}{d\Omega} = \frac{d\Sigma}{d\Omega} - \left.\frac{d\Sigma}{d\Omega}\right|_{H_0 \to \infty}, \quad (5.37)$$

which, besides eliminating the nuclear SANS contribution ($\propto |\widetilde{N}|^2$), corresponds to the total magnetic SANS cross section (at a specific field) minus the total cross section at a very large (saturating) magnetic field; in other words, $d\Sigma_{\mathrm{SM}}/d\Omega \propto \partial \sigma_{\mathrm{SM}}/\partial \Omega$ describes the deviation from the uniformly magnetized particle case.

The perpendicular SANS cross section [eqn (5.36)], as it is visible in the top row of Fig. 5.7, is dominated by the saturation term $J_1^2 \sin^2\theta/\mathrm{k}^2$, which masks the effects of the vortex-center displacement. The vortex is a low-field configuration, which implies that the subtraction of the saturated SANS cross section significantly distorts the cross-section images. This can be understood by noting that the saturated state is characterized by a maximum of magnetic poles (surface charges) on the outer boundary of the dot. The divergence (jump) of the magnetization on a scale of the cylinder diameter $D_{\mathrm{cyl}} = 2R$ then gives rise to a large magnetic SANS signal at small momentum transfers. By contrast, the magnetic scattering due to the vortex state [Fig. 5.5(a)], which is related to magnetic structure on a scale smaller than D_{cyl} and characterized by small magnetic charges, shows up at larger q.

The saturation term itself is determined by the dot shape, and for circular dots depends only on the dot's size R (entering the definition of k). That is why, to reveal the finer details of the SANS cross section, it is advantageous to add back the saturation term to $\partial \sigma_{\mathrm{SM}}^{\perp}/\partial \Omega$. The in-this-way "corrected" cross sections are shown in the bottom row of Fig. 5.7, where the symmetry breaking due to the vortex-center displacement now becomes more clearly visible. The corrected cross sections can be represented as a sum of two terms of zero and second order in b, which are shown separately in Fig. 5.8. Larger vortex displacement means more weight on the second-order term in this sum. In Fig. 5.9, the azimuthally averaged total perpendicular magnetic SANS cross section $\partial \sigma_{\mathrm{M}}^{\perp}/\partial \Omega = \partial \sigma_{\mathrm{SM}}^{\perp}/\partial \Omega + J_1^2/(2\mathrm{k}^2)$, the spin-misalignment SANS cross section $\partial \sigma_{\mathrm{SM}}^{\perp}/\partial \Omega$, and the saturated cross section $\partial \sigma_{\mathrm{sat}}^{\perp}/\partial \Omega = J_1^2/(2\mathrm{k}^2)$ are compared.

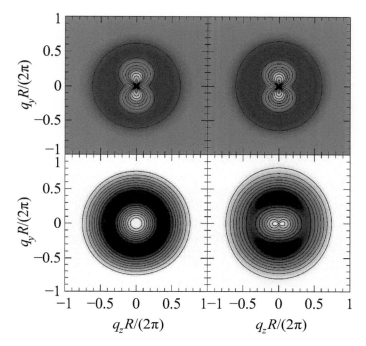

Fig. 5.7: Perpendicular magnetic SANS cross section of a ferromagnetic disc containing a centered (left, **b** = 0) and a displaced (right, **b** = 0.4) magnetic vortex ($\mathbf{k_0} \perp \mathbf{H_0}$). The vortex displacement produces the magnetization $M_z/M_s = \frac{2}{3} \times 0.4 \cong 0.27$, which, reading from Fig. 5.6, roughly corresponds to $H_0 = 0.6\,\text{kOe}$ for the sample from [609]. The top row shows the spin-misalignment SANS cross section as it is commonly defined with the saturated magnetic term subtracted [eqn (5.36)]. The bottom row displays the total magnetic SANS cross section with the magnetic saturation term $J_1^2 \sin^2 \theta / \mathbf{k}^2$ added back. After [467].

Apart from just computing and adding back the saturation term in the cross section, another way to exclude it and to highlight the effects of the vortex-center displacement during the SANS-image analysis is to subtract the zero-order terms altogether. This can be achieved by considering the following combination of spin-misalignment cross-section values:

$$\frac{\partial \sigma_2}{\partial \Omega} = \frac{\partial \sigma_{\text{SM}}}{\partial \Omega} - \frac{\partial \sigma_{\text{SM}}}{\partial \Omega}\bigg|_{H_0 \to 0}, \qquad (5.38)$$

in which only the second- and higher-order terms in the vortex-center displacement parameter **b** remain. This combination of SANS cross sections is expected to have the structure which is shown in the right half of Fig. 5.8. Departure from this simple dependency might reveal higher-order effects and may shed new light on the details of the vortex-core deformation during the magnetization process. This can be a valuable input to help decide which model of vortex displacement better describes the magne-

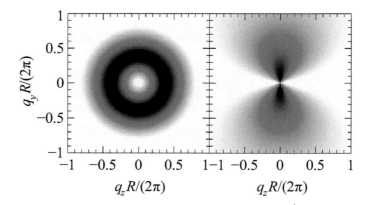

Fig. 5.8: Zero-order (left) and second-order (right) terms in b of the perpendicular magnetic SANS cross section of a ferromagnetic thin disc containing a magnetic vortex ($\mathbf{k}_0 \perp \mathbf{H}_0$). The zero-order term is displayed with the magnetic-saturation contribution added back (as in the bottom row of Fig. 5.7), otherwise its structure is masked by the saturation term. The second-order term is independent of this addition. After [467].

tization process: the uniform translation [611], the conformal mode [602], or the mode with no magnetic charges on the cylinder's side faces [603].

A small external field applied along the cylinder's axis (which is parallel to the incoming neutron beam) does not lead to a vortex-center displacement and does not change the symmetry of the magnetization distribution. This implies that the parallel SANS cross section is isotropic. For the case of vanishing field (b = 0) and neglecting the vortex core, the spin-misalignment SANS cross section in the parallel scattering geometry can be expressed algebraically as ($\mathbf{k}_0 \parallel \mathbf{H}_0$) [467]:

$$\frac{\partial \sigma_{\mathrm{SM}}^{\parallel}}{\partial \Omega} = \frac{\pi^2 \left(J_1 H_0 - J_0 H_1 \right)^2}{4k^2} - \frac{J_1^2}{k^2}, \tag{5.39}$$

which uses the same notation as eqn (5.36), except that now $q = \left(q_x^2 + q_y^2 \right)^{1/2}$. It has the shape of a series of concentric rings with the first maximum strongly dominating the others. The second term in eqn (5.39) originates from subtracting the magnetically saturated state, and the first term coincides with the first term in eqn (5.36) for the perpendicular cross section. When in both spin-misalignment SANS cross sections, eqns (5.36) and (5.39), the respective saturation term is added back (to yield the respective total magnetic SANS), then their subtraction directly yields the second-order contribution in b.

Experimental SANS studies on nanodot arrays are still rare. Roshchin et al. [610] reported polarized grazing-incidence SANS data on 20 nm-thick iron dots with diameters smaller than 100 nm. The authors have provided evidence for the appearance of the magnetic vortex state, and for a transition from the vortex state to a single domain with decreasing dot diameter and increasing applied field. From the measurements of the out-of-plane magnetization of the vortex core, a vortex core size of

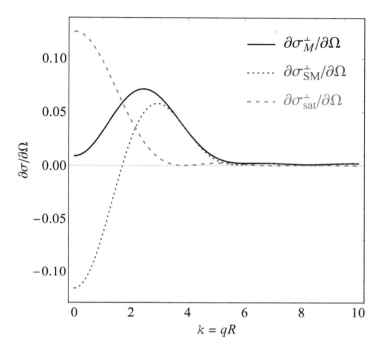

Fig. 5.9: Comparison of the azimuthally averaged total perpendicular magnetic SANS cross section $\partial\sigma_M^\perp/\partial\Omega = \partial\sigma_{SM}^\perp/\partial\Omega + J_1^2/(2k^2)$, the spin-misalignment SANS cross section $\partial\sigma_{SM}^\perp/\partial\Omega$ [eqn (5.36)] (both at a field corresponding to $b = 0.4$), and of the cross section in the saturated state $\partial\sigma_{sat}^\perp/\partial\Omega = J_1^2/(2k^2)$ (see inset) ($\mathbf{k_0} \perp \mathbf{H_0}$).

19 ± 4 nm was deduced for 65 nm-diameter dots. Regarding spin-polarized neutron scattering, the displaced non-centrosymmetric vortex structure is expected to show up as a polarization-dependent contribution to the spin-flip cross section. Since the nuclear coherent scattering is non-spin-flip, the finer details of the vortex can be investigated by carrying out polarization-analysis experiments.

In the context of SANS by magnetic vortices, it is also worth mentioning the paper by Mirebeau et al. [468], who studied the re-entrant spin glass $Ni_{0.81}Mn_{0.19}$ in single-crystalline form. Their main experimental results are summarized in Fig. 5.10. Supported by Monte Carlo simulations of a two-dimensional lattice structure with competing ferromagnetic and antiferromagnetic exchange interactions, they have suggested the existence of vortex-like chiral spin structures in $Ni_{0.81}Mn_{0.19}$, in this way revisiting earlier interpretations of field-dependent SANS data [122, 612–614]. For the analysis of their magnetic neutron data, which exhibit a maximum at some characteristic q-value [compare Fig. 5.9 and Fig. 5.10(b)], expressions similar to eqn (5.36) were derived. The peak position q_{max} is field dependent [Fig. 5.10(c)] and the related defect size $r_d = \pi/q_{max}$ ($\sim$ typical radius of a vortex in their theory) is inversely proportional to the square root of the applied field [Fig. 5.10(e)], following the same trend as the micromagnetic exchange length l_H [eqn (3.66)]. Interestingly, Mirebeau et al. [468] find that the number of defects or scattering centers N_d increases with increasing field and

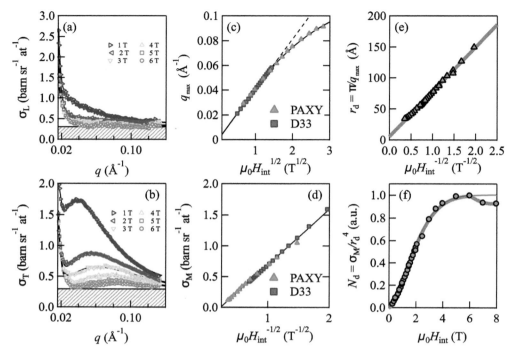

Fig. 5.10: Magnetic SANS results on the re-entrant spin glass $Ni_{0.81}Mn_{0.19}$. Evolution of the longitudinal (a) and transverse (b) magnetic neutron scattering cross sections as a function of magnetic field. In both panels, the hatched areas represent the q-independent background contribution (see [468] for details). Field dependence of the peak position q_{max} (c) and integrated scattering intensity σ_M (d) as obtained from the analysis of SANS data taken on the instruments D33 (blue squares) and PAXY (red triangles). (e) Field dependence of the defect size $r_d = \pi/q_{max}$ computed using the data of (c). (f) Field dependence of the number of scattering centers N_d. After [468].

saturates at $\sim 6\,T$ [Fig. 5.10(f)]. The work supports the existence of scaling laws, which describe the position, width, and intensity of the magnetic SANS signal.

5.2   Complex systems

5.2.1   Ferrofluids

Ferrofluids are stable colloidal suspensions of magnetic nanoparticles, most often iron oxides (e.g., magnetite Fe_3O_4 or maghemite γ-Fe_2O_3), in a liquid carrier material (e.g., water, toluene, kerosene, or various oils). In many studies, the shape of the nanoparticles is nearly spherical or cylindrical with a typical characteristic dimension between $\sim 5-100$ nm. Due to their technological relevance and also from the fundamental science point of view, ferrofluids are the subject of an intense worldwide research effort (e.g., [615–617]). Magnetic neutron scattering and, in particular, the SANS method is very powerful for characterizing and analyzing the static and dynamic magnetic microstructure of ferrofluids in the bulk and on the relevant nanometer length scale (see, e.g., [25, 36, 50, 51, 61, 459, 463–466, 618–677] and references therein). By contrast, in order to avoid multiple scattering effects, the application of x-ray and light scattering techniques to the study of ferrofluids is usually limited to thin and dilute samples. To achieve a stable ferrofluid suspension and to prevent the formation of larger particle agglomerates due to the magnetic dipole-dipole interaction or the van der Waals interaction, the particles are stabilized by either electric charges or by coating the particles with long-chained organic molecules (surfactants) [616].

In addition to the average particle size, there are at least two further characteristic length scales which determine the properties of a ferrofluid: the average interparticle distance and the range of the interparticle interaction potential [25]. The main interparticle interactions to be taken into account are: (i) the isotropic and short-range steric repulsion of particles; (ii) the anisotropic and long-range magnetodipolar interaction; (iii) the van der Waals interaction; and (iv) hydrodynamic interactions. The latter is caused by the motion of the particles in the liquid and is only important when the dynamical properties of non-dilute ferrofluids are studied [678]. In addition to these interactions, larger particles can be magnetized nonuniformly, e.g., due to the presence of surface anisotropy or inhomogeneous demagnetizing fields. Obviously, the internal spin structure and the magnetodipolar interaction depend explicitly on an applied magnetic field $\mathbf{H}_0$. For nonspherical particles, the hydrodynamic interaction also depends on $\mathbf{H}_0$ due to the change of the particle orientation degree when $\mathbf{H}_0$ is changed [678]. The complexity of the various long, medium, and short-range, isotropic and anisotropic, and repulsive and attractive interactions between the nanoparticles in a ferrofluid render an analytical treatment of the magnetic SANS cross section extremely difficult. Experimental data are therefore frequently analyzed in terms of the superparamagnetic particle-matrix approach described in Section 5.1.2, using the well-known concepts of nuclear SANS such as the Guinier or Porod laws. Based largely on the detailed and extensive review by Avdeev and Aksenov [51], selected characteristic features of magnetic SANS on ferrofluids are discussed in the following.

The application of SANS to the study of complex magnetic liquids largely profits from the possibility to vary the scattering-length density (SLD) of the solvent relative to the SLD of the particle phase. This allows one to tune the contrast and, hence, the scattering cross section; for instance, in the simple case of a light water/heavy water solvent mixture, the SLD of the solvent can be adjusted according to [660]:

$$\rho_{\text{sol}} = u_1 \rho_{\text{D}_2\text{O}} + (1 - u_1) \rho_{\text{H}_2\text{O}}, \tag{5.40}$$

where

$$\rho_{\text{D}_2\text{O}} = +6.34 \times 10^{10} \, \text{cm}^{-2} \quad \text{and} \quad \rho_{\text{H}_2\text{O}} = -0.56 \times 10^{10} \, \text{cm}^{-2} \tag{5.41}$$

denote the SLDs of heavy and light water, respectively, and u_1 is the volume fraction of D_2O. Figure 5.11 depicts, for a spherical core-shell particle, the various situations which can arise. We emphasize that we are here only considering the nuclear SANS contribution, and that the subscript "nuc" in the various SLDs is suppressed. The core-shell form factor can be written as [51]:

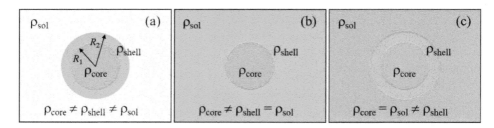

Fig. 5.11: Sketch illustrating the scattering-length density (SLD) contrast between a spherical core-shell particle and the solvent in a magnetic liquid. Only the core of the particle is assumed to be magnetic; ρ_{sol}, ρ_{shell}, and ρ_{core} denote, respectively, the nuclear SLDs of the solvent, particle shell, and particle core. (a) General case of $\rho_{sol} \neq \rho_{shell} \neq \rho_{core}$; (b) $\rho_{sol} = \rho_{shell} \neq \rho_{core}$ and the shell becomes invisible; (c) $\rho_{sol} = \rho_{core} \neq \rho_{shell}$ and the shell is highlighted.

$$F^2(q) = [(\rho_{core} - \rho_{shell})V_{s1}\Xi(qR_1) + (\rho_{shell} - \rho_{sol})V_{s2}\Xi(qR_2)]^2, \tag{5.42}$$

where ρ_{core} and ρ_{shell} are the nuclear SLDs of the core and of the surfactant shell, V_{s1} and V_{s2} are the sphere volumes which are defined by the radii of the core (R_1) and of the core plus shell (R_2), and $\Xi(x) = 3j_1(x)/x$. The surfactant shell thickness is given by $R_2 - R_1$. The general case where all the contrasts in eqn (5.42) contribute to the nuclear scattering is shown in Fig. 5.11(a). When the SLD of the solvent is matched to the SLD of the shell [$\rho_{sol} = \rho_{shell}$, Fig. 5.11(b)], then the nuclear SANS cross section is governed by $\rho_{core} - \rho_{sol}$, while for $\rho_{sol} = \rho_{core}$ [Fig. 5.11(c)] the SANS signal contains information about the shell.

These simple considerations emphasize the enormous potential of the contrast-variation technique for soft matter studies. The method has originally been advanced by Stuhrmann (e.g., [679, 680]) for the case of non-interacting, identical, and non-magnetic nanoparticles (the so-called basic functions approach). It has been further expanded and adapted by Avdeev [655] to include the effect of polydispersity and the magnetic scattering contribution due to non-interacting superparamagnetic single-domain particles.

Figure 5.12 shows unpolarized zero-field SANS data of magnetite nanoparticles dispersed in a protonated and in a deuterated solvent (benzene). Both ferrofluids are assumed to consist of weakly interacting particles. A significantly different q-dependence of both SANS cross sections can be observed, which is a consequence of the different nuclear scattering contrasts related to the different SLDs of the solvent. For the analysis of both data sets a model based on eqn (5.20) was employed (solid lines in Fig. 5.12). The parameters of the particle-size distribution $f(R)$ of the D-solvent ferrofluid were held fixed in the neutron-data analysis [629]. Then, asssuming the absence of a magnetically "dead" surface layer on the particles, the total nuclear and magnetic SANS cross section could be well described by a model of non-interfering polydisperse core-shell particles, where the thickness and the SLD of the surfactant shell were taken as adjustable parameters. On the other hand, for the description of the magnetite ferrofluid in the H-solvent, a fit to non-interfering polydisperse spheres

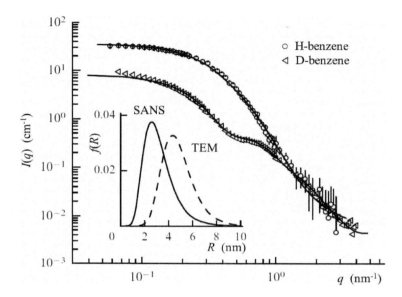

Fig. 5.12: Experimental SANS curves of magnetite ferrofluids in H and D-solvents (benzene). No magnetic field is applied. The parameters of the particle-size distribution function $f(R)$ are varied in H-benzene and fixed in D-benzene, where the effective thickness and the SLD of the surfactant shell are varied (no magnetically "dead" surface assumed). The solid lines represent fits to a set of non-interfering polydisperse sphere (H-benzene) and core-shell (D-benzene) form factors. The inset compares the $f(R)$ obtained by SANS and transmission electron microscopy (TEM). After [51].

provided a good description of the SANS data, i.e., the scattering contribution of the shell has been neglected when using the H-solvent. Moreover, for the H-solvent sample, it turned out that the fitting analysis is insensitive to the magnetic scattering term, i.e., the agreement between data and fit was even improved if the magnetic term is not taken into account [629, 636]. In fact, Avdeev and Aksenov [51] conclude that for unpolarized neutrons and standard ferrofluid types (e.g., iron oxides in water, toluene, or benzene), magnetic scattering in the case of H-solvents is negligibly small as compared to the nuclear SANS. A similar conclusion has been reached by Gazeau et al. [634] in their neutron study on biocompatible ionic ferrite-based ferrofluids. The discrepancy between the $f(R)$ obtained from SANS and transmission electron microscopy (TEM) (see inset in Fig. 5.12) has been attributed (i) to the different sensitivity of both methods to the size range being investigated, and (ii) it has been conjectured that the TEM sample preparation procedure may result in a different particle aggregate structure, effectively altering the size distribution of the remaining nanoparticles [51].

The effect of nanoparticle volume fraction x_p on the unpolarized zero-field SANS cross section of a benzene-based and a pentanol-based magnetite ferrofluid is displayed in Fig. 5.13. The analytical description of SANS data on concentrated ferrofluid systems, e.g., in terms of the $d\Sigma/d\Omega = P(q)S(q)$ approach discussed briefly in Section 2.9.4, is generally difficult due to polydispersity and due to the presence of the

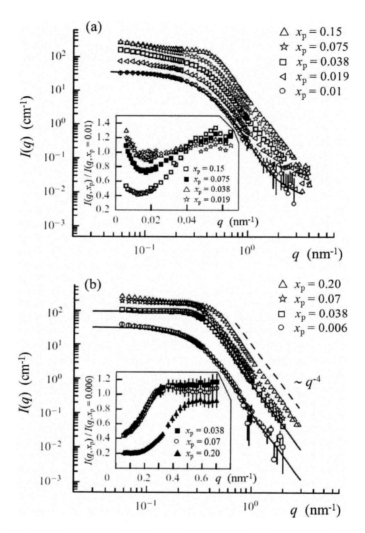

Fig. 5.13: Experimental SANS data of (a) benzene- and (b) pentanol-based ferrofluids (both H-solvents) at different magnetite volume fractions x_p. No magnetic field is applied. Solid lines: simulations (at minimal volume fraction) based on a set of non-interfering polydisperse spheres. The insets depict the effective "nuclear" structure factors at increased concentrations. After [51].

long-range and anisotropic magnetodipolar interaction between the particles. An often-used approach to obtain the effective structure factor is to divide the SANS cross section for a given concentration by the SANS signal of a dilute sample ($x_p \rightarrow 0$), which is assumed to represent single-particle scattering (see, e.g., [39,627,628,631,634,666,667]). The in-this-way obtained $S(q)$ can be seen in the insets of Fig. 5.13. The main difference between the benzene and the pentanol-based ferrofluid is that in the former ferrofluid the surfactant consists of a single layer of oleic acid, while in the latter a surfactant double layer should stabilize the ferrofluid and prevent the particles from aggregation. The behavior of $S(q)$ in the limit of low q, when the system is probed on macroscopic length scales, provides information on thermodynamical quantities and on the nature of the interaction potential (repulsive versus attractive). This is embodied in the compressibility relation [264],

$$\lim_{q \to 0} S(q) = \rho_{\mathrm{p}} k T \kappa_{\mathrm{T}} = -\rho_{\mathrm{p}} k T \frac{1}{V} \left(\frac{\partial V}{\partial p} \right)_{T}, \tag{5.43}$$

where $\rho_{\mathrm{p}} = N_{\mathrm{p}}/V$ is the particle density, and κ_{T} denotes the isothermal compressibility of the liquid, which describes the fractional decrease in volume per unit increase in pressure at constant temperature [349]. Repulsive interparticle interactions result in $S(0) < 1$, whereas attractive interactions give $S(0) > 1$ [39]. Returning to the discussion of the data shown in Fig. 5.13, it is seen that the difference in the chemistry and structure of the particle + surfactant phase is reflected in the behavior of the effective $S(q)$ at low momentum transfers: for the benzene-based ferrofluid, $S(q)$ in the limit of $q \to 0$ increases with decreasing x_{p} to values larger than unity [inset in Fig. 5.13(a)], which is indicative of an attractive interaction between the ferrofluid particles, while the $S(q)$ for the pentanol-based ferrofluid [Fig. 5.13(b)] decreases with increasing x_{p} and takes on values smaller than unity at low q. The observed difference in the character of the interaction suggests the complete screening of attractive interactions by a thicker surfactant shell in a doubly stabilized magnetite ferrofluid.

Hayter and Pynn [622] have provided an analytical expression for the structure factor of a saturated ferrofluid, where the spherical particles interact through magnetic dipolar and attractive central potentials. In a pioneering study [459], they have experimentally confirmed their predictions by performing one of the first polarization-analysis experiments on a concentrated core-shell-type Co ferrofluid (see Fig. 5.14). The structure factors parallel and perpendicular to the applied magnetic field were determined by comparison to the predominantly form-factor-only scattering of a diluted system. As can be seen in Fig. 5.14, the theory (solid and dashed lines) agrees very well with the experimental $S(q)$ data and the deviations at small q are believed to result from slight polydispersity present in the sample. The fluctuation scattering has been obtained from spin-flip scattering intensities and it contains the contribution due to misaligned $x-y$ magnetization components [25]. Reducing the field increases the magnitude of the fluctuation scattering.

Mériguet et al. [666] have investigated the effect of particle volume fraction, surface coating, pH-value, and ionic strength on the interparticle interactions in maghemite (γ-Fe_2O_3) ferrofluids via the measurement of the structure factor in zero field and under an applied magnetic field. The results of the Paris group impressively demonstrate that the interactions in ferrofluids can be chemically tuned. Figure 5.15 depicts the dependency of $S(q)$ on the pH-value, the salt concentration C_{S} ($Na^+NO_3^-$), and the applied field for a ferrofluid with an average particle diameter of 10.5 nm and a particle volume fraction of $x_{\mathrm{p}} = 1.5\%$. At zero field, a clear evolution of the structure factor $S(q)$ from a repulsive interparticle interaction (on average) at pH $= 3$ and $C_{\mathrm{S}} = 10^{-3}$ M

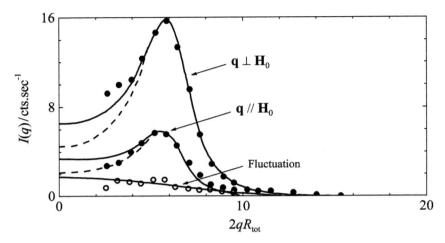

Fig. 5.14: Measured values of the coherent non-spin-flip scattering (filled circles) and fluctuation scattering (open circles) for a concentrated Co ferrofluid (particle volume fraction: 0.27) in an applied magnetic field of 0.5 T at 298 K. The abscissa is scaled by the total particle diameter $2R_{tot}$. Solid and dashed lines: theory prediction for $S(q)$ corresponding, respectively, to the inclusion or exclusion of an attractive component in the central part of the interaction potential. After [459].

[Fig. 5.15(a)] to an attractive interaction (on average) [Fig. 5.15(b) and (c)] can be observed. Application of a field of $H = 800$ kA/m gives rise to an anisotropic structure factor. In Fig. 5.15(a), the structure factors $S_\parallel(q)$ and $S_\perp(q)$ to the field remain those of a repulsive system and are almost the same as in zero field. For the pH = 3.7 and $C_S = 0.13$ M sample [Fig. 5.15(c)], the interactions remain those of an attractive system in both directions, while in the intermediate situation [Fig. 5.15(b)] the application of a field induces a repulsion along the direction of the field and leaves the interaction along the perpendicular direction attractive.

The formation of cluster and chain structures consisting of several nanoparticles (both with and without the application of an external magnetic field) has been reported by a number of authors (e.g., [51, 637, 664, 667, 668]). For theoretical studies on the formation of dipolar-field-induced particle chains that are aligned along the applied-field direction see [618, 619]. Such clusters leave their fingerprint in the scattering curve at small q, while the individual particles scatter into the large-q region. This is illustrated in Fig. 5.16 for light-water-based magnetite ferrofluids with different combinations of double-stabilizing surfactants. The zero-field SANS curves can be roughly subdivided into two regimes: the interval $q \lesssim 0.5$ nm^{-1}, within which the SANS cross section is described by a power-law dependence, corresponding to structure sizes larger than $\sim$10 nm. In [51], this feature has been interpreted as small-angle scattering on fractal clusters with a size of at least 120 nm (given that $q_{min} \cong 0.05$ nm^{-1} and $2\pi/q_{min} \cong 120$ nm). Scattering at $q \gtrsim 0.5$ nm^{-1} is attributed to the structural units composing the clusters. The data in this range cannot be analytically described by a dilute set of polydisperse homogeneous spheres. The points on the q-axis where

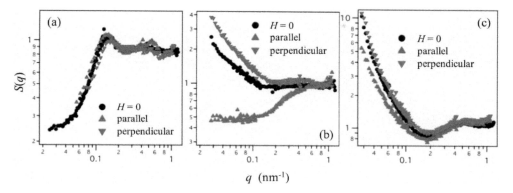

Fig. 5.15: Zero-field and under-field ($H = 800\,\mathrm{kA/m}$) structure factors of a maghemite (γ-Fe$_2$O$_3$) ferrofluid with a particle volume fraction of $x_\mathrm{p} = 1.5\%$ in three different states of interaction. (a) pH = 3 and $C_\mathrm{S} = 10^{-3}\,\mathrm{M}$; (b) pH = 3.5 and $C_\mathrm{S} = 6.5 \times 10^{-2}\,\mathrm{M}$; (c) pH = 3.7 and $C_\mathrm{S} = 0.13\,\mathrm{M}$. After [666].

deviation from power-law behavior occurs [indicated by the arrows in Fig. 5.16(a)] are used to estimate the characteristic size of the structural units. These depend on the nature of the sample and the values range between 18–25 nm, larger than the size of the individual magnetite particles. As shown in Fig. 5.16(b), the cluster structures are sensitive to temperature. Increasing the temperature from 25 °C to 70 °C results in a change of the slope at small q, which is interpreted as the disintegration of the clusters. Reducing the temperature back to 25 °C leads again to partial cluster formation.

Wiedenmann et al. [635, 641–643, 647, 653] have observed the formation of pseudocrystalline structures in concentrated Co ferrofluids. For concentrations above about 1 vol% Co, interparticle interactions are induced by an external magnetic field, which gives rise to an ordering of Co core-shell particles in hexagonal planes with their magnetic moments aligned along the [110] direction. In addition to the hexagonal sheets, the existence of segments of dipolar chains are also reported. Similar clustering and field-induced ordering phenomena were also observed in iron-oxide-based ferrofluids [657, 674]. The pseudocrystalline structures form when the dipolar interaction between the particles is strengthened by using strongly magnetic materials (using the pure elements instead of their oxides) or by increasing the particle size (resulting in an increase of the magnetic moment) [51]. Chain formation under the action of an applied field is also of relevance for the understanding of the magnetoviscous effect, which describes the increase of a ferrofluids' viscosity with increasing field. Pop and Odenbach et al. [639, 645, 649] have established the correlation between the microstructure and the magnetoviscous effect by performing shear-rate and magnetic-field-dependent SANS experiments on magnetite and cobalt-based ferrofluids using a specially designed rheometer (see also [681]). An analogous electroviscous effect related to electric-field-induced pattern formation and particle aggregation has also been studied using the SANS technique [671, 676, 682].

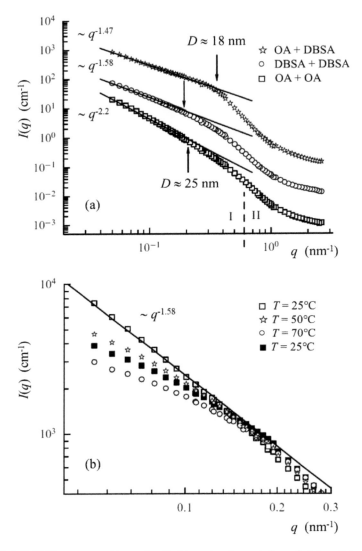

Fig. 5.16: (a) SANS cross sections for light-water-based ferrofluids with different surfactant combinations in a double stabilizing layer (OA = oleic acid; DBSA = dode-
cylbenzene sulphonic acid). No magnetic field is applied. Solid curves are power-law
dependencies corresponding to scattering on fractal clusters. Arrows indicate points on
the scattering curves at which the power-law dependencies break down. These points
are used to estimate the size of structural units within the clusters. (b) Temperature-
dependent changes of the scattering curve of a DBSA + DBSA sample. The figure
depicts the q-range within which noticeable changes occur. The temperature is first
increased from $25\,°C$ to $70\,°C$ and then reduced back to room temperature. After [51].

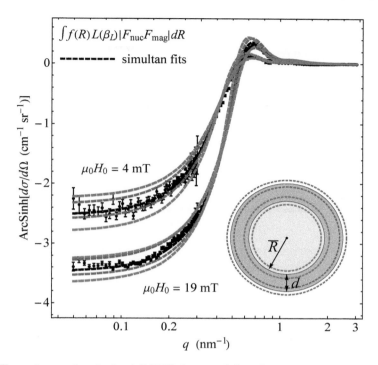

Fig. 5.17: Experimental polarized SANS data and fitted scattering cross sections for a Co-based ferrofluid ($\mathbf{k}_0 \perp \mathbf{H}_0$). The dashed black lines are the result of a simultaneous fit of eqn (5.21) for different external fields. The inset depicts the core-shell particle model with confidence intervals for the core radius (green dashed line) and the shell thickness (pink dashed line). After [64].

As an example for the application of the Langevin formalism (see Section 5.1.2) to the analysis of SANS data on ferrofluids, Fig. 5.17 depicts the difference of the two SANSPOL cross sections for a Co-based magnetic liquid at several external fields [64]. The data were fitted to the nuclear-magnetic cross term [eqn (5.21)] under the constraint that all structural contributions such as form factors are field independent and the overall intensity scales with the magnetic contrast via the Langevin function only. For clarity, Fig. 5.17 shows only two applied fields, but the model was fitted with four different fields (including the saturation field) to obtain the final confidence intervals. In principle, such an analysis is already possible with unpolarized neutrons, but as one can see from eqn (5.18), the non-saturated parts in the cross section give rise to an isotropic magnetic SANS contribution, leading to larger error margins in the nuclear-magnetic scattering separation [64].

TISANE. Based on a proposal by Gähler and Golub [683], Wiedenmann et al. [646] have implemented the TISANE technique on a time-of-flight spectrometer and paved the way for its later usage on SANS instruments (TISANE = Time Dependent SANS Experiments [684]). The chopper-based TISANE method opens up the way to kinetic

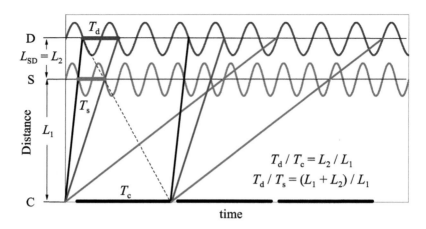

Fig. 5.18: Distance-time diagram for the TISANE mode. A wide velocity band of neutrons emerges from successive chopper openings (C) causing frame overlap at the sample (S). Neutrons from different chopper pulses which probe the same oscillation state are observed at the same time channel at the detector (D), provided that eqn (5.45) is satisfied. $T_c = \nu_c^{-1}$ is the repetition time of the chopper, $T_s = \nu_s^{-1}$ is the time period with which the scattering properties of the sample are modulated, and $T_d = \nu_d^{-1}$ is the signal period of the detector. After [646].

studies. It represents an improvement of the conventional time-resolved SANS technique (using a continuous neutron beam), which is limited to about 300 Hz time resolution by the neutron time-of-flight spread resulting from the wavelength distribution of the incident neutrons [665]. Stroboscopic TISANE allows one to probe nanoscale magnetism up to the $\sim 10\,\mu s$ regime, which permits the investigation of magnetization dynamics of nanoparticles in oscillating magnetic fields [50, 646, 672, 685, 686], the dynamics of vortex [687] and skyrmion lattices [688], or systems out of equilibrium. An example for a nonmagnetic application of the TISANE method are CO_2-microemulsions, which exhibit strong pressure-dependent properties. Using time-resolved SANS, Sottmann et al. [689, 690] have investigated the kinetics of pressure-induced structural changes in super or near-critical CO_2-microemulsions upon periodic pressure jumps of adjustable amplitude.

The working principle of TISANE is depicted in Fig. 5.18. For a more detailed treatment, including a mathematical analysis and a discussion of its limitations, see [684]. In TISANE, the sample is subjected to a periodic perturbation of its scattering properties, for instance via an oscillating magnetic field:

$$\mathbf{B}(t) = \mathbf{B}_0 \cos(2\pi\nu_s t) + \mathbf{B}_{st}, \tag{5.44}$$

where $B_0 = |\mathbf{B}_0|$ and ν_s, denote, respectively, the amplitude and frequency of the time-dependent field, and $\mathbf{B}_{st}$ is an optional superposed static field. The directions of $\mathbf{B}_0$ and $\mathbf{B}_{st}$ can be parallel and perpendicular to each other as well as parallel and perpendicular to the incident neutron beam. In the magnetic TISANE studies published so far, B_0^{max} is typically a few 10 mT and ν_s takes on values up to, say,

Fig. 5.19: (left) Counter-rotating double-disk chopper system for TISANE (© B. Jarry/ILL) and (right) a two-axis Helmholtz coil system allowing one to generate a rotating (kHz) magnetic field at the sample position. Image courtesy of Dirk Honecker, ISIS Neutron and Muon Source, Rutherford Appleton Laboratory, Didcot, United Kingdom.

20 kHz. The chopper is located at a distance L_1 from the sample and produces a pulsed polychromatic beam at a frequency of ν_c. Figure 5.19 depicts some of the main components of a (magnetic) TISANE setup. The detector, located at $L_{SD} = L_2$ from the sample, is characterized by a data-acquisition frequency ν_d. By the time the neutrons reach the sample position, there is an extreme frame overlap, i.e., neutrons of different wavelengths emerging from different chopper pulses arrive at the sample. At a certain distance downstream of the sample, where naturally the detector is placed, all the wavelengths reorder themselves. The chopper frequency ν_c is locked in with the oscillation frequency ν_s of the magnetic field and with the data-acquisition frequency ν_d of the detector to fulfill the TISANE frequency condition [684]:

$$\nu_d = \nu_s - \nu_c \quad \text{with} \quad \nu_c = \nu_s \frac{L_2}{L_1 + L_2}, \tag{5.45}$$

which must be satisfied in order to acquire with high resolution the scattering signal of frequency ν_d at the detector ($T_{d,s,c} = \nu_{d,s,c}^{-1}$).

The limitation of the conventional method can be seen from the following estimate. The neutron time-of-flight time $t_{tof} = L_2 \lambda \frac{m_n}{h}$ is given by:

$$t_{tof}[\text{ms}] = L_2 \lambda \times 2.52778, \tag{5.46}$$

where L_2 is in m and λ in nm. The spread Δt_{tof} of the neutron arrival times at the detector, i.e., the difference in travel times between the slowest and fastest neutrons in the distribution, can then be expressed as:

$$\Delta t_{tof} = \frac{\Delta \lambda}{\lambda} t_{tof} = \Delta \lambda L_2 \times 2.52778. \tag{5.47}$$

Assuming $\lambda = 1.0$ nm, $\Delta\lambda/\lambda = 0.10$, and $L_2 = 2-20$ m, the time spread equals $\Delta t_{tof} = 0.5-5$ ms [684]. This value needs to be compared with the period of the oscillatory magnetic field $T_s = \nu_s^{-1}$; for instance, $T_s = 3.3-20$ ms for $\nu_s = 50-300$ Hz, which is already comparable to Δt_{tof}. Increasing ν_s results in a progressively increased smearing of the oscillations in the detector response. Therefore, in order to probe the sample with frequencies in the $1-10$ kHz regime, the time resolution of the continuous technique is no longer sufficient and must be replaced by the pulsed TISANE technique [659].

Figure 5.20 provides a direct comparison between the conventional time-resolved and the TISANE technique. The sample in the study of Wiedenmann et al. [646] was a concentrated Co ferrofluid, consisting of 6 vol.% Co nanoparticles (core radius: 4.4 nm) coated with a surfactant shell (thickness: 1.9 nm) and dispersed in a viscous oil. It can be seen that the pulsed TISANE technique largely extends the frequency range of the continuous mode, which under the conditions of [646] is limited to ~300 Hz, up to ~1300 Hz. The corresponding characteristic timescale could be related to the Brownian rotational diffusion of individual core-shell nanoparticles. TISANE therefore provides access to the submillisecond timescale, which at least partially closes the gap between inelastic neutron scattering techniques, muon-spin spectroscopy, and Mössbauer spectroscopy ($\sim 10^{-12}-10^{-5}$ s) [155,593]. Compared to ac magnetic susceptibility measurements ($\sim 10^{-5}-10^{-1}$ s), it has the advantage of providing additional information on spatial nanometer-scale nuclear and magnetic correlations.

5.2.2   Magnetic steels

Despite centuries of research, steel, which has an enormous economical and societal relevance, is still one of the most complicated and investigated functional materials. Magnetic SANS, along with other techniques such as scanning and transmission electron microscopy, or atom-probe tomography, has played and continues to play a decisive role for the microstructure determination of this important class of materials; in particular, SANS is used to determine the size distribution as well as the corresponding volume fractions of precipitates, which govern to a large extent the mechanical properties. Analysis of the so-called A-ratio provides an indication for the chemical composition and the magnetization of the precipitates relative to the matrix. In this section, we highlight some prototypical applications of the SANS method for analyzing steels. The research literature regarding SANS on magnetic steels and on related metallic alloys is vast and any realistic attempt to provide an encyclopedic listing is beyond the scope of this book. For early reviews on this topic see [20,33,34]; for more recent accounts we refer to [64,691,692].

The SANS analysis on magnetic steels is based on the two-phase particle-matrix concept and assumes the sample to be fully saturated by a strong applied magnetic field. For unpolarized neutrons and $\mathbf{k}_0 \perp \mathbf{H}_0$, the total nuclear and magnetic SANS cross section at saturation can be expressed as [compare eqn (2.110)]:

$$\frac{d\Sigma}{d\Omega}(q,\theta) = \frac{d\Sigma_{nuc}}{d\Omega}(q) + \frac{d\Sigma_M}{d\Omega}(q)\sin^2\theta, \tag{5.48}$$

where θ denotes the angle enclosed by the direction of the applied field and the momentum transfer vector $\mathbf{q}$. Equation (5.48), which assumes isotropic nuclear and magnetic form factors, forms the basis for the separation of nuclear and magnetic SANS by means of sector averaging along the horizontal ($\parallel \mathbf{H}_0$) and vertical directions on the

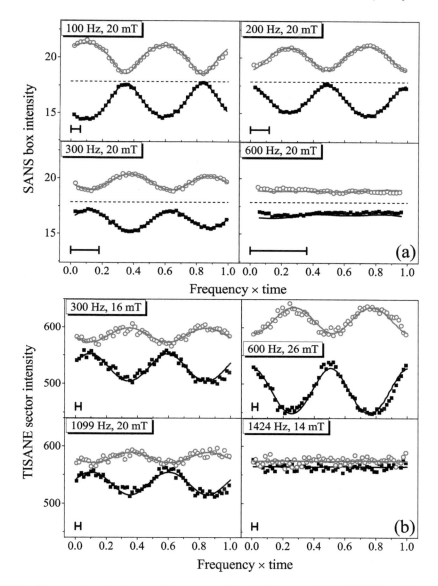

Fig. 5.20: Comparison between the conventional time-resolved SANS mode and the TISANE technique. Shown is the frequency dependence of the scattering intensities integrated over a q-range of $0.4 \pm 0.3 \, \text{nm}^{-1}$. Bars indicate the experimental resolution. (a) Conventional SANS mode. (b) TISANE mode. Closed symbols: horizontal sector averages. Open symbols: vertical sector averages. Insets specify the frequencies and the field amplitudes. Solid lines: fits to a "superparamagnetic" model function. Dashed lines in (a): static SANS mode at zero field. After [646].

detector (compare Appendix A). The accordingly separated $d\Sigma_{nuc}/d\Omega$ (or $d\Sigma_M/d\Omega$) is then further used to obtain the size distribution of the scatterers (e.g., precipitates) by solving the inverse problem which is embodied by an equation of the type of eqn (5.19). This procedure implicitly assumes the absence of interparticle interferences (dilute limit). However, in order to calculate the size distribution in absolute units, the nuclear and magnetic contrasts between scatterer (S) and matrix (Ma) must be known. This leads to the following equations [64]:

$$(\Delta\rho)_{nuc} = \langle\rho_{nuc}\rangle_S - \langle\rho_{nuc}\rangle_{Ma}$$

$$= \sum_{i\in S} \rho_{a,i,S}\, b_{i,S} - \sum_{i\in Ma} \rho_{a,i,Ma}\, b_{i,Ma}\,, \tag{5.49}$$

$$(\Delta\rho)_{mag} = \langle\rho_{mag}\rangle_S - \langle\rho_{mag}\rangle_{Ma}$$

$$= \sum_{i\in S} \rho_{a,i,S}\, b_{m,i,S} - \sum_{i\in Ma} \rho_{a,i,Ma}\, b_{m,i,Ma}\,, \tag{5.50}$$

where ρ_{nuc} and ρ_{mag} denote, respectively, the nuclear and magnetic scattering-length density, and $\rho_{a,i}$, b_i, and $b_{m,i}$ are, respectively, the atomic density, and the nuclear and magnetic atomic scattering length of atomic species i. The above parameters determine the A-ratio, which was originally introduced as the ratio of the scattering cross sections perpendicular ($\theta = 90°$) and parallel ($\theta = 0°$) to the saturating magnetic field [compare eqn (5.48)] [693]:

$$A = \frac{d\Sigma_\perp/d\Omega}{d\Sigma_\parallel/d\Omega} = 1 + \frac{d\Sigma_M/d\Omega}{d\Sigma_{nuc}/d\Omega} = 1 + \frac{(\Delta\rho)_{mag}^2}{(\Delta\rho)_{nuc}^2}. \tag{5.51}$$

Note that A (the symbol should not be confused with the exchange-stiffness constant) is in general a function of q. Compare also the similar ratio of scattering intensities, eqn (5.4), introduced by Ernst, Schelten, and Schmatz [451, 452] for the analysis of the spin structure of nanoparticles. However, as implied by eqn (5.51), if the magnetic and nuclear scatterers are the same objects, A can be expressed as the ratio of contrasts. Given the matrix structure and composition, eqns (5.49)–(5.51) describe a relationship between a measurable quantity of SANS and the structure and composition of the scatterers. The problem of extracting the composition of the scatterers from experimental A-data is obviously under-determined, except for the simplest systems and for nonmagnetic scatterers (magnetic holes). Alternatively, if the assumption of a dilute assembly of scatterers is violated, the Porod invariant P can be used to calculate the volume fraction x_p of scatterers [compare eqn (2.69)] [290]:

$$P = \frac{1}{2\pi^2} \int_0^\infty \frac{d\Sigma}{d\Omega}(q)q^2 dq = x_p(1 - x_p)(\Delta\rho)^2, \tag{5.52}$$

where the subscripts "nuc" and "M" have been omitted. Equation (5.52) is applicable independent of the structure of the two-phase medium and the degree of dilution of the system. Care has to be exercised for the proper extrapolation of the integrand.

We emphasize that the described two-phase approach relies on the (dilute) steel sample being in the magnetically saturated state. Deviations from saturation and the impact of the related spin-misalignment scattering have been studied by Bischof et al. [121]. These authors carried out magnetic-field-dependent SANS measurements on precipitate-containing martensitic steels, which were exposed to different heat treatments. When the sample is not fully magnetized, then magnetic scattering terms $\propto |\widetilde{M}_y|^2 \cos^2\theta$ give rise to a "contamination" of the nuclear scattering channel [compare, e.g., eqn (2.93) and Appendix A]. If this is not properly taken into account, large errors may result owing to spin-misalignment scattering. SANS studies on the precipitation and decomposition reaction in complex alloys (e.g., Fe–Cr, Ni–Al, Fe–Cu) also rely on the separation of nuclear and magnetic scattering contributions by means of a saturating field (e.g., [694–700]). Furthermore, it is important to remark that spin-polarized neutrons allow the sign of the product $(\Delta\rho)_{\text{nuc}}(\Delta\rho)_{\text{mag}}$ to be determined. This is impossible using unpolarized SANS and it may help to identify the type of scatterers in some cases [701, 702].

The main purpose of SANS in steel research is to characterize nanometer-sized precipitates and voids. Such nanofeatures can be the result of the synthesis procedure (e.g., [691, 703–715]) or they can form during the lifetime of the steel due to radiation-induced processes or thermal aging (e.g., [716–725]). As an example for the latter, we show in Fig. 5.21(a) the effect of neutron flux on the formation of Cu-rich precipitates in reactor pressure vessel steels. Advancing embrittlement of neutron-irradiated reactor pressure vessel steels is obviously an issue of utmost significance for safe reactor operation. While the dominant embrittlement mechanisms were gradually understood [727], special long-term irradiation effects are also of specific interest [728]. For the case of neutron-irradiated materials, the respective un-irradiated sample is usually taken as a reference, and the outcome of the data analysis is then the distribution of irradiation-induced nanofeatures. It is seen in Fig. 5.21(a) that the increase of the peak radius of the volume-related size distribution of scatterers by a factor of about 2 is the most prominent effect of reducing the flux by a factor of 35 at constant fluence. The total volume fraction of scatterers is approximately independent of flux. Figure 5.21(b) shows that the volume fraction of irradiation-induced solute atom clusters (containing Cu, Ni, and Mn) increases at increasing fluence, whereas the mean radius of the clusters is approximately constant. This is at difference with the effect of flux at constant accumulated fluence [Fig. 5.21(a)], for which the mean cluster size decreases at increasing flux, whereas the volume fraction was found to be approximately constant. Components of future fission reactors and fusion devices will have to withstand more severe conditions than those depicted in Fig. 5.21, including neutron exposures of up to 200 displacements per atom (dpa), a factor of 1000 more than exposures typical for the reactor pressure vessel of running nuclear power plants. Fe–$(8 - 18\%)$Cr steels are candidate materials for such extreme conditions.

Figure 5.22 displays the effect of neutron irradiation on the A-ratio of a $Fe_{1-x}Cr_x$ alloy $(x = 0.125)$. Fe–Cr alloys attract interest as model systems, since they allow for the exploration of the irradiation behavior of Cr steels. Cr contents in excess of the solubility limit of Cr in Fe below $500\,^{\circ}\text{C}$ give rise to a phase separation into a Fe-rich α-phase and a Cr-rich α'-phase. The α-phase is ferromagnetic, while the α'-phase is usually considered to be nonmagnetic. The formation of α' is responsible for the so-called $475\,^{\circ}\text{C}$ embrittlement during thermal aging and for the embrittlement during

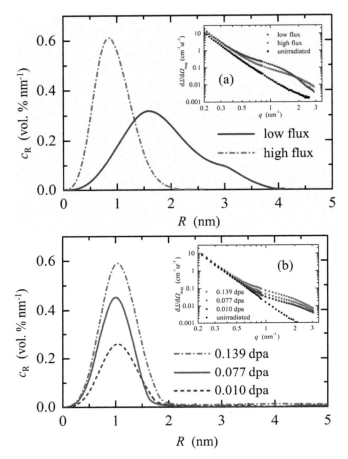

Fig. 5.21: (a) Effect of neutron flux (for equal neutron exposure of 0.032 dpa) on the size distribution c_R of irradiation-induced Cu-rich precipitates for a reactor pressure vessel weld (0.22 wt% Cu, 1.1 wt% Mn, 1.1 wt% Ni) [720]. The measured magnetic SANS cross section is shown in the inset. Low and high fluxes correspond, respectively, to 0.087×10^{-9} dpa/s and 3.05×10^{-9} dpa/s (factor: 35). After [64]. (b) Size distribution of irradiation-induced solute atom clusters in A533B-1 reactor pressure vessel steel (IAEA reference steel JRQ) as a function of neutron fluence (i.e., the product of neutron flux and irradiation time at constant flux). The measured magnetic SANS cross section is shown in the inset. After [726].

neutron irradiation of $\gtrsim 9\%$ Cr steels. α- and α'-phases are both bcc with only minor differences of the lattice parameters. Bergner et al. [729] derived the volume fraction of α' from SANS experiments performed in saturation on industrial-purity $Fe_{1-x}Cr_x$ ($x = 0.125$; neutron-irradiated at 300 °C). They found that a quasi-steady state of Cr precipitation had been reached. The measured A-ratio for this particular composition is independent of the magnitude of q. The average values are $A = 2.07 \pm 0.05$ and

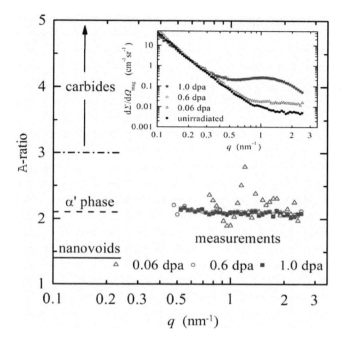

Fig. 5.22: Effect of neutron exposure on the A-ratio of neutron-irradiated $Fe_{1-x}Cr_x$ ($x = 0.125$) [729]. The measured magnetic scattering is displayed in the inset. The un-irradiated reference was subtracted before calculating $A(q)$. Theoretical A-ratios of Cr-rich carbide precipitates and nanovoids are indicated as well. After [64].

$A = 2.05 \pm 0.05$ for neutron exposures of 0.6 and 1.0 dpa, respectively (see Fig. 5.22). These results are consistent with the small-angle scattering from α'-particles dispersed in a α-matrix, for which a range of A-ratios from 2.09 to 2.13 was reported [719]. The A-ratio calculated for pure Cr precipitates in α-Fe, $A = 2.03$ [719], is also in the experimentally observed range of values. By contrast, neither nanovoids nor the known types of Cr-rich carbides [719, 729] are able to explain the experimental data.

The discussed examples provide a flavor of the potential of magnetic SANS for elucidating the microstructure of magnetic steels. A decisive point for the validity of the neutron-data analysis, which is based on the separation of nuclear and magnetic SANS using a large applied field, appears to be the question whether the saturated magnetization state has been reached "sufficiently close". Magnetic-field-dependent SANS measurements are required in order to address this issue.

5.2.3 Spin glasses, amorphous metals, and critical SANS

Spin glasses are a special class of magnetic alloys which are characterized by terms such as disorder, randomness, competing interactions, and frustration [730–738]. In contrast to conventional ferromagnets and antiferromagnets, spin glasses do not exhibit long-range ordering of magnetic moments. Instead, below the so-called freezing temperature

T_f, ideal spin glasses when cooled from the paramagnetic regime cooperatively enter a perfectly disordered state in which all the spins (or spin clusters) are aligned into random directions. Among the various experimental techniques which are employed for elucidating the spin structure and the complex magnetic interactions in spin glasses, magnetic SANS stands out, since with it the spin correlations on the nanometer length scale and within the bulk of the sample can be probed.

Spin glasses can be found in crystalline or amorphous structural state. We will start describing the experimental findings of archetypal crystalline spin glasses consisting of magnetic transition-metal impurities in noble metal hosts. Prototypical examples for such systems are binary alloys of composition $A_{1-x}B_x$, where magnetic B-ions (e.g., Fe, Mn, Ni) are substituted in a nonmagnetic host A-matrix (e.g., Ag, Au, Cu, Pt) [733]. At low concentration ($x \ll 1$), the magnetic ions are randomly distributed in the host matrix (so-called site randomness*) and interact with each other via the long-range and oscillatory Ruderman–Kittel–Kasuya–Yosida (RKKY) exchange interaction mediated by the conduction electrons [740–742]. Asymptotically, for large distances between the magnetic moments ($r \to \infty$), the indirect RKKY exchange between two ions takes on the form [733]

$$J(r) = J_0 \frac{\cos(2k_F r + \varphi_0)}{(k_F r)^3}, \qquad (5.53)$$

where J_0 and φ_0 are constants, and k_F is the Fermi wave number of the host metal. Equation (5.53), which has been derived in the free-electron approximation, shows that for a given magnetic impurity there will be some ferromagnetic interactions ($J > 0$) with its neighboring spins, while other interactions are of antiferromagnetic type ($J < 0$). Hence, no spin alignment can be found that is satisfactory to all the exchange bonds. This scenario—random distances between the magnetic moments in the nonmagnetic metallic matrix plus distance-dependent RKKY-type interactions between them—leads to competing ferromagnetic and antiferromagnetic interactions and to concomitant frustration. As a consequence, the magnetic ground state of the system is multi-degenerate and no true long-range magnetic order is established, i.e., no magnetic Bragg peaks are seen in a diffraction measurement when crossing T_f. It should be emphasized that the RKKY Hamiltonian contains the basic necessary ingredients of disorder and frustration and is able to reproduce some elementary key features of canonical spin glasses. However, the real microscopic physics is often much more complicated and involves other interactions too (e.g., magnetic anisotropy, dipole-dipole interaction, DMI) [734]. This is particularly true for larger concentrations of magnetic impurities, when cluster spin glasses (so-called mictomagnetism) and long-range-ordered inhomogeneous states form; see Fig. 2.9 in the book by Mydosh [735] for the different types of magnetic behavior which may occur in a spin glass as a function of impurity concentration.

The temperature-concentration magnetic phase diagram of one of the most scrutinized spin-glass systems, $Au_{1-x}Fe_x$, is displayed in Fig. 5.23(a) [743]. For small

---

*In addition to site randomness, there are also systems which exhibit bond randomness, e.g., the ordered crystalline material $Rb_2Cu_{1-x}Co_xF_4$ [739]. In this case, the bonds between nearest-neighbor ions can have different values of the exchange (varying between $+J$ and $-J$), and these bonds are randomly distributed throughout the sample [155].

iron concentration, this system directly enters the spin-glass state on cooling from a high-temperature paramagnetic phase. Decreasing the temperature slows down the fluctuations of the individual spins which become progressively correlated and tend to form locally correlated units (clusters) [155]. On approaching the freezing temperature T_f, the correlations among the spins grow, and at T_f the sample enters a cooperative state by taking on one of its many (nearly equivalent) ground states. One of the signatures of this process is a peak in the SANS intensity. This is displayed in Fig. 5.23(b) [732, 744], which shows the temperature dependence of the forward scattering intensity of an $Au_{90}Fe_{10}$ alloy at selected q-values. The observed q-dependence of T_f [see inset in Fig. 5.23(b)] has been interpreted as an indication that there is not a unique freezing temperature in these alloys, a finding which has been controversially discussed in the literature [744–746]. For higher iron concentrations, the behavior of $Au_{1-x}Fe_x$ can be different, e.g., there can be a transition from a paramagnetic to a superparamagnetic to a cluster-glass state, or the system can exhibit a sequence from a paramagnetic to a ferromagnetic to a low-temperature spin-glass-like phase (so-called re-entrant behavior). It must be emphasized that the nature of the spin glass freezing transition is not fully understood yet and still the subject of current research; see, e.g., the articles by Parisi [736], Sherrington [747], and Mydosh [738] for accounts on spin glasses and a perspective on open problems.

Besides the above-mentioned archetypal crystalline spin glasses, there are many other materials classes which may exhibit spin-glass or spin-glass-like behavior; for instance, high-temperature superconductors, heavy-fermion systems, Heusler alloys, spinels and pyrochlore oxides, or amorphous metals [734, 738]. Amorphous metals can be synthesized by the extremely rapid quenching of metal melts, so that the liquid state is "frozen in". These systems are therefore also called metallic glasses [748–750]. Their magnetism has many parallels to that of nanocrystalline magnets. While some of these amorphous materials form spin glasses due to their random topology at an atomic scale and due to competing ferromagnetic and antiferromagnetic exchange interactions between neighboring atoms, others may be described as random anisotropy magnets. In the latter case, the notable difference to the nanocrystalline systems resides in the fact that the orientation of the local easy-axis directions varies randomly at the scale of the interatomic spacing, rather than the grain size. In addition to random anisotropy, the magnetodipolar interaction as well as the DMI [411–413] may be of relevance for determining the spin structure of metallic glasses. Descriptions within the framework of random anisotropy models are otherwise closely similar, and the application of this approach to metallic glasses [454, 751–754] preceded its adaptation to nanocrystalline magnets [755, 756].

Magnetic SANS has greatly contributed to the understanding of the magnetic properties of spin glasses, amorphous metals, and complex alloys in general, since it becomes possible to measure the evolving spatial spin correlations on a mesoscopic length scale and as a function of several parameters such as temperature, applied magnetic field, or alloy composition (see, e.g., [122, 468, 528, 612–614, 744–746, 757–791] and references therein). SANS research into spin glasses peaked during the 1970s−1990s. For a SANS review on rare-earth-based amorphous alloys see [30]; for a more comprehensive treatment of inelastic and elastic magnetic neutron scattering from spin glasses and

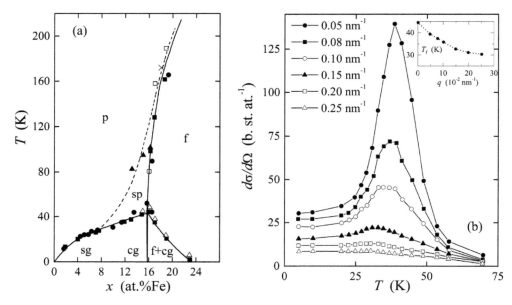

Fig. 5.23: (a) Magnetic phase diagram of $Au_{1-x}Fe_x$ solid solutions (p = paramagnetic; sp = superparamagnetic; sg = spin glass; cg = cluster glass; f = ferromagnetic). After [743]. (b) Forward scattering intensity of a $Au_{90}Fe_{10}$ alloy as a function of temperature for a series of q-values. Lines are a guide to the eyes. Inset: peak temperatures T_f as a function of q. The point at $q = 0$ is obtained from ac magnetic susceptibility measurements. After [732, 744].

amorphous magnets, we refer the reader to the book by Hicks [201]. In the following, we will discuss the SANS cross-section expressions that are commonly used to analyze the (q, T, H_0)-dependence of experimental scattering data on such materials.

Generally, for ferromagnets, it is found that the q-dependence of the scattering cross section at temperatures above the ferromagnetic ordering temperature T_C is well described by a Lorentzian (L) function [540, 541]:

$$\frac{d\Sigma}{d\Omega} = \frac{w_0}{\kappa^2 + q^2}. \tag{5.54}$$

This expression for $d\Sigma/d\Omega$ corresponds to a spin-spin correlation function of the Ornstein–Zernike type (in three spatial dimensions and for not-too-small r),

$$\langle \mathbf{M}(0)\, \mathbf{M}(r) \rangle \propto \frac{\exp\left(-\kappa r\right)}{r}, \tag{5.55}$$

where κ^{-1} denotes the correlation length $\xi = \kappa^{-1}$ [cf. the analogous expression for nuclear scattering, eqn (2.73)]. As the temperature is approaching T_C from above, the sizes of the correlated regions start to grow and $\kappa \to 0$, corresponding to the onset of infinite-range ferromagnetic order. Within the theory of critical phenomena, the behavior of the correlation parameter is described by the scaling law $\xi(T) \sim |t|^{-\nu}$, where

$t = (T - T_C)/T_C$ is the reduced temperature, and the critical scaling exponent $\nu = 1/2$ for mean-field theory. For the initial susceptibility and the spontaneous magnetization (below T_C), one finds, respectively, $\chi(T) \sim |t|^{-\gamma}$ with $\gamma = 1$ and $m(T) \sim (-t)^{\beta}$ with $\beta = 1/2$ (for mean-field theory) [154]. Early SANS studies addressing the critical behavior of the $3d$ transition metals can be found e.g., in [238,792–800]. For temperatures below T_C, fits with

$$\frac{d\Sigma}{d\Omega} = \frac{w_1}{\kappa^2 + q^2} + \frac{w_2}{(\kappa^2 + q^2)^2}, \tag{5.56}$$

have been found to approximate the data well [767, 772], when the amplitudes w_i and the correlation length κ^{-1} are used as adjustable parameters. It has been pointed out, however, that in some instances eqn (5.56) is difficult to distinguish from other functional forms which may provide similarly good fit (e.g., [773, 774]). In other instances, the κ in eqn (5.56) were refined independently, and the addition of an extra Lorentzian-squared (L^2) term with yet another κ was required to fit the data [779].

Equation (5.56) with the same κ in the L and L^2 term emerges from an analysis of the Ising model in the presence of random fields [801–803]. In that analysis, the squared Lorentzian arises from the static magnetic microstructure with an exponentially decaying correlation function [cf. eqn (2.72)], whereas the simple Lorentzian represents scattering due to spin waves and residual spin fluctuations. Del Moral and Cullen [454] have provided an extended spin-wave theory for the magnetic correlations in random anisotropy magnets in the quasi-saturated regime. Besides corroborating the origins of the L and L^2 terms in eqn (5.56), one of their main results is the extension of eqn (5.56) to obtain new SANS cross sections of $L^{1/2}$ and $L^{3/2}$ types. Boucher and Chieux [30] provide a detailed discussion of the various scattering laws and related power-law exponents encountered in the SANS of rare-earth-based amorphous alloys. The application of eqns (5.54) and (5.56) to amorphous Tb–Fe alloys is summarized in Fig. 5.24. It is seen that they provide a good fit to the data, and the correlation length κ^{-1} in the remanent state exhibits the expected peak at $T = T_C$ [772].

Before continuing with the discussion of magnetic SANS on spin glasses and amorphous alloys, we briefly touch upon an important point which may complicate the analysis and understanding of critical-scattering data—the estimation of the effect of inelasticity [804]. When no analysis of the energy of the scattered neutrons is carried out, then $d\Sigma/d\Omega$ represents the cross section integrated over all final neutron energies. If the inelastic scattering contributions are small compared to the incident neutron energy ($\hbar\omega \ll E_0$), then the energy-integrated $d\Sigma/d\Omega$ is proportional to the spatial Fourier transform of the instantaneous spin-spin correlation function [76,86]. In this so-called static approximation (sometimes also denoted as the quasi-static or quasi-elastic approximation), $d\Sigma/d\Omega$ can be related to the wave-vector-dependent static susceptibility $\chi_{st}(\mathbf{q})$, which is a quantity of clear theoretical significance that can be compared with model calculations [196]. Moreover, from an experimental point of view, it is also easier to measure the differential rather than the partial differential scattering cross section [541]. Under the above assumptions and neglecting Bragg diffraction, one can write [86, 196, 201]:

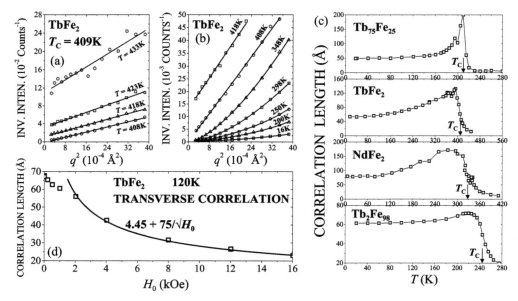

Fig. 5.24: (a) Inverse intensity versus q^2 for TbFe$_2$ above and close to $T_C = 409\,\mathrm{K}$ showing the Lorentzian (L) line shape (solid lines). (b) Scattering at lower temperatures illustrating departures from the Lorentzian line shape below T_C. The lines are the fit to the L plus L^2 cross section for $T \lesssim 408\,\mathrm{K}$. (c) Spin correlation lengths $\xi = \kappa^{-1}$ for amorphous Tb$_{75}$Fe$_{25}$, TbFe$_2$, NdFe$_2$, and Tb$_2$Fe$_{98}$ showing the quenching of long-range order at and below the transition temperatures. (d) Field dependence of the correlation length corresponding to the transverse part of the cross section. The solid line is the theoretically predicted $1/\sqrt{H_0}$ dependence. After [768].

$$\frac{d\Sigma}{d\Omega} \propto |f(\mathbf{q})|^2 kT\chi_{\mathrm{st}}(\mathbf{q}), \tag{5.57}$$

where the atomic magnetic form factor $f(\mathbf{q}) \cong 1$ in the small-angle region. The magnetic neutron scattering cross section in the static approximation, eqn (5.57), is also useful for comparing magnetic neutron data with bulk susceptibility measurements [201]. The magnitude of the scattering vector for inelastic scattering is given by eqn (1.9), which is rewritten here for convenience (reciprocal lattice vector: $\mathbf{G} = 0$):

$$q = \sqrt{k_0^2 + k_1^2 - 2k_0 k_1 \cos\psi}$$

$$= \sqrt{\frac{2m_{\mathrm{n}}}{\hbar^2}} \sqrt{2E_0 - \hbar\omega - 2\sqrt{E_0(E_0 - \hbar\omega)}\cos\psi}, \tag{5.58}$$

where $E_0 = \hbar^2 k_0^2/(2m_{\mathrm{n}})$ is the initial neutron energy, $E_1 = \hbar^2 k_1^2/(2m_{\mathrm{n}})$ is the final neutron energy, $\hbar\omega = E_0 - E_1$ is the energy transfer between the neutron and the sample, and ψ is the scattering angle. In the static approximation, which amounts to neglecting the ω-dependence of the momentum transfer vector for constant E_0 and scattering angle, the spin fluctuations appear almost static to the incident neutrons and information about the dynamics is lost [76, 86]. Near a critical point, however, inelastic contributions to the total energy-integrated cross section may become appreciable [541], so that it is necessary to quantify their influence. The most direct way to do this consists of experimentally measuring the distribution of final neutron energies; for instance, via time-of-flight energy analysis, as has been done by Burke and Rainford in their extensive study on the evolution of magnetic order in Cr–Fe alloys [764–766]. These authors found [766] that the width of the quasi-elastic line ($\sim$40 μeV) is much smaller than the energy of the 11.8 Å neutrons (588 μmeV) used in their experiment. Inelasticity corrections can also be assessed by numerically integrating the partial differential scattering cross section using different models for the spin dynamics (see, e.g., [805–808] and references therein for further details).

Many of the findings for the amorphous magnets are consistent with the results for the elastic magnetic SANS from random-anisotropy-type nanocrystalline ferromagnets, as discussed in Sections 4.4 and 6.5. First, it is well documented that under a small applied field the maximum magnetic scattering from the amorphous systems occurs parallel to the field direction (e.g., [122, 528, 767]). Figure 5.25 depicts an example for the case of the Fe–Al spin-glass system. This is in agreement with the observations for the nanoscaled magnets [e.g., Fig. 6.19(a)], where the origin of field-parallel scattering is related to transversal (misaligned) spins. Second, as can be seen in Fig. 5.24(d), the correlation length of eqn (5.56) varies with the applied magnetic field as $\kappa^{-1} \propto H_0^{-1/2}$ below T_C [754]. This agrees with the behavior of the micromagnetic exchange length $l_H \propto H_0^{-1/2}$ (cf. Section 6.5), consistent with the closely similar description of the two types of materials [454, 751–756]. Third, at not-too-small fields, the asymptotic form of the elastic spin-misalignment scattering is given by (compare the discussion in Section 4.3.5) [528]:

$$\frac{d\Sigma_{SM}}{d\Omega} \cong \frac{w_3}{q^2} + \frac{w_4}{q^4}. \tag{5.59}$$

To see this, note that the random fields and the random variations in the magnitude of the magnetic moments of an amorphous magnet may vary at the scale of the interatomic distance d_{NN}. Since the accessible q-values of typical SANS experiments are much smaller than $2\pi/d_{NN}$, both the anisotropy-field interference function $S_H(q)$ and the magnetostatic function $S_M(q)$ are almost q-independent over a large part of the experimental SANS q-range. Consequently, the asymptotic behavior of the related spin-misalignment scattering $d\Sigma_{SM}/d\Omega = S_H R_H + S_M R_M$ is determined by the asymptotic behavior of both response functions $R_H \propto q^{-4}$ and $R_M \propto q^{-2}$ [compare eqns (4.16) and Fig. 4.2]. Fourth, it is noteworthy that Rhyne and Glinka [767] report that the magnitude of the Lorentzian contribution to eqn (5.56), which is expected to describe energy-integrated scattering due to spin waves, becomes insignificant at the higher fields. This can be readily understood since, according to Section 2.3, a large applied field results in the creation of a spin-wave gap and thus in the suppression of spin-wave scattering in SANS.

As a general comment, the Lorentzian (L) and Lorentzian-squared (L^2) terms of eqn (5.56) represent a special case of the more general results discussed in this book

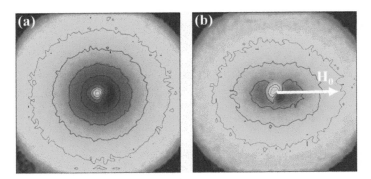

Fig. 5.25: Two-dimensional SANS intensity distribution of the amorphous spin glass $Fe_{70}Al_{30}$ at $T = 9\,K$. (a) $\mu_0 H_0 = 0\,T$; (b) $\mu_0 H_0 = 0.1\,T$. Image courtesy of Dina Mergia (Ph.D. dissertation, University of Athens, Greece, 1997).

[compare the discussion related to eqn (5.59)]. While eqn (5.56) applies in the idealized limiting case where the random fields vary at a scale too small to be resolved by SANS, the equations of Sections 4.2–4.4 admit that the "random" anisotropy comprises correlation of the easy-axis direction and magnitude on much larger length scales (restricted by the validity range of the continuum theory of micromagnetics). In real amorphous magnets, and beyond the concept of randomness on the atomic scale, one may expect that the anisotropy is indeed characterized by a spectrum of correlation lengths that includes also longer-range features; for instance, due to medium-range chemical or topological order, or due to magnetoelastic coupling. Such features might extend to within the range of resolution of SANS studies. As the analysis of field-dependent SANS discussed in Sections 4.2–4.4 applies analogously to nanocrystalline as well as to amorphous systems, it could reveal such features. This type of study has not been reported so far. Yet another outstanding experiment is the study of field-dependent SANS in the early stages of crystallization of a magnetic glass. Here, magnetic SANS is expected to provide a sensitive probe of the emergence of correlated easy-axis directions in the crystal nuclei. While magnetic SANS has been employed for studying the precipitation of a nonmagnetic (Cu) phase from an amorphous magnetic alloy [809], the potential of field-dependent SANS studies of crystallization for identifying the nucleation of a magnetic phase remains to be exploited.

Finally, it is noted that scattering laws similar to eqn (5.56) have been found to provide a good phenomenological description of experimental magnetic SANS data from various magnetic systems in different scientific contexts; for instance, interacting assemblies of magnetic nanoparticles, which exhibit pronounced temperature and magnetic-field-dependent correlations (e.g., [460,564,810–813]), are sometimes also described by a combination of L and L^2 terms. In the study of Bellouard et al. [564], fits to temperature-dependent scattering data have indicated a progressive alignment of the spins near the surface and an emerging correlation between the net magnetization directions of neighboring Fe nanoparticles as the temperature is decreased. Oku et al. [811] have studied Mn–Zn-ferrite nanoparticles coated with a nonmagnetic surfactant layer and randomly dispersed in a solid polystyrene matrix (particle volume fraction: 4%). For $6\,\mathrm{K} \leq T \leq 260\,\mathrm{K}$, they report the appearance of strong interparticle magnetic correlations, which are monitored by fitting the magnetic SANS cross section to a combination of L and L^2 terms. The magnetic correlation length increases with decreasing temperature down to $\sim 40\,\mathrm{K}$, but long-range order is not attained. For $T \lesssim 40\,\mathrm{K}$, the range of the correlations decreases with decreasing temperature, an observation which is interpreted in terms of random-field effects. Further examples for materials classes where L and L^2 terms find usage are e.g., magnetic shape-memory alloys, perovskite manganites, and pyrochlore oxides (see the following Section 5.2.4).

5.2.4 Some further applications of magnetic SANS

The literature on magnetic SANS is abundant and very diverse. In the following short sections, we address several materials classes where the magnetic SANS technique finds application. The aim is to give a rough idea on the main research directions and to cite some relevant papers in the respective field.

Flux-line lattices, skyrmions, and long-range-ordered noncollinear spin structures. These systems are characterized by the existence of a long-range-ordered periodic magnetic structure with a "lattice constant" that is usually much larger than the wavelength of cold neutrons. Therefore, the concepts of diffuse magnetic SANS which are developed in this book are generally not suitable for the description of the magnetic neutron scattering cross section, but the well-known formalism of neutron diffraction applies. SANS experiments on flux-line lattices, skyrmion crystals, and long-range-ordered noncollinear structures use SANS instruments as diffractometers. For the specific example of a magnetic structure with a single propagation vector of $\mathbf{k}_{\mathrm{mag}}$ (e.g., the various helical, conical, or oscillatory structures in the heavy rare-earth metals [814]), the magnetic cross section is proportional to $\delta(\mathbf{q} - [\mathbf{G} \pm \mathbf{k}_{\mathrm{mag}}])$ [76], where $\mathbf{G}$ is a reciprocal lattice vector (compare Fig. 1.2). Magnetic Bragg diffraction occurs in this case for $\mathbf{q} = \mathbf{G} \pm \mathbf{k}_{\mathrm{mag}}$, i.e., each Bragg peak is surrounded by two satellite reflections. As the direct beam $[\mathbf{G} = (000)]$ corresponds to a ferromagnetic zone center, the ferromagnetic components of any noncollinear magnetic structure may be conveniently studied by SANS (e.g., [815–825]).

Flux-line lattices in type-II superconductors were among the first topics which were studied using two-axis diffractometers and SANS instruments [826–834]. In fact, the experimental proof for the existence of the theoretically predicted Abrikosov vortex lattice of type-II superconductors [835] was performed in 1964 by Cribier et al. [826] using neutron diffraction (well before their successful real-space imaging). Since then, SANS has provided important information on e.g., the structure and symmetry of the vortex lattice, the penetration depth, the nature of the pairing and the pairing mechanism, and on the superconducting energy gap (e.g., [836–850]). For further reading, see the review by Eskildsen, Forgan, and Kawano-Furukawa [53] and Section IX in [64].

Some of the type-II superconductors such as niobium and vanadium exhibit a so-called intermediate mixed state (IMS) between the field-free Meissner phase and the Abrikosov vortex-lattice phase (also denoted as the Shubnikov phase) [851–858]. The IMS phase consists of isolated Shubnikov domains with a typical size in the micrometer regime, which are embedded in the Meissner region. While most of the SANS studies focus on the properties of the vortex lattice, more recent investigations address the domain formation and the domain structure of the IMS as a function of temperature, applied field, sample purity, pinning, or sample shape. As shown by Mühlbauer et al. [855–858], for the investigation of the IMS, the combination of conventional SANS and USANS/VSANS with the real-space neutron grating interferometry method has proven to be very useful in obtaining unique insight into the hierarchical properties of vortex-matter domains on a wide range of length scales from about 10 nm to 10 μm.

Skyrmions and skyrmion crystals may be considered as the magnetic analog of superconducting vortex lattices. Both types of vortex matter—skyrmion crystals and flux-line lattices—represent nanoscale magnetic objects that emerge from chemically and structurally homogeneous samples due to collective ordering phenomena of the electronic system. Magnetic skyrmions are topologically nontrivial whirls of a continuous magnetization vector field. Depending on the mechanisms which generate skyrmions, their size may range from about 1 nm to 1 μm [382]. A DMI-stabilized skyrmion-crystal phase was first discovered in the cubic non-centrosymmetric helimagnet MnSi in 2009 by Mühlbauer et al. [374]. Inspired by this finding, the observation of magnetic skyrmions in a large variety of other compounds, well beyond the B20 family, has established their existence as a generic phenomenon of materials that promote chiral magnetic interactions (see, e.g., [257, 375, 382, 385–387, 390, 392, 393, 395–398, 859–869] and references therein); see also Section VIII in [64] for a review of neutron studies on skyrmions and noncollinear spin structures.

Heavy-fermion compounds. Elliott [870] and Ginzburg and Maleyev [871] have developed a theory of magnetic neutron scattering by conduction electrons in a metal. It is predicted that the magnetic SANS cross section is sizable if the effective electron mass $m_e^\star \gtrsim 100\,m_e$, where m_e is the free-electron mass. Such high values of $m_e^\star/m_e$ can e.g., be found in heavy-fermion compounds. SANS on heavy-fermion quasi-particles is anisotropic, absent for $\mathbf{q} \parallel \mathbf{B}$ and strongest for $\mathbf{q} \perp \mathbf{B}$, where $\mathbf{B}$ is the magnetic induction inside the sample, and it is expected to increase with an applied magnetic field. At $q = 0.15\,\mathrm{nm}^{-1}$ and zero field, the magnitude of the SANS cross section is estimated to be of the order of $d\Sigma/d\Omega \cong 0.5 \times 10^{-2}\,\mathrm{cm}^{-1}$. The main features of this type of weak neutron-electron scattering have been demonstrated by Kopitsa et al. [872] on a CeRu$_2$Si$_2$ single crystal at 0.85 K.

Invar alloys. Iron-nickel alloys with a Ni concentration of ∼35 at.% exhibit an anomalously small thermal expansion over a wide range of temperature. The origin of this so-called Invar effect is still not fully resolved. It is believed to be related to the presence of magnetic inhomogeneities and noncollinear magnetization deep in the ferromagnetic ordered state. Therefore, SANS is very well suited to study these systems (see, e.g., [873–880] and references therein). Komura et al. [873, 874] employed the Guinier law to estimate the size and temperature variation of magnetic cluster structures in Fe$_{65}$Ni$_{35}$ single crystals. Similarly, Chamberod et al. [875] report the existence of small clusters associated with a magnetization inhomogeneity. Grigoriev et al. [877] studied the magnetic phase transition in Invar alloys by combining SANS with neutron depolarization measurements. For carbon-doped Fe$_{75}$Ni$_{25}$, they report the coexistence of two length scales above the critical point: the smaller length scale, as seen by the SANS method, has been attributed to the "usual" critical fluctuations and their shape was well described by a Lorentzian, while much larger magnetic inhomogeneities, accessible via the neutron depolarization technique, were found above T_C and were roughly modeled by a squared Lorentzian. In later SANS studies on $Fe_{70}Ni_{30}$ [878] and $Fe_{65}Ni_{35}$ [879], a spatial distribution of the Curie temperature T_C has been introduced in order to explain the experimental data. Based on the early polarization analysis work by Menshikov and Schweizer [876], Stewart et al. [880] have employed uniaxial polarization analysis on a high-purity $Fe_{65}Ni_{35}$ single crystal to prove the existence of dominant but uncorrelated longitudinal spin fluctuations. These coexist with transversal spin components, which are related to spin clusters with a size of ~13 nm. In addition to SANS, high-resolution neutron spin-echo spectroscopy on a polycrystalline ingot indicated the presence of slow dynamical spin fluctuations in the ordered ferromagnetic state at $q \cong 0.7\,nm^{-1}$ (comparable to the SANS experiment).

Magnetic recording media. Magnetic SANS and, in particular, polarized SANS has significantly contributed to the understanding of the structure of magnetic recording media (e.g., [470, 565, 881–894]). Many of these materials exist in the form of complex thin-film architectures, wherein the magnetically active recording layer, which has a thickness of typically only a few nanometers, is buried. SANS experiments are therefore extremely challenging due to the small magnetic volume that is probed and due to the large amount of nuclear background scattering from the substrate and the underlayers of the medium. Based on a Stoner–Wohlfarth model allowing for thermally activated magnetization switching, Lister et al. [891] could describe the grain-size-dependent magnetic reversal process in a perpendicular $CoCrPt-SiO_x$-based recording layer. Strong evidence for an increase in magnetic anisotropy with grain diameter was found. Oku et al. [569] studied spherical $Fe_{16}N_2$ nanoparticles (for potential magnetic recording tape applications) using polarized SANS and successfully analyzed the data with a core-shell model. Dufour et al. [895] employed unpolarized SANS and polarization analysis to study the domain structure and the magnetization-reversal process in an exchange-biased $DyFe_2/YFe_2$ superlattice. They report that the reversal occurs via the formation of 10–100s nm-sized magnetic domains, which are found to be arranged in a quasi-periodic manner in the plane of the sample, and that the reversal mechanism involves the rotation of the magnetization in and out of the sample plane.

Magnetic shape-memory alloys. Ni–Mn-based Heusler alloys exhibit unique physical properties such as magnetocaloric, magnetoresistance, exchange bias, and, in particular, magnetic shape-memory effects [896–898]. These features originate from the strongly coupled magnetic and structural degrees of freedom and occur in stoichiometric as well as in nonstoichiometric alloys. Depending on the composition, Ni–Mn–X (X: Al, Ga, In, Sn, Sb) Heusler alloys display a martensitic phase transition from a high-temperature cubic $L2_1$ austenite phase to a low-temperature tetragonal $L1_0$ martensite phase. The near-stoichiometric Heusler alloys Ni_2MnX have a ferromagnetic ground state, while the off-stoichiometric $Ni_{50}Mn_{50-x}In_x$ alloys with $0 \leq x \leq 15$ show the presence of antiferromagnetic correlations [896]. Magnetic SANS is employed on these systems for studying the highly complex magnetic ordering and mesoscale inhomogeneity across the martensitic phase transformation (e.g., [298,594,899–907]); for instance, polarized SANS was used to study the effect of Si, Cr, Ni, C, and N alloying on the nuclear and magnetic homogeneity of Fe–Mn-based shape-memory alloys [901,903]. Runov et al. [899,900,902,904] were the first to investigate the nuclear and magnetic microstructure of polycrystalline Ni_2MnGa and single-crystalline $Ni_{49.1}Mn_{29.4}Ga_{21.5}$ by means of temperature (15–400 K) and magnetic-field-dependent (up to 4.5 kOe) polarized SANS, neutron depolarization, and neutron diffraction. The spin dynamics was also probed via the so-called left-right asymmetry method. Their main conclusions are that all the structural phase transformations in the Ni–Mn–Ga alloy system proceed via mesoscopically inhomogeneous phases, and that the structural changes (including changes in the lattice modulation) are accompanied by changes in the spin dynamics [904]. Bhatti et al. [594,905] scrutinized the complex magnetism of $Ni_{50-x}Co_xMn_{40}Sn_{10}$ polycrystalline alloys with $x = 6$–8. The temperature dependence of the unpolarized zero-field SANS data (within 30–600 K) across $T_C = 425$ K and the martensitic transition at $\sim$380–390 K were analyzed by a combination of a Porod, Gaussian, and Lorentzian scattering function. These are expected to model, respectively, the scattering from large-scale structures (e.g., magnetic domains or crystal grains), nanosized spin clusters, and critical fluctuations. In agreement with conclusions drawn from magnetometry data, these authors observe the formation of nanoscopic spin clusters with a mean center-to-center spacing of $\sim$12 nm at low temperatures. Benacchio et al. [298] carried out a polarization analysis on field-annealed $Ni_{50}Mn_{45}In_5$ and found evidence for the formation of nanoprecipitates with magnetically disordered regions. Such structures have been postulated to exist based on the outcome of integral measurement techniques [908,909].

In the context of magnetoresponsive shape-memory alloys, one may also mention the SANS study of Laver et al. [910,911] on a magnetostrictive $Fe_{81}Ga_{19}$ single crystal. Magnetic-field-dependent unpolarized and spin-polarized SANS (with and without compressive strain) revealed the existence of shape-anisotropic nanosized precipitates with an average distance of $\sim$15 nm and a magnetization which is different to that of the matrix phase. The role of these heterogeneities for the magnetostriction process in Fe–Ga alloys has been discussed. Zhang et al. [912] performed field-dependent SANS on a $Fe_{81}Ga_{19}$ galfenol single crystal and analyzed the two-dimensional SANS patterns in terms of a spherical-harmonics expansion. This allowed conclusions to be drawn regarding the role of isotropic and anisotropic magnetization processes. Ke et al. [913] employed SANS to study the effect of Tb-doping on the stress states of melt-quenched and melt-spun polycrystalline Fe–Ga alloys. They report the presence of Tb-rich nanoscale precipitates with sizes between 4–16 nm and power-law exponents of the zero-field total SANS cross section between 6–7, which was attributed to the ferromagnetic domain structure.

Nd–Fe–B-based permanent magnets.   A chronological assessment of magnetic SANS studies on Nd–Fe–B-based permanent magnets probably starts with the work of Fujii et al. [914], who investigated the role of domain walls and grain boundaries on the magnetic microstructure of sintered $Nd_{15}Fe_{77}B_8$ and $Nd_{15}Fe_{76}Al_1B_8$. Despite this pioneering and promising approach of using the SANS method for studying Nd–Fe–B-based magnets, only relatively little SANS work has been carried out on this materials class. Takeda et al. [915] continued with SANS research on sintered Nd–Fe–B by analyzing the temperature dependence of SANS patterns, putting special attention on the correlation between the average structure and the coercivity. Moreover, fractal magnetic domains on multiple length scales have been reported in a $Nd_2Fe_{14}B$ single crystal [916], the signature of magnetic volumes charges was observed in sintered isotropic Nd–Fe–B [487], multiple magnetic scattering has been reported for Nd–Fe–B nanocomposites [917], and the magnetic microstructure and magnetization-reversal process of sintered (isotropic and textured) and nanocrystalline (isotropic and hot-deformed, grain-boundary doped and undoped) Nd–Fe–B has been investigated [526, 533, 918–928] (see also Sections 4.4.2 and 6.5.1 for further details).

Perovskite manganites. Magnetic perovskite materials with the general formula $R_{1-x}A_xMnO_3$ (R = rare-earth cation; A = alkali or alkaline-earth cation) exhibit an extremely rich phase diagram, which is characterized by the intricate interplay between structural, transport, and magnetic properties [929–932]. Depending on the doping, these materials may show antiferromagnetic-insulating or ferromagnetic-conducting behavior; for instance, the perovskite manganite $LaMnO_3$ is an antiferromagnetic insulator, which on doping with Ca, Sr, or Ba becomes a ferromagnetic metal. Much interest in the study of these materials has emerged from the observation that the electrical resistivity decreases dramatically in the vicinity of T_C when a magnetic field is applied. Based on results obtained on the mixed-valence compound $La_{1-x}Sr_xMnO_3$, magnetoresistance ratios of 10^{12} in fields of a few Tesla have been reported [64], and the term colossal magnetoresistance (CMR) has been coined [929–932]. Since many of the phenomena which are observed in the manganites are related to spatial inhomogeneities in the electronic and magnetic properties on a nanometer and/or micrometer length scale, the SANS method is perfectly suited to provide information on this so-called electronic or magnetic phase separation. Quite commonly, for the analysis of SANS data, the familiar concepts from nuclear SANS such as the Guinier or Porod approximation, or a combination of Lorentzian and Lorentzian-squared scattering functions, find application to study the temperature and field dependence of the magnetic correlations. There exist numerous SANS investigations on perovskite-type materials (see, e.g., [933–961] and references therein). We refer to Section VII in the review [64], which provides an excellent overview of SANS work on perovskite manganites and cobaltites, and on other complex magnetic alloys.

Pyrochlore oxides. Pyrochlore oxides are a class of magnetically frustrated crystalline materials, many of which have the composition $R_2B_2O_7$, where R is a rare-earth ion, and B is usually a transition-metal ion [962]. Frustration in these compounds is geometrical in origin rather than induced by positional disorder [963]. Their magnetic properties reveal many parallels to the ones of spin glasses, spin ices, and spin liquids. Consequently, SANS is used to monitor the temperature evolution of the magnetic correlations (e.g., [936, 964–969]). A detailed SANS study on the frustrated pyrochlore oxides $Y_2Mn_2O_7$, $Ho_2Mn_2O_7$, and $Yb_2Mn_2O_7$ by Greedan et al. [967] reports that the magnetic properties of these systems are very similar to that seen for more conventional spin glasses. The obtained correlation lengths κ_1^{-1}, related to the Lorentzian term in the cross section, and κ_2^{-1} (Lorentzian-squared) are unequal and in general

$\kappa_2^{-1} > \kappa_1^{-1}$. While κ_1^{-1} remains finite, reaching maximum values which range from $10-20$ Å depending on the compound, κ_2^{-1} shows a very strong temperature dependence and reaches large values of >500 Å for Y and Ho and appears to saturate near 400 Å for Yb. Based on these results, it was concluded that the magnetic microstructure of these materials consists of coexisting randomly spin-canted domains (κ_2^{-1}) and smaller ferromagnetic clusters (κ_1^{-1}). Lynn et al. [936] carried out a neutron study of the magnetic correlations, phase transition, and the long-wavelength spin dynamics of the colossal magnetoresistive pyrochlore $Tl_2Mn_2O_7$. Comparison of their data to results obtained on the doped $LaMnO_3$ manganite led them to conclude that $Tl_2Mn_2O_7$ possibly belongs to a new class of CMR systems with a different underlying magnetoresistive mechanism. Buhariwalla et al. [968] investigated long-wavelength correlations in ferromagnetic titanate pyrochlores. In the compound $Yb_2Ti_2O_7$ they observed "rods" of diffuse scattering extending along $\langle 111 \rangle$ directions in reciprocal space, which were interpreted as arising from domain boundaries between finite-sized ferromagnetic domains. The anisotropic feature is absent in $Ho_2Ti_2O_7$. Quite similarly, Scheie et al. [969] report magnetic-field and temperature-dependent SANS data on single-crystalline $Yb_2Ti_2O_7$, which exhibit additional streaks along other crystallographic directions. They interpret the spin structure of this material in terms of a mixed magnetic state, which comprises antiferromagnetically coupled regions within an otherwise ferromagnetic ground state.

6

REAL-SPACE ANALYSIS

In the majority of cases, magnetic SANS data are analyzed in reciprocal space by fitting a particular model to the experimental SANS cross section (see, e.g., Section 4.4). An alternative real-space approach to analyzing SANS data is the computation of the correlation function of the system. This can be accomplished by direct or indirect Fourier transformation. Correlation functions provide intuitive guidance in real space about the magnetization distribution and the behavior of characteristic nuclear and magnetic length scales. The present chapter introduces the correlation function of the spin-misalignment SANS cross section (Section 6.2), discusses its properties based on micromagnetic model calculations (Sections 6.3 and 6.4), and relates them to experimental data on a number of hard and soft magnetic materials (Section 6.5). In preparation of this task and in order to develop an intuitive feeling for magnetization correlations, we will first analyze the magnetic-field-dependent behavior of the magnetization around a spherical inclusion in an anisotropy-field-free matrix (Section 6.1).

6.1 Magnetization profile around a spherical inclusion

Based on linearized micromagnetic theory, the magnetization profiles and autocorrelation functions of the spin misalignment for various models of the magnetic anisotropy field of a spherical inclusion were computed in [443]. To be more specific, a single isolated spherical nanoparticle (core-shell particle), which is embedded in an infinitely extended magnetic matrix, is considered. The particle is characterized by its magnetic anisotropy field $\mathbf{H}_\mathrm{p}(\mathbf{r})$, whereas the matrix is assumed to be anisotropy-field-free. For different spatial profiles of $\mathbf{H}_\mathrm{p}$ (uniform sphere, uniform core-shell, power-law decay, linear increase, exponential decay), the magnetization response around the defect and the corresponding correlation functions were calculated. Figure 6.1 displays the assumed models for $\mathbf{H}_\mathrm{p}(\mathbf{r})$. The theory assumes uniform values for the exchange-stiffness constant A and the saturation magnetization M_s, i.e., jumps in M_s at the particle-matrix interface are not taken into account. Note also that the validity of the approach is restricted to large applied fields (approach-to-saturation regime) in order to guarantee that the normalized transversal magnetization component is small ($M_\mathrm{p}/M_\mathrm{s} \ll 1$, see Fig. 6.2). Moreover, we would like to emphasize that the magnetostatic field $\mathbf{H}_\mathrm{d}$ due to non-vanishing volume charges $\nabla \cdot \mathbf{M} \neq 0$ has been neglected. Only the exchange interaction, magnetic anisotropy, and the Zeeman interaction are considered in these simple model calculations; see Fig. 2 in [443] for a comparison of the cases with and without magnetostatic interaction.

The applied-field dependence of the reduced transversal (perpendicular to $\mathbf{H}_0$) magnetization $M_\mathrm{p}/M_\mathrm{s}$ is shown in Fig. 6.2. These magnetization profiles were com-

Magnetic Small-Angle Neutron Scattering: A Probe for Mesoscale Magnetism Analysis. Andreas Michels, Oxford University Press (2021). © Andreas Michels. DOI: 10.1093/oso/ 9780198855170.003.0006

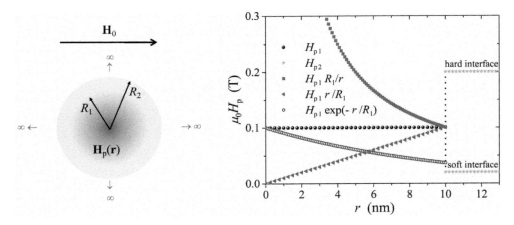

Fig. 6.1: Models for the spatial structure of the magnetic anisotropy field $\mathbf{H}_p(\mathbf{r})$ of a spherical particle (core-shell), which is embedded in an anisotropy-field-free, infinitely extended magnetic matrix. The following functional dependencies for $H_p(r)$ were assumed: (uniform sphere) $H_p(r) = H_{p1} = $ constant for $r \leq R_1$; (uniform core-shell particle) $H_p(r) = H_{p1}$ for $r \leq R_1$ and $H_p(r) = H_{p2}$ for $R_1 \leq r \leq R_2$; (power-law decay) $H_p(r) = H_{p1}(R_1/r)$; (linear increase) $H_p(r) = H_{p1}(r/R_1)$; (exponential decay) $H_p(r) = H_{p1} \exp(-r/R_1)$. Depending on whether the anisotropy field of the shell, H_{p2}, is larger or smaller than H_{p1}, the interface is denoted as a hard or soft interface. For illustration purposes, we have chosen $R_1 = 10\,\mathrm{nm}$, $R_2 = 13\,\mathrm{nm}$, $\mu_0 H_{p1} = 0.1\,\mathrm{T}$, and $\mu_0 H_{p2} = 0.2\,\mathrm{T}$ $(0.02\,\mathrm{T})$. After [443].

puted by numerically solving the following equation:

$$
\frac{M_p}{M_s}(r) = \sqrt{\frac{2}{\pi}} \int_0^\infty \frac{\widetilde{H}_p(q)}{H_{\mathrm{eff}}(q, H_i)} \frac{\sin(qr)}{qr} q^2 dq, \tag{6.1}
$$

where $\widetilde{H}_p(q)$ denotes the Fourier transform of the magnetic anisotropy profile $H_p(r)$ (Fig. 6.1), and H_{eff} is the effective magnetic field [eqn (3.65)]. As expected, increasing the applied field suppresses both the magnitude and the range of the transversal spin fluctuations. The M_p/M_s profiles reveal that the perturbation which is caused by the anisotropy field of the particle is largest at the center of the dominant defect, and then M_p/M_s decays smoothly at the larger distances. While the transversal magnetization for the uniform-sphere model [Fig. 6.2(a)] and for the exponential decay case [Fig. 6.2(c)] are qualitatively similar, the shapes of the other two M_p/M_s curves are significantly different: the core-shell particle [Fig. 6.2(b)] exhibits a peak in M_p/M_s, which is due to the presence of the hard magnetic shell at $10\,\mathrm{nm} \leq r \leq 13\,\mathrm{nm}$, and for the power-law decay case [Fig. 6.2(d)], we find an almost linear decrease of M_p/M_s at small r and for not-too-large fields. Note that the M_p/M_s data for a soft interface (i.e., for $H_{p1}/H_{p2} > 1$) are qualitatively similar to the uniform-sphere case [443]. The magnetization profiles which are displayed in Fig. 6.2 suggest that the spatial structure of the perturbing magnetic anisotropy field has a strong impact on the magnetization

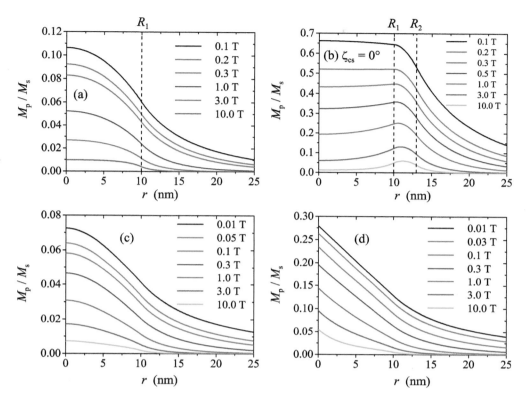

Fig. 6.2: Normalized transversal magnetization component M_p/M_s as a function of the distance r from the center of the inclusion (at $r = 0$). The M_p/M_s curves describe the magnetization response to the magnetic anisotropy field profiles shown in Fig. 6.1. Values of the applied magnetic field are indicated in the insets and increase from top to bottom, respectively. (a) Uniform-sphere model ($R_1 = 10$ nm, $\mu_0 H_{p1} = 0.1$ T); (b) uniform core-shell model with a hard interface ($R_1 = 10$ nm, $R_2 = 13$ nm, $H_{p1}/H_{p2} = 0.1$); $\zeta_{cs} = 0°$ denotes the angle between the anisotropy-field directions of the core and the shell; (c) exponential decay; (d) power-law decay. The dashed vertical lines in (a) and (b) indicate, respectively, the radius R_1 of the uniform particle and the region $R_1 \leq r \leq R_2$ of the magnetically hard shell. Magnetic materials parameters of Ni were assumed ($A = 8.2$ pJ/m; $M_s = 500$ kA/m). After [443].

distribution and, eventually, on the associated magnetic SANS cross section; see [443] for the corresponding autocorrelation functions.

The results in Fig. 6.2 also demonstrate the special role of the micromagnetic exchange length $l_H(H_i) = \sqrt{2A/(\mu_0 M_s H_i)}$ [eqn (3.66)], which is contained in the expression for H_{eff} in eqn (6.1). This length scale can be taken as the spatial resolution limit of the magnetization with respect to local defect-induced perturbations [556]. Variations in the magnetic anisotropy field on a characteristic microstructural length scale—given in the above examples by the radius R_1 of the inclusion [Fig. 6.2(a), (c), and (d)] or by the thickness $\Delta R = R_2 - R_1$ of the shell [Fig. 6.2(b)]—can be followed by the magnetization only when $l_H \lesssim R_1, \Delta R$. At the largest field, l_H is of the order of a few nm, e.g., $l_H(\mu_0 H_i = 10\,\mathrm{T}) \cong 1.8\,\mathrm{nm}$ for Ni [compare also Fig. 3.5(a)], with the consequence that sharp variations in M_p/M_s on a scale of the order of l_H can be resolved. This can be clearly seen, for instance in Fig. 6.2(a), where M_p/M_s at the highest field is almost constant inside the particle ($r \leq R_1$), and then M_p/M_s decays relatively steeply in the near vicinity of the interface ($r = R_1$). Decreasing the field, e.g., $l_H(0.1\,\mathrm{T}) \cong 18\,\mathrm{nm}$, results in the "loss of information", i.e., the details about the perturbation at the interface are smeared out. Similarly, the perturbing effect of the magnetically hard shell in Fig. 6.2(b) is only seen at the largest field values, where it gives rise to a peak feature in M_p/M_s. At the lower fields, when l_H is much larger than the shell thickness $\Delta R = 3\,\mathrm{nm}$, the peak transforms into a broad shoulder.

Regarding the experimental determination of the spin-misalignment correlation length, which may be taken as a parameter describing the field-dependent extension (size) of the spin perturbation, it should be emphasized that polycrystalline magnetic materials generally do contain many different microstructural defects, each of which may exhibit its own characteristic magnetization profile with respect to magnitude and decay length. Experimental correlation lengths represent therefore an average over a large number of (different) lattice imperfections. In Section 6.5, we will see that, for many ferromagnets, the experimental correlation lengths can be described by a field-dependent contribution, related to the exchange length l_H, and by a field-independent term, which characterizes the size of the defect.

6.2   Correlation function of the spin-misalignment SANS cross section

In analogy to the autocorrelation function of the nuclear excess scattering-length density [eqn (2.64)]

$$\delta N(\mathbf{r}) = N(\mathbf{r}) - \langle N \rangle, \tag{6.2}$$

one can define the autocorrelation function of the spin misalignment as [303, 442, 443]:

$$C_{\mathrm{SM}}(\mathbf{r}) = \frac{1}{V} \int \delta\mathbf{M}(\mathbf{r}')\delta\mathbf{M}(\mathbf{r}' + \mathbf{r})d^3r', \tag{6.3}$$

where

$$\delta\mathbf{M}(\mathbf{r}) = \mathbf{M}(\mathbf{r}) - \langle \mathbf{M} \rangle \tag{6.4}$$

denotes the deviation of the local magnetization vector field $\mathbf{M}(\mathbf{r})$ from the mean magnetization $\langle \mathbf{M} \rangle$. Alternatively, $C_{\mathrm{SM}}(\mathbf{r})$ can be expressed as:

$$C_{\mathrm{SM}}(\mathbf{r}) = \frac{1}{V} \int \left| \widetilde{\delta\mathbf{M}}(\mathbf{q}) \right|^2 \exp\left(i\mathbf{q} \cdot \mathbf{r} \right) d^3q, \tag{6.5}$$

where $\widetilde{\delta\mathbf{M}}(\mathbf{q})$ is the Fourier transform of $\delta\mathbf{M}(\mathbf{r})$. In the high-field limit, $\langle \mathbf{M} \rangle$ is nearly parallel to the applied magnetic field $\mathbf{H}_0 \parallel \mathbf{e}_z$ with $|\langle \mathbf{M} \rangle| \cong M_s$, so that $\delta\mathbf{M}(\mathbf{r}) \cong \{M_x(\mathbf{r}), M_y(\mathbf{r}), 0\}$ and

$$C_{\mathrm{SM}}(\mathbf{r}) = \frac{1}{V} \int \left(|\widetilde{M}_x(\mathbf{q})|^2 + |\widetilde{M}_y(\mathbf{q})|^2 \right) \exp\left(i\mathbf{q} \cdot \mathbf{r} \right) d^3 q. \tag{6.6}$$

Comparison of eqns (2.59)–(2.65) with eqns (6.3)–(6.6) reveals an important difference between nuclear and magnetic scattering [besides the fact that $\delta N(\mathbf{r})$ is a scalar and $\delta \mathbf{M}(\mathbf{r})$ a vector function]: while the nuclear SANS cross section is directly proportional to the magnitude square of the Fourier transform of the nuclear scattering-length density, $d\Sigma_{\mathrm{nuc}}/d\Omega \propto |\widetilde{N}|^2$, the function $|\widetilde{\delta \mathbf{M}}|^2$, being proportional to the Fourier transform of C_{SM}, does not represent the experimentally measurable quantity $d\Sigma_{\mathrm{SM}}/d\Omega$ [compare, e.g., eqn (4.5)]. Rather, the magnetic SANS cross section corresponds to the Fourier transform of the autocorrelation function of the Halpern–Johnson vector [compare eqn (1.35)], which is the component of the magnetization perpendicular to the scattering vector.

The function C_{SM} can also be related to the approach-to-saturation regime of a magnetization curve. This follows from the fact that the component of the magnetization along the field direction, M_z, can be approximated as ($M_x, M_y \ll M_{\mathrm{s}}$):

$$M_z = M_{\mathrm{s}} \sqrt{1 - \frac{M_x^2 + M_y^2}{M_{\mathrm{s}}^2}} \cong M_{\mathrm{s}} \left(1 - \frac{1}{2} \frac{M_x^2 + M_y^2}{M_{\mathrm{s}}^2} \right). \tag{6.7}$$

Since the value of C_{SM} at the origin,

$$C_{\mathrm{SM}}(\mathbf{r} = 0) = \frac{1}{V} \int \left[M_x^2(\mathbf{r}) + M_y^2(\mathbf{r}) \right] d^3 r = \langle (\delta M)^2 \rangle, \tag{6.8}$$

equals the mean-square magnetization fluctuation $\langle (\delta M)^2 \rangle$, where $\langle (...) \rangle = V^{-1} \int (...) \, d^3 r$ denotes a volume average, comparison to eqn (6.7) suggests that

$$\frac{\langle M_z \rangle}{M_{\mathrm{s}}} \cong 1 - \frac{1}{2} \frac{\langle (\delta M)^2 \rangle}{M_{\mathrm{s}}^2} = 1 - \frac{1}{2} \frac{C_{\mathrm{SM}}(\mathbf{r} = 0)}{M_{\mathrm{s}}^2}, \tag{6.9}$$

which links the autocorrelation function to the macroscopic magnetization. Comparisons between magnetization and neutron data based on eqn (6.9) have been carried out in [303, 449] for the cases of nanocrystalline Tb and Gd.

As we have discussed before, $|\widetilde{\delta \mathbf{M}}|^2$ does not correspond to the experimentally measurable quantity $d\Sigma_{\mathrm{SM}}/d\Omega$. With a view towards analyzing experimental data, one may therefore define the correlation function $C(\mathbf{r})$ of the spin-misalignment SANS cross section $d\Sigma_{\mathrm{SM}}/d\Omega$ and its normalized version $c(\mathbf{r})$ as [312]:

$$C(\mathbf{r}) = \frac{1}{8\pi^3} \int \frac{d\Sigma_{\mathrm{SM}}}{d\Omega}(\mathbf{q}) \exp\left(i\mathbf{q} \cdot \mathbf{r} \right) d^3 q, \tag{6.10}$$

$$c(\mathbf{r}) = \frac{C(\mathbf{r})}{C(\mathbf{r} = 0)} = \frac{\int \frac{d\Sigma_{\mathrm{SM}}}{d\Omega}(\mathbf{q}) \exp\left(i\mathbf{q} \cdot \mathbf{r} \right) d^3 q}{\int \frac{d\Sigma_{\mathrm{SM}}}{d\Omega}(\mathbf{q}) d^3 q}. \tag{6.11}$$

We emphasize that the integral of $d\Sigma_{\mathrm{SM}}/d\Omega$ over the reciprocal space, corresponding to

$$C(\mathbf{r} = 0) = \frac{1}{8\pi^3} \int \frac{d\Sigma_{SM}}{d\Omega}(\mathbf{q}) d^3q, \tag{6.12}$$

does not provide an obvious invariant of the spin-misalignment SANS. This is in contrast to nuclear SANS, where the Porod invariant $\mathbf{P} = (2\pi^2)^{-1} \int_0^\infty \frac{d\Sigma_{nuc}}{d\Omega} q^2 dq = \langle N^2 \rangle - \langle N \rangle^2$ yields the mean-square density fluctuation [compare eqns (2.69) and (5.52)]. We also remind the reader that $d\Sigma_{SM}/d\Omega$ at a particular internal magnetic field H_i can be obtained by subtracting the total nuclear and magnetic scattering at a saturating field from the measurement of the total $d\Sigma/d\Omega$ at the particular H_i. By using the analytical results in Section 4.2 for the spin-misalignment SANS cross section, the correlation function can be computed. Using the final expressions for the azimuthally averaged $\frac{d\Sigma_{SM}}{d\Omega} = \frac{d\Sigma_{SM}}{d\Omega}(q)$ in the $\mathbf{k}_0 \perp \mathbf{H}_0$ and $\mathbf{k}_0 \parallel \mathbf{H}_0$ geometries [eqns (4.14) and (4.21)], eqn (6.11) can be expressed as:

$$c(r) = \frac{\int\limits_0^\infty \frac{d\Sigma_{SM}}{d\Omega}(q) j_0(qr) q^2 dq}{\int\limits_0^\infty \frac{d\Sigma_{SM}}{d\Omega}(q) q^2 dq}, \tag{6.13}$$

where $j_0(z) = \sin(z)/z$ denotes the zeroth-order spherical Bessel function. We refer to Appendix E, where the three- and two-dimensional Fourier transforms of functions $h(x, y, z)$ and $g(x, y)$ are derived in spherical polar coordinates [970, 971].

Inserting the analytical expression for the spin-misalignment SANS cross section in the perpendicular scattering geometry, eqn (4.14), into eqn (6.13) yields for $c(r)$:

$$c(r) = \frac{\int\limits_0^\infty [S_H(q)R_H(q, H_i) + S_M(q)R_M(q, H_i)] j_0(qr) q^2 dq}{\int\limits_0^\infty [S_H(q)R_H(q, H_i) + S_M(q)R_M(q, H_i)] q^2 dq}, \tag{6.14}$$

where the micromagnetic response functions $R_H(q, H_i)$ and $R_M(q, H_i)$ are given by eqns (4.10) and (4.11). It is straighforward to adapt eqn (6.14) to other scattering geometries or to correlation functions of the type of eqn (6.21). $S_H(q)$ and $S_M(q)$ may be described by form-factor functions, which characterize the size and shape of spatial regions over which the magnetic anisotropy field (ξ_H) and the saturation magnetization (ξ_M) are homogeneous (see Section 6.3). Structure factors can of course also be included. We refer to Fig. 6.17 for an example where $c(r)$ data, which were computed from experimental $d\Sigma_{SM}/d\Omega$ data, have been analyzed with this approach [312].

In a SANS experiment, only the components of the momentum-transfer vector $\mathbf{q}$ perpendicular to the incident-beam direction $\mathbf{k}_0$ are effectively probed. For $\mathbf{k}_0 \perp \mathbf{H}_0$ ($q_x \cong 0$) this implies that $\frac{d\Sigma_{SM}}{d\Omega} \cong \frac{d\Sigma_{SM}}{d\Omega}(q_y, q_z)$, whereas $\frac{d\Sigma_{SM}}{d\Omega} \cong \frac{d\Sigma_{SM}}{d\Omega}(q_x, q_y)$ for $\mathbf{k}_0 \parallel \mathbf{H}_0$ ($q_z \cong 0$). We remind that $\mathbf{H}_0 \parallel \mathbf{e}_z$ for both scattering geometries (compare Fig. 2.1). With this in mind, it is useful to also consider the following correlation function ($\mathbf{k}_0 \perp \mathbf{H}_0$) [226]:

$$c_\perp(y,z) = \frac{\int\limits_{-\infty}^{+\infty}\int\limits_{-\infty}^{+\infty} \frac{d\Sigma_{\mathrm{SM}}}{d\Omega}(q_y, q_z)\cos(q_y y + q_z z)dq_y dq_z}{\int\limits_{-\infty}^{+\infty}\int\limits_{-\infty}^{+\infty}\frac{d\Sigma_{\mathrm{SM}}}{d\Omega}(q_y, q_z)dq_y dq_z}. \tag{6.15}$$

An analogous expression holds for the parallel scattering geometry. Due to the fact that $\frac{d\Sigma_{\mathrm{SM}}}{d\Omega} = \frac{d\Sigma_{\mathrm{SM}}}{d\Omega}(q_y, q_z)$, the $c_\perp(y, z)$ that is computed according to eqn (6.15) represents a projection of the three-dimensional correlation function $c(x, y, z)$ along the direction of the incident neutron beam. This is in analogy to nuclear SANS, where the two-dimensional SANS cross section $\frac{d\Sigma_{\mathrm{nuc}}}{d\Omega}(q_y, q_z)$ and the two-dimensional projected correlation function $C_{\mathrm{N}\perp}(y, z)$ are related in the following way [185, 226, 972]:

$$\frac{d\Sigma_{\mathrm{nuc}}}{d\Omega}(q_y, q_z) \cong \int\limits_{-\infty}^{+\infty}\left[\int C_{\mathrm{N}}(x, y, z)\exp\left(-i\mathbf{q}\cdot\mathbf{r}\right)d^3r\right]\delta(q_x)dq_x$$

$$= \int\limits_{-\infty}^{+\infty}\int\limits_{-\infty}^{+\infty}\int\limits_{-\infty}^{+\infty} C_{\mathrm{N}}(x, y, z)\exp\left(-i[q_y y + q_z z]\right)dx dy dz$$

$$= \int\limits_{-\infty}^{+\infty}\int\limits_{-\infty}^{+\infty} C_{\mathrm{N}\perp}(y, z)\exp\left(-i[q_y y + q_z z]\right)dy dz \tag{6.16}$$

with

$$C_{\mathrm{N}\perp}(y, z) = \int\limits_{-\infty}^{+\infty} C_{\mathrm{N}}(x, y, z)dx. \tag{6.17}$$

In the above expressions, $C_{\mathrm{N}}(x, y, z)$ denotes the three-dimensional autocorrelation function of the nuclear excess scattering-length density [compare eqn (2.64) and Appendix C], and the delta function $\delta(q_x)$ takes account of the fact that in small-angle approximation the Ewald sphere may be approximated as being flat (neglecting its curvature).

Equation (6.15), or likewise eqn (6.16), can be transformed into polar coordinates, which results in:

$$c_\perp(r, \phi) = \frac{\int\limits_0^\infty\int\limits_0^{2\pi} \frac{d\Sigma_{\mathrm{SM}}}{d\Omega}(q, \theta)\cos(qr\cos(\theta - \phi))q d\theta dq}{\int\limits_0^\infty\int\limits_0^{2\pi}\frac{d\Sigma_{\mathrm{SM}}}{d\Omega}(q, \theta)q d\theta dq}, \tag{6.18}$$

where the angle ϕ specifies the orientation of $\mathbf{r} = (r, \phi)$ in the y–z-plane. By introducing the nth-order Bessel function (see p.19f in [973]),

$$J_n(z) = \frac{1}{2\pi}\int\limits_0^{2\pi}\cos(n\alpha - z\sin(\alpha))d\alpha = \frac{1}{2\pi}\int\limits_\zeta^{2\pi+\zeta}\cos(n\alpha - z\sin(\alpha))d\alpha, \tag{6.19}$$

where n is an integer and the last equation is valid for any angle ζ, we can obtain an average of $c_\perp(r, \phi)$ over all angles ϕ in the detector plane:

$$
c_\perp(r) = \frac{1}{2\pi} \int_0^{2\pi} c_\perp(r, \phi) d\phi = \frac{\int_0^\infty \int_0^{2\pi} \frac{d\Sigma_{\mathrm{SM}}}{d\Omega}(q, \theta) J_0(qr) q d\theta dq}{\int_0^\infty \int_0^{2\pi} \frac{d\Sigma_{\mathrm{SM}}}{d\Omega}(q, \theta) q d\theta dq}. \tag{6.20}
$$

The integrations in eqn (6.20) with respect to the angle θ can be taken either analytically using the theoretical expression for $d\Sigma_{\mathrm{SM}}/d\Omega$ (compare Section 4.2) or numerically using experimental data. It follows that (see also Appendix E)

$$
c_\perp(r) = \frac{\int_0^\infty \frac{d\Sigma_{\mathrm{SM}}}{d\Omega}(q) J_0(qr) q dq}{\int_0^\infty \frac{d\Sigma_{\mathrm{SM}}}{d\Omega}(q) q dq}, \tag{6.21}
$$

which differs from eqn (6.13). Both expressions for the correlation function [eqns (6.13) and (6.21)] will be compared to each other in the next section (see Fig. 6.7).

In many cases, it is desirable to estimate the correlation length l_C of the spin-misalignment SANS cross section. Figure 6.3 illustrates the meaning of l_C, which specifies the range over which perturbations in the spin structure around lattice defects (e.g., pores, grain boundaries, dislocations, vacancies) are transmitted by the exchange interaction into the surrounding crystal lattice. A convenient definition to determine the spin-misalignment correlation length l_C is

$$
C(r = l_C) = C(r = 0) \exp(-1), \tag{6.22}
$$

where $C(0)$ denotes the extrapolated value of the correlation function at the origin. Equation (6.22) yields the exact correlation length for exponentially decaying correlations. Note, however, that this definition does not imply that the correlations do decay exponentially. In fact, spin-misalignment correlations generally exhibit a more complex behavior (see, e.g., Fig. 6.14). Equation (6.22) is merely a convenient way to define a characteristic length which can be related to the magnetic microstructure and which can be computed model-independently. An alternative route to extracting a spin-misalignment length is the computation of moments of the correlation function; for instance, for exponentially decaying $c(r)$ the above definition and $l_C = \int_0^\infty c(r) dr$ are equivalent.

In several studies [49, 442–444, 526], it was found that $l_C(H_i)$ data can be well described by the following relation:

$$
l_C(H_i) = \mathcal{L} + \sqrt{\frac{2A}{\mu_0 M_s (H_i + H^\star)}}, \tag{6.23}
$$

where the field-independent parameter $\mathcal{L}$ is of the order of the defect size, and the second field-dependent term on the right-hand side represents a modified exchange

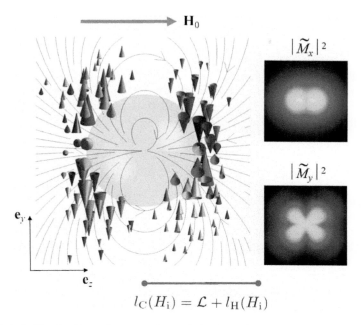

Fig. 6.3: Sketch illustrating the meaning of the correlation length l_C of the spin mis-alignment. (left) Computed spin misalignment around a spherical nanopore in a fer-romagnetic iron matrix (two-dimensional cut-out of a three-dimensional simulation). Shown is the magnetization component $\mathbf{M}_\perp(\mathbf{r})$ perpendicular to $\mathbf{H}_0 \parallel \mathbf{e}_z$. Thickness of arrows is proportional to the magnitude of $\mathbf{M}_\perp$. Solid grey lines: magnetodipo-lar field distribution. The correlation length $l_C = \mathcal{L} + l_H$ is a measure for the size of the inhomogeneously magnetized region around the defect. l_C consists of a field-independent contribution $\mathcal{L}$, which is related to the structural size of the defect, and of a field-dependent exchange length l_H, which transmits the perturbation at the pore-matrix interface into the surrounding crystal lattice. (right) Corresponding magneti-zation Fourier components $|\widetilde{M}_x(\mathbf{q})|^2$ and $|\widetilde{M}_y(\mathbf{q})|^2$ projected into the plane $q_x = 0$. After [476].

length l_H of the field. The quantity H^* takes account of magnetostatic and magnetic anisotropy fields. Equation (6.23) is a phenomenological prediction based on micro-magnetic theory, which embodies the convolution relationship between the magnetic anisotropy-field microstructure $\mathbf{H}_p(\mathbf{r})$, the spatial variation of the saturation magneti-zation $M_s(\mathbf{r})$, and micromagnetic response functions which decay with l_H [441,456]. In general, the correlation length $\mathcal{L}$ depends on the characteristic length scales of $\mathbf{H}_p(\mathbf{r})$ (ξ_H) and $M_s(\mathbf{r})$ (ξ_M), i.e., $\mathcal{L} = \mathcal{L}(\xi_H, \xi_M)$; for instance, for a statistically isotropic poly-crystalline single-phase magnetic material with a uniform M_s and where each grain is assumed to be a single crystal with magnetocrystalline anisotropy only, the corre-lation length $\mathcal{L} \cong \xi_H$ is sensibly related to the average crystallite size [443,444]. It must also be emphasized that in many of the results which are discussed here, we set

$\xi_H = \xi_M = \mathcal{L} = R$ and use the sphere form factor with radius R for both functions $S_H(q)$ and $S_M(q)$. However, one should keep in mind that $\mathcal{L}$ has a general meaning, and that this microstructural length scale should not necessarily be identified with a sphere radius.

The field $H^\star$ in eqn (6.23) models the influence of the magnetodipolar interaction and of the magnetic anisotropy [526]. For soft magnetic materials with low crystalline anisotropy and at large applied magnetic fields (when the magnetostatic interaction may be negligible), one may ignore the field $H^\star$, so that eqn (6.23) simplifies to:

$$l_C(H_i) = \mathcal{L} + l_H(H_i) = \mathcal{L} + \sqrt{\frac{2A}{\mu_0 M_s H_i}}. \tag{6.24}$$

The latter equation has been found to excellently describe the field-dependent spin-misalignment correlations in nanocrystalline Co and Ni (see Section 6.5.2) [442]. By contrast, for uniaxial hard magnets, the anisotropy field $H_K = 2K_u/M_s$, which for a $Nd_2Fe_{14}B$ single crystal is about $8\,T$ at $300\,K$ [974], is expected to suppress the extent of spin inhomogeneities. Likewise, jumps ΔM of the magnitude of the magnetization at internal phase boundaries, which in Fe-based nanocomposites can be as large as $1.5\,T$ [446], give rise to magnetic stray-field torques that produce spin disorder in the surrounding magnetic phase (see Fig. 6.3). Such perturbations are also characterized by small-sized gradients in the magnetization distribution (see Fig. 6.10). At $H_i = 0$ and for $H^\star = H_K = 2K_u/(\mu_0 M_s)$, eqn (6.23) reduces to:

$$l_C = \mathcal{L} + \sqrt{\frac{A}{K_u}}, \tag{6.25}$$

where $\sqrt{A/K_u}$ is the domain-wall parameter.

Before discussing selected experimental data in Section 6.5, we will in the following investigate—within the framework of the micromagnetic SANS theory—the basic properties of the correlation function, such as its angular anisotropy, field dependence, influence of scattering geometry ($\mathbf{k}_0 \perp \mathbf{H}_0$ and $\mathbf{k}_0 \parallel \mathbf{H}_0$), or its dependency on the ratio $H_p/\Delta M$ [312]. We remind the reader that the ratio $H_p/\Delta M$ reflects the relative contribution of terms $S_H R_H$ and $S_M R_M$ to $d\Sigma_{SM}/d\Omega$ [compare eqn (4.14)]. Such model calculations provide a useful guideline for the study of experimental data.

6.3 Basic properties of the correlation function of the spin-misalignment SANS

In this section, we study the basic characteristics of the spin-misalignment correlation function by employing the results of the micromagnetic SANS theory (see Sections 3.2, 3.3, and 4.2). We focus on the $\mathbf{k}_0 \perp \mathbf{H}_0$ scattering geometry, since for $\mathbf{k}_0 \parallel \mathbf{H}_0$ all $d\Sigma_{SM}/d\Omega$ are isotropic. By contrast, for $\mathbf{k}_0 \perp \mathbf{H}_0$, both the spin-misalignment SANS cross section $d\Sigma_{SM}(q_y, q_z)/d\Omega$ and the spin-misalignment correlation function $c_\perp(y, z)$ are generally anisotropic functions, so that the value of the correlation length that is determined from $c_\perp(y, z)$ also depends on the orientation on the two-dimensional detector. Moreover, for both functions $S_H(q\xi_H) = 8\pi^3 V^{-1} b_H^2 \tilde{H}_p^2(q\xi_H)$ and $S_M(q\xi_M) =$

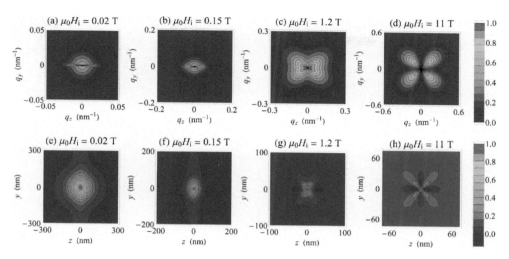

Fig. 6.4: (a)–(d) Contour plots of the normalized $d\Sigma_{SM}/d\Omega(q_y, q_z)$ [eqn (4.5)] at internal magnetic fields as indicated ($\mathbf{k}_0 \perp \mathbf{H}_0$; $\mathbf{H}_0$ is horizontal; $\xi_H = \xi_M = 5$ nm). (e)–(h) Corresponding two-dimensional correlation functions $c_\perp(y, z)$ [eqn (6.15)]. Note the changing q- and r-scales. After [312].

$8\pi^3 V^{-1} b_H^2 \widetilde{M}_z^2(q\xi_M)$ in the expression for $d\Sigma_{SM}/d\Omega$ we use the sphere form factor [compare eqns (3.82)–(3.84)]. We remind that ξ_H and ξ_M represent, respectively, the correlation lengths of the magnetic anisotropy field $\mathbf{H}_p(\mathbf{r})$ and of the spatial profile $M_s(\mathbf{r})$ of the saturation magnetization. In most of the results that follow, we have set $\xi_H = \xi_M = \mathcal{L} = R$. Note also that the factor $8\pi^3 V^{-1} b_H^2$ which appears in the definitions of S_H and S_M cancels in the normalization procedure [compare, e.g., eqn (6.15)].

The lower row of Fig. 6.4 displays a series of $c_\perp(y, z)$ [computed according to eqn (6.15)] at selected internal magnetic fields H_i. The upper row in Fig. 6.4 shows the corresponding $d\Sigma_{SM}/d\Omega$ (neglecting the Dzyaloshinskii–Moriya interaction; $l_D = 0$). While the $d\Sigma_{SM}/d\Omega$ at small fields [Fig. 6.4(a) and (b)] are enhanced parallel to the applied-field direction, the correlation functions exhibit maxima in the direction perpendicular to the field. The range of the correlations extends to several hundreds of nanometers [Fig. 6.4(e) and (f)]. Increasing the field results in the suppression of the correlations and at the largest field $d\Sigma_{SM}/d\Omega$ possesses a nearly fourfold angular anisotropy with maxima along the detector diagonals and minima along the horizontal and vertical axes [Fig. 6.4(d)], which translate into the corresponding extrema in the $c_\perp(y, z)$ [Fig. 6.4(h)].

Figure 6.5 illustrates the anisotropy of the correlations. In Fig. 6.5(a) we depict the correlation function along different directions: while the correlation length at 1.2 T varies only relatively little with direction (from 8.8 nm to 10.9 nm for the particular chosen set of parameters [312]), the functional dependencies of the $c_\perp(y, z)$ are significantly different, with the correlation function along the horizontal z-direction becoming negative at $r \cong 18$ nm. The curves in Fig. 6.5 were obtained by solving eqn (6.18) for $\phi = 0, \pi/4, \pi/2$. In nuclear SANS, negative values of the distance distribution func-

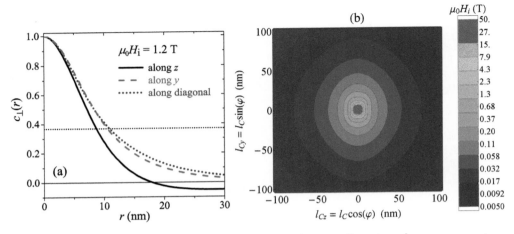

Fig. 6.5: (a) $c_\perp(r)$ [eqn (6.18)] along different real-space directions [same parameters as in Fig. 6.4(g)]. Dotted horizontal line: $c_\perp(r) = \exp(-1)$. (b) Contour plot revealing the in-plane (ϕ) variation of l_C for several values of the internal magnetic field H_i. Logarithmic color scale for the field is used. After [312].

tion $p(r)$ are attributed to distances that connect regions with the opposite sign of the scattering-length density more frequently than regions with the same sign [10]. However, for magnetic SANS, such an easily accessible interpretation of the correlation function $c_\perp(r)$ of the spin-misalignment SANS cross section in terms of a specific magnetization distribution is not straightforward. This is mainly related to the above-mentioned fact that $c_\perp(r)$ does not directly represent the correlations in the magnetic microstructure, as does the autocorrelation function C_{SM} [eqn (6.3)], but also includes the magnetodipolar interaction of the neutrons with the sample via the trigonometric functions and the cross term in the magnetic SANS cross section [compare eqn (4.5)]. The anisotropy of the correlations is further depicted in Fig. 6.5(b), where we show a contour plot of l_C for several values of H_i. This graph reveals a relatively weak anisotropy of l_C. At small fields, the correlations along the vertical (y) direction decay on a larger length scale than along the horizontal (z) direction. With increasing field, the anisotropy becomes less pronounced.

An experimental example demonstrating the angular anisotropy of the two-dimensional spin-misalignment scattering cross section and of the related correlation function and correlation lengths is shown in Fig. 6.6. These data were obtained on a soft magnetic bulk metallic glass [528], where magnetostatic scattering contributions $\propto S_M R_M$ are negligible. For small values of r, corresponding to large $c_\perp$-values ($c_\perp \gtrsim 0.8$), the anisotropy of the correlation function shows a maximum along the vertical direction, in agreement with an horizontal elongation in $\mathbf{q}$-space and in qualitative agreement with the results shown in Fig. 6.5.

Figure 6.7 provides a comparison of the correlation functions of the spin-misalignment SANS cross section [eqns (6.13) and (6.21)] and the autocorrelation function of the spin misalignment [eqn (6.6)]. At small internal fields, the results

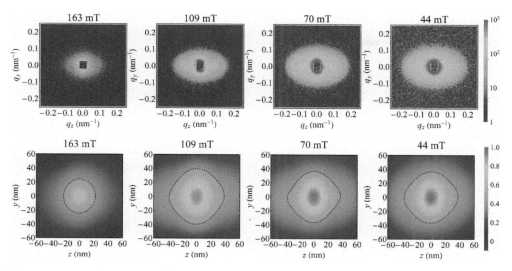

Fig. 6.6: (upper row) Field dependence of the two-dimensional spin-misalignment SANS cross section $d\Sigma_{SM}/d\Omega$ of an as-cast bulk metallic glass ($Fe_{70}Mo_5Ni_5P_{12.5}B_{2.5}C_5$) at selected applied magnetic fields (see insets). $\mathbf{H}_0$ is horizontal in the plane of the detector. (lower row) Corresponding numerically computed two-dimensional correlation functions $c_\perp(y, z)$. Dashed contour lines: $c_\perp(y, z) = \exp(-1)$. After [528].

for $c(r)$, $c_\perp(r)$, and the derived $l_C(H_i)$ differ considerably, whereas for larger fields (here: $\mu_0 H_i \gtrsim 1\,T$) both equations yield almost the same correlation lengths. Equation (6.13) and the autocorrelation function [eqn (6.6)] give nearly identical correlation lengths over the whole field range.

The field dependence of the correlations in both scattering geometries is illustrated in Figs. 6.8 and 6.9. The dotted horizontal lines indicate the values of the correlation length l_C of the spin misalignment. Increasing H_i results in both geometries in the suppression of transversal spin-misalignment fluctuations and in a concomitant reduction of the $c(r)$ and reduced l_C-values. At small fields, l_C may take on values of the order of 100 nm, which decrease to values of the order of the assumed particle size, here $\mathcal{L} = R = 5\,nm$, for fields larger than a few Tesla [see also the dotted horizontal line in Fig. 6.9(b)]. For the chosen limiting case of $H_p/\Delta M \to \infty$ (magnetic anisotropy only), the difference between the $c(r)$ and the $l_C(H_i)$ in the two scattering geometries is only minor (see Fig. 6.9). However, noting that the $d\Sigma_{SM}/d\Omega$ and, hence, the $c(r)$ in the parallel geometry are independent of $\widetilde{M}_z$-fluctuations [compare eqn (4.21)], and with reference to Fig. 6.10, we see that the difference between the two scattering geometries increases with decreasing value of $H_p/\Delta M$.

The effect of the ratio $H_p/\Delta M$ on the correlation function and on the l_C-values is shown in Fig. 6.10 for the transversal scattering geometry. Perturbations in the spin microstructure which are dominated by fluctuations of the magnetic anisotropy field ($H_p/\Delta M \gg 1$) decay on a larger length scale than magnetostatically dominated

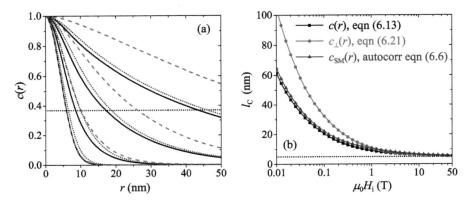

Fig. 6.7: (a) Comparison between the correlation functions of the spin-misalignment SANS cross section, $c(r)$ and $c_\perp(r)$, and the autocorrelation function of the spin misalignment $c_{SM}(r)$. Solid lines: $c(r)$ [eqn (6.13)]. Dashed lines: $c_\perp(r)$ [eqn (6.21)]. Dotted lines: $c_{SM}(r)$ [eqn (6.6)] ($\mathbf{k}_0 \perp \mathbf{H}_0$; $H_p/\Delta M = 1$; $\xi_H = \xi_M = \mathcal{L} = 5$ nm). Values of H_i (in T) increase from top to bottom: 0.02, 0.15, 1.2, 11. Dotted horizontal line: $c(r) = \exp(-1)$. (b) Corresponding $l_C(H_i)$ (log-linear scale) (lines are a guide to the eyes). Dotted horizontal line: $l_C = \mathcal{L} = 5$ nm. After [312].

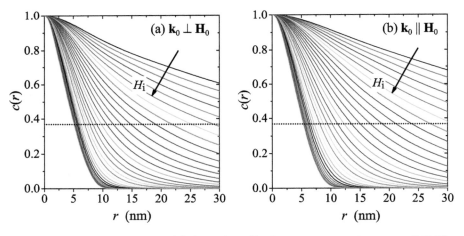

Fig. 6.8: Correlation functions $c(r)$ [eqn (6.13)] of the spin-misalignment SANS cross section at several internal-field values H_i for (a) $\mathbf{k}_0 \perp \mathbf{H}_0$ and (b) $\mathbf{k}_0 \parallel \mathbf{H}_0$. H_i increases, respectively, from 0.01 T to 100 T on a logarithmic scale, i.e., $\mu_0 H_i^j = 10^{4\frac{j}{j_{max}} - 2}$ T, where $j_{max} = 30$ and $j = 0, \ldots, 30$ ($H_p/\Delta M = 1$). The arrows specify the direction of increasing H_i. Dotted horizontal lines in (a) and (b): $c(r) = \exp(-1)$. After [312].

($H_p/\Delta M \ll 1$) perturbations. We also see that for anisotropy-field-dominated correlations, eqn (6.24) provides an excellent description of the data [solid lines in Fig. 6.9(b) and Fig. 6.10(b)] [312]. At large fields, when the spin-misalignment SANS cross section is small and the exchange length l_H takes on values of a few nm, l_C reflects, irrespective

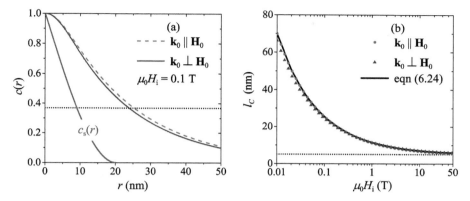

Fig. 6.9: (a) Comparison of the $c(r)$ [eqn (6.13)] for the two scattering geometries ($\mu_0 H_i = 0.1$ T; $H_p/\Delta M \to \infty$). Dotted horizontal line: $c(r) = \exp(-1)$. The autocorrelation function $c_s(r)$ of a single uniform sphere with radius $R = 10$ nm [eqn (6.27)] is drawn for comparison. (b) Comparison of the field dependence of the spin-misalignment correlation length l_C for the two scattering geometries ($H_p/\Delta M \to \infty$) (log-linear scale). Solid line: eqn (6.24). Dotted horizontal line: $l_C = \mathcal{L} = 5$ nm. After [312].

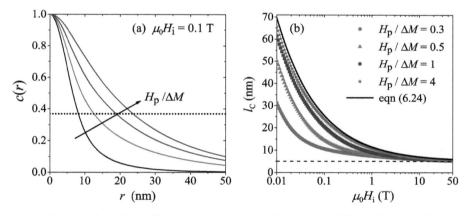

Fig. 6.10: (a) $c(r)$ [eqn (6.13)] for several values of the ratio $H_p/\Delta M$ at $\mu_0 H_i = 0.1$ T ($\mathbf{k}_0 \perp \mathbf{H}_0$). $H_p/\Delta M$-values: 0.004, 0.4, 0.8, 4. The arrow specifies the direction of increasing $H_p/\Delta M$. Dotted horizontal line in (a): $c(r) = \exp(-1)$. (b) Field dependence of the spin-misalignment correlation length l_C for different values of $H_p/\Delta M$ (log-linear scale). Solid line: eqn (6.24). Dashed horizontal line: $l_C = \mathcal{L} = 5$ nm. After [312].

of the value of $H_p/\Delta M$, the size of the (in this case, spherical) defect.

The behavior of the correlation function at small r relates to the properties of the SANS cross section at large q. For nuclear particle scattering, the final slope of the scattering curve has been analyzed by Porod [290]; in particular, for small r, the correlation function computed according to eqn (6.13) can be expanded into a power

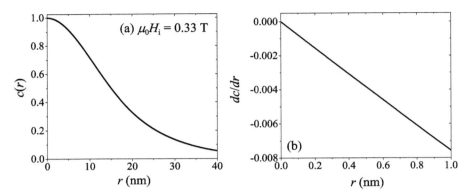

Fig. 6.11: (a) Spin-misalignment correlation function $c(r)$ [eqn (6.13)] at $\mu_0 H_i = 0.33\,\text{T}$. (b) Corresponding first-order r-derivative of $c(r)$.

series,

$$c(r) = 1 + a_1 r + a_2 r^2 + a_3 r^3 + \dots, \tag{6.26}$$

where the coefficients a_i, termed "differential" parameters by Porod, are related to the fine structure of the particle. For the example of a single uniform sphere with radius R, one finds the well-known expression (valid for $r \leq 2R$)

$$c_s(r) = 1 - \frac{3r}{4R} + \frac{r^3}{16R^3}, \tag{6.27}$$

from which one can recognize that the first derivative of $c_s(r)$ evaluated at $r = 0$ is related to the surface-to-volume ratio $\mathcal{S}/V$ of the particle, i.e., $c_s'(0) = a_1 = -3/(4R) = -\mathcal{S}/(4V)$ [compare eqn (2.74) and Fig. 6.9(a)]. The latter statement is generally valid for uniform particles with a sharp boundary [11].

Inspection of the micromagnetic correlation functions which are computed based on eqn (6.13) reveals that the $c(r)$ enter the origin with zero slope—in contrast to the uniform particle case (see, e.g., Figs. 6.11 and 6.14). As discussed by Porod [290], this feature can be interpreted by the absence of a sharp interface and by the presence of a nonuniform continuous magnetic scattering-length density variation. To be more precise, within the framework of the micromagnetic SANS theory [312, 414, 439, 444, 467, 488], the magnetic microstructure in real space, $\mathbf{M}(\mathbf{r})$, corresponds to a complicated convolution product between the magnetic anisotropy-field microstructure, the saturation-magnetization profile, and micromagnetic functions [compare eqns (3.68) and (3.69)]. As a result, smoothly varying magnetization profiles are at the origin of the related spin-misalignment scattering. This statement does of course not preclude the existence of sharp interfaces in the nuclear grain microstructure of a magnetic material; for instance, there may well exist sharp particle-matrix interfaces in a magnetic nanocomposite, but the corresponding spin distribution which decorates these interfaces (and which gives rise to $d\Sigma_{\text{SM}}/d\Omega$) might be continuous and smooth over the defects. Therefore, consistent with the absence of a sharp interface in the magnetic microstructure (compare Figs. 6.2 and 6.3), we note that the correlation functions of bulk ferromagnets computed based on eqn (6.13) enter the origin $r = 0$ with zero slope [312]. Consequently, the linear term in an expansion of the form of eqn (6.26) vanishes, and [290]

$$c(r) = 1 - br^2 + \ldots, \tag{6.28}$$

where $b > 0$ (compare Fig. 6.14). This particular form for $c(r)$, valid for small r and characteristic for fluctuating systems, can be motivated by expanding the magnetization fluctuation $\delta \mathbf{M}(\mathbf{r})$ in the expression for the autocorrelation function $C_{\mathrm{SM}}(\mathbf{r})$ [eqn (6.3)] into a Taylor series. Considering for simplicity only one component of $\delta \mathbf{M}(\mathbf{r})$, say M_x, and, due to statistical isotropy in the plane perpendicular to the applied field, treating it as a function of only one variable, say ζ, one can in a simplified way write for the autocorrelation function [290]:

$$C_{\mathrm{SM}}(\zeta) = \frac{1}{V^{1/3}} \int_{-\infty}^{\infty} M_x(\zeta') M_x(\zeta' + \zeta) d\zeta'$$

$$\simeq \frac{1}{V^{1/3}} \int_{-\infty}^{\infty} M_x(\zeta') \left\{ M_x(\zeta') + \frac{dM_x}{d\zeta'}\zeta + \frac{1}{2}\frac{d^2 M_x}{d\zeta'^2}\zeta^2 + \ldots \right\} d\zeta', \tag{6.29}$$

where $V^{1/3}$ is some characteristic sample dimension. The first term in the curly brackets evaluates to $\overline{M_x^2}$, and contributes to the mean-square magnetization fluctuation [eqn (6.8)]. Integration by parts shows that the second (linear) term vanishes for a statistically isotropic material, and that the coefficient of the third term is negative, since the second derivative of a function must be negative near a maximum. The final (small ζ) result for the autocorrelation function is [290]:

$$C_{\mathrm{SM}}(\zeta) \simeq \overline{M_x^2} - \frac{1}{2}\overline{\left(\frac{dM_x}{d\zeta'}\right)^2}\zeta^2, \tag{6.30}$$

which is of the functional form of eqn (6.28). A similar theoretical estimation for the behavior of the spin-misalignment correlation function can be accomplished by expanding the spherical Bessel function ($j_0(z) \cong 1 - \frac{1}{6}z^2 + \ldots$) in the enumerator of eqn (6.13) and by taking into account the asymptotic q-dependency of $d\Sigma_{\mathrm{SM}}/d\Omega$. Moreover, for systems which exhibit large variations of the magnetic materials parameters at defect sites, e.g., a strongly reduced exchange interaction at grain boundaries, the magnetization distribution over the defect may show a quasi-discontinuity [165, 361, 364, 975]. It would be of interest to work out the analytical behavior of the correlation function for this situation.

With reference to the small-angle scattering on homogeneous particles, where the leading term in the asymptotic expansion of the cross section (Debye–Porod law) is proportional to $c'(0)$ [compare eqn (2.75)] [11, 290], the result eqn (6.30) is then compatible with the absence of an asymptotic q^{-4}-behavior of the spin-misalignment scattering. This is in agreement with theoretical predictions and experimental observations (see Section 4.3.5). We remind that the $c_\perp(r)$ which is computed according to eqn (6.21) is equivalent to the projection of the general $c(x, y, z)$ along the incident-beam (x) direction [compare also eqns (6.15) and (6.16)]. In this case, the slope of the correlation function at the origin vanishes also for any homogeneous three-dimensional particle with a sharp interface (see Appendix C).

6.4   Total correlation function

In the micromagnetic model calculations of the previous Section 6.3, we have discussed the Fourier transform of $d\Sigma_{SM}/d\Omega = S_H R_H + S_M R_M$, which experimentally may be obtained by the subtraction of the (nuclear and magnetic) residual scattering cross section $d\Sigma_{res}/d\Omega$, measured at saturation, from the SANS data at lower fields. In the following, we will briefly consider the effect of a field-independent $d\Sigma_{res}/d\Omega$ on the correlation function and correlation length; in other words, we consider the *total* correlation function, which is simply the Fourier transform of the total unpolarized $d\Sigma/d\Omega = d\Sigma_{res}/d\Omega + d\Sigma_{SM}/d\Omega$ [eqns (2.93) and (2.94)].

For a statistically isotropic ferromagnet in the $\mathbf{k}_0 \perp \mathbf{H}_0$ scattering geometry, the azimuthally averaged $d\Sigma_{res}/d\Omega$ is given by eqn (4.13), which is rewritten here:

$$\frac{d\Sigma_{res}}{d\Omega} = \frac{8\pi^3}{V} \left(|\widetilde{N}|^2 + \frac{1}{2} b_H^2 |\widetilde{M}_s|^2 \right). \tag{6.31}$$

Equation (4.15) in Section 4.2 shows that, within the micromagnetic theory, the problem of the total correlation function essentially reduces to the study of the impact of the field-independent nuclear SANS background on the correlations. However, since in experimental situations it is often more convenient to measure $d\Sigma_{res}/d\Omega$, rather than the nuclear SANS cross section $d\Sigma_{nuc}/d\Omega = \frac{8\pi^3}{V} |\widetilde{N}|^2$, we concentrate here on $d\Sigma_{res}/d\Omega$ and model this term by the sphere form-factor function containing a single effective correlation length ξ_{res}:

$$\frac{d\Sigma_{res}}{d\Omega} = A_{res} P(q \xi_{res}). \tag{6.32}$$

We emphasize that ξ_{res} may be different from the length scales which characterize the magnetic anisotropy field (ξ_H) and the spatial structure of the saturation-magnetization profile (ξ_M). As before, we assume for $S_H(q\xi_H)$ and $S_M(q\xi_M)$ the sphere form factor [eqns (3.82)–(3.84)], and the amplitude parameter A_{res} in eqn (6.32) may be taken as a measure for the strength of $d\Sigma_{res}/d\Omega$ relative to $d\Sigma_{SM}/d\Omega$.

Figure 6.12 depicts some results for the total correlation function $c_{tot}(r)$ [computed based on eqn (6.13)] and the correlation length $l_C^{tot}(H_i)$. In Fig. 6.12(a) and (b), we have set $\xi_{res} = 10\,\mathrm{nm}$, $\xi_H = \xi_M = 5\,\mathrm{nm}$, and $H_p = \Delta M$, which then implies that $S_H = S_M$ [compare eqns (3.82)–(3.84)]. As expected, an increase of the magnitude of $d\Sigma_{res}/d\Omega$ results in a suppression of the range of the field-dependent correlations. Since $d\Sigma_{SM}/d\Omega$ vanishes for large fields, the value of l_C^{tot} as $H_i \to \infty$ is determined by ξ_{res} (unless $A_{res} = 0$). Since the small-angle scattering is proportional to the squared particle volume, a relatively small A_{res} is already sufficient to significantly cut down the field dependence of l_C^{tot}. Figure 6.12(c) shows $l_C^{tot}(H_i)$ for $\xi_{res} = 10\,\mathrm{nm}$, $\xi_M = 25\,\mathrm{nm}$, and $\xi_H = 40\,\mathrm{nm}$. When $A_{res} = 0$, corresponding to the case of pure spin-misalignment scattering (no

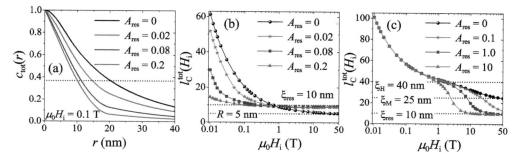

Fig. 6.12: (a) Total correlation function $c_{\text{tot}}(r)$ [computed based on eqn (6.13)] at $\mu_0 H_i = 0.1\,\text{T}$ and for several values of the amplitude parameter A_{res} ($\mathbf{k_0} \perp \mathbf{H_0}$; $H_p/\Delta M = 1$; $S_H = S_M$). The structural correlation length of $d\Sigma_{\text{res}}/d\Omega$ was set to $\xi_{\text{res}} = 10\,\text{nm}$ [eqn (6.32)], while the characteristic length scales of S_H and S_M were both set to $\xi_H = \xi_M = R = 5\,\text{nm}$. Dotted horizontal line: $c_{\text{tot}}(r) = \exp(-1)$. (b) Corresponding field dependency of the correlation length l_C^{tot} (log-linear scale). Lines are a guide to the eyes. Dotted horizontal lines: $l_C^{\text{tot}} = R = 5\,\text{nm}$ and $l_C^{\text{tot}} = \xi_{\text{res}} = 10\,\text{nm}$. (c) $l_C^{\text{tot}}(H_i)$ for $\xi_{\text{res}} = 10\,\text{nm}$, $\xi_M = 25\,\text{nm}$, and $\xi_H = 40\,\text{nm}$ (log-linear scale).

field-independent background), one can see that $l_C^{\text{tot}}(H_i \to \infty)$ approaches the smallest of the two correlation lengths ξ_H and ξ_M.

6.5 Selected experimental results

6.5.1 Nanocrystalline Nd–Fe–B-based permanent magnets

In the following, we discuss the prototypical real-space SANS data analysis for the example of a nanocrystalline Nd–Fe–B-based permanent magnet. We consider a two-phase nanocomposite, which consists of hard magnetic $Nd_2Fe_{14}B$ particles (size: $\sim 22\,\text{nm}$) and Fe_3B crystallites (size: $\sim 29\,\text{nm}$) [526, 533]. It is important to mention that for this particular alloy the difference ΔM in the saturation magnetizations of the $Nd_2Fe_{14}B$ phase and the Fe_3B crystallites is rather small, $\mu_0 \Delta M \cong 0.01\,\text{T}$ [538]. Consequently, the related longitudinal magnetic SANS cross section $\propto |\widetilde{M}_z|^2 \propto (\Delta M)^2$ is negligible as compared to the nuclear SANS $\propto |\widetilde{N}|^2$.

Figure 6.13(a) displays the total unpolarized $d\Sigma/d\Omega$ of the Nd–Fe–B nanocomposite. A strong field dependence between the largest applied field of $10\,\text{T}$ and the coercive field of $\mu_0 H_c = -0.55\,\text{T}$ is observable. Since nuclear SANS is field independent and since SANS due to $|\widetilde{M}_z|^2$ fluctuations is negligible for this alloy, it is evident that the dominant contribution to $d\Sigma/d\Omega$ is due to transversal spin misalignment. In order to obtain the corresponding spin-misalignment SANS cross section [see Fig. 6.13(b)], the $d\Sigma/d\Omega$ at $10\,\text{T}$ was subtracted from the $d\Sigma/d\Omega$ at lower fields [compare eqn (4.1)]. The resulting $d\Sigma_{\text{SM}}/d\Omega$ is of comparable magnitude as $d\Sigma/d\Omega$, but possesses a strikingly different q-dependency; in particular, the shoulder in $d\Sigma/d\Omega$ at about $q = 0.2\,\text{nm}^{-1}$ and at the highest fields is absent in $d\Sigma_{\text{SM}}/d\Omega$. Possible origins for the shoulder in $d\Sigma/d\Omega$ are interparticle interferences and/or diffusion zones around the particles, as discussed by Hermann and Heinemann et al. [315–317]. The different shapes of $d\Sigma/d\Omega$

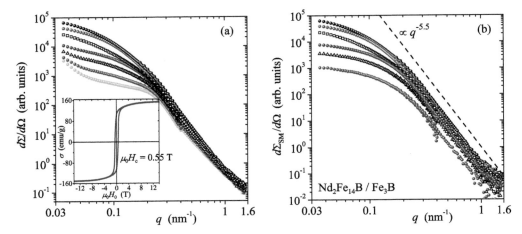

Fig. 6.13: (a) Azimuthally averaged total SANS cross section $d\Sigma/d\Omega$ of $Nd_2Fe_{14}B/Fe_3B$ as a function of momentum transfer q and applied magnetic field H_0 ($T = 300\,K$) ($\mathbf{k}_0 \perp \mathbf{H}_0$) (log-log scale). The field values follow the course of a hysteresis loop, starting from a large positive field and then reducing the field to negative values. Solid circles ($\bullet$): applied-field values (in Tesla) decrease from bottom to top: 10, 6, 1, -0.25, -0.55. ($\square$): $-1\,T$. ($\triangle$): $-3\,T$. Inset: room-temperature magnetization curve of $Nd_2Fe_{14}B/Fe_3B$. (b) Applied-field dependence of the spin-misalignment SANS cross section $d\Sigma_{SM}/d\Omega$ of nanocrystalline $Nd_2Fe_{14}B/Fe_3B$. Solid circles ($\bullet$): field values (in Tesla) decrease from bottom to top: 6, 1, -0.25, -0.55. ($\square$): $-1\,T$. ($\triangle$): $-3\,T$. The $d\Sigma_{SM}/d\Omega$ data displayed in (b) were obtained by subtracting the $10\,T$ data shown in (a) from the $d\Sigma/d\Omega$ at lower fields. Dashed line: $d\Sigma_{SM}/d\Omega \propto q^{-5.5}$. After [526].

and $d\Sigma_{SM}/d\Omega$ are also reflected in different asymptotic power-law exponents m. While the spin-misalignment SANS cross section is characterized by power-law exponents which range between $m \sim 5-6$ at all fields investigated [see Fig. 4.23(b)], the total unpolarized SANS reveals significantly lower values for m, which approach the Porod value of $m = 4$ at $10\,T$ [526].

Fourier transformation of the $d\Sigma_{SM}/d\Omega$ data according to eqn (6.13) yields the correlation function $C(r)$ of the spin misalignment (see Fig. 6.14). The field-dependent correlations in Fig. 6.14 do not decay exponentially, consistent with the absence of an $m = 4$ power-law exponent in $d\Sigma_{SM}/d\Omega$. Furthermore, the $C(r)$ seem to approach the origin $r = 0$ with zero slope (dotted line in Fig. 6.14), which is in agreement with the notion of magnetic SANS from continuous magnetization profiles and the absence of a sharp interface in the magnetic microstructure [compare eqn (6.28)] [290].

The values of the correlation length l_C are plotted in Fig. 6.15 as a function of the applied magnetic field, which is usually the control parameter in magnetic SANS experiments. For the Nd–Fe–B nanocomposite (with $\Delta M \cong 0$), one can expect that l_C describes the spatial extent of magnetization inhomogeneities, mainly within the soft magnetic Fe_3B grains, that are caused by the jump in the magnetic materials parameters (exchange constant, direction and magnitude of magnetic anisotropy) at

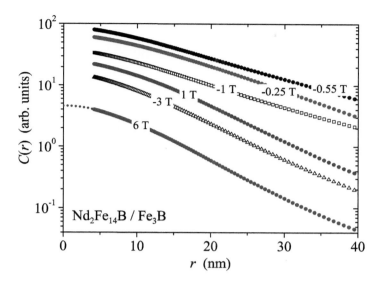

Fig. 6.14: Field dependence of the (unnormalized) correlation function $C(r)$ of the spin misalignment of nanocrystalline $Nd_2Fe_{14}B/Fe_3B$ [computed according to eqn (6.13)] (log-linear scale). The field values follow the course of a hysteresis loop, starting from a large positive field and then reducing the field to negative values [as in Fig. 6.13(b), see insets]. Dotted line (extrapolating the 6 T data to $r = 0$): $C(r) = 4.58 - 0.043\,r^2$. After [526].

the interface between the $Nd_2Fe_{14}B$ particles and the surrounding Fe_3B crystallites. As can be seen in Fig. 6.15, l_C approaches a constant value of about 12.5 nm at the largest positive fields and increases with decreasing applied field to take on a maximum value of about 18.5 nm at the experimental coercive field of $\mu_0 H_c = -0.55$ T. Further increase of H_0 towards more negative values results again in a decrease of l_C towards ~12.5 nm. From the fit of the $l_C(H_0)$ data to eqn (6.23) (solid line in Fig. 6.15), we obtain $\mathcal{L} = 10.9$ nm (close to the experimental average grain radius of the $Nd_2Fe_{14}B$ phase) and $\mu_0 H^\star = +0.60$ T, which is close to the absolute value of the experimental coercive field. At the remanent state, the penetration depth of the spin disorder into the Fe_3B phase amounts to ~5−6 nm.

The here-used procedure of subtracting the total unpolarized SANS scattering at a field close to saturation from data at lower fields (Fig. 6.13) suggests that it may not always be necessary to resort to complex polarization-analysis experiments in order to obtain the magnetic (spin-flip) SANS cross section. Ignoring nuclear-spin-dependent SANS and the chiral scattering term, then the comparison of the spin-flip SANS cross section $d\Sigma^{+-}/d\Omega = d\Sigma^{-+}/d\Omega$ [eqn (2.100)] with the spin-misalignment SANS cross section $d\Sigma_{SM}/d\Omega$ [eqns (4.3) and (4.5)] reveals that the subtraction procedure yields, except for the longitudinal magnetic term $|\widetilde{M}_z|^2 \sin^2\theta \cos^2\theta$, a combination of (difference) Fourier components that is very similar to the spin-flip SANS cross section; in other words, under the above assumptions, $d\Sigma_{SM}/d\Omega$ depends on the transversal magnetization Fourier components, as does $d\Sigma^{+-}/d\Omega$, albeit with different trigonometric

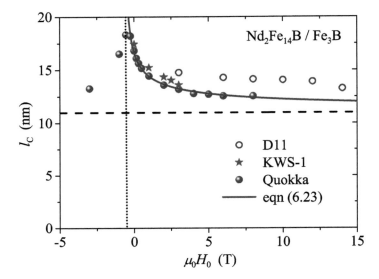

Fig. 6.15: Applied-field dependence of the correlation length l_C of the spin misalignment of nanocrystalline $Nd_2Fe_{14}B/Fe_3B$. Solid line: fit of the data to eqn (6.23), where $\mathcal{L} = 10.9\,nm$ and $\mu_0 H^\star = +0.60\,T$ are treated as adjustable parameters, and the quantities $A = 12.5\,pJ/m$ and $\mu_0 M_s = 1.6\,T$ are held fixed. In addition to $l_C(H_0)$ data obtained at the instrument Quokka (ANSTO), results obtained at the SANS instruments KWS-1 (JCNS) and D11 (ILL) are also shown. Dashed horizontal line: average radius of the $Nd_2Fe_{14}B$ particles ($R = 11\,nm$). Dotted vertical line: coercive field $\mu_0 H_c = -0.55\,T$. After [526].

weights. If the nuclear particle microstructure of the material under study does not change with the applied field, then the subtraction procedure might be a practicable alternative to time-consuming and low-intensity polarization-analysis experiments.

6.5.2 Nanocrystalline soft magnets

The SANS response of nanocrystalline soft magnets is extremely strong, which is mainly related to their highly disordered nanometer-scale spin structure. On a scale of a few nanometers, the direction and/or the strength of the magnetic anisotropy field changes randomly giving rise to a contrast for magnetic SANS. It is of crucial importance to realize that for these systems—even in the approach-to-saturation regime—spin-misalignment scattering is strongly dependent on the applied magnetic field. As an example, Fig. 6.16(a) shows the unpolarized (nuclear and magnetic) $d\Sigma/d\Omega$ of nanocrystalline Co with an average crystallite size of $D_{gs} = 9.5 \pm 3.0\,nm$ [441, 442]. We note that the sample under study is a fully dense polycrystalline bulk metal with a nanometer grain size, not nanoparticles in a matrix. The SANS cross section at the smallest momentum transfers varies by more than three orders of magnitude between $5\,mT$ and $1800\,mT$. Even in the saturation regime [compare hysteresis loop in Fig. 6.16(b)], $d\Sigma/d\Omega$ exhibits an extraordinarily large field variation. The origin of the large field dependence of $d\Sigma/d\Omega$ near saturation is not related to a macroscopic

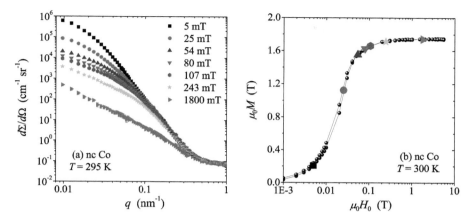

Fig. 6.16: (a) Azimuthally averaged unpolarized SANS cross section $d\Sigma/d\Omega$ of nanocrystalline Co metal (average grain size: $D_{gs} = 9.5 \pm 3.0\,\text{nm}$) as a function of momentum transfer q and at selected applied magnetic fields (see inset) ($\mathbf{k_0} \perp \mathbf{H_0}$) (log-log scale). (b) ($\bullet$) Room-temperature magnetization curve of nanocrystalline Co (log-linear scale) [441]. The $M(H_0)$-values where the SANS measurements shown in (a) have been carried out are indicated by the large colored symbols. The differences in the demagnetizing fields between the SANS and VSM samples are negligible. After [59].

magnetic domain structure (see, e.g., Fig. 1.5), but to the failure of the spins to completely align along the external field; in other words, it is due to spin misalignment. Closer inspection of the $d\Sigma/d\Omega$ data in Fig. 6.16(a) reveals that, with increasing field, a characteristic length scale (wavelength) is evolving towards larger q-values (i.e., q-scaling is lost). This observation provides evidence for the existence of a characteristic magnetic-field-dependent length scale $l_C(H_i)$ in the system (compare also Fig. 4.4).

A comparison between experiment and theory is depicted in Fig. 6.17, where the spin-misalignment correlation functions $C(r)$ of nanocrystalline Co and Ni together with global fits based on eqn (6.13) are displayed. The expression for $d\Sigma_{SM}/d\Omega$ used in eqn (6.13) is given by the micromagnetic eqn (4.14), which provides a very good global description of the field-dependent correlations. From the analysis, values for the exchange-stiffness constant A, the ratio $H_p/\Delta M$, and the anisotropy-field correlation length ξ_H can be estimated (see [312] for further details).

Figure 6.18 depicts the results for $l_C(H_i)$ for nanocrystalline Co and Ni [442]. The field variation of l_C for Co [Fig. 6.18(a)] confirms the conclusions that were already drawn by the inspection of the corresponding $d\Sigma/d\Omega$ and $C(r, H_i)$ (Figs. 6.16 and 6.17), namely, a pronounced dependence on the applied field with values ranging from about 94 nm at the lowest field to about 15 nm at the highest field. The values for l_C are (at all fields) larger than those expected for uniformly magnetized grains of the experimental size 9.5 nm [dotted line in Fig. 6.18(a)]. By contrast, the field dependence of l_C for Ni with an average crystallite size of $D_{gs} = 49$ nm [Fig. 6.18(b)] indicates the presence of magnetization inhomogeneities on a scale smaller than D_{gs}. Here, l_C varies from about 45 nm at 39 mT to 13 nm at 1790 mT.

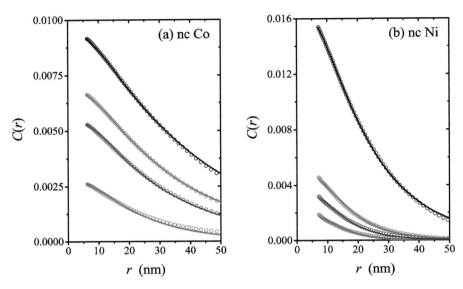

Fig. 6.17: Comparison between experimental $C(r)$ data (unnormalized) and theoretical prediction. (∘) Correlation functions of the spin-misalignment SANS cross section of nanocrystalline Co (a) and nanocrystalline Ni (b) with average crystallite sizes of $D_{gs} = 9.5\,\text{nm}$ (Co) and $D_{gs} = 49\,\text{nm}$ (Ni). $C(r)$ data are taken from [442]. Solid lines: global fits based on eqn (6.13) using eqn (4.14) for $d\Sigma_{SM}/d\Omega$. Values of the internal magnetic field H_i (in mT) from top to bottom: (a) 54, 80, 107, 243; (b) 190, 570, 800, 1240. After [312].

The predictions for $l_C(H_i)$ by eqn (6.24) are plotted as solid lines in Fig. 6.18(a) and (b) using the known values of the materials and microstructural parameters (see [442] for details). Note that there are no free parameters. It can be seen that eqn (6.24) provides an excellent description of the Co data, while the agreement for Ni is at least qualitative. The observation in Fig. 6.18(a) that $l_C(H_i)$ deviates from a pure $H_i^{-1/2}$-law at large fields and approaches a constant value close to the experimental grain size of $D_{gs} = 9.5\,\text{nm}$ suggests that in nanocrystalline Co the dominant microstructural defect is the anisotropy field of each individual crystallite, i.e., the typical length scale over which the anisotropy field in the Co sample is uniform is of the order of D_{gs}, and nonuniformities on a smaller scale, originating e.g., from the defect cores of grain boundaries or stacking faults, play only a minor role. Contrary to what is found for Co, the data for Ni suggest the existence of nonuniformities in the anisotropy field on a scale smaller than the average crystallite size. Here, the average distance between twin faults in the sample, $L_{twin} = 4.3\,\text{nm}$ [dashed line in Fig. 6.18(b)], was used as the characteristic length scale $\mathcal{L}$ of the microstructure in eqn (6.24). The deviation between the data points and the theoretical prediction indicates a wide distribution in the characteristic length scales (e.g., grain size, distance between twin faults) that are associated with the individual microstructural anisotropy fields.

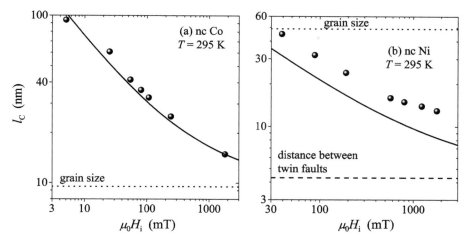

Fig. 6.18: Field dependence of the correlation length $l_C(H_i)$ of the spin misalignment for nanocrystalline Co (a) and Ni (b) (log-log scale). Dotted lines in (a) and (b): average grain sizes. Dashed line in (b): average distance between twin faults. Solid lines in (a) and (b): eqn (6.24). After [442].

The results in Figs. 6.16−6.18 suggest that the magnetic microstructure of soft magnetic nanocrystalline Ni and Co is dominated by static nanometer-scale magnetization fluctuations, whose amplitude and wavelength increase continuously as the field is lowered. The existence, at small applied magnetic fields, of structure in the spin system which extends over many neighboring grains is a central result of random-anisotropy-type models of the remanent state of nanocrystalline soft magnets (compare Herzer's random anisotropy model [976]). Eventually, the continuous increase in l_C is expected to break down when the coercive field is reached and a macroscopic domain structure with domain sizes of the order of several micrometers forms. In this situation, the dominant magnetic SANS contrast may originate from the jump of the direction of the magnetization between neighboring domains, and the characteristic length scale may be too large to be resolved with conventional SANS. Moreover, the existence of long-range gradients in the magnetization distribution demonstrates once more that discontinuous hard-sphere models are not appropriate to describe magnetic SANS from such materials.

6.5.3 Nanocrystalline hard magnets

Terbium. A different behavior of the spin-misalignment correlations is obtained in nanocrystalline Tb ($4f^8$ for Tb^{3+}; L = 3, S = 3, J = 6), which is prepared by the inert-gas-condensation method [303]. Here, perturbations in the spin structure are mainly restricted to developing on an intraparticle length scale. Before presenting the experimental SANS results, we will briefly summarize some relevant magnetic properties of Tb [120]. Single-crystalline Tb possesses a ferromagnetic ordering temperature of $T_C = 220\,\text{K}$ and a zero-temperature magnetic moment of $9.34\,\mu_B$. Below T_C, the magnetic moments are confined to the basal plane of the hcp lattice by an extremely

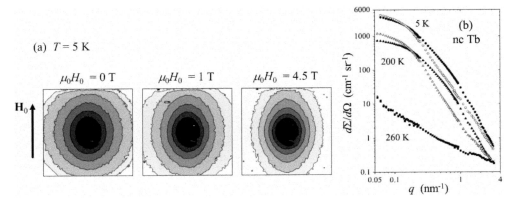

Fig. 6.19: (a) Grey-scale-coded map of the experimental scattering intensity of nanocrystalline Tb (grain size: $D_{gs} = 9 \pm 2\,\text{nm}$) recorded on the two-dimensional detector at $T = 5\,\text{K}$ and at three different magnetic fields as indicated in the figure. The field direction is vertical, and pixels in the corners have momentum transfer of $q \cong 1.3\,\text{nm}^{-1}$. Subsequent shadings denote doubling of the intensity with darker regions corresponding to higher intensity. (b) Experimental differential scattering cross sections $d\Sigma/d\Omega$ of nanocrystalline Tb at different temperatures as indicated (log-log scale). Closed symbols: $\mu_0 H_0 = 0\,\text{T}$. Open symbols: $\mu_0 H_0 = 4.5\,\text{T}$. After [303].

large magnetic anisotropy [977]; for instance, at $T = 5\,\text{K}$ and at $T = 200\,\text{K}$ estimated applied fields of, respectively, $62\,\text{T}$ and $20\,\text{T}$ are required to overcome the anisotropy and to align the moments along the c-axis [303]. Therefore, the application of a typical laboratory magnetic field ($\mu_0 H_0 \sim 5\,\text{T}$) will preferentially realign the magnetization within the basal plane of each crystallite, rather than producing noticeable out-of-plane orientation. Furthermore, the relatively low ordering temperature of Tb allows one to correct for the combined nuclear and magnetic SANS due to porosity by performing SANS measurements well above T_C [303]. Pore scattering is of particular importance in inert-gas-condensed nanocrystalline materials [978].

Figure 6.19 shows the SANS cross section of nanocrystalline Tb with an average crystallite size of $D_{gs} = 9\,\text{nm}$ at several applied magnetic fields and temperatures. Remarkable are three observations. (i) The SANS signal in the paramagnetic temperature regime at $T = 260\,\text{K}$ and at the smallest momentum transfers is about two orders of magnitude smaller than the combined nuclear and magnetic signal in the ferromagnetic regime [Fig. 6.19(b)]. This fact clearly demonstrates the dominance of the magnetic SANS contribution, in particular, due to spin misalignment, which gives rise to enhanced magnetic scattering along the field direction [see Fig. 6.19(a)]. (ii) Whereas in nanocrystalline Ni and Co the application of an external field of the order of typically $2\,\text{T}$ results in an alignment of the magnetization along the field direction and in a concomitant decrease in the magnetic SANS signal by almost three orders of magnitude [compare Fig. 6.16(a)], the field dependence of the SANS cross section for Tb is relatively weak. (iii) Unexpectedly, at $q \cong 0.2\,\text{nm}^{-1}$, one observes a crossover of the scattering cross sections measured at zero field and at $4.5\,\text{T}$. Obviously, increasing the applied field leads in nanoscaled Tb to an increase of the scattering contrast at the smallest q. As discussed below, the appearance of a crossover in the scattering curves can be explained with the existence of two characteristic length scales.

Figure 6.20 depicts the magnetic field dependence of the characteristic length l_C of the spin misalignment of nanocrystalline Tb at 5 K and at 200 K. The values for l_C at 5 K exhibit a relatively weak field dependence with $1.5\,\mathrm{nm} \lesssim l_C \lesssim 2.0\,\mathrm{nm}$, practically at the resolution limit of the SANS instrument. At $T = 200\,\mathrm{K}$, the correlation length varies more significantly with the applied field. After a slight decrease at small fields, the characteristic length increases to a value of about 5 nm at 4.5 T. All values for l_C are smaller than the average crystallite size of $D_{gs} = 9\,\mathrm{nm}$. In agreement with the results of x-ray diffraction and magnetization measurements [303], this suggests that the origin of spin misalignment is related to disorder from within the particles.

The initial (at small H_0) negative field response of l_C in Tb is in qualitative agreement with the experimental observations in nanocrystalline Ni and Co, although the l_C-values of nanocrystalline Tb are significantly smaller (compare Fig. 6.18). However, in contrast to the soft magnetic systems, l_C of Tb at 200 K exhibits a minimum and a positive field-response for applied fields larger than 1 T. Weissmüller et al. [303] explain the l_C-behavior and the related crossover of the scattering curves in Fig. 6.19(b) as follows: the analysis of magnetization, x-ray diffraction, and SANS measurements has provided evidence that large inhomogeneous lattice strains (with a root-mean-square microstrain of about 0.5% in the Tb study [303]) result via magnetoelastic coupling in spin-misalignment fluctuations on an intraparticle length scale. The suppression of this structure in the spin system by moderate applied magnetic fields leads to the decrease in l_C at small H_0, to the concomitant decrease in the scattering cross section at large q, and to an increase in the magnitude of the net magnetic moment of each grain, while leaving its misalignment relative to the field direction constant. Since the large c-axis anisotropy of Tb ($K_2 \cong 1.7 \times 10^7\,\mathrm{Jm^{-3}}$ at 200 K [303, 979]) prevents a tilting of the moments out of the easy plane, the increase in the applied field leaves the direction of the net magnetization of each crystallite essentially unchanged. At small q, roughly $2\pi/D_{gs}$ and below, the scattering contrast is then simply proportional to the (squared) net moments of the grains. This quantity increases with increasing field, as argued above. The different field response of $d\Sigma/d\Omega$, positive at small q and negative at large q, then manifests itself in the crossover.

The above-sketched behavior for l_C can be further supported by making use of a well-known result from nuclear SANS theory, the autocorrelation function $c_s(r)$ of a uniform sphere [eqn (6.27)]. By using the definition $l_C = -(d\ln c_s/dr)_{r\to 0}^{-1}$ [eqn (2.74)] and $D_{gs} = 9\,\mathrm{nm}$, one obtains $l_C = \frac{2}{3}D_{gs} = 6\,\mathrm{nm}$ [303]. Indeed, the spin-misalignment length at 200 K approaches a value of $l_C \cong 5\,\mathrm{nm}$ at the largest applied fields (Fig. 6.20). At 5 K, l_C still exhibits a slight tendency to increase at the largest fields. Obviously, here, the larger magnetic anisotropy at low temperature ($K_2 \cong 5.65 \times 10^7\,\mathrm{Jm^{-3}}$ at 5 K [303, 979]) prevents a homogeneous magnetization within the basal planes, even at $\mu_0 H_0 = 4.5\,\mathrm{T}$.

Gadolinium. As with the related rare-earth metal Tb, the magnetism of Gd originates from the strongly localized electronic spins of the $4f$ shell, which are indirectly coupled via the Ruderman–Kittel–Kasuya–Yosida (RKKY) exchange interaction. However, due

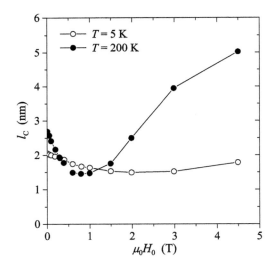

Fig. 6.20: Spin-misalignment correlation length l_C of nanocrystalline Tb (grain size: $D_{gs} = 9\,\text{nm}$) versus the applied magnetic field H_0 at $T = 5\,\text{K}$ and at $T = 200\,\text{K}$. Lines are a guide to the eyes. After [303].

to the absence of a $4f$ angular momentum ($4f^7$ for Gd^{3+}; $\text{L} = 0$, $\text{S} = \text{J} = 7/2$), Gd exhibits, in contrast to Tb, a comparatively low magnetocrystalline anisotropy, and thus takes on a prominent position among the heavy rare-earth elements. In the following, we will briefly summarize the results of magnetic SANS experiments on nanocrystalline Gd in the ferromagnetic temperature regime, which provide strong evidence for the existence of grain-boundary-induced spin disorder. For the Gd study [447, 449, 980], the low-capturing isotope ^{160}Gd (enrichment: 98.6%) was used as a starting material in the inert-gas-condensation process [543–545].

Grain-size-dependent magnetization isotherms of Gd at $T = 5\,\text{K}$ are displayed in Fig. 6.21(a). It is seen that a reduction of the grain size D_{gs} results in a considerable decrease of the macroscopic mean magnetization with respect to the coarse-grained state; e.g., at $\mu_0 H_i = 1\,\text{T}$, we find a relative reduction of about 20% for the smallest grain size of 14 nm. Even at fields as high as 9 T, the effect of nanocrystallinity on the macroscopic magnetization is still significant. By stepwise annealing the nanocrystalline samples to the coarse-grained state, we recover the single-crystal saturation magnetization value of $\mu_0 M_s = 2.69\,\text{T}$, and a scaling law for the relative magnetization reduction $\Delta M/M \propto D_{gs}^{-1}$ is found [see Fig. 6.21(c)]. This is particularly remarkable since the volume fraction of grain boundaries follows approximately the same grain-size dependence [981]. Therefore, we suggest that incomplete saturation due to spin disorder within the grain-boundary region may be responsible for the observed D_{gs}^{-1}-behavior; in other words, these results indicate the presence of a reduced effective magnetization in the grain-boundary phase, which may e.g., be a consequence of competing or frustrated interactions between the $4f$ moments. The estimated value for the relative reduction of the grain-boundary magnetization (at 9 T) with respect to the bulk of the grains amounts to 26% [449].

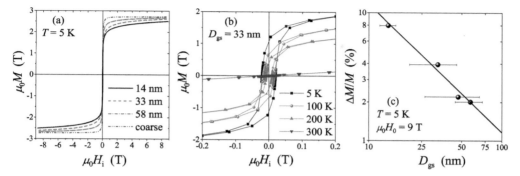

Fig. 6.21: Grain-size-dependent magnetization data of nanocrystalline Gd. (a) Magnetization isotherms of Gd at $T = 5\,\text{K}$. The coarse-grained reference sample was obtained by annealing from the nanocrystalline as-prepared state. (b) Hysteresis loops of the $D_{gs} = 33\,\text{nm}$ sample for temperatures of $T = 5, 100, 200$, and $300\,\text{K}$. Lines are a guide to the eyes. (c) Relative reduction $\Delta M/M := (M_{cg} - M_{nc})/M_{cg}$ (data points) of the macroscopic magnetization as a function of the average grain size D_{gs} at $\mu_0 H_0 = 9\,\text{T}$ and $T = 5\,\text{K}$ (log-log scale). M_{cg} is the experimental magnetization of coarse-grained Gd ($D_{gs} > 100\,\text{nm}$), which (at 9 T and 5 K) is practically identical to the single-crystal value for the saturation magnetization, $\mu_0 M_s = 2.69\,\text{T}$, and M_{nc} denotes the corresponding magnetization value of nanocrystalline Gd. Solid line: fit to a scaling law $\Delta M/M \propto D_{gs}^{-1}$. The quantity ΔM in the above expression should not be confused with the magnetization jump at internal particle-matrix interphases ($\widetilde{M}_z \propto \Delta M$), which is of relevance e.g., in nanocomposites [compare eqn (3.74)]. After [449, 980].

The effect of temperature on the hysteresis loop is displayed in Fig. 6.21(b) for a Gd sample with an average grain size of $D_{gs} = 33\,\text{nm}$. The data exhibit the usual reduction of remanence and coercivity with increasing temperature. As may be expected from the single-crystal value of $T_C = 293\,\text{K}$ for the Curie temperature of Gd [120], the data measured at the highest temperature of $T = 300\,\text{K}$ do not show a hysteresis. Note also that T_C is additionally shifted towards lower temperatures in the nanocrystalline material [545, 551].

The results of magnetometry presented in this section unambiguously demonstrate that nanocrystallinity has a strong impact on the magnetization isotherm of Gd (and likewise for Tb [303]). This becomes particularly clear from the observed dependence of the coercive field and the macroscopic magnetization on the average crystallite size. The reduction of the macroscopic magnetization in the nanocrystalline material for magnetic fields up to 9 T and low temperature and, in particular, its variation according to $\Delta M/M \propto D_{gs}^{-1}$, points towards the special role of the grain boundaries as the dominant source of spin disorder. Such a scaling law is characteristic for the volume fraction of atoms located in the core regions of grain boundaries.

The correlation function $C(r)$ at several applied fields and at $T = 78\,\text{K}$ is displayed in Fig. 6.22(a). $C(r)$ is a monotonically decaying function of the distance r, and an increase of the applied field results in the suppression of the magnitude of spin-misalignment fluctuations and in an overall decrease of the range of the correlations.

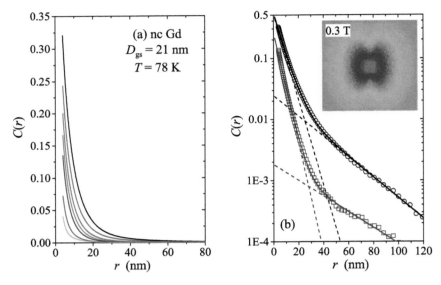

Fig. 6.22: (a) Correlation function $C(r)$ of the spin misalignment [eqn (6.13), un-normalized] of nanocrystalline (nc) ^{160}Gd at $T = 78\,\mathrm{K}$ ($D_{\mathrm{gs}} = 21 \pm 6\,\mathrm{nm}$). Values of the applied magnetic field from top to bottom (in mT): 0, 10, 30, 60, 100, 300, 600. (b) $C(r)$ from (a) on a log-linear scale at $0\,\mathrm{mT}$ (open circles) and $100\,\mathrm{mT}$ (open squares). It is seen that the data contain two characteristic length scales, as indicated by the dashed lines. Solid lines are fits to a sum of two decaying exponentials. Inset: clover-leaf anisotropy at $0.3\,\mathrm{T}$ ($\mathbf{H}_0$ is vertical). After [449, 980].

Closer inspection of the $C(r)$ data suggests the existence of at least two characteristic length scales in the spin system [compare Fig. 6.22(b)]. In order to extract these characteristic length scales, the correlation function (at a particular field) has been analyzed by a sum of two decaying exponentials, $C(r) = C_1 \exp(-r/L_1) + C_2 \exp(-r/L_2)$, where L_1 and L_2 denote the spin-misalignment lengths, which can be taken as a measure for the characteristic distance over which spin inhomogeneities decay.

Figure 6.23 depicts the results for the field dependencies of L_1 and L_2. It is seen that L_1 [Fig. 6.23(a)] ranges from about $4\,\mathrm{nm}$ to $6\,\mathrm{nm}$, which is about five times the structural width of a grain boundary [982] but still significantly smaller than the grain size of the nanocrystalline Gd sample, $D_{\mathrm{gs}} = 21\,\mathrm{nm}$. The length scale L_1 decreases with increasing applied field. In agreement with the observation of the clover-leaf-type anisotropy in the SANS signal of nanocrystalline Gd [see inset in Fig. 6.22(b)], which has been explained with the existence of a jump in the magnetization $\mathbf{M}$ at grain boundaries [447], it seems plausible to relate the short-range length scale L_1 to the defect character of the grain boundaries. The atomic-site mismatch (disorder) that is localized in the grain-boundary core of width $\sim 1\,\mathrm{nm}$ seems to induce spin misalignment which is transmitted by the exchange interaction into the interior of the adjacent crystallites. Obviously, the grain-boundary core is decorated by a gradient in $\mathbf{M}$ which decays within L_1. Since the exchange integral in Gd depends sensitively on

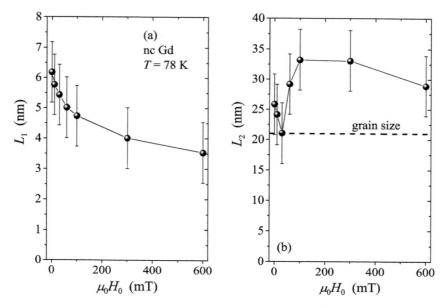

Fig. 6.23: Field dependence of the correlation lengths of the spin misalignment L_1 (a) and L_2 (b) of nanocrystalline (nc) [160]Gd. L_1 and L_2 were obtained by fitting the $C(r)$ data from Fig. 6.22(a) with a sum of two decaying exponentials. Lines are a guide to the eyes. Dashed horizontal line in (b): average crystallite size D_{gs} of the nanocrystalline [160]Gd sample. After [980].

interatomic distances [983,984], the atomic-site disorder at internal interfaces may also give rise to exchange-weakened bonds or even to antiferromagnetic couplings. Such a scenario [985] may then explain, at least partly, the reduction of the Curie transition temperature in bulk Gd at small crystallite size [545,551].

The second length scale L_2 varies between about 20 nm and 35 nm. Correlating L_2 with an additional-but-different regime of spin misalignment requires identifying an additional and plausible source of spin disorder. We assign the random magnetic anisotropy which is present in the bulk of the Gd grains to manifest the second source term. For uniformly magnetized grains with discontinuous jumps in the orientation of **M**, the characteristic length scale is expected to be of the order of D_{gs}, as was found in nanocrystalline Co [442]. Therefore, it seems plausible that the value of L_2 compares with D_{gs} [dashed line in Fig. 6.23(b)]. However, the variation of L_2 with field is beyond what can be explained by the neutron analysis.

The results for $C(r)$ of nanocrystalline Gd indicate the presence of two characteristic length scales in the spin structure. In agreement with grain-size-dependent magnetization data and the observation of a clover-leaf-type anisotropy in the magnetic scattering cross section, we attribute L_1 to the spin disorder which is caused by the defect cores of the grain boundaries, whereas L_2 is related to the magnetic anisotropy of the individual crystallites. As shown previously for the itinerant band magnets Co and Ni [441, 442], the atomic-site disorder of grain boundaries is not reflected in an associated spin disorder seen in the SANS signal. In those systems, the random magnetic anisotropy is the dominant source of spin misalignment. However, in nanocrystalline Gd it has become possible to unravel the spin disorder at or across internal interfaces mediated by the highly position-sensitive RKKY interaction.

7
MICROMAGNETIC SIMULATIONS

As extensively discussed in the previous chapters, necessary prerequisite for the quantitative analysis of elastic magnetic SANS data is the knowledge of the Fourier components of the static magnetization vector field $\mathbf{M}(\mathbf{r})$ of the sample under study. The continuum theory of micromagnetics [134, 154, 344] provides the proper framework for the computation of $\mathbf{M}(\mathbf{r})$. The total magnetic energy in such computations, $E_{\mathrm{tot}}[\mathbf{M}(\mathbf{r})]$, consists of the standard energy contributions due to the externally applied magnetic field, magnetic anisotropy, isotropic exchange, possibly the Dzyaloshinskii–Moriya interaction (DMI), and the magnetodipolar interaction energy. However, the solution of Brown's equations of micromagnetics amounts to the solution of a set of nonlinear partial differential equations with complex boundary conditions (see Section 5.1.3), a task which cannot be done analytically for most practically relevant problems. Analytical closed-form expressions for the magnetic SANS cross section are limited to the approach-to-saturation regime [1, 5, 312, 414, 439, 441, 444, 467, 488], in which the micromagnetic equations can be linearized (see Sections 3.2, 3.3, 4.2, 4.6, and 5.1.4). Consequently, numerical minimization of $E_{\mathrm{tot}}[\mathbf{M}(\mathbf{r})]$ should be carried out. The use of numerical techniques allows one to take into account the full nonlinearity of Brown's equations, without resorting to the high-field approximation.

In this chapter, we summarize the main results of the micromagnetic SANS simulation methodology which has been developed by Berkov and Erokhin [58, 473–478]. These authors have carried out extensive simulation studies on a wide range of magnetic microstructures. We begin in Section 7.1 with a short overview of existing micromagnetic SANS studies. In Section 7.2, we highlight the strength and the power of the simulation package developed by Berkov and Erokhin by discussing prototypical sample microstructures in simulations. Section 7.3 furnishes some useful information on the simulation methodology and on the implementation of the different energy contributions. Section 7.4 features the most important simulation results for the magnetic SANS cross section, with Section 7.4.1 focusing on nanocomposite bulk magnets and Section 7.4.2 summarizing the state-of-the-art on magnetic nanoparticles. We close this chapter with an outlook on simulation studies of magnetic SANS (Section 7.5).

7.1 Brief overview of micromagnetic SANS simulations

Numerical micromagnetic simulations of SANS experiments are very rare to date [470–472]. The first full-scale simulations of SANS measurements on a two-phase system have been reported by Ogrin et al. [470]. These authors have modeled the magnetization configuration of a longitudinal magnetic recording media film (see Fig. 7.1). Based on the experimental characterization of this material, a two-phase model was

Magnetic Small-Angle Neutron Scattering: A Probe for Mesoscale Magnetism Analysis. Andreas Michels, Oxford University Press (2021). © Andreas Michels. DOI: 10.1093/oso/ 9780198855170.003.0007

built, where each magnetic grain consisted of a hard magnetic grain core and an essentially paramagnetic grain shell having a very high susceptibility. The so-called Object Oriented MicroMagnetic Framework (OOMMF) code employing the standard finite-difference method has been used [986], so that a very fine discretization scheme ($0.3 \times 0.3 \times 0.3 \, \text{nm}^3$ cells) had to be applied in order to reproduce the spherical shape of grain cores with a required accuracy. For this reason, only a rather limited number of grains ($\sim$40−60) could be simulated. In addition, the exchange interaction both between the grains and within the soft magnetic matrix (represented by the merging of grain shells) was neglected. Still, using several adjustable parameters, a good agreement between the simulated SANS intensity profile and the experimental neutron data was achieved.

In [471], the OOMMF code was employed in order to simulate the field dependence of the magnetic SANS cross section of a Fe–Si model system and the results were compared to the experimental data of VITROPERM samples [535]. It was concluded that the scattering behavior at large applied fields is dominated by the local anisotropy-field fluctuations which could be well reproduced by the simulations. Interestingly, the authors of [471] also report micromagnetic simulations of magnetic force microscopy (MFM) images. For nanocrystalline hcp Co with an average grain size of 20 nm, their results suggest that the magnetodipolar field resulting from the local magnetic fluctuations induced by the grains should lead to an observable contrast in MFM images.

Zighem et al. [472] have developed a numerical procedure with which it becomes possible to calculate the three-dimensional magnetic form factor of nanosized objects (spheres, cylinders, flat disks) from realistic magnetization distributions obtained by micromagnetic calculations. In their simulations, form factors obtained by Fourier transforming the internal, external, and total magnetic induction field as well as the magnetization were compared.

Grigoriev et al. [479–485] have investigated the spin textures of three-dimensional Ni- and Co-based inverse opal-like structures (IOLS) by using a combination of polarized small-angle neutron diffraction and micromagnetic simulations. For their computations, the finite-element-based code NMag has been employed [987, 988]. IOLS can be considered as a three-dimensional network of ferromagnetic nano-objects (quasi-cubes and quasi-tetrahedra), which are connected to each other via relatively long and thin crosspieces ("legs") [483]. The highly complex microstructure of these lattices, with reported unit-cell sizes in the range between about 640−845 nm, gives rise to characteristic Bragg peaks and to a strong diffuse SANS signal due to the mesoscale spin disorder within the unit cell. Owing to their similarity to spin-ice pyrochlore compounds such as $Dy_2Ti_2O_7$ and $Ho_2Ti_2O_7$, the physics of IOLS has been interpreted in terms of Ising-like magnetic moments situated at the vertices of tetrahedra-shaped structural units.

7.2 Sample microstructures

Illustrations of typical sample microstructures which are used in the micromagnetic SANS simulations of Berkov and Erokhin [58, 473–478] are depicted in Fig. 7.2. The simulation volume of these microstructures is usually discretized into cubical mesh el-

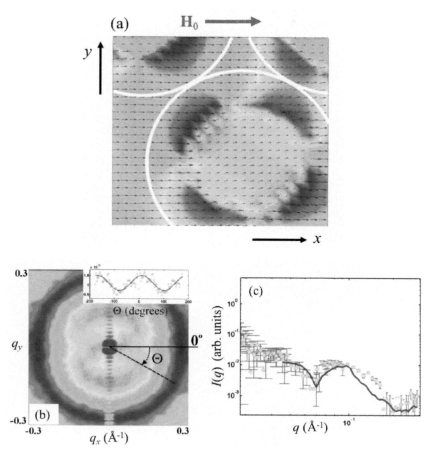

Fig. 7.1: Micromagnetic simulation of magnetic SANS from a Co–Cr–Pt–B-based longitudinal recording medium ($\mu_0 H_0 = 1.45\,\mathrm{T}$). (a) Simulated magnetic flux-density distribution inside and outside the grain cores (compare also to Fig. 7.3). White solid lines indicate grain boundaries. (b) Calculated magnetic scattering intensity $I(\mathbf{q})$ from the magnetic flux-density distribution shown in (a). Inset shows a 90° phase dependence of $I(\theta)$ at $q = 0.05\,\mathrm{\mathring{A}}^{-1}$. (c) Comparison between experimental (data points) and simulated (solid line) magnetic scattering. After [470].

ements with a size of about 1–$2\,\mathrm{nm^3}$, which may be taken as the "resolution limit" of the magnetization vector. The choice for this particular mesh-element size is motivated by the magnitude of the governing micromagnetic exchange lengths; for instance, the magnetostatic exchange length $l_\mathrm{M} = \sqrt{2A/(\mu_0 M_\mathrm{s}^2)}$, which reflects the competition between exchange (A) and self-magnetostatic (M_s) energies, is of the order of 3–$10\,\mathrm{nm}$ for many magnetic materials (compare also to Fig. 3.5) [134]. A particular strength of the simulation approach directly becomes evident by inspecting Fig. 7.2: it is possi-

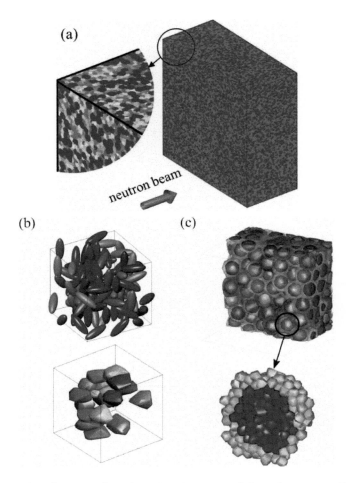

Fig. 7.2: Examples for sample microstructures used in micromagnetic simulations of magnetic SANS. (a) Two-phase nanocomposite sample consisting of a magnetic particle phase (blue polyhedrons) which is embedded into a magnetic matrix phase (yellow-orange-red polyhedrons) (simulation volume: $V = 250 \times 600 \times 600 \, \text{nm}^3$) [58]. (b) Examples of the spatial distribution of hard magnetic crystallites (soft crystallites not shown) with different aspect ratios of corresponding ellipsoids of revolution [477]. (c) Core-shell microstructure used in micromagnetic simulations of a Nd–Fe–B-based nanocomposite and example of a polyhedron mesh-element distribution in a single grain; blue (yellow) elements correspond to the core (shell). The typical mesh-element size is 2 nm [478]. After [64].

ble to take site-dependent magnetic parameters (saturation magnetization, magnetic anisotropy) and interactions (exchange, DMI, magnetodipolar field) into account. This allows one to study most of the relevant classes of magnetic materials (e.g., magnetic nanocomposites, porous ferromagnets, single-phase magnets, magnetic nanoparticles).

Needless to say that site dependency accounts for the polycrystalline nature of magnetic materials, which implies the presence of microstructural defect regions where the magnetic materials parameters may differ from their (single crystalline) bulk values.

Let us discuss this scenario for the case of a two-phase magnetic nanocomposite. Generally, the sources of the magnetodipolar field $\mathbf{H}_d$ are spatially inhomogeneous magnetization distributions, either in orientation and/or in magnitude [cf. eqn (3.27)]. For magnetic nanocomposites, the most prominent magnetic volume charges $\rho_V(\mathbf{r}) = -\nabla \cdot \mathbf{M}(\mathbf{r})$ (see Section 3.1.6) are related to the nanoscale variations in the magnetic materials parameters at the phase boundary between particles and matrix, e.g., variations in the magnetization, anisotropy, or exchange interaction. Such jumps in the magnetic materials parameters may give rise to an inhomogeneous spin structure which decorates each nanoparticle. Figure 7.3 displays the computed real-space magnetization distribution around two single-domain nanoparticles. In this example, the difference in the saturation magnetizations of particle and matrix phase was set to 1.5 T, which is typical for the Fe-based two-phase alloy NANOPERM [446]. It is seen that the jump in M_s gives rise to a large stray-field torque producing spin disorder in the surrounding matrix. We refer to Fig. 1.8 for the corresponding spin distribution around a spherical pore in an iron matrix (see also Fig. 1.9). With reference to the later discussion of the SANS images, we note that the symmetry of the spin structure in Fig. 7.3 replicates the symmetry of the cross term $CT = -(\widetilde{M_y}\widetilde{M_z^*} + \widetilde{M_y^*}\widetilde{M_z})$ in the SANS cross section [compare eqn (7.10) and Fig. 7.9]. In the presence of an applied magnetic field, the stray field and associated magnetization configuration around each nanoparticle "look" similar (on the average), thus giving rise to dipolar correlations which add up to a positive-definite CT contribution to $d\Sigma_M/d\Omega$. In Fourier space, the signature of these dipolar correlations is a clover-leaf-shaped scattering pattern with maxima approximately along the detector diagonals. This angular anisotropy becomes best visible by displaying the spin-misalignment SANS cross section $d\Sigma_{SM}/d\Omega$, which is obtained by subtracting the scattering at saturation (see, e.g., Figs. 7.12 and 7.13). Note, however, that for statistically isotropic polycrystalline microstructures, clover-leaf-type anisotropies only become visible in $d\Sigma_{SM}/d\Omega$ for $\mathbf{k}_0 \perp \mathbf{H}_0$. For $\mathbf{k}_0 \parallel \mathbf{H}_0$, the $d\Sigma_{SM}/d\Omega$ of such systems is isotropic (compare, e.g., Figs. 4.31 and 7.6).

Another example for the geometrical flexibility of the methodology developed by Berkov and Erokhin is the micromagnetic modeling of an Nd–Fe–B nanocomposite, where a core-shell particle microstructure for the description of Nd–Fe–B grains was implemented [478]. In this model, each 20 nm-sized grain is supposed to consist of a magnetically hard core with parameters as for bulk Nd–Fe–B, which is surrounded by a shell with different magnetic parameters. The core-shell model should take into account changes in the magnetic parameters of the Nd–Fe–B crystallites near their surface, which may be imperfect due to the synthesis process. A typical core-shell microstructure used in the simulations is shown in Fig. 7.2(c). In order to resolve the magnetization distribution inside the shell, the mesh-element size was set to 2 nm. The mesh-element structure of a grain is also presented in Fig. 7.2(c). This relatively fine discretization scheme entails longer computational times [as compared to the two-phase nanocomposite system displayed in Fig. 7.2(a)], so that the magnetization-reversal behavior of a sample consisting of only 260 core-shell nanograins was simulated (sample

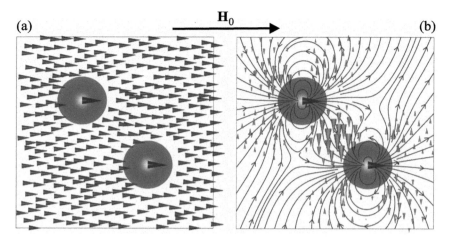

Fig. 7.3: Result of a micromagnetic simulation for the two-dimensional spin distribution around two selected Fe nanoparticles (blue circles, size: $\sim$10 nm), which are assumed to be in a single-domain state ($\mu_0 H_0 = 0.3$ T). (a) Magnetization distribution in both phases. Note that $\mu_0 \Delta M = \mu_0 (M_{\text{hard}} - M_{\text{soft}}) = 1.5$ T. In order to highlight the spin misalignment in the soft magnetic phase, panel (b) displays the magnetization component $\mathbf{M}_\perp$ perpendicular to $\mathbf{H}_0$ (red arrows). Thickness of arrows is proportional to the magnitude of $\mathbf{M}_\perp$. Blue lines: dipolar field distribution. After [58].

volume: $115 \times 115 \times 115$ nm^3). Figure 7.4 depicts the magnetization distribution of the core-shell-type Nd–Fe–B-based nanocomposite at selected points on the hysteresis curve (approach-to-saturation, remanence, coercivity). In the high-field regime, the shells have a larger magnetization along the field direction, since the materials parameters (in particular, the anisotropy constant) are reduced relative to the core regions. This situation prevails down to the remanent state, where qualitatively a similar spin distribution is encountered. However, at the coercive field, one can see that the shells reverse their magnetization easier than the cores, which again can be related to the reduced materials parameters in the shell region. Altogether, the results in Figs. 7.3 and 7.4 clearly reveal that large mesoscale gradients in the spin distribution are common phenomena in nanocomposites, and that these represent a strong contrast for magnetic SANS.

7.3 Simulation methodology and implementation of energy contributions

7.3.1 Methodology

In view of the rather early stage of applying micromagnetic algorithms for SANS investigations, it is certainly fair to say that there is a huge potential for improving the simulation methodology. That is why we restrict ourselves in this section only to the main features. The majority of results which are presented later on in Section 7.4.1 were obtained on nanocomposite magnets [58, 473–478], so that the following discussion

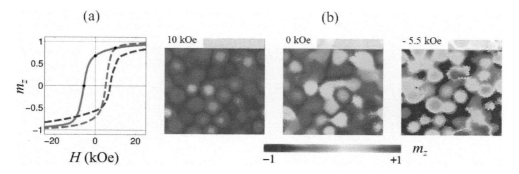

Fig. 7.4: (a) Simulated upper part of the hysteresis loop (green line) and (b) color-coded magnetization distributions (two-dimensional cross section out of a three-dimensional simulation) of a core-shell-type Nd–Fe–B-based nanocomposite at selected points on the hysteresis curve (approach-to-saturation, remanence, coercivity). Contributions to the hysteresis loop originating only from the cores (blue dashed line) and from the shells (red dashed line) are also shown as lower parts in (a). In this particular example, the anisotropy and exchange constants of the shell, as well as the exchange constant between neighboring shells, are reduced relative to the core values (core diameter: $d_{core} = 14.1$ nm). After [478].

is largely focused on this class of magnetic materials (e.g., Fe-based soft magnets, Nd–Fe–B-based permanent magnets, nanopores or precipitates in steels). It is emphasized that nanocomposites are one of the most complicated objects from the point of view of numerical simulations. The main difficulty is that they consist of at least two phases, and the boundaries between these phases are complicated curved surfaces. A typical example is a hard-soft nanocomposite consisting of magnetically hard (i.e., having a large magnetocrystalline anisotropy) crystal grains surrounded by a magnetically soft matrix.

In order to perform accurate and efficient simulations of multiphase magnetic materials, Berkov and Erokhin have developed an algorithm generating a polyhedron mesh with the following properties [64]. (i) It should be possible to represent each hard magnetic crystallite as a single finite element, because the magnetization inside such a crystallite is essentially homogeneous.† (ii) The mesh should allow for an arbitrarily fine discretization of the soft magnetic matrix in-between the hard grains in order to account for the possible large variations of the magnetization direction between the hard grains. (iii) The shape of the meshing polyhedrons should be as close as possible to spherical to ensure a good quality of the spherical dipolar approximation for the calculation of the magnetodipolar interaction energy, even for the nearest-neighboring mesh elements.

A polyhedron mesh satisfying all these requirements and consisting of a relatively large number of finite elements (up to $N \sim 10^6$) can be generated using a physical method, which is based on the model of randomly positioned spheres interacting via

---

†If required, this constraint can be dropped and the hard magnetic grains can be subdivided into finer mesh elements in order to resolve their internal spin structure.

(a)　　　　　　　　　　　　　　　(b)

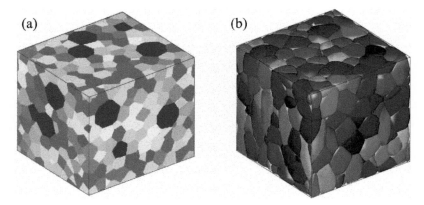

Fig. 7.5: Schematical representation of nanocomposite "samples", which are used for the micromagnetic SANS simulations. (a) Isotropic sample microstructure based on the model of randomly positioned spheres (blue is hard magnetic phase, yellow-orange-red-magenta is soft magnetic phase). After [58]. (b) Anisotropic sample microstructure (red colors are hard magnetic phase, blue colors are soft magnetic phase). Image courtesy of Sergey Erokhin, General Numerics Research Lab, Jena, Germany.

a short-range repulsive potential (see [58] for details). The whole algorithm can be viewed as a means to discretize the sample volume into polyhedrons having nearly spherical shape. This is due to the fact that the polyhedrons "inherit" the spatial structure obtained by the positioning of closely packed spheres. Figure 7.5(a) features an example of a generated sample microstructure based on the model of randomly positioned spheres. The circumstance that the shape of the volume occupied by each finite element, which may be assumed to be homogeneously magnetized, is nearly spherical allows one to use the spherical dipole approximation—equivalent to the point dipole approximation—for the evaluation of the magnetodipolar interaction energy between the various finite elements. Moreover, the micromagnetic methodology for nanocomposite modeling employs periodic boundary conditions, which are routinely applied in simulations of extended thin films and bulk materials in order to eliminate strong finite-size effects. It is also emphasized that the simulation package has been extended to include hard-phase crystallites with a highly anisotropic shape [see Fig. 7.5(b)] [477].

7.3.2 Energy contributions

In the micromagnetic simulations, all four standard contributions to the total magnetic Gibbs free energy are taken into account, i.e., energy in the external magnetic field, energy of the magnetocrystalline anisotropy, exchange-stiffness energy (the antisymmetric DMI can also be implemented), and the magnetodipolar interaction energy. The system energy due to the presence of an external magnetic field and the energy of the magnetocrystalline anisotropy (which can be of uniaxial "u" and/or cubic "c" symmetry) are calculated in the standard way, namely:

$$E_{\mathrm{z}} = -\mu_0 \sum_{i=1}^{N} \boldsymbol{\mu}_i \cdot \mathbf{H}_0, \tag{7.1}$$

$$E_{\mathrm{mc}}^{\mathrm{u}} = -\sum_{i=1}^{N} K_{\mathrm{u}i} \Delta V_i \left(\mathbf{m}_i \cdot \mathbf{n}_i\right)^2, \tag{7.2}$$

$$E_{\mathrm{mc}}^{\mathrm{c}} = \sum_{i=1}^{N} K_{\mathrm{c}i} \Delta V_i \left(m_{i,x'}^2 m_{i,y'}^2 + m_{i,x'}^2 m_{i,z'}^2 + m_{i,y'}^2 m_{i,z'}^2\right), \tag{7.3}$$

where $\mathbf{H}_0$ denotes the external field, $\boldsymbol{\mu}_i = \boldsymbol{\mu}(\mathbf{r}_i)$ and ΔV_i are the magnetic moment and the volume of the i-th finite element (polyhedron), and $\mathbf{m}_i = \boldsymbol{\mu}_i/\mu_i$ denotes the unit magnetization vector. Both the anisotropy constants $K_{\mathrm{u}i}$, $K_{\mathrm{c}i}$ and the directions of the anisotropy axes $\mathbf{n}_i$ can be site-dependent, as required for a polycrystalline material. The symbols $m_{i,x'}$ (and so on) in eqn (7.3) represent the components of the unit magnetization vectors in the local coordinate system that is attached to the cubic anisotropy axes. Higher-order anisotropy contributions can also be taken into account.

The evaluation of the exchange-energy contribution requires a much more sophisticated approach than in the standard finite-difference method, because the continuous integral version of this energy contains magnetization gradients [cf. eqn (3.6)]:

$$E_{\mathrm{ex}} = \int_V A(\mathbf{r}) \left[(\nabla m_x)^2 + (\nabla m_y)^2 + (\nabla m_z)^2\right] d^3 r, \tag{7.4}$$

where A denotes the (site-dependent) exchange-stiffness constant and V is the sample volume. Finding an approximation to eqn (7.4) for a disordered system based on some interpolation procedure preserving the smooth behavior of the magnetization components—required for the correct evaluation of derivatives in eqn (7.4)—is a highly complicated task. For a regular cubic grid with a cell size a and cell volume $\Delta V = a^3$, it can be shown rigorously (see the detailed proof in [989]) that the integral in eqn (7.4) can be approximated by the sum

$$E_{\mathrm{ex}} = -\frac{1}{2} \sum_{i=1}^{N} \sum_{j \subset \mathrm{n.n.}(i)} \frac{2 A_{ij} \Delta V}{a^2} \left(\mathbf{m}_i \cdot \mathbf{m}_j\right). \tag{7.5}$$

Here, A_{ij} denotes the exchange-stiffness constant between cells i and j, and the notation $j \subset \mathrm{n.n.}(i)$ means that the inner summation is performed over the nearest neighbors of the ith cell only. We note in passing that the Heisenberg-like eqn (7.5) is valid only when the angles between neighboring moments are not too large. Neglecting this condition can lead to completely unphysical results [989].

For a disordered system of finite elements having different volumes, different distances between the element centers, and different numbers of nearest neighbors for each element, eqn (7.5) can obviously not be used. Therefore, the following expression for E_{ex}, which is analogous to eqn (7.5), has been introduced [58]:

$$E_{\mathrm{ex}} = -\frac{1}{2} \sum_{i=1}^{N} \sum_{j \subset \mathrm{n.n.}(i)} \frac{2 A_{ij} \Delta \overline{V}_{ij}}{r_{ij}^2} \left(\mathbf{m}_i \cdot \mathbf{m}_j\right), \tag{7.6}$$

where $\Delta\overline{V}_{ij} = (\Delta V_i + \Delta V_j)/2$, r_{ij} is the distance between the centers of the i-th and the j-th finite elements with volumes ΔV_i and ΔV_j, and A_{ij} is the exchange constant. Finally, eqn (7.6) should be corrected by taking into account that the number of nearest neighbors for different finite elements may be different. This can be accomplished by introducing the correction factor $6/n_{av}$, where n_{av} is the average number of nearest neighbors for the particular random realization of the disordered finite-element system [58]. The accuracy of this simple correction method is surprisingly good, as it has been shown by tests presented in [58, 474].

The energy of the long-range magnetodipolar interaction between magnetic moments and the corresponding contribution to the total effective magnetic field are computed using the point-dipole approximation as [154]:

$$E_{\text{dip}} = -\frac{1}{2}\mu_0 \sum_{i=1}^{N} \boldsymbol{\mu}_i \cdot \mathbf{h}_i = -\frac{1}{2}\mu_0 \sum_{i=1}^{N} \boldsymbol{\mu}_i \sum_{j \neq i} \frac{3\mathbf{e}_{ij}\left(\mathbf{e}_{ij} \cdot \boldsymbol{\mu}_j\right) - \boldsymbol{\mu}_j}{r_{ij}^3}, \qquad (7.7)$$

i.e., the magnetic moments of the finite elements are treated as point dipoles located at the polyhedron centers ($\mathbf{e}_{ij} = \mathbf{r}_{ij}/r_{ij}$). This approximation is equivalent to the approximation of spherical dipoles, i.e., it would be exact for spherical finite elements. Hence, for a discretized system, computational errors are introduced, because the finite elements are polyhedrons. However, these errors are small, because the shape of the polyhedrons is close to spherical, due to the special algorithm employed for the generation of the mesh. If necessary, the errors can be significantly reduced further, by taking into account higher-order terms in the multipole expansion. The summation in eqn (7.7) is performed by the so-called particle-mesh Ewald method. The specific implementation of the lattice-based Ewald method for the magnetodipolar interaction for regular and disordered systems of magnetic particles is beyond the scope of this book and described in [990, 991]. The major advantage of the lattice Ewald version is the possibility to use fast Fourier transformation for computing the long-range part of the total magnetodipolar field.

The minimization of the total magnetic energy, obtained as the sum of all the contributions described above, is carried out using a highly optimized version of a gradient method employing the dissipation part of the Landau–Lifshitz equation of motion for magnetic moments [compare eqn (7.13)] [359, 992]. The basic step of this procedure consists of updating the magnetization configuration as:

$$\mathbf{m}_i^{\text{new}} = \mathbf{m}_i^{\text{old}} - \Delta t \left(\mathbf{m}_i^{\text{old}} \times \left[\mathbf{m}_i^{\text{old}} \times \mathbf{h}_i^{\text{eff}}\right]\right), \qquad (7.8)$$

where $\mathbf{h}_i^{\text{eff}} = \mathbf{H}_i^{\text{eff}}/M_s$ is the reduced effective field, evaluated in the standard way as the negative functional energy derivative over the magnetic-moment projections [344], and Δt is the step in the minimization method. After a certain number of successful basic steps, a specific extrapolation of magnetic-moment projections is performed, which leads to a strong acceleration of the minimization process. For the termination of the energy minimization, the local torque criterion is used: the iteration process is stopped, if the maximal torque acting on magnetic moments is smaller than some prescribed value, i.e.,

$$\max_{\{i\}} \left|\mathbf{m}_i \times \mathbf{h}_i^{\text{eff}}\right| < \epsilon. \qquad (7.9)$$

This condition is more appropriate than the alternative criterion of a sufficiently small energy difference between two subsequent steps. Typically, a value of $\epsilon = 10^{-4} - 10^{-3}$ is found to be small enough to ensure the convergence of the minimization procedure.

After the equilibrium magnetization distribution $M_{x,y,z}(\mathbf{r})$ has been calculated using the equation-of-motion minimization technique, the magnetic SANS cross section can be computed in terms of the Fourier components $\widetilde{M}_{x,y,z}(\mathbf{q})$ of the magnetization configuration [compare, e.g., eqns (7.10) and (7.11)]. The Fourier components of $\mathbf{M}(\mathbf{r})$ for a disordered system can be obtained in the most efficient way by mapping (interpolating) the spin distribution onto a regular grid, which allows the usage of the fast Fourier transformation technique [474]. In order to compare the numerical results with experiment, it is necessary to appropriately evaluate the scattering cross section in the plane of the detector. For this purpose, three approaches have been followed in the literature. (i) In [58,473–475], an averaging procedure which takes into account magnetic fluctuations also along the direction of the incident neutron beam has been employed. Without loss of generality, we restrict the following considerations to the transversal scattering geometry ($\mathbf{k}_0 \perp \mathbf{H}_0$), where $\mathbf{k}_0 \parallel \mathbf{e}_x$ and $\mathbf{H}_0 \parallel \mathbf{e}_z$. The computed (mapped on a regular grid) three-dimensional magnetization distribution of the sample $\mathbf{M}(x, y, z)$ is divided into thin slices $i = 1, \ldots, N_x$ with a typical thickness of 2.5 nm. The slicing of the sample is done along the direction of the incident beam. This results in a set of magnetization distributions $\mathbf{M}^{(i)}(y, z) = \mathbf{M}(x_i = \text{constant}, y, z)$, which are then summed up and divided by the number N_x of slices in order to obtain the averaged real-space quantities. Subsequent two-dimensional Fourier transformation yields the $\widetilde{\mathbf{M}}(q_y, q_z)$, which are used to compute the SANS cross section. (ii) In [476], the three-dimensional spin structure $\mathbf{M}(x, y, z)$ is also subdivided into thin slices $i = 1, \ldots, N_x$ along the direction of the incident beam. However, in contrast to case (i), the resulting $\mathbf{M}^{(i)}(y, z)$ are first Fourier-transformed to yield the $\widetilde{\mathbf{M}}^{(i)}(q_y, q_z)$, and the squared Fourier coefficients $|\widetilde{M}_x^{(i)}|^2$ (and so on) are summed up and divided by N_x; in other words, in (ii) the averaging is done in Fourier space, whereas in (i) it is done in real space. (iii) In [64], three-dimensional Fourier transformation is applied to $\mathbf{M}(x, y, z)$ to numerically obtain the corresponding $\widetilde{\mathbf{M}}(q_x, q_y, q_z)$. Then, data on the (q_y, q_z) plane at $q_x = 0$ is used, corresponding to one single slice (compare also Fig. 7.10). Procedure (i) results for $\mathbf{k}_0 \perp \mathbf{H}_0$ in an isotropic $|\widetilde{M}_x|^2$ Fourier component [see Fig. 7.8(a)], as does procedure (iii) (see Fig. 7.10) and the analytical theory of Section 3.2, while procedure (ii) gives an anisotropic $|\widetilde{M}_x|^2$ (which is enhanced along the field direction [476]). Since $|q_x| = q \sin(\psi/2)$, future work should therefore establish the appropriate range of q_x-values (number of slices) for computing the Fourier components.

7.4 Decrypting the SANS cross section

7.4.1 Nanocomposite bulk magnets

In the following, we will discuss micromagnetic simulations of magnetic SANS from two-phase nanocomposites [58]. Most of the results were obtained on the sample microstructure which is depicted in Fig. 7.2(a) and which is characteristic for Fe-based two-phase nanocomposite magnets from the NANOPERM family of alloys [993]. This material consists of a distribution of crystalline iron nanoparticles in a soft magnetic

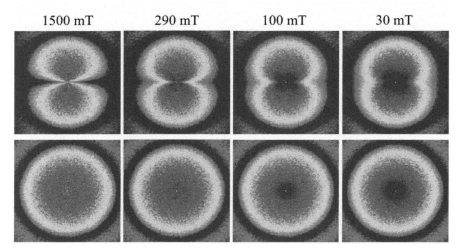

Fig. 7.6: Applied-field dependence of the total magnetic SANS cross section $d\Sigma_{\mathrm{M}}/d\Omega$ for $\mathbf{k}_0 \perp \mathbf{H}_0$ [eqn (7.10), upper row] and for $\mathbf{k}_0 \parallel \mathbf{H}_0$ [eqn (7.11), lower row]. The external magnetic field $\mathbf{H}_0 \parallel \mathbf{e}_z$ is applied horizontally in the plane of the detector for $\mathbf{k}_0 \perp \mathbf{H}_0$ ($q_x = 0$, upper row) and normal to the detector plane for $\mathbf{k}_0 \parallel \mathbf{H}_0$ ($q_z = 0$, lower row). Pixels in the corners of the images have $q \cong 1.2\,\mathrm{nm}^{-1}$. After [58].

amorphous matrix. In the micromagnetic simulations (see, e.g., Fig. 7.3), the sizes of the "hard" magnetic Fe particles are distributed according to a Gaussian function centered at about $10\,\mathrm{nm}$, and the particle volume fraction equals 40%, as in the experimental study [446]. For later reference, the following materials parameters were employed: saturation magnetizations $M_{\mathrm{hard}} = 1750\,\mathrm{kA/m}$ and $M_{\mathrm{soft}} = 550\,\mathrm{kA/m}$, anisotropy constants $K_{\mathrm{hard}} = 4.6 \times 10^4\,\mathrm{J/m^3}$ and $K_{\mathrm{soft}} = 1.0 \times 10^2\,\mathrm{J/m^3}$, and an exchange-stiffness constant of $A = 2.0 \times 10^{-12}\,\mathrm{J/m}$ for exchange interactions both within the soft phase and between the hard and soft phases. The difference in the saturation magnetizations between particle and matrix phase gives rise to a large jump of the magnetization magnitude at particle-matrix interfaces, $\mu_0 \Delta M = \mu_0(M_{\mathrm{hard}} - M_{\mathrm{soft}}) = 1.5\,\mathrm{T}$. The nuclear SANS contribution is ignored in these simulations. Note, however, that for polycrystalline texture-free magnetic nanocomposites, the nuclear SANS signal is independent of the applied magnetic field and isotropic, and its magnitude is generally small compared to the here-relevant spin-misalignment scattering [49].

Figure 7.6 displays the applied-field dependence of the two-dimensional magnetic SANS cross sections (including the respective $|\widetilde{M}_z|^2$ terms) for the perpendicular and parallel scattering geometries. For $\mathbf{k}_0 \perp \mathbf{H}_0$, the unpolarized elastic magnetic SANS cross section equals

$$\frac{d\Sigma_{\mathrm{M}}}{d\Omega}(\mathbf{q}) = \frac{8\pi^3}{V} b_{\mathrm{H}}^2 \left(|\widetilde{M}_x|^2 + |\widetilde{M}_y|^2 \cos^2\theta + |\widetilde{M}_z|^2 \sin^2\theta \right.$$
$$\left. -(\widetilde{M}_y \widetilde{M}_z^* + \widetilde{M}_y^* \widetilde{M}_z) \sin\theta \cos\theta \right), \tag{7.10}$$

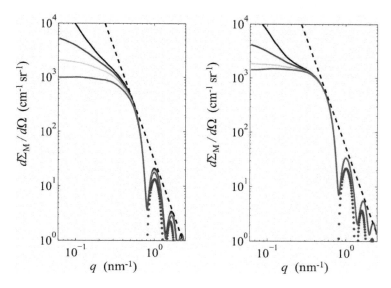

Fig. 7.7: Solid lines: azimuthally averaged $d\Sigma_{\mathrm{M}}/d\Omega$ as a function of scattering vector q for $\mathbf{k}_0 \perp \mathbf{H}_0$ (left image) and for $\mathbf{k}_0 \parallel \mathbf{H}_0$ (right image) (data have been smoothed) (log-log scale). Field values (in mT) from top to bottom, respectively: 30, 100, 290, 1500. Dashed lines in both images: $d\Sigma_{\mathrm{M}}/d\Omega \propto q^{-4}$. Solid circles in both images represent part of the form factor of a sphere with a radius of $R = 5.7\,\mathrm{nm}$. After [58].

whereas for $\mathbf{k}_0 \parallel \mathbf{H}_0$ we have

$$\frac{d\Sigma_{\mathrm{M}}}{d\Omega}(\mathbf{q}) = \frac{8\pi^3}{V} b_{\mathrm{H}}^2 \left(|\widetilde{M}_x|^2 \sin^2\theta + |\widetilde{M}_y|^2 \cos^2\theta + |\widetilde{M}_z|^2 \right. $$
$$\left. -(\widetilde{M}_x \widetilde{M}_y^* + \widetilde{M}_x^* \widetilde{M}_y)\sin\theta\cos\theta \right). \tag{7.11}$$

The corresponding azimuthally averaged data can be seen in Fig. 7.7. While $d\Sigma_{\mathrm{M}}/d\Omega$ for $\mathbf{k}_0 \parallel \mathbf{H}_0$ is isotropic (i.e., θ-independent) over the whole field and momentum-transfer range (Fig. 7.6, lower row), it is highly anisotropic for $\mathbf{k}_0 \perp \mathbf{H}_0$ (Fig. 7.6, upper row). At a saturating applied magnetic field of $\mu_0 H_0 = 1.5\,\mathrm{T}$, where the normalized sample magnetization is larger than 99.9%, the anisotropy of $d\Sigma_{\mathrm{M}}/d\Omega$ is for $\mathbf{k}_0 \perp \mathbf{H}_0$ clearly of the $\sin^2\theta$-type, i.e., elongated normal to $\mathbf{H}_0$. This is because magnetic scattering due to transversal spin misalignment is small for fields very close to saturation and the dominant scattering contrast arises from nanoscale jumps of the longitudinal magnetization at phase boundaries. On decreasing the field, the transversal magnetization components increase in magnitude as long-range spin misalignment develops at the smallest q. The SANS pattern for $\mathbf{k}_0 \perp \mathbf{H}_0$ essentially remains of the $\sin^2\theta$-type at lower fields, although a more complicated anisotropy builds up at small q. As can be seen in Fig. 7.7, $d\Sigma_{\mathrm{M}}/d\Omega$ at small q increases by more than one order of magnitude as the field is decreased from 1.5 T to 30 mT. Asymptotically, at large q, the power-law dependence of $d\Sigma_{\mathrm{M}}/d\Omega$ can be described by $d\Sigma_{\mathrm{M}}/d\Omega \propto q^{-4}$.

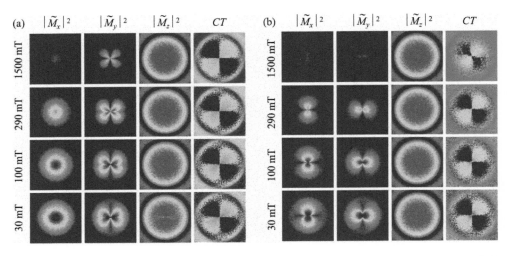

Fig. 7.8: Results of the micromagnetic simulations for the Fourier components of the magnetization. The images represent projections of the respective functions into the plane of the detector, i.e., $q_x = 0$ for $\mathbf{k}_0 \perp \mathbf{H}_0$ (a) and $q_z = 0$ for $\mathbf{k}_0 \parallel \mathbf{H}_0$ (b). From left column to right column, respectively: $|\widetilde{M}_x|^2$, $|\widetilde{M}_y|^2$, $|\widetilde{M}_z|^2$, and $CT = -(\widetilde{M}_y \widetilde{M}_z^* + \widetilde{M}_y^* \widetilde{M}_z)$ (a) and $CT = -(\widetilde{M}_x \widetilde{M}_y^* + \widetilde{M}_x^* \widetilde{M}_y)$ (b). In the first three columns from left, red color corresponds, respectively, to high intensity and blue color to low intensity. In the fourth column, blue color corresponds to negative and orange-yellow color to positive values of the CT. Pixels in the corners of the images have $q \cong 1.2 \, \text{nm}^{-1}$. After [58].

In agreement with the nature of the underlying microstructure, $\sim$10 nm-sized single-domain iron particles embedded in a nearly saturated matrix, one can describe the oscillations of $d\Sigma_\text{M}/d\Omega$ at large q and H_0 by the form factor of a sphere with a radius of $R = 5.7$ nm (solid circles in Fig. 7.7). We remind that the shape of the particles in the micromagnetic algorithm is not strictly spherical.

Figure 7.8 shows for both scattering geometries the projections of the magnetization Fourier components $|\widetilde{M}_x|^2$, $|\widetilde{M}_y|^2$, $|\widetilde{M}_z|^2$, and of the cross terms $CT = -(\widetilde{M}_y \widetilde{M}_z^* + \widetilde{M}_y^* \widetilde{M}_z)$ and $CT = -(\widetilde{M}_x \widetilde{M}_y^* + \widetilde{M}_x^* \widetilde{M}_y)$ into the plane of the two-dimensional detector at the same external-field values as in Fig. 7.6. It can be seen in Fig. 7.8(a) that for $\mathbf{k}_0 \perp \mathbf{H}_0$ both $|\widetilde{M}_x|^2$ and $|\widetilde{M}_z|^2$ are isotropic over the displayed (q, H_0)-range, while the Fourier component $|\widetilde{M}_y|^2$ reveals the clover-leaf anisotropy. For $\mathbf{k}_0 \parallel \mathbf{H}_0$ [Fig. 7.8(b)], $|\widetilde{M}_x|^2$ and $|\widetilde{M}_y|^2$ are both strongly anisotropic with characteristic maxima in the plane perpendicular to $\mathbf{H}_0$, while $|\widetilde{M}_z|^2$ is isotropic, as expected for isotropic microstructures. When for $\mathbf{k}_0 \parallel \mathbf{H}_0$ all Fourier components are multiplied by the corresponding trigonometric functions and summed up [according to eqn (7.11)], then the resulting $d\Sigma_\text{M}/d\Omega$ becomes isotropic (Fig. 7.6, lower row).

The cross terms for both scattering geometries vary in sign between the quadrants on the detector (Fig. 7.8). The respective CT is positive in the upper right quadrant of the detector ($0° < \theta < 90°$), negative in the upper left quadrant ($90° < \theta < 180°$), and so on. When both CTs are multiplied by $\sin\theta \cos\theta$, the corresponding contribution

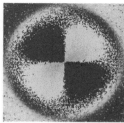

Fig. 7.9: (left image) $CT = -(\widetilde{M}_y\widetilde{M}_z^* + \widetilde{M}_y^*\widetilde{M}_z)$ for $\mathbf{k}_0 \perp \mathbf{H}_0$ ($\mu_0 H_0 = 0.29$ T). Blue color corresponds to negative and orange-yellow color to positive values of the CT. (right image) $CT \sin\theta\cos\theta$ [compare eqn (7.10)]. Red color corresponds to high intensity and blue color to low intensity. All other settings are as in Fig. 7.8(a). After [58].

to $d\Sigma_M/d\Omega$ becomes positive-definite for all angles θ [compare eqns (7.10) and (7.11) and Fig. 7.9]. Since for the perpendicular scattering geometry the CT is related to correlated M_y-M_z fluctuations, this observation suggests that the CT appears to be of special relevance for strongly inhomogeneous multiphase nanocomposite magnets. This is contrary to the common assumption that the CT averages to zero for statistically isotropic polycrystalline microstructures.

It is worth emphasizing that both Fourier components $|\widetilde{M}_x|^2(q_x, q_y, q_z)$ and $|\widetilde{M}_y|^2(q_x, q_y, q_z)$ do coincide with respect to rotation around the q_z-axis, because the only symmetry breaking in the system is due to the external magnetic field ($\mathbf{H}_0 \parallel \mathbf{e}_z$). Figure 7.10(a) and (b) depict three-dimensional images of the Fourier component $|\widetilde{M}_x|^2(q_x, q_y, q_z)$, derived from the spatial magnetization distribution $M_x(x, y, z)$, at selected values of q_x and q_y. It is seen in Fig. 7.10(a) that $|\widetilde{M}_x|^2$ is isotropic in the plane $q_x = 0$ [compare to Fig. 7.8(a)], while it becomes strongly anisotropic at larger q_x. In the q_x-q_z-plane, $|\widetilde{M}_x|^2$ also exhibits the clover-leaf pattern [see Fig. 7.10(b)].

In Fig. 7.11, we show for the $\mathbf{k}_0 \perp \mathbf{H}_0$ scattering geometry the azimuthally averaged total $d\Sigma_M/d\Omega$ along with the azimuthally averaged individual scattering contributions to $d\Sigma_M/d\Omega$, i.e., the azimuthal average of terms $\frac{8\pi^3}{V}b_H^2|\widetilde{M}_x|^2$, $\frac{8\pi^3}{V}b_H^2|\widetilde{M}_y|^2\cos^2\theta$, $\frac{8\pi^3}{V}b_H^2|\widetilde{M}_z|^2\sin^2\theta$, and $-\frac{8\pi^3}{V}b_H^2(\widetilde{M}_y\widetilde{M}_z^* + \widetilde{M}_y^*\widetilde{M}_z)\sin\theta\sin\theta$ [compare eqn (7.10)]. At saturation [$\mu_0 H_0 = 1.5$ T, Fig. 7.11(c)], both transversal scattering contributions, i.e., terms $\propto |\widetilde{M}_x|^2$ and $\propto |\widetilde{M}_y|^2$, are small relative to the other terms and the main contribution to the total $d\Sigma_M/d\Omega$ originates from longitudinal magnetization fluctuations, i.e., from terms $\propto |\widetilde{M}_z|^2$. Note that the CT is the product of a transversal and the longitudinal magnetization Fourier component. On decreasing the field [Fig. 7.11(b) and (a)], the transversal Fourier components and the CT become progressively more important, in particular at small q.

The finding (for $\mathbf{k}_0 \perp \mathbf{H}_0$) that—on the detector plane—$|\widetilde{M}_x|^2$ and $|\widetilde{M}_z|^2$ are isotropic and $|\widetilde{M}_y|^2 = |\widetilde{M}_y|^2(q, \theta)$ is highly anisotropic provides a straightforward explanation for the experimental observation of the clover-leaf anisotropy in the SANS data of the nanocrystalline two-phase alloy NANOPERM [446]. The simulation results for the spin-misalignment SANS cross section $d\Sigma_{SM}/d\Omega \propto (|\widetilde{M}_x|^2 + |\widetilde{M}_y|^2\cos^2\theta +$

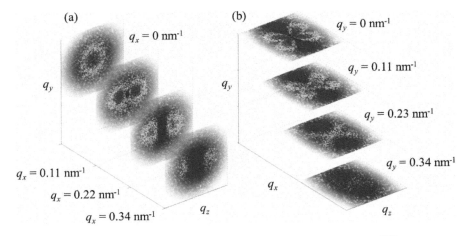

Fig. 7.10: Two-dimensional cuts (shifted for a better visibility) of $|\widetilde{M}_x|^2(q_x, q_y, q_z)$ at selected points in the q_y-q_z-plane (a) and in the q_x-q_z-plane (b). After [64].

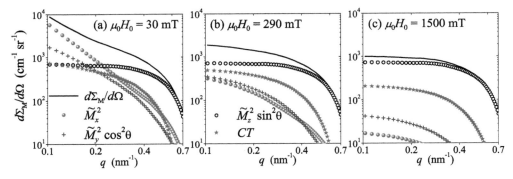

Fig. 7.11: Azimuthally averaged total magnetic SANS cross section $d\Sigma_{\mathrm{M}}/d\Omega$ (solid line) and azimuthally averaged individual scattering contributions to $d\Sigma_{\mathrm{M}}/d\Omega$ [eqn (7.10)] as a function of the scattering vector q and applied magnetic field H_0 (see inset) (log-log scale) ($\mathbf{k}_0 \perp \mathbf{H}_0$). After [58].

$CT \sin\theta \cos\theta$) (see Fig. 7.12), where the scattering at saturation ($\mu_0 H_0 = 1.5$ T) has been subtracted, agree qualitatively very well with the experimental data [473, 474]. Clover-leaf-type anisotropies in $d\Sigma_{\mathrm{SM}}/d\Omega$ have also been reported for a number of other materials, including precipitates in steels [121], nanocrystalline Gd [447, 449], and nanoporous Fe [448]. The maxima in the spin-misalignment SANS cross section depend for $\mathbf{k}_0 \perp \mathbf{H}_0$ on q and H_0, and may appear at angles θ significantly smaller than 45°. This becomes evident in Fig. 7.13, where we show polar plots of the simulated $d\Sigma_{\mathrm{SM}}/d\Omega$ at selected q and H_0. The simulation results in Fig. 7.13 agree qualitatively with the analytical SANS theory, which predicts a field-induced transition from anisotropy-field-dominated to magnetostatic scattering that is accompanied by a change in the angular anisotropy of $d\Sigma_{\mathrm{SM}}/d\Omega$ (compare Fig. 4.3, upper row of Fig. 4.7, and Fig. 4.9).

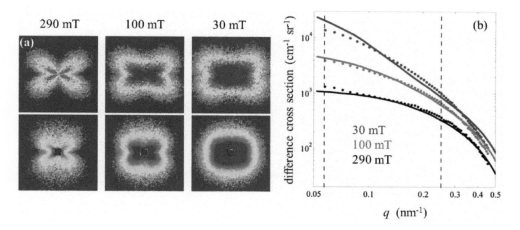

Fig. 7.12: (a) Comparison between simulation (upper row) and experimental data (lower row) for the spin-misalignment SANS cross section $d\Sigma_{\rm SM}/d\Omega \propto (|\widetilde{M}_x|^2 + |\widetilde{M}_y|^2 \cos^2\theta + CT\sin\theta\cos\theta)$ of NANOPERM at different external fields as indicated ($\mathbf{k}_0 \perp \mathbf{H}_0$). Pixels in the corners of the images have $q \cong 0.64\,{\rm nm}^{-1}$. $\mathbf{H}_0$ is horizontal in the plane. (b) (•) Corresponding azimuthally averaged data. Field values (in mT) from top to bottom: 30, 100, 290. Solid lines: results of the micromagnetic simulations (data have been smoothed). Vertical dashed lines indicate the region where the clover-leaf anisotropy is observed. Since the experimental data [446] were not obtained in absolute units, they were scaled with a factor for comparison with the simulated data. After [58].

We refer to Fig. 4.10 in Section 4.3.2 which highlights the impact of the magnetodipolar interaction on the magnetic SANS cross section. When the dipolar interaction is ignored in the micromagnetic computations, all Fourier components are isotropic at all q and H_0 investigated (data for $|\widetilde{M}_x|^2$ and $|\widetilde{M}_z|^2$ are not shown in Fig. 4.10). This observation demonstrates that for any realistic description of experimental magnetic SANS data from nonuniformly magnetized structures, the long-range and anisotropic magnetostatic interaction has to be taken into account.

As mentioned above, not only variations in the magnetization magnitude, but also variations in the direction and/or magnitude of the magnetic anisotropy field $\mathbf{H}_{\rm K}$ and variations in the magnitude of the exchange coupling may give rise to dipolar correlations. The micromagnetic simulation package allows one to vary the magnetic materials parameters of both phases of the nanocomposite. Hence, it becomes possible to study the impact of such situations on the magnetic SANS cross section. In order to investigate variations in $\mathbf{H}_{\rm K}$, which are by construction naturally included into the micromagnetic algorithm (random anisotropy), the spin distribution for the situation that $M_{\rm hard} = M_{\rm soft} = M_{\rm s}$ (i.e., $\Delta M = 0$) was computed, but for different values of $M_{\rm s}$. Figure 7.14 reveals that a clover-leaf-type pattern in $|\widetilde{M}_y|^2$ develops with increasing saturation-magnetization value $M_{\rm s}$, i.e., with increasing strength of the magnetodipolar interaction. As jumps in the magnetization at phase boundaries are now excluded

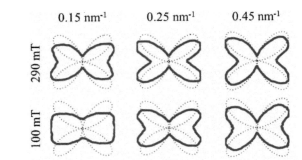

Fig. 7.13: Polar plots of the simulated spin-misalignment SANS cross section $d\Sigma_{\mathrm{SM}}/d\Omega \propto (|\widetilde{M}_x|^2 + |\widetilde{M}_y|^2 \cos^2\theta + CT\sin\theta\cos\theta)$ at different combinations of momentum transfer q and applied magnetic field H_0 (see insets) ($\mathbf{k}_0 \perp \mathbf{H}_0$). Data have been smoothed. Dotted lines ($\propto \sin^2\theta\cos^2\theta$) serve as a guide to the eyes. After [58].

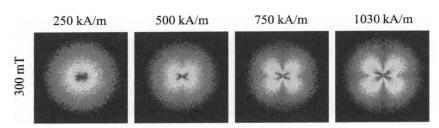

Fig. 7.14: Fourier component $|\widetilde{M}_y|^2(\mathbf{q})$ at $\mu_0 H_0 = 300\,\mathrm{mT}$ and for $M_{\mathrm{hard}} = M_{\mathrm{soft}} = M_{\mathrm{s}}$ (i.e., $\Delta M = 0$) ($\mathbf{k}_0 \perp \mathbf{H}_0$). M_{s} increases from left to right (see insets). $K_{\mathrm{hard}} = 4.6 \times 10^4\,\mathrm{J/m}^3$, $K_{\mathrm{soft}} = 1.0 \times 10^2\,\mathrm{J/m}^3$, and random variations in easy-axis directions from particle to particle are assumed. After [475].

as possible sources for perturbations in the spin structure, it is straightforward to conclude that nanoscale fluctuations in $\mathbf{H}_{\mathrm{K}}$ give rise to inhomogeneous magnetization states, which decorate each nanoparticle and which look similar to the structure shown in Fig. 7.14. This finding strongly suggests that the origin of the clover-leaf pattern in $d\Sigma_{\mathrm{SM}}/d\Omega$ of nanomagnets is not only related to variations in the magnetization magnitude but also due to variations in the magnitude and direction of the magnetic anisotropy field.

7.4.2 Magnetic nanoparticles

As discussed in Section 5.1, for magnetic nanoparticle assemblies the conventional theory of magnetic SANS, embodied by eqn (5.1), assumes uniformly magnetized nanoparticles (macrospin model). However, there exist many experimental and theoretical studies which suggest that this assumption is violated; for instance, the presence of crystalline defects, surface anisotropy, or, more generally speaking, the interplay between various magnetic interactions (exchange, magnetic anisotropy, magnetostatics, external field) may lead to nonuniform spin structures. Numerical micromagnetic simulations provide a guide to the analysis of experimental data. Here, we will discuss some selected results on an isolated cobalt nanowire [162]. For results on inhomogeneously magnetized spherical iron nanoparticles, we refer to Section 1.6.2 and to [163].

The simulations on the Co cylinder were carried out using the GPU-based open-source software package MuMax3 [166, 167], which can calculate the space and time-dependent magnetic microstructure of nanosized and micron-sized ferromagnets. Mumax3 employs a finite-difference discretization scheme of space using a two-dimensional or three-dimensional grid of orthorhombic cells. In the micromagnetic simulations all standard contributions to the total magnetic Gibbs free energy can be taken into account, i.e., energy in the external magnetic field, magnetodipolar interaction energy, energy of the magnetocrystalline anisotropy, isotropic and symmetric exchange energy, as well as the antisymmetric DMI. In the following, we discuss the spin structures and related magnetic SANS cross sections in the remanent state as a function of the aspect ratio D_{cyl}/L_{cyl} of the cylinder. The length of the nanorod is fixed at $L_{cyl} = 500\,\text{nm}$ and the diameter is varied between $D_{cyl} = 30, 60$, and $90\,\text{nm}$. The long axis of the nanorod is parallel to the incident neutron beam. As is well known, for such shape-anisotropic particles, the long-range magnetodipolar interaction gives rise to magnetic shape anisotropy, which tends to align the magnetic moments along the nanorod's long axis. On top of that, we consider an uniaxial magnetocrystalline anisotropy axis $\mathbf{K}_u$, which is directed along the cylinder's diameter (perpendicular to the long axis). Consequently, the system behaves effectively like a biaxial magnetic material and the spin structure reflects the competition between both interactions; in particular, as D_{cyl} increases at constant L_{cyl} and at constant magnitude and direction of $\mathbf{K}_u$.

Computed real-space spin structures of the $D_{cyl} = 30\,\text{nm}$ and the $D_{cyl} = 90\,\text{nm}$ nanowires are depicted in Fig. 7.15. Shown are the $\mathbf{m}(\mathbf{r})$ in the remanent state. Note also that the nanorod was initially saturated by an applied field directed along the z-direction (along the diameter), which coincides with the uniaxial magnetic easy-axis direction. For $D_{cyl} = 30\,\text{nm}$ and a corresponding relatively small ratio of $D_{cyl}/L_{cyl} = 0.06$ [Fig. 7.15(a)], the spin structure is determined by the magnetic shape anisotropy. When $H_0 = H_z$ is reduced starting from saturation, the magnetization rotates towards the shape-anisotropy axis. At remanence, the magnetization is then predominantly oriented along the x-direction, with some minor spin misalignment along the z-direction at the nanowire's end faces, and a vanishingly small magnetization component along the y-direction. The localized spin inhomogeneities at the end faces are related to the combined effects of the inhomogeneous dipole field of the rod, which tends to align the moments parallel to the surface of the nanorod's ends, the exchange interaction, which tries to keep the ferromagnetic order, and the magnetocrystalline anisotropy, which prefers the magnetization along the $\pm z$-direction. When D_{cyl} increases, at constant L_{cyl} and K_u, the dipolar interaction favors more complex magnetic structures such as domain wall or vortex states [162], which minimize the magnetic flux through the surface of the system. An example for a vortex-like

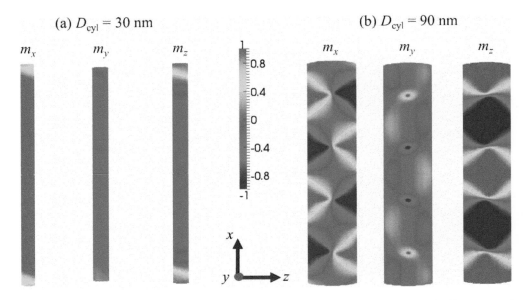

Fig. 7.15: Color-coded maps of the computed real-space spin structures of a Co nanorod with a length of $L_{cyl} = 500$ nm and diameters of (a) $D_{cyl} = 30$ nm and (b) $D_{cyl} = 90$ nm (remanent state). The magnetization distributions $\mathbf{m}(x, y, z)$ are identical in all three images in (a) and (b), respectively. The color scale encodes the different Cartesian components of $\mathbf{m}$. The uniaxial magnetocrystalline anisotropy axis ($K_u = 4.5 \times 10^5$ J/m^3) is directed along the $\pm z$-direction, which coincides with the direction of the prior saturation field. After [162].

configuration is displayed in Fig. 7.15(b), where the $\mathbf{m}(\mathbf{r})$ of the $D_{cyl} = 90$ nm nanorod is shown ($D_{cyl}/L_{cyl} = 0.18$). Indeed, the spin texture which stabilizes the system at remanence is a combination of different vortices (in those structures the magnetization is confined in the x–z-plane) with their cores aligned parallel and antiparallel to the y-direction; in other words, the vortex cores are oriented perpendicular to the plane that is spanned by the uniaxial and shape-anisotropy axes.

The Fourier components of the magnetization and the magnetic SANS cross sections of the cylinders' spin structures are displayed in Figs. 7.16 and 7.17 at the remanent state. This representation decrypts the magnetic SANS cross sections and highlights the magnitudes and q-dependencies of the different scattering contributions. The data for the individual Fourier components in Fig. 7.16 also reveal that these functions may explicitly depend on the angle θ in the detector plane, as is the case with the nanocomposite bulk magnets (see Section 7.4.1). Based on the real-space observations in Fig. 7.15, the corresponding reciprocal-space SANS cross section and the magnetization Fourier components of the $D_{cyl} = 30$ nm nanorod can be understood: $d\Sigma_M/d\Omega$ is dominated by the $|\widetilde{M}_x|^2$ Fourier component and contains a weak $|\widetilde{M}_z|^2$ contribution. The cross term $CT = -(\widetilde{M}_y\widetilde{M}_z^* + \widetilde{M}_y^*\widetilde{M}_z)$ is negligible, since the $\widetilde{M}_y$ component is very small. The azimuthally averaged $d\Sigma_M/d\Omega$ of the $D_{cyl} = 30$ nm nanorod can be well

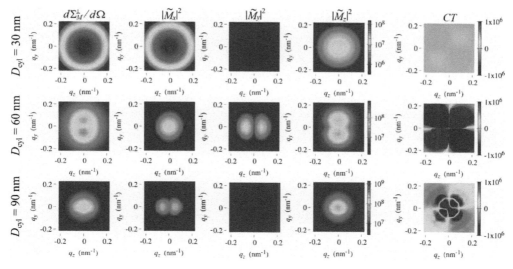

Fig. 7.16: Results of micromagnetic simulations for the total magnetic SANS cross section and the Fourier components of the magnetization of a single Co nanorod in the remanent state ($\mathbf{k}_0 \perp \mathbf{H}_0$). The long axis of the nanorod is parallel to the incident neutron beam. The images represent projections of $d\Sigma_\mathrm{M}/d\Omega$, $|\widetilde{M_x}|^2$, $|\widetilde{M_y}|^2$, $|\widetilde{M_z}|^2$ (all on a logarithmic color scale) and of the cross term $CT = -(\widetilde{M_y}\widetilde{M_z^*} + \widetilde{M_y^*}\widetilde{M_z})$ (linear color scale) into the plane of the two-dimensional detector ($q_x = 0$). $\mathbf{H}_0$ is horizontal in the plane. $D_\mathrm{cyl} = 30\,\mathrm{nm}$ (upper row); $D_\mathrm{cyl} = 60\,\mathrm{nm}$ (middle row); $D_\mathrm{cyl} = 90\,\mathrm{nm}$ (lower row). After [162].

described by eqn (5.1) [compare solid line in Fig. 7.17(a)], which assumes a uniformly magnetized particle.

Increasing the nanorod's diameter results in the emergence of a variety of magnetization switching processes and complex inhomogeneous states (see, e.g., [994–998] for further reading). While for $D_\mathrm{cyl} = 30\,\mathrm{nm}$ the shape anisotropy determines the magnetic configuration at remanence [Fig. 7.15(a)], changing the diameter to $D_\mathrm{cyl} = 60\,\mathrm{nm}$ ($D_\mathrm{cyl}/L_\mathrm{cyl} = 0.12$) reduces the strength of the shape anisotropy and the competition with the magnetocrystalline anisotropy leads to a change in the switching mechanism from essentially coherent rotation to the nucleation and propagation of a transversal domain wall [162]. The domain oriented along the x-direction grows by propagating a transversal domain wall at the lateral boundaries of the cylinder. Therefore, the remanent state is characterized by this transversal domain wall. The related SANS cross section (middle row in Fig. 7.16) can be seen as a combination of all the three magnetization Fourier components. Note also that the CT is now increased in magnitude as compared to the $D_\mathrm{cyl} = 30\,\mathrm{nm}$ case (due to an increased $\widetilde{M_y}$ contribution). Notably, the circular symmetry of the magnetic SANS cross section observed at remanence for the smallest diameter has now turned into a slightly vertically elongated pattern due to the existence of magnetization components perpendicular to the rod's axis, which directly result from the transversal domain-wall configuration.

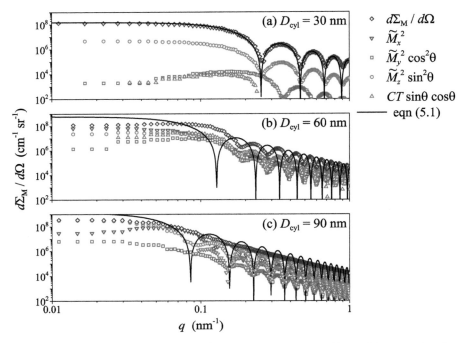

Fig. 7.17: 2π-azimuthally averaged remanent-state data (from Fig. 7.16) of a Co nanorod of 500 nm length and different diameters D_{cyl} ($\mathbf{k}_0 \perp \mathbf{H}_0$) (log-log scale). (a) $D_{cyl} = 30$ nm; (b) $D_{cyl} = 60$ nm; (c) $D_{cyl} = 90$ nm. Solid lines in (a)−(c): magnetic contribution to eqn (5.1) using the circular-disc form factor $F^2(q) = 4J_1^2(qR)/(qR)^2$. Note that $d\Sigma_M/d\Omega$ and $|\widetilde{M}_x|^2$ nearly overlap in (a). After [162].

When the diameter is further increased to $D_{cyl} = 90$ nm [Fig. 7.15(b)], there is again a change in the mechanism from the nucleation and propagation of transversal domain-wall-like structures to vortex-like configurations. Here, the $|\widetilde{M}_y|^2$ Fourier component is small in magnitude and the SANS cross section (lower row in Fig. 7.16) is dominated by $|\widetilde{M}_x|^2$ and $|\widetilde{M}_z|^2$, resulting in a distorted pattern with a slight elongation along the field direction. By comparison of the magnetization Fourier components along the rod (x) and the magnetocrystalline (z) axes, it can be seen that the shape anisotropy still contributes to the effective anisotropy, however, its strength is lowered as compared to the anisotropy created parallel to the rod's diameter, i.e., increasing the diameter reinforces the effect of the magnetocrystalline anisotropy. The CT contributes to $d\Sigma_M/d\Omega$ only with a low signal, which is related to the small volume fraction of vortex cores [small $\widetilde{M}_y$ contribution, compare Fig. 7.15(b)]. Interestingly, at the smallest q, the CT has inverted its sign compared to the $D_{cyl} = 30$ nm and $D_{cyl} = 60$ nm results [162], which may reflect an asymmetry of the magnetic configuration. In monocrystalline hcp Co nanowire arrays with $D_{cyl}/L_{cyl} = 0.225$, Vivas et al. [999] and Ivanov et al. [1000] have demonstrated that, at remanence, the magnetic structure can be composed of multiple stable magnetic vortex domains with different chirality. The pronounced form-factor oscillations in $d\Sigma_M/d\Omega$, visible for the $D_{cyl} = 30$ nm nanorod [Fig. 7.17(a)], are progressively washed out with increasing diameter [Fig. 7.17(b) and (c)]. This is due to the formation of a complex multi-domain structure and the concomitant existence of different length scales (periodicities) characterizing the Fourier components $\widetilde{M}_{x,y,z}$.

In order to grasp the deviation between eqn (5.1) and the spin-misalignment SANS into one single parameter, we introduce the function

$$\eta_M(H_0) = \frac{d\Sigma_M}{d\Omega}(q^*, H_0) \left/ \frac{d\Sigma_M}{d\Omega}(q^*, H_0 \to \infty) \right., \tag{7.12}$$

which describes the normalized variation of the azimuthally averaged magnetic SANS cross section $d\Sigma_M/d\Omega$ with the applied magnetic field. Figure 7.18 shows the results for $\eta_M(H_0)$ obtained by computing $d\Sigma_M/d\Omega$ of the Co nanorod for both scattering geometries and for several H_0-values between positive and negative saturation with and without magnetocrystalline anisotropy. For the computation of η_M, the average intensity at the respective characteristic q-value of $q^* = 2\pi/D_{cyl}$ was used. The behavior of $\eta_M(H_0)$ for a uniformly magnetized shape-anisotropic nanoparticle and for the perpendicular and parallel scattering geometries can be understood within the context of the Stoner–Wohlfarth model [164]: for $\mathbf{k}_0 \parallel \mathbf{H}_0$ (corresponding to the case that the rod's long axis is parallel to $\mathbf{H}_0$), the magnetization curve exhibits the well-known rectangular shape, implying that $\mathbf{M}$ points always along the z-direction. There is, of course, a switching from the $+z$ to the $-z$ direction at the nucleation field, but this feature appears to be too "sharp" to be resolved with the SANS technique (for the single particle). As a consequence, the scattering vector $\mathbf{q}$ is always perpendicular to $\mathbf{M}$, so that the expectation value of the $\sin^2 \alpha$ factor in eqn (5.1) equals unity and the function $\eta_M(H_0)$ does not depend on the field. The corresponding scattering pattern is isotropic. For $\mathbf{k}_0 \perp \mathbf{H}_0$ (corresponding to the case that the rod's long axis is perpendicular to $\mathbf{H}_0$), the Stoner–Wohlfarth model predicts a linear paramagnetic-like decrease from saturation (along the horizontal z-direction) to a zero remanent magnetization, where now (due to the shape anisotropy) the magnetization points along the wire axis (x-direction). The expectation value of $\sin^2 \alpha$ changes from a value of $1/2$ at saturation to a value of 1 at $H_0 = 0$; in other words, $\eta_M(H_0)$ increases from 1 to a value of 2.

The above-described idealized Stoner–Wohlfarth scenario is fairly well reproduced by the simulations for the smallest diameter of $D_{cyl} = 30$ nm ($-\blacktriangledown-$ data in Fig. 7.18). The very small deviations from $\eta_M = 2$ in the perpendicular geometry [$K_u \neq 0$; Fig. 7.18(c)] and from $\eta_M = 1$ in the parallel geometry [$K_u \neq 0$; Fig. 7.18(d)] are attributed to the small spin-misalignment scattering that is related to the spin disorder at the end faces of the cylinder [compare Fig. 7.15(a)]. Neglecting the transversal magnetocrystalline anisotropy in the simulations results in a quasi-single-domain $D_{cyl} = 30$ nm cylinder (Fig. 1.10). For the two largest diameters (60 nm and 90 nm), we observe a significant deviation from the Stoner–Wohlfarth prediction, and therefore also from eqn (5.1), due to the emerging spin-disorder scattering ($-\bullet-$ and $-\blacksquare-$ data in Fig. 7.18; compare Fig. 7.15). While for both $K_u = 0$ cases and for $K_u \neq 0$ in the parallel geometry we observe the trend that the spin-inhomogeneity parameter $\eta_M(H_0)$ increases with increasing D_{cyl}/L_{cyl} ratio, the behavior of $\eta_M(H_0)$ is different

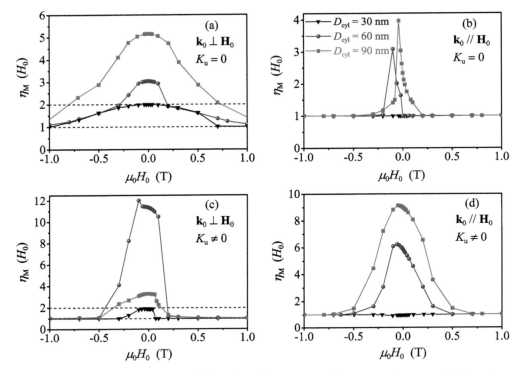

Fig. 7.18: The function $\eta_{\mathrm{M}}(H_0)$ [eqn (7.12)] computed for several nanorod diameters D_{cyl} [see inset in (b)]. Lines are a guide to the eyes. (a) $\mathbf{k}_0 \perp \mathbf{H}_0$ and zero magnetocrystalline anisotropy ($K_{\mathrm{u}} = 0$); (b) $\mathbf{k}_0 \parallel \mathbf{H}_0$ and $K_{\mathrm{u}} = 0$; (c) $\mathbf{k}_0 \perp \mathbf{H}_0$ and $K_{\mathrm{u}} = 4.5 \times 10^5\ \mathrm{J/m^3}$; (d) $\mathbf{k}_0 \parallel \mathbf{H}_0$ and $K_{\mathrm{u}} = 4.5 \times 10^5\ \mathrm{J/m^3}$. Dashed lines in (a) and (c): Stoner–Wohlfarth limit. After [162].

for $K_{\mathrm{u}} \neq 0$ in the perpendicular geometry [Fig. 7.18(c)]. Here, the $D_{\mathrm{cyl}} = 90\,\mathrm{nm}$ rod exhibits a field variation which is much smaller than the one of the $D_{\mathrm{cyl}} = 60\,\mathrm{nm}$ cylinder. This observation indicates that the magnitude of the spin-misalignment scattering of nanorods depends sensitively on their domain structure, which is determined by the nanoparticle's geometry (ratio $D_{\mathrm{cyl}}/L_{\mathrm{cyl}}$) and by the magnetic interactions, including the applied-field direction (compare also the Fourier images in Fig. 7.16).

7.5 Summary and outlook on simulation studies

Figures 7.8, 7.11, 7.16, and 7.17 embody the power of the micromagnetic simulation methodology: besides taking into account the full nonlinearity of Brown's static equations of micromagnetics, the approach allows one to study the dependency of the individual magnetization Fourier components $\widetilde{M}_{x,y,z}$ on the relevant parameters, such as the applied magnetic field $\mathbf{H}_0$ and, most importantly, the momentum-transfer vector $\mathbf{q}$; in other words, it becomes possible to decrypt the magnetic SANS cross section. Simulations ideally complement and guide neutron experiments, in which a weighted sum of the $\widetilde{M}_{x,y,z}$ is generally measured [compare eqns (7.10) and (7.11)]. This fact often hampers the straightforward interpretation of recorded SANS data. While it is in principle possible to determine some Fourier components, e.g., through the application of a saturating magnetic field or by exploiting the neutron-polarization degree of freedom via SANSPOL or POLARIS methods (e.g., [332, 1001–1009]), it is difficult to unambiguously determine a particular scattering contribution without contamination by unwanted Fourier components; for instance, when the applied field is not large enough to completely saturate the sample ($\mathbf{k}_0 \perp \mathbf{H}_0$), then the scattering of unpolarized neutrons along the field direction does not represent the pure nuclear SANS, but contains also the magnetic SANS due to the misaligned spins (see Appendix A) [121]. Moreover, the flexibility of the micromagnetic package developed by Berkov and Erokhin in terms of microstructure variation (e.g., particle-size distribution, texture, site-dependent magnetic materials parameters and interactions) suggests that the combination of full-scale three-dimensional micromagnetic simulations with experimental magnetic-field-dependent SANS data continues to provide fundamental insights into the magnetic SANS of a wide range of magnetic materials. It is also rather straightforward to switch on and off certain magnetic interactions in the simulations and to test in this way their impact on the neutron scattering. The micromagnetic simulations underline the importance of the magnetodipolar interaction for understanding magnetic SANS; in particular, the clover-leaf-shaped angular anisotropy in $d\Sigma_\mathrm{M}/d\Omega$—which was previously believed to be exclusively related to nanoscale jumps in the magnetization magnitude at internal interfaces—is of relevance for all bulk nanomagnets with spatially fluctuating magnetic parameters.

Regarding the magnetic SANS on nanoparticle systems, significant deviations from the macrospin-based model [eqn (5.1)] due to internal spin disorder of different origin are observed. In order to account for these inhomogeneous spin structures and to understand the associated complex scattering patterns, micromagnetic simulations, e.g., using the MuMax3 software package [166, 167], provide the key tool set for progressing the understanding of magnetic SANS on nanoparticles. The inclusion of interparticle interactions in the micromagnetic computations, including the particle arrangement on regular lattices, will also allow for a comparison of simulation results to experimental data on nanowire arrays, which are usually dominated by the strong scattering signal due to the structure factor $S(q)$ of the array [586–589, 1010]. As with the bulk ferromagnets, the magnetic microstructure and SANS of nanoparticles is affected by microstructural defects (e.g., vacancies and antiphase boundaries [177]) and the inclusion of these features into the micromagnetic SANS description is a future task. Furthermore, a systematic study of the effect of inhomogeneous nanoparticle magnetization on the correlation function $c(r)$ or, likewise, on the pair-distance distribution function $p(r) = r^2 c(r)$ is also missing [163, 298]. In this regard, the compilation of a library of SANS cross sections and associated correlation functions for different nanoparticle sizes and shapes, size-distribution functions, packing densities of nanoparticles (strength of the dipolar interaction), different types of magnetic anisotropy (bulk and surface anisotropy), or defect structures, etc., would be highly desirable. Atomistic calculations [1011] are also promising, since they would allow to test the limits of applicability of the continuum micromagnetic approach.

The inclusion of nuclear small-angle scattering into micromagnetic SANS software packages would be a further advantage, which will allow for the computation of polarization-dependent nuclear-magnetic interference terms, in this way providing a more complete picture. In view of the further development of efficient micromagnetic algorithms and the continuously increasing computing power (see, e.g., the discussion in [167]), it is to be expected that the total nuclear and magnetic scattering cross section of true macroscopic samples, with dimensions of the order of several microns, can be simulated. As a starting point, one may evaluate the nuclear SANS cross section based on the computed SANS for the completely saturated state, which is the Fourier image of the saturation-magnetization profile. For simple cases, e.g., a two-phase system with identical nuclear and magnetic form factors, the nuclear SANS is proportional to the computed magnetic SANS at saturation.

As discussed in Section 5.2.1, progress in SANS instrumentation regarding time-resolved data-acquisition procedures (TISANE = Time Dependent SANS Experiments) opens up the way to kinetic studies [646, 665, 683, 684]. TISANE allows one to probe magnetism up to the ~10 μs regime, which permits the investigation of magnetization dynamics in nanoparticle assemblies [50, 646, 672, 685, 686], the dynamics of vortex [687] and skyrmion lattices [688], or systems out of equilibrium. Therefore, the analytical and numerical extension of the present static micromagnetic approach [eqns (5.26)–(5.33)] to study magnetization dynamics using SANS would be desirable. This can be accomplished by solving the so-called Landau–Lifshitz–Gilbert (LLG) equation for $\mathbf{M}(\mathbf{r}, t)$, which reads [359, 992]:

$$\frac{d\mathbf{M}}{dt} = -\frac{\gamma_0}{1 + \alpha_\mathrm{d}^2} \left(\mathbf{M} \times \mathbf{H}_\mathrm{eff}\right) - \frac{\gamma_0}{M_\mathrm{s}} \frac{\alpha_\mathrm{d}}{1 + \alpha_\mathrm{d}^2} \left(\mathbf{M} \times [\mathbf{M} \times \mathbf{H}_\mathrm{eff}]\right), \qquad (7.13)$$

where γ_0 denotes the gyromagnetic ratio, α_d is a phenomenological dimensionless damping parameter, and the effective magnetic field $\mathbf{H}_\mathrm{eff}$ contains the relevant magnetic interactions. We refer to the review by Berkov [992] for a discussion of the origin and limitations of the standard LLG equation. Bertotti et al. [1012, 1013] have provided an analytical treatment of the LLG equation and found exact solutions for the nonlinear large motion of the magnetization vector in a body with uniaxial symmetry subject to a circularly polarized field. Their work may serve as a guideline for future dynamic magnetic SANS studies. In general, we believe that the combination of experimental scattering data with large-scale numerical computations is a promising approach for the future resolution of three-dimensional mesoscale magnetization structures [163].

Appendix A

Caveat on the separation of nuclear and magnetic SANS from unpolarized SANS data

The unpolarized nuclear and magnetic SANS cross section is given by

$$\frac{d\Sigma}{d\Omega}(\mathbf{q}) = \frac{8\pi^3}{V}b_{\mathrm{H}}^2\left(b_{\mathrm{H}}^{-2}|\widetilde{N}|^2 + |\widetilde{\mathbf{Q}}|^2\right), \tag{A.1}$$

which reads

$$\frac{d\Sigma}{d\Omega}(\mathbf{q}) = \frac{8\pi^3}{V}b_{\mathrm{H}}^2\left(b_{\mathrm{H}}^{-2}|\widetilde{N}|^2 + |\widetilde{M}_x|^2 + |\widetilde{M}_y|^2\cos^2\theta + |\widetilde{M}_z|^2\sin^2\theta\right.$$
$$\left. -(\widetilde{M}_y\widetilde{M}_z^* + \widetilde{M}_y^*\widetilde{M}_z)\sin\theta\cos\theta\right) \tag{A.2}$$

for the perpendicular scattering geometry ($\mathbf{k}_0 \perp \mathbf{H}_0$). At complete magnetic saturation ($M_x = M_y = 0$ and $M_z = M_{\mathrm{s}}$), eqn (A.2) reduces to

$$\frac{d\Sigma}{d\Omega}(\mathbf{q}) = \frac{8\pi^3}{V}b_{\mathrm{H}}^2\left(b_{\mathrm{H}}^{-2}|\widetilde{N}|^2 + |\widetilde{M}_{\mathrm{s}}|^2\sin^2\theta\right), \tag{A.3}$$

where $\widetilde{M}_{\mathrm{s}}(\mathbf{q})$ denotes the Fourier transform of $M_{\mathrm{s}}(\mathbf{r})$. Equation (A.3) is often employed for separating nuclear from magnetic SANS by taking sector averages of the two-dimensional $d\Sigma/d\Omega$ along the horizontal direction ($\theta = 0°$), yielding

$$\frac{d\Sigma}{d\Omega} \cong \frac{d\Sigma_{\mathrm{nuc}}}{d\Omega} = \frac{8\pi^3}{V}|\widetilde{N}|^2, \tag{A.4}$$

and along the vertical direction ($\theta = 90°$), providing

$$\frac{d\Sigma}{d\Omega} \cong \frac{d\Sigma_{\mathrm{nuc}}}{d\Omega} + \frac{d\Sigma_{\mathrm{M}}}{d\Omega} = \frac{8\pi^3}{V}\left(|\widetilde{N}|^2 + b_{\mathrm{H}}^2|\widetilde{M}_{\mathrm{s}}|^2\right). \tag{A.5}$$

In the following, it is demonstrated that the above separation should be considered with great care; in particular, if the value of the applied magnetic field is not large enough to saturate the sample. For only partially saturated specimens, the transversal Fourier components in eqn (A.2) also give rise to a scattering contribution parallel to $\mathbf{H}_0$ (compare, e.g., the term $|\widetilde{M}_y|^2\cos^2\theta$), so that the supposed $d\Sigma_{\mathrm{nuc}}/d\Omega$, determined by an average along the $\theta = 0°$ direction, is contaminated by these contributions. This scenario is depicted in Fig. A.1, which shows the two-dimensional total $d\Sigma/d\Omega$ and the horizontally averaged ($\theta = 0°$) SANS cross sections of the Fe-based nanocrystalline alloy NANOPERM at a series of applied fields. The hysteresis loop of the NANOPERM sample is depicted in Fig. A.2.

The magnetization data reveal that at least the two highest fields in the SANS experiment, 1500 mT and 290 mT, fall into the approach-to-saturation regime. The supposed nuclear SANS signal which is determined at these two fields is significantly different [Fig. A.1(e)], and it can be safely assumed that the "$d\Sigma_{nuc}/d\Omega$" would still continue to decrease if the applied field would be increased beyond 1500 mT. This observation highlights the enormous sensitivity of the magnetic SANS method for detecting small transversal spin misalignment (compare also Fig. 6.16). We also refer to the study by Hennion et al. [122] on re-entrant spin glasses, where in the saturation regime a large field-dependent SANS cross section was observed with an enhanced intensity parallel to the field direction.

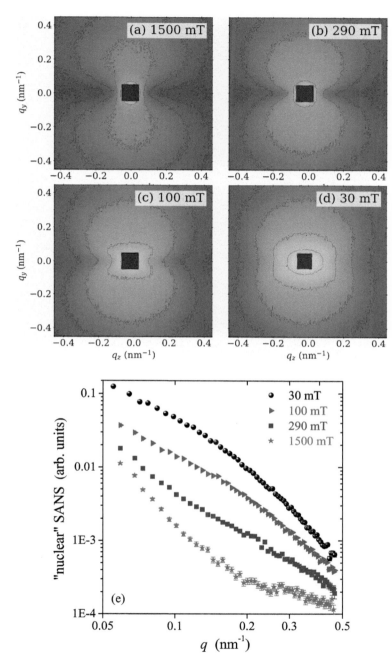

Fig. A.1: (a)–(d) Total unpolarized $d\Sigma/d\Omega$ of NANOPERM ($Fe_{89}Zr_7B_3Cu_1$) at selected applied magnetic fields (see insets) ($\mathbf{k}_0 \perp \mathbf{H}_0$). $\mathbf{H}_0$ is horizontal. (e) Corresponding supposed "nuclear" SANS cross section, determined by azimuthally averaging the two-dimensional data along the horizontal ($\theta = 0°$) direction ($\pm 10°$ sector averages) (log-log scale).

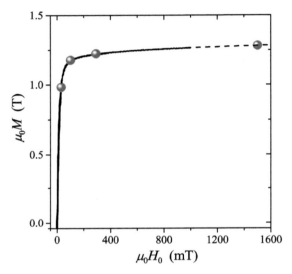

Fig. A.2: Room-temperature magnetization curve of NANOPERM. The $M(H_0)$-values where the SANS measurements shown in Fig. A.1 have been carried out are approximately indicated by the large data points. The experimental magnetization data have been extrapolated to the SANS data point at 1500 mT (dashed line). The differences in the demagnetizing fields between the SANS and magnetometry samples are negligible for the present qualitative consideration. Data courtesy of Kiyonori Suzuki, Monash University, Clayton, Australia.

Appendix B
Magnetic materials parameters

A quantity of utmost importance for magnetic SANS is the magnetic scattering-length density contrast $(\Delta\rho)^2_{\mathrm{mag}}$. Assuming a saturated two-phase particle-matrix-type microstructure, we have [eqn (2.124)]:

$$(\Delta\rho)^2_{\mathrm{mag}} = b^2_{\mathrm{H}} \left(\Delta M\right)^2, \tag{B.1}$$

where $b_{\mathrm{H}} = 2.91 \times 10^8\,\mathrm{A^{-1}m^{-1}}$, and

$$\Delta M = M^{\mathrm{p}}_{\mathrm{s}} - M^{\mathrm{m}}_{\mathrm{s}} \tag{B.2}$$

denotes the difference in the saturation magnetizations between the particle and the matrix phase. Table B.1 lists the room-temperature M_{s}-values of some magnetic materials. Note that for use in eqn (B.1) these values should be converted to A/m (e.g., for iron $\mu_0 M_{\mathrm{s}} = 2.185\,\mathrm{T} \cong 1739\,\mathrm{kA/m}$), so that $(\Delta\rho)^2_{\mathrm{mag}}$ comes in units of $\mathrm{m^{-4}}$.

Another magnetic parameter which is of great relevance for magnetic SANS on nanoparticles, in particular in relation to spin-misalignment scattering (compare, e.g., Fig. 1.11), is the so-called critical single-domain size D_{c}. This quantity marks the transition between the single-domain and the two-domain state. For a single spherical particle, D_{c} can be roughly estimated by means of [119,165]:

$$D_{\mathrm{c}} \cong \frac{72\sqrt{AK_1}}{\mu_0 M^2_{\mathrm{s}}}, \tag{B.3}$$

where A denotes the exchange-stiffness constant, and K_1 is the first-order anisotropy constant. Table B.1 displays D_{c}-values which were computed using eqn (B.3). It is seen that D_{c} ranges between about $10\,\mathrm{nm}$ for soft magnetic materials and $1\,\mu\mathrm{m}$ for very hard magnets.

Table B.1 Magnetic materials parameters of some selected materials (taken from [119, 165, 1014]). M_s = saturation magnetization at room temperature; T_C = Curie temperature; $K_{1,2}$ = first- and second-order anisotropy constants at room temperature; D_c = critical single-domain diameter for a corresponding spherical particle.

Material	$\mu_0 M_s$ (T)	T_C (K)	K_1 (MJ/m^3)	K_2 (MJ/m^3)	D_c (nm)
Fe	2.185	1043	0.048	−0.01	9.7
Co	1.79	1403	0.453	0.145	55.5
Ni	0.62	627	−0.0045	−0.0025	22.6
Gd	2.69	293	−	−	−
Ni$_3$Fe	1.1	873	0.00012	−	−
Nd$_2$Fe$_{14}$B	1.61	588	4.3	0.65	210
Pr$_2$Fe$_{14}$B	1.56	565	5.6	≈ 0	−
SmCo$_5$	1.05	993	17	−	1170
Sm$_2$Co$_{17}$	1.29	1070	4.2	−	420
Sm$_2$Fe$_{17}$N$_3$	1.56	749	8.6	1.9	−
BaFe$_{12}$O$_{19}$	0.48	723	0.32	< 0.1	62
SrFe$_{12}$O$_{19}$	0.46	733	0.35	−	−
Fe$_3$O$_4$	0.60	858	−0.011	−0.003	83
γ-Fe$_2$O$_3$	0.47	863	−0.0046	−	−
CrO$_2$	0.56	390	0.025	−	180
MnFe$_2$O$_4$	0.52	573	−0.0028	−	−
CoFe$_2$O$_4$	0.50	793	0.270	−	−
NiFe$_2$O$_4$	0.34	858	−0.0069	−	−
MnBi	0.73	633	0.90	−	480
FePt	1.43	750	6.6	−	340

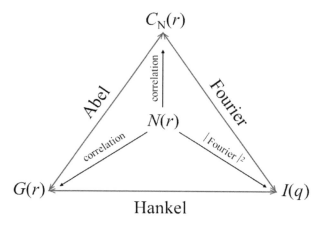

Fig. C.1: Relationship between the Fourier, Hankel, and Abel transforms for an isotropic nuclear scattering-length density profile $N(r)$. Note that $G(r)$ can be found by calculating the autocorrelation function of the projection of $N(r)$. After [289, 343].

Appendix C
Fourier–Hankel–Abel cycle

In this Appendix, we recall the basic integral-transform relationships between the nuclear SANS cross section $\frac{d\Sigma_{\mathrm{nuc}}}{d\Omega} = I(q)$, the autocorrelation function $C_{\mathrm{N}}(r)$ of the nuclear scattering-length density $N(r)$, and the projected correlation function $G(r)$, which is accessible in a SESANS experiment. This is known as the Fourier–Hankel–Abel (FHA) cycle, which is schematically depicted in Fig. C.1. All the involved functions are assumed to be isotropic. Applying successively the three transforms, or a cyclic order of them ($\mathcal{H} \circ \mathcal{F} \circ \mathcal{A} = \mathcal{F} \circ \mathcal{A} \circ \mathcal{H} = \mathcal{A} \circ \mathcal{H} \circ \mathcal{F} = \mathcal{I}$), yields the original function [1015]. Here, $\mathcal{I}$ is the identity operator, and $\circ$ denotes the composition of operators, e.g., $H \circ F \circ Au = H[F(Au)]$, where u is some function. We refer to the papers by Andersson et al. [289] and Kohlbrecher and Studer [343] for detailed discussions of the FHA cycle in relation to SANS techniques.

In Section 2.4.3, we have seen that $C_{\mathrm{N}}(r)$ and $I(q)$ are related via the following set of Fourier transforms [compare eqns (2.66) and (2.67)]:

$$C_{\mathrm{N}}(r) = \frac{1}{2\pi^2 r} \int_0^\infty I(q) \sin(qr) q \, dq \qquad (\text{C.1})$$

and

$$I(q) = \frac{4\pi}{q} \int_0^\infty C_\mathrm{N}(r)\sin(qr)rdr, \tag{C.2}$$

where $C_\mathrm{N}(r)$ is the autocorrelation function of $N(r)$,

$$C_\mathrm{N}(r) = \frac{4\pi}{V} \int_0^\infty N(|\mathbf{r}'|)N(|\mathbf{r}'+\mathbf{r}|)r'^2 dr'. \tag{C.3}$$

By means of the convolution theorem, it can be shown that $I(q)$ is related to the Fourier transform of $N(r)$ according to [compare eqns (2.58) and (2.59)]:

$$I(q) = \frac{1}{V} \left| \frac{4\pi}{q} \int_0^\infty N(r)\sin(qr)rdr \right|^2. \tag{C.4}$$

Using the results of Appendix E, namely that the 2D Fourier transform of a radially symmetric function is equal to the zero-order Hankel transform of that function [eqns (E.40) and (E.41)], we can relate $I(q)$ to the projected correlation function $G(r)$ [compare also eqns (6.15)–(6.17) and eqn (6.21)]:

$$I(q) = \int_0^\infty G(r)J_0(qr)rdr, \tag{C.5}$$

$$G(r) = \int_0^\infty I(q)J_0(qr)qdq. \tag{C.6}$$

The function G corresponds to the $C_{\mathrm{N}\perp}$ of Section 6.2. Finally, we make use of a well-known result of SESANS theory, i.e., the autocorrelation function $C_\mathrm{N}(r)$ is related to the projected $G(r)$ via an Abel transform [289, 343]:

$$G(r) = 4\pi \int_r^\infty \frac{C_\mathrm{N}(\delta)\delta}{\sqrt{\delta^2 - r^2}} d\delta, \tag{C.7}$$

$$C_\mathrm{N}(r) = -\frac{1}{2\pi^2} \int_r^\infty \frac{G'(\delta)}{\sqrt{\delta^2 - r^2}} d\delta, \tag{C.8}$$

where the latter equation is valid for $C_\mathrm{N}(r)$, which decays to zero faster than r^{-1} (at large r), and $G'(z) = dG/dz$. For the sphere with $V_\mathrm{s} = \frac{4\pi}{3}R^3$, we have [eqn (2.71)]:

$$C_\mathrm{N}(r) = \begin{cases} 1 - \frac{3r}{4R} + \frac{r^3}{16R^3} & \text{if } 0 \le r \le 2R \\ 0 & \text{if } r \ge 2R \end{cases} \tag{C.9}$$

and

$$I(q) = 9V_{\rm s} \frac{[\sin(qR) - qR\cos(qR)]^2}{(qR)^6},$$ (C.10)

so that eqn (C.6), or eqn (C.7), yields

$$G(r) = \begin{cases} \pi \dfrac{6R(8R^2 + r^2)\sqrt{4R^2 - r^2} + 3r^2(16R^2 - r^2)\ln\left(\frac{r}{2R + \sqrt{4R^2 - r^2}}\right)}{32R^3} & \text{if } 0 \le r \le 2R \\ 0 & \text{if } r \ge 2R. \end{cases}$$ (C.11)

$G(r)$ with $G(0) = 3\pi R$ is plotted in Fig. C.2 together with $C_{\rm N}(r)$. It can be seen that the slope of $C_{\rm N}(r)$ is finite at the origin $[C'_{\rm N}(0) = -3/(4R)]$, while $G'(0) = 0$, which is true for any three-dimensional particle with a sharp interface [289].

The latter statement can be proven as follows [1016] by considering Debye–Bueche-type correlation functions $C_{\rm N}(r)$ of uniform particles with a sharp interface [300]. For such systems, $C_{\rm N}(r)$ has the following properties [11]:

$$C_{\rm N}(r) = \begin{cases} C_{\rm N}(r) & \text{if } r \le D \\ 0 & \text{if } r \ge D, \end{cases}$$ (C.12)

where D is the largest dimension in a particle, and

$$\left.\frac{dC_{\rm N}}{dr}\right|_{r=0} = C'_{\rm N}(0) \ne 0.$$ (C.13)

Inserting eqn (C.2) into eqn (C.6) and changing the order of integration yields:

$$G(r) = 4\pi \int_0^D \left(\int_0^\infty J_0(qr)\sin(q\delta)dq \right) C_{\rm N}(\delta)\delta d\delta.$$ (C.14)

Making use of eqn (6.671.7) in [1017],

$$\int_0^\infty J_0(qr)\sin(q\delta)dq = \begin{cases} \frac{1}{\sqrt{\delta^2 - r^2}} & \text{if } 0 < r < \delta \\ 0 & \text{if } 0 < \delta < r, \end{cases}$$ (C.15)

we have (for $0 < r < \delta$):

$$\frac{G(r)}{4\pi} = \int_r^D \frac{C_{\rm N}(\delta)\delta}{\sqrt{\delta^2 - r^2}}d\delta,$$ (C.16)

which equals eqn (C.7). Integration by parts gives:

$$\frac{G(r)}{4\pi} = C_{\rm N}(\delta)\sqrt{\delta^2 - r^2}\Big|_r^D - \int_r^D C'_{\rm N}(\delta)\sqrt{\delta^2 - r^2}d\delta,$$ (C.17)

where the first term on the right-hand side of eqn (C.17) vanishes due to $C_{\rm N}(D) = 0$. The first-order r-derivative of the remaining expression equals:

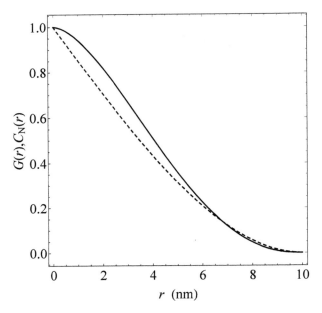

Fig. C.2: Solid line: normalized $G(r)$ [eqn (C.11)] with $R = 5\,\text{nm}$. Dashed line: $C_N(r)$ [eqn (C.9)].

$$\frac{d}{dr}\frac{G(r)}{4\pi} = r \int_{r}^{D} \frac{C'_N(\delta)}{\sqrt{\delta^2 - r^2}}\,d\delta. \tag{C.18}$$

Substituting $t = \delta/r$, the integral transforms into:

$$\int_{r}^{D} \frac{C'_N(\delta)}{\sqrt{\delta^2 - r^2}}\,d\delta = \int_{1}^{D/r} \frac{C'_N(rt)}{\sqrt{t^2 - 1}}\,dt. \tag{C.19}$$

As $r \to 0$, the asymptotic behavior of the integral can be approximated by:

$$\int_{1}^{D/r} \frac{C'_N(rt)}{\sqrt{t^2 - 1}}\,dt \cong \int_{1}^{D/r} \frac{C'_N(0)}{\sqrt{t^2 - 1}}\,dt = C'_N(0)\ln\left(\frac{D + \sqrt{D^2 - r^2}}{r}\right), \tag{C.20}$$

which implies that

$$\frac{d}{dr}\frac{G(r)}{4\pi} \cong rC'_N(0)\ln\left(\frac{D + \sqrt{D^2 - r^2}}{r}\right), \tag{C.21}$$

demonstrating that the slope of $G(r)$ vanishes at the origin $r = 0$, and at $r = D$ [compare eqn (C.18)] [1018], provided that $C_N(r)$ fulfills eqns (C.12) and (C.13).

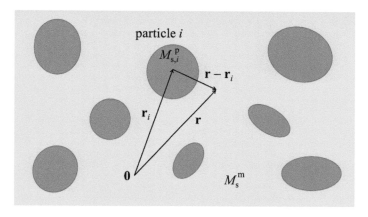

Fig. D.1: Sketch illustrating the distribution of N_p homogeneous spherically symmetric particles with saturation magnetizations $M_{s,i}^p$ in a homogeneous matrix of saturation magnetization M_s^m.

Appendix D

Fourier transform of the local saturation magnetization in a material containing ferromagnetic spherical inclusions

The following derivation is taken from Schlömann [335]. We consider a saturated microstructure consisting of a distribution of $i = 1, \ldots, N_p$ homogeneous particles (fixed rigidly in space) in a homogeneous matrix of saturation magnetization M_s^m. The particles have a saturation magnetization $M_{s,i}^p$ and a volume $V_{p,i}$ (see Fig. D.1). The aim is to compute the Fourier transform

$$\widetilde{M}_s(\mathbf{q}) = \frac{1}{(2\pi)^{3/2}} \int M_s(\mathbf{r}) \exp\left(-i\mathbf{q} \cdot \mathbf{r}\right) d^3r \tag{D.1}$$

of the local saturation magnetization

$$M_s(\mathbf{r}) = \frac{1}{(2\pi)^{3/2}} \int \widetilde{M}_s(\mathbf{q}) \exp\left(i\mathbf{q} \cdot \mathbf{r}\right) d^3q, \tag{D.2}$$

where the integrals are, respectively, taken over all space. With reference to Fig. D.1, we see that $M_s(\mathbf{r})$ can be expressed as a sum of convolution products:

$$M_s(\mathbf{r}) = M_s^m + \sum_{i=1}^{N_p} (M_{s,i}^p - M_s^m) f_i(\mathbf{r}) \otimes \delta(\mathbf{r} - \mathbf{r}_i), \qquad (D.3)$$

where

$$\delta(\mathbf{r} - \mathbf{r}_i) = \begin{cases} 0 & \text{if } \mathbf{r} \neq \mathbf{r}_i \\ \infty & \text{if } \mathbf{r} = \mathbf{r}_i \end{cases} \qquad (D.4)$$

represents Dirac's delta function, and

$$\frac{1}{(2\pi)^{3/2}} \int \delta(\mathbf{r} - \mathbf{r}_i) \exp(-i\mathbf{q} \cdot \mathbf{r}) \, d^3r = \frac{\exp(-i\mathbf{q} \cdot \mathbf{r}_i)}{(2\pi)^{3/2}} \qquad (D.5)$$

is the Fourier transform of $\delta(\mathbf{r} - \mathbf{r}_i)$. The function $f_i(\mathbf{r})$ is defined as:

$$f_i(\mathbf{r}) = \begin{cases} 1 & \text{if } \mathbf{r} \in V_{p,i} \\ 0 & \text{if } \mathbf{r} \notin V_{p,i}, \end{cases} \qquad (D.6)$$

with its Fourier transform being

$$\frac{1}{(2\pi)^{3/2}} \int f_i(\mathbf{r}) \exp(-i\mathbf{q} \cdot \mathbf{r}) \, d^3r = \frac{1}{(2\pi)^{3/2}} \int_{V_{p,i}} \exp(-i\mathbf{q} \cdot \mathbf{r}) \, d^3r = \frac{F_i(\mathbf{q})}{(2\pi)^{3/2}}, \quad (D.7)$$

where $F_i(\mathbf{q})$ is the particle form factor with $F_i(\mathbf{q} = 0) = V_{p,i}$. For a sphere

$$F_i(qR_i) = 3V_{s,i} \frac{j_1(qR_i)}{qR_i}. \qquad (D.8)$$

The convolution in eqn (D.3) maps the functions $f_i(\mathbf{r})$ onto the particle positions $\mathbf{r}_i$. Using the convolution theorem, the Fourier transform of $M_s(\mathbf{r})$ is then obtained as:

$$\widetilde{M_s}(\mathbf{q}) = (2\pi)^{3/2} M_s^m \delta(\mathbf{q}) + \sum_{i=1}^{N_p} \frac{M_{s,i}^p - M_s^m}{(2\pi)^{3/2}} F_i(\mathbf{q}) \exp(-i\mathbf{q} \cdot \mathbf{r}_i), \qquad (D.9)$$

where

$$\delta(\mathbf{q}) = \frac{1}{(2\pi)^3} \int \exp(-i\mathbf{q} \cdot \mathbf{r}) \, d^3r. \qquad (D.10)$$

Moreover, by assuming that all $M_{s,i}^p = M_s^p$, it is easily verified from eqn (D.9) that

$$\frac{(2\pi)^{3/2}}{V} \widetilde{M_s}(\mathbf{q} = 0) = \langle M_s \rangle = \frac{1}{V} \int M_s(\mathbf{r}) d^3r = M_s^m(1 - x_p) + M_s^p x_p \qquad (D.11)$$

yields the macroscopic mean magnetization $\langle M_s \rangle$, which can be measured with a magnetometer; $x_p = \sum_i V_{p,i}/V$ denotes the volume fraction of the particle phase. Note that Dirac's delta function is the identity element of the algebra of convolutions, i.e., $\delta \otimes f = f \otimes \delta = f$. At $\mathbf{q} \neq 0$, the magnitude square of $\widetilde{M}_s(\mathbf{q})$ evaluates to:

$$|\widetilde{M}_s(\mathbf{q})|^2 = \frac{1}{8\pi^3} \sum_{i=1}^{N_P} \sum_{j=1}^{N_P} (M_{s,i}^P - M_s^m)(M_{s,j}^P - M_s^m) F_i(\mathbf{q}) F_j^*(\mathbf{q}) \exp\left(-i\mathbf{q} \cdot [\mathbf{r}_i - \mathbf{r}_j]\right).$$

(D.12)

Since the locations of the particles are not actually known, it is reasonable to take an average of $|\widetilde{M}_s(\mathbf{q})|^2$ over all possible particle positions and to assume that this average value is representative of the actual sample under consideration. When the volume fraction of the inclusions is small ($x_p \ll 1$), each inclusion can be located with equal probability at any point within V. In this situation, the average of $\exp\left(-i\mathbf{q} \cdot [\mathbf{r}_i - \mathbf{r}_j]\right)$ vanishes unless $i = j$, and we have:

$$|\widetilde{M}_s(\mathbf{q})|_{av}^2 = \frac{1}{8\pi^3} \sum_{i=1}^{N_P} (M_{s,i}^P - M_s^m)^2 |F_i^2(\mathbf{q})|;$$

(D.13)

see [335] for an estimate of the error involved when going from eqn (D.12) to eqn (D.13).

Appendix E
Three- and two-dimensional Fourier transforms in polar coordinates

The following derivations are taken from the articles by Baddour [970, 971]. In this Appendix, Fourier transforms are denoted with a capital letter and a tilde ($\tilde{\ }$) above it, while for Hankel transforms an overhat ($\hat{\ }$) on a lower-case letter is used.

E.0.1 Three-dimensional case

The 3D Cartesian forward Fourier transform of a function $f(\mathbf{r}) = f(x, y, z)$ can be defined as:

$$\tilde{F}(\mathbf{q}) = \tilde{F}(q_x, q_y, q_z) = \frac{1}{(2\pi)^{3/2}} \int\limits_{-\infty}^{+\infty} \int\limits_{-\infty}^{+\infty} \int\limits_{-\infty}^{+\infty} f(\mathbf{r}) \exp\left(-i\mathbf{q} \cdot \mathbf{r}\right) dx dy dz, \qquad (\text{E}.1)$$

where $\mathbf{r} = \{x, y, z\}$ and $\mathbf{q} = \{q_x, q_y, q_z\}$. By applying the usual coordinate transformations from Cartesian to spherical coordinates, $r^2 = x^2 + y^2 + z^2$, $\alpha = \arctan([x^2 + y^2]^{1/2}/z)$, $\beta = \arctan(y/x)$, $q^2 = q_x^2 + q_y^2 + q_z^2$, $\theta = \arctan([q_x^2 + q_y^2]^{1/2}/q_z)$, $\eta = \arctan(q_y/q_x)$, the position vector $\mathbf{r}$ and the wave vector $\mathbf{q}$ can be expressed in terms of their spherical polar coordinates, respectively, (r, α, β) and (q, θ, η), where $0 \leq \alpha, \theta \leq \pi$ denote the polar angles and $0 \leq \beta, \eta \leq 2\pi$ are the corresponding azimuthal angles. Equation (E.1) then transforms into:

$$\tilde{F}(\mathbf{q}) = \tilde{F}(q, \theta, \eta) = \frac{1}{(2\pi)^{3/2}} \int\limits_{0}^{2\pi} \int\limits_{0}^{\pi} \int\limits_{0}^{\infty} f(r, \alpha, \beta) \exp\left(-i\mathbf{q} \cdot \mathbf{r}\right) r^2 \sin\alpha \, dr d\alpha d\beta, \qquad (\text{E}.2)$$

where $\mathbf{q} \cdot \mathbf{r} = qr[\cos\alpha \cos\theta + \sin\alpha \sin\theta \cos(\beta - \eta)]$. In order to evaluate this integral, we use the well-known expansion of the complex exponential function in terms of spherical harmonics $Y_{l,m}$ [161],

$$\exp\left(i\mathbf{q} \cdot \mathbf{r}\right) = 4\pi \sum_{l=0}^{\infty} \sum_{m=-l}^{+l} i^l j_l(qr) Y_{l,m}(\alpha, \beta) Y_{l,m}^*(\theta, \eta), \qquad (\text{E}.3)$$

$$\exp\left(-i\mathbf{q} \cdot \mathbf{r}\right) = 4\pi \sum_{l=0}^{\infty} \sum_{m=-l}^{+l} (-i)^l j_l(qr) Y_{l,m}^*(\alpha, \beta) Y_{l,m}(\theta, \eta), \qquad (\text{E}.4)$$

where $j_l(z)$ denotes the spherical Bessel function of order l, and the asterisk ($*$) marks the complex-conjugated quantity. We also assume that $f(r, \alpha, \beta)$ can be expanded into a series of spherical harmonics,

$$f(r, \alpha, \beta) = \sum_{l=0}^{\infty} \sum_{m=-l}^{+l} f_{l,m}(r) Y_{l,m}(\alpha, \beta), \qquad (E.5)$$

where the expansion coefficients are given by

$$f_{l,m}(r) = \int_0^{2\pi} \int_0^{\pi} f(r, \alpha, \beta) Y^*_{l,m}(\alpha, \beta) \sin \alpha d\alpha d\beta. \qquad (E.6)$$

Inserting eqns (E.4) and (E.5) into eqn (E.2) and making use of the orthonormality relation,

$$\int_0^{2\pi} \int_0^{\pi} Y_{l,m}(\alpha, \beta) Y^*_{l',m'}(\alpha, \beta) \sin \alpha d\alpha d\beta = \delta_{ll'} \delta_{mm'}, \qquad (E.7)$$

where δ_{ij} is the Kronecker delta function, we obtain

$$\widetilde{F}(q, \theta, \eta) = \sum_{l=0}^{\infty} \sum_{m=-l}^{+l} (-i)^l \left(\sqrt{\frac{2}{\pi}} \int_0^{\infty} f_{l,m}(r) j_l(qr) r^2 dr \right) Y_{l,m}(\theta, \eta) \qquad (E.8)$$

$$= \sum_{l=0}^{\infty} \sum_{m=-l}^{+l} (-i)^l \hat{f}_{l,m}(q) Y_{l,m}(\theta, \eta), \qquad (E.9)$$

where $\hat{f}_{l,m}(q)$ (the term in round brackets) represents the lth-order spherical Hankel transform of $f_{l,m}(r)$. We remind that for a function $h(r)$, the nth-order spherical Hankel transform and its inverse are given by the set of equations [1019]:

$$\hat{h}_n(q) = \sqrt{\frac{2}{\pi}} \int_0^{\infty} h(r) j_n(qr) r^2 dr \qquad (E.10)$$

and

$$h(r) = \sqrt{\frac{2}{\pi}} \int_0^{\infty} \hat{h}_n(q) j_n(qr) q^2 dq. \qquad (E.11)$$

The inverse transform to eqn (E.1),

$$f(\mathbf{r}) = f(x, y, z) = \frac{1}{(2\pi)^{3/2}} \int_{-\infty}^{+\infty} \int_{-\infty}^{+\infty} \int_{-\infty}^{+\infty} \widetilde{F}(\mathbf{q}) \exp\left(i\mathbf{q} \cdot \mathbf{r}\right) dq_x dq_y dq_z, \qquad (E.12)$$

can be found in a similar manner. We expand $\tilde{F}(\mathbf{q}) = \tilde{F}(q, \theta, \eta)$ in a series of spherical harmonics,

$$\tilde{F}(q, \theta, \eta) = \sum_{l=0}^{\infty} \sum_{m=-l}^{+l} \tilde{F}_{l,m}(q) Y_{l,m}(\theta, \eta), \tag{E.13}$$

where

$$\tilde{F}_{l,m}(q) = \int_0^{2\pi} \int_0^{\pi} \tilde{F}(q, \theta, \eta) Y_{l,m}^*(\theta, \eta) \sin \theta d\theta d\eta, \tag{E.14}$$

and use the expansion eqn (E.3) and eqn (E.7) to finally obtain

$$f(\mathbf{r}) = f(r, \alpha, \beta) = \sum_{l=0}^{\infty} \sum_{m=-l}^{+l} i^l \left(\sqrt{\frac{2}{\pi}} \int_0^{\infty} \tilde{F}_{l,m}(q) j_l(qr) q^2 dq \right) Y_{l,m}(\alpha, \beta). \tag{E.15}$$

Note that $\tilde{F}_{l,m}(q)$ is not the Fourier transform of $f_{l,m}(r)$. The integral in round brackets in eqn (E.15) denotes the lth-order inverse spherical Hankel transform of $\tilde{F}_{l,m}(q)$. Comparing eqns (E.9) and (E.13), we have

$$\tilde{F}_{l,m}(q) = (-i)^l \hat{f}_{l,m}(q), \tag{E.16}$$

which relates the coefficients $\tilde{F}_{l,m}(q)$ in the expansion of $F(q, \theta, \eta)$ into spherical harmonics to the lth-order spherical Hankel transform of $f_{l,m}(r)$.

When the functions $f(\mathbf{r})$ and $\tilde{F}(\mathbf{q})$ are radially symmetric, i.e., $f = f(r)$ and $\tilde{F} = \tilde{F}(q)$, then eqn (E.2) and its inverse transform can be written as:

$$\tilde{F}(q) = \frac{1}{(2\pi)^{3/2}} \int_0^{\infty} f(r) \left(\int_0^{2\pi} \int_0^{\pi} \exp\left(-i\mathbf{q} \cdot \mathbf{r}\right) \sin \alpha d\alpha d\beta \right) r^2 dr, \tag{E.17}$$

$$f(r) = \frac{1}{(2\pi)^{3/2}} \int_0^{\infty} \tilde{F}(q) \left(\int_0^{2\pi} \int_0^{\pi} \exp\left(i\mathbf{q} \cdot \mathbf{r}\right) \sin \theta d\theta d\eta \right) q^2 dq. \tag{E.18}$$

The expressions in the round brackets are related to the solid-angle average of the function $\exp(i\mathbf{q} \cdot \mathbf{r})$, i.e., $\langle \exp(i\mathbf{q} \cdot \mathbf{r}) \rangle = (4\pi)^{-1} \int_{\Omega} \exp(i\mathbf{q} \cdot \mathbf{r}) d\Omega = \sin(qr)/(qr) = j_0(qr)$. This can be seen by taking $\mathbf{q}$ in eqn (E.17) and $\mathbf{r}$ in eqn (E.18) along the respective polar axis and by using the integral definition of the spherical Bessel function of zero order [973]

$$j_0(z) = \frac{1}{2} \int_0^{\pi} \exp\left(-iz \cos \alpha\right) \sin \alpha d\alpha. \tag{E.19}$$

Equations (E.17) and (E.18) can then be expressed as zero-order spherical Hankel transforms:

$$\widetilde{F}(q) = \sqrt{\frac{2}{\pi}} \int_0^\infty f(r) j_0(qr) r^2 dr, \tag{E.20}$$

$$f(r) = \sqrt{\frac{2}{\pi}} \int_0^\infty \widetilde{F}(q) j_0(qr) q^2 dq. \tag{E.21}$$

Equations (E.20) and (E.21) reveal that the 3D Fourier transform of a radially symmetric function and its inverse Fourier transform are related to each other via zero-order spherical Hankel transformation.

E.0.2   Two-dimensional case

The 2D Cartesian forward Fourier transform of a function $f(\mathbf{r}) = f(x, y)$ can be defined as:

$$\widetilde{F}(\mathbf{q}) = \widetilde{F}(q_x, q_y) = \frac{1}{2\pi} \int_{-\infty}^{+\infty} \int_{-\infty}^{+\infty} f(\mathbf{r}) \exp\left(-i\mathbf{q} \cdot \mathbf{r}\right) dx dy, \tag{E.22}$$

where $\mathbf{r} = \{x, y\}$ and $\mathbf{q} = \{q_x, q_y\}$. The inverse transform is then given by

$$f(\mathbf{r}) = \frac{1}{2\pi} \int_{-\infty}^{+\infty} \int_{-\infty}^{+\infty} \widetilde{F}(\mathbf{q}) \exp\left(i\mathbf{q} \cdot \mathbf{r}\right) dq_x dq_y. \tag{E.23}$$

As with the 3D case, we express the position vector $\mathbf{r}$ and the wave vector $\mathbf{q}$ in terms of their (planar) polar coordinates, respectively, (r, ϕ) and (q, θ), where $0 \leq \phi, \theta \leq 2\pi$. Equations (E.22) and (E.23) then transform into [970]

$$\widetilde{F}(q, \theta) = \frac{1}{2\pi} \int_0^{2\pi} \int_0^\infty f(r, \phi) \exp\left(-iqr \cos[\theta - \phi]\right) r dr d\phi, \tag{E.24}$$

$$f(r, \phi) = \frac{1}{2\pi} \int_0^{2\pi} \int_0^\infty \widetilde{F}(q, \theta) \exp\left(iqr \cos[\theta - \phi]\right) q dq d\theta, \tag{E.25}$$

where $\mathbf{q} \cdot \mathbf{r} = qr \cos(\theta - \phi)$. By using the following so-called Jacobi–Anger expansion of the complex exponential in terms of the nth-order Bessel function $J_n(z)$ [534],

$$\exp\left(i\mathbf{q} \cdot \mathbf{r}\right) = \sum_{n=-\infty}^\infty i^n J_n(qr) \exp\left(in\phi\right) \exp\left(-in\theta\right), \tag{E.26}$$

$$\exp\left(-i\mathbf{q} \cdot \mathbf{r}\right) = \sum_{n=-\infty}^\infty i^{-n} J_n(qr) \exp\left(-in\phi\right) \exp\left(in\theta\right), \tag{E.27}$$

one can further evaluate eqns (E.24) and (E.25). For this purpose, we expand $f(r, \phi)$ and $\tilde{F}(q, \theta)$ into Fourier series,

$$f(r, \phi) = \sum_{n=-\infty}^{\infty} f_n(r) \exp(in\phi), \tag{E.28}$$

$$\tilde{F}(q, \theta) = \sum_{n=-\infty}^{\infty} \tilde{F}_n(q) \exp(in\theta), \tag{E.29}$$

where the $f_n(r)$ and $\tilde{F}_n(q)$ are found from

$$f_n(r) = \frac{1}{2\pi} \int_0^{2\pi} f(r, \phi) \exp(-in\phi)\, d\phi, \tag{E.30}$$

$$\tilde{F}_n(q) = \frac{1}{2\pi} \int_0^{2\pi} \tilde{F}(q, \theta) \exp(-in\theta)\, d\theta. \tag{E.31}$$

By inserting eqns (E.27) and (E.28) into eqn (E.24) and using the identity

$$\int_0^{2\pi} \exp(i[m-n]\phi)\, d\phi = 2\pi \delta_{mn}, \tag{E.32}$$

we obtain for the forward transform

$$\tilde{F}(q, \theta) = \frac{1}{2\pi} \int_0^{2\pi} \int_0^{\infty} \sum_{m=-\infty}^{\infty} f_m(r) \exp(im\phi) \sum_{n=-\infty}^{\infty} i^{-n} J_n(qr) \exp(-in\phi) \exp(in\theta)\, r dr d\phi$$

$$= \sum_{n=-\infty}^{\infty} i^{-n} \exp(in\theta) \int_0^{\infty} f_n(r) J_n(qr) r dr. \tag{E.33}$$

Comparing eqn (E.33) with eqn (E.29), we find for the Fourier coefficients of the Fourier domain expansion

$$\tilde{F}_n(q) = i^{-n} \int_0^{\infty} f_n(r) J_n(qr) r dr. \tag{E.34}$$

Similarly, for the inverse transform we find

$$f(r, \phi) = \sum_{n=-\infty}^{\infty} i^n \exp(in\phi) \int_0^{\infty} \tilde{F}_n(q) J_n(qr) q dq \tag{E.35}$$

and

$$f_n(r) = i^n \int_0^\infty \widetilde{F}_n(q) J_n(qr) q \, dq. \tag{E.36}$$

Equations (E.34) and (E.36) demonstrate that the nth terms in the Fourier-series expansions of the original functions are related to each other via nth-order Hankel transformation.

When the functions $f(\mathbf{r})$ and $\widetilde{F}(\mathbf{q})$ are radially symmetric (in 2D also denoted as circularly symmetric), i.e., $f = f(r)$ and $\widetilde{F} = \widetilde{F}(q)$, then eqns (E.24) and (E.25) can be written as:

$$\widetilde{F}(q) = \int_0^\infty f(r) \left(\frac{1}{2\pi} \int_0^{2\pi} \exp\left(-iqr\cos[\theta - \phi]\right) d\phi \right) r \, dr, \tag{E.37}$$

$$f(r) = \int_0^\infty \widetilde{F}(q) \left(\frac{1}{2\pi} \int_0^{2\pi} \exp\left(iqr\cos[\theta - \phi]\right) d\theta \right) q \, dq. \tag{E.38}$$

Using the integral definition of the Bessel function of zero order (see p.19f in [973]),

$$J_0(z) = \frac{1}{2\pi} \int_0^{2\pi} \exp\left(-iz\cos[\theta - \phi]\right) d\phi = \frac{1}{2\pi} \int_0^{2\pi} \exp\left(iz\cos[\phi - \theta]\right) d\theta, \tag{E.39}$$

eqns (E.37) and (E.38) can be expressed as zero-order Hankel transforms [1019]:

$$\widetilde{F}(q) = \int_0^\infty f(r) J_0(qr) r \, dr, \tag{E.40}$$

$$f(r) = \int_0^\infty \widetilde{F}(q) J_0(qr) q \, dq. \tag{E.41}$$

The transform pair [eqns (E.40) and (E.41)] shows that the 2D Fourier transform of a radially symmetric function and its inverse Fourier transform are related to each other via zero-order Hankel transformation. Due to the symmetric inclusion of the factor $(2\pi)^{-1}$ in the definition of the Fourier transforms [eqns (E.22) and (E.23)], this factor vanishes in the final result for $\widetilde{F}(q)$ and $f(r)$ (as compared to [970]). It is also of interest to note that for the special case considered here, the Fourier transformation can be seen as an operator acting in the 2D regime, while the Hankel transform is an operator in the 1D regime [970]. We emphasize that spin-echo-based SANS techniques such as spin-echo SANS (SESANS) and spin-echo modulated SANS (SEMSANS), as well as dark-field imaging (DFI), are directly sensitive to the projected correlation function, corresponding to $f(r)$, which by means of eqn (E.40) can be connected to the SANS cross section, corresponding to $\widetilde{F}(q)$ (see Appendix C for further details) [289, 343, 1020–1025].

Appendix F
Magnetic units

Throughout this book, we use the International System of Units (SI) for magnetic quantities and fields. In this Appendix, we briefly recall only the most relevant relations and we refer to the textbook by Jackson [161] and to [1026] for more detailed information on unitology. Of central importance for magnetic SANS is the magnetization vector field $\mathbf{M}$, which is related to the magnetic field strength $\mathbf{H}$ (sometimes also called the magnetizing force) and to the magnetic induction $\mathbf{B}$ (or magnetic flux density) via the defining materials equation

$$\mathbf{B} = \mu_0 \left(\mathbf{H} + \mathbf{M} \right) = \mu_0 \mathbf{H} + \mathbf{J}, \tag{F.1}$$

where $\mu_0 = 4\pi \times 10^{-7}\,\mathrm{Tm/A} = 4\pi \times 10^{-7}\,\mathrm{H/m}$ denotes the permeability of free space (or vacuum permeability), and

$$\mathbf{J} = \mu_0 \mathbf{M} \tag{F.2}$$

is the magnetic polarization. We emphasize that generally

$$\mathbf{H} = \mathbf{H}_0 + \mathbf{H}_\mathrm{d}, \tag{F.3}$$

where $\mathbf{H}_0$ is the externally applied magnetic field (e.g., due to an electric current flowing down a copper wire or a permanent magnet) and $\mathbf{H}_\mathrm{d}$ represents the magnetostatic field (or magnetodipolar field), which results from nonzero divergences of the magnetization (either in magnitude and/or in orientation). The dimensionless susceptibility χ may be defined as:

$$\mathbf{M} = \chi \mathbf{H}, \tag{F.4}$$

which is related to the permeability μ via the relation [compare eqn (F.1)]

$$\mu = \mu_0 (1 + \chi) = \mu_0 \mu_\mathrm{r}, \tag{F.5}$$

where μ_r is the dimensionless relative permeability. Table F.1 lists some selected magnetic quantities and their conversion from the SI to the (older) centimeter–gram–second (CGS) system.

Table F.1 SI units of some selected magnetic quantities and their conversion to the CGS system. Gaussian units are the same as CGS emu for magnetic properties (see also Appendix A in [155]). T = Tesla; G = Gauss; A = Ampère; m = meter; Oe = Oersted; emu = electromagnetic unit; cm = centimeter; kg = kilogram; g = gram; J = Joule; H = Henry; Wb = Weber; Mx = Maxwell. Note that the designation "emu" is not a unit.

Quantity	Symbol	SI	CGS	Conversion
Magnetic flux density, Magnetic induction	B	T	G	10^{-4} T$\,\hat{=}\,$1 G
Magnetic field strength	H	A/m	Oe	$\frac{10^3}{4\pi}$ A/m$\,\hat{=}\,$1 Oe
Volume magnetization	M	A/m	emu/cm^3	10^3 A/m$\,\hat{=}\,$1 emu/cm^3
Mass magnetization	σ, M	Am^2/kg	emu/g	1 Am^2/kg$\,\hat{=}\,$1 emu/g
Magnetic polarization	J	T	emu/cm^3	$4\pi10^{-4}$T$\,\hat{=}\,$1 emu/cm^3
Magnetic moment	μ, m	Am^2 (= J/T)	emu (= erg/G)	10^{-3} Am^2$\,\hat{=}\,$1 emu
Susceptibility	χ	dimensionless	emu/cm^3/Oe	$4\pi\chi^{\mathrm{SI}}\,\hat{=}\,\chi^{\mathrm{CGS}}$
Permeability	μ	H/m (= Tm/A)	dimensionless	$4\pi10^{-7}$ H/m$\,\hat{=}\,$1 G/Oe
Magnetic flux	Φ	Wb	Mx	10^{-8} Wb$\,\hat{=}\,$1 Mx
Demagnetizing factor	N, N_{d}	dimensionless	dimensionless	$4\pi N_{\mathrm{d}}^{\mathrm{SI}}\,\hat{=}\,N_{\mathrm{d}}^{\mathrm{CGS}}$
Energy	E	J	erg	10^{-7} J$\,\hat{=}\,$1 erg
Volume energy density, Energy product	ω	J/m^3	erg/cm^3	10^{-1} J/m^3$\,\hat{=}\,$1 erg/cm^3
Anisotropy constant	K	J/m^3	erg/cm^3	10^{-1} J/m^3$\,\hat{=}\,$1 erg/cm^3
Exchange-stiffness constant	A	J/m	erg/cm	10^{-5} J/m$\,\hat{=}\,$1 erg/cm

Appendix G
Fundamental constants

Avogadro constant	$N_A = 6.0221 \times 10^{23}\,\mathrm{mol}^{-1}$
Bohr magneton	$\mu_B = 9.2740 \times 10^{-24}\,\mathrm{Am}^2$ or J/T
Bohr radius	$a_0 = 0.5292 \times 10^{-10}\,\mathrm{m}$
Boltzmann constant	$k = 1.3807 \times 10^{-23}\,\mathrm{J/K}$
Classical electron radius	$r_e = 2.8179 \times 10^{-15}\,\mathrm{m}$
Electron mass	$m_e = 9.1094 \times 10^{-31}\,\mathrm{kg}$
Elementary charge	$e = 1.6022 \times 10^{-19}\,\mathrm{C}$
Neutron magnetic moment	$\mu_n = 9.6624 \times 10^{-27}\,\mathrm{Am}^2$ or J/T
Neutron mass	$m_n = 1.6749 \times 10^{-27}\,\mathrm{kg}$
Nuclear magneton	$\mu_N = 5.0508 \times 10^{-27}\,\mathrm{Am}^2$ or J/T
Permeability of vacuum	$\mu_0 = 4\pi \times 10^{-7}\,\mathrm{Tm/A}$
Permittivity of vacuum	$\epsilon_0 = 8.8542 \times 10^{-12}\,\mathrm{F/m}$
Planck constant	$h = 6.6261 \times 10^{-34}\,\mathrm{Js}$
Planck constant divided by 2π	$\hbar = 1.0546 \times 10^{-34}\,\mathrm{Js}$
Proton magnetic moment	$\mu_p = 1.4106 \times 10^{-26}\,\mathrm{Am}^2$ or J/T
Proton mass	$m_p = 1.6726 \times 10^{-27}\,\mathrm{kg}$
Speed of light in vacuum	$c = 2.9979 \times 10^8\,\mathrm{m/s}$

Values taken from [1026].

Appendix H
Glossary of symbols

A	Exchange-stiffness constant
$\mathcal{A} = 2qR_{\mathrm{HS}}$	Argument of the hard-sphere structure factor
$\mathtt{A}(q)$	A-ratio used in the analysis of magnetic steels in Section 5.2.2
a	Lattice constant
a_0	Bohr radius
a_1, a_2, a_3	Expansion coefficients of the nuclear correlation function ("differential" parameters)
$\mathbf{B}$	Magnetic induction
$\mathtt{b}$	Vortex-center displacement parameter
b	Nuclear scattering length
$b_\pm$	Compound neutron-nucleus scattering lengths
$\hat{b}$	Nuclear spin-dependent scattering length operator
$b_{\mathrm{coh}} = \overline{b}$	Nuclear coherent scattering length
b_{H}	Magnetic scattering length [eqn (1.33)]
b_{inc}	Nuclear incoherent scattering length
b_{m}	Magnetic scattering length
$C(r)$	Correlation function of the spin-misalignment SANS cross section
C_{ijkl}	Elasticity tensor
$C_{\mathrm{N}}(r)$	Autocorrelation function of the nuclear scattering-length density
$C_{\mathrm{N}\perp}(r)$	Projected autocorrelation function of the nuclear scattering-length density
$C_{\mathrm{SM}}(r)$	Autocorrelation function of the spin misalignment
CT	$= -(\widetilde{M}_y \widetilde{M}_z^* + \widetilde{M}_y^* \widetilde{M}_z)$ cross term in $d\Sigma_{\mathrm{M}}/d\Omega$ for $\mathbf{k}_0 \perp \mathbf{H}_0$
	$= -(\widetilde{M}_x \widetilde{M}_y^* + \widetilde{M}_x^* \widetilde{M}_y)$ cross term in $d\Sigma_{\mathrm{M}}/d\Omega$ for $\mathbf{k}_0 \parallel \mathbf{H}_0$
c	Speed of light in free space
$c(r)$	Normalized correlation function of the spin-misalignment SANS cross section [eqn (6.13)]
$c_\perp(r)$	Normalized projected correlation function of the spin-misalignment SANS cross section [eqn (6.21)]
$c_{\mathrm{s}}(r)$	Autocorrelation function of a uniform sphere
$c_{\mathrm{tot}}(r)$	Normalized total correlation function

D	Diameter of a sphere (particle)
D	Dzyaloshinskii–Moriya constant
D	Spin-wave stiffness constant
D_{c}	Critical single-domain size
D_{cyl}	Diameter of a circular cylinder
D_{ds}	Magnetic domain size
D_{gs}	Average crystallite (grain) size
d	Characteristic pair distance
$d\Sigma/d\Omega$	Macroscopic differential SANS cross section
$d\Sigma_{\mathrm{M}}/d\Omega$	Magnetic SANS cross section
$d\Sigma_{\mathrm{nuc}}/d\Omega$	Nuclear SANS cross section
$d\Sigma_{\mathrm{nuc}}^{\mathrm{coh}}/d\Omega$	Nuclear coherent scattering cross section
$d\Sigma_{\mathrm{nuc}}^{\mathrm{inc}}/d\Omega$	Nuclear incoherent scattering cross section
$d\Sigma_{\mathrm{res}}/d\Omega$	Residual nuclear and magnetic SANS cross section (measured at saturation)
$d\Sigma_{\mathrm{SM}}/d\Omega$	Spin-misalignment SANS cross section
$d\sigma/d\Omega$	Differential scattering cross section
$d^2\sigma/d\Omega dE_1$	Partial or double differential scattering cross section
$d\Omega$	Element of solid angle
E_0	Initial energy of neutron
E_1	Final energy of neutron
E_{dmi}	Dzyaloshinskii–Moriya energy
E_{ex}	Exchange energy
E_{m}	Magnetostatic (or dipolar interaction) energy
E_{mc}	Magnetocrystalline anisotropy energy
E_{me}	Magnetoelastic anisotropy energy
E_{tot}	Total magnetic energy
E_{z}	Zeeman energy
e	Elementary charge
$F(\mathbf{q})$	Particle form factor
$F_{\mathrm{mag}}(\mathbf{q})$	Magnetic particle form factor
$F_{\mathrm{nuc}}(\mathbf{q})$	Nuclear particle form factor
F_{R}	Flipping ratio
F_{Z}	Special function in Section 4.2.5
$f(\mathbf{q})$	Atomic magnetic form factor
$f_{\mathrm{N}}(\psi)$	Nuclear scattering amplitude
$f(R)$	Particle-size distribution function
$\mathbf{G}$	Reciprocal lattice vector
G	Gibbs free energy
$G(r)$	Projected autocorrelation function of the nuclear scattering-length density
g	Landé factor
g_{A}	Special function in Section 4.2.5
g_{MS}	Special function in Section 4.2.5

$g(r)$	Static pair-correlation function
$g_{x,y}(\mathbf{r})$	Partial derivatives of the magnetoelastic coupling energy density with respect to the magnetization
$\widetilde{g}_{x,y}(\mathbf{q})$	Fourier transform of $g_{x,y}(\mathbf{r})$
$\mathbf{H}_0$	Externally applied magnetic field
H_{c2}	External magnetic field value required to transform a helix into a ferromagnetic collinear fully polarized state [eqn (2.23)]
$\mathbf{H}_d$	Magnetostatic field
$\mathbf{H}_{dmi}$	Dzyaloshinskii–Moriya field
$\mathbf{H}_{eff}(\mathbf{r})$	Effective magnetic field in the balance-of-torques equation [eqn (3.40)]
$H_{eff}(q, H_i)$	Effective magnetic field [eqn (3.65)]
$\mathbf{H}_{ex}$	Exchange field
$\mathbf{H}_i$	Internal magnetic field
$\mathbf{H}_p$	Magnetic anisotropy field
h	Planck constant
$\hbar$	Planck constant divided by 2π
$h_i = H_i/M_s$	Reduced magnetic field in Section 4.2.5
$\hbar\omega_{dmi}$	Dispersion relation for Dzyaloshinskii–Moriya magnets
$\hbar\omega_{ph}$	Dispersion relation for long-wavelength phonons
$\hbar\omega_{sw}$	Dispersion relation for long-wavelength spin waves
I	Nuclear spin quantum number
$I(q)$	Scattering intensity
J	Exchange integral
J	Total angular momentum number
$J_m(z)$	Bessel function of order m
$J_s = \mu_0 M_s$	Saturation polarization
$j_n(z)$	Spherical Bessel function of order n
K	Wave-vector magnitude inside matter
K_c	Cubic magnetic anisotropy constant
K_s	Surface anisotropy constant
K_u	Uniaxial magnetic anisotropy constant
k	Boltzmann constant
$\mathbf{k} = qR$	Argument of Bessel and Struve functions in Section 5.1.4
k_F	Fermi wave number
$\mathbf{k}_s$	Wave vector of helix [eqn (2.23)]
$\mathbf{k}_0$	Wave vector of incident neutrons
$\mathbf{k}_1$	Wave vector of scattered neutrons
L	Lorentzian function
L	Orbital angular momentum number
$\mathcal{L}$	Length scale characterizing the size of a microstructural defect

L_{col}	Collimation length
L_{cyl}	Length of cylinder
L_{mfp}	Neutron mean free path
L_{SD}	Sample-to-detector distance
L_{vol}	Volume-weighted average column length
l_C	Correlation length of the spin misalignment
l_C^{tot}	Total correlation length
$l_{coh}^{\parallel}$	Longitudinal coherence length
$l_{coh}^{\perp}$	Transversal coherence length
l_D	Exchange length of the Dzyaloshinskii–Moriya interaction
l_H	Exchange length of the field
l_K	Domain-wall thickness
l_M	Magnetostatic exchange length
l_σ	Exchange length of the stress
$\mathbf{M}$	Magnetization vector field
$\langle \mathbf{M} \rangle$	Average magnetization
$\widetilde{\mathbf{M}}$	Fourier transform of the magnetization
M_{nuc}	Mass of the nucleus
$M_s = |\mathbf{M}|$	Saturation magnetization
m	Asymptotic power-law exponent
$\mathbf{m} = q s$	Argument of the functions g_A and g_{MS} in Section 4.2.5
m_e	Electron mass
$m_e^\star$	Effective electron mass
m_n	Neutron mass
m_p	Proton mass
N	Number of nuclei or magnetic moments in a sample
$N(\mathbf{r})$	Nuclear scattering-length density function
$\langle N \rangle$	Average nuclear scattering-length density
N_d	Demagnetizing factor
N_p	Number of particles in a sample
$\mathbf{n}$	Unit vector normal to surface
n	Index of refraction
$\mathbf{P}$	Polarization of incident neutron beam
P	Porod invariant
P_n	Nuclear spin polarization
p	Polarizer efficiency
$p(r)$	Distance distribution function
p_A	Analyzer efficiency
$\mathbf{p}_n$	Neutron momentum

$\widetilde{\mathbf{Q}}$	Halpern–Johnson vector, magnetic interaction vector, or magnetic scattering vector
Q_p	Invariant of the paramagnetic SANS cross section
$\mathbf{q}$	Wave vector, momentum-transfer vector, or scattering vector
$\hat{\mathbf{q}}$	Unit scattering vector
q_ph	Magnitude of phonon wave vector
q_sw	Magnitude of magnon wave vector
R	Radius of a sphere
R_0	Median of log-normal distribution function
$R(q)$	Ratio of nuclear to longitudinal magnetic SANS [eqn (2.118)]
$R(\mathbf{q}, \langle \mathbf{q} \rangle)$	Resolution function of SANS instrument
R_dip	Ratio of Fourier components $|\widetilde{M}_y|^2$ with and without the dipolar interaction
R_G	Radius of gyration
R_GH	Radius of gyration of the magnetic anisotropy field
R_GM	Radius of gyration of the saturation-magnetization profile
$R_\mathrm{GSM}(H_\mathrm{i})$	Radius of gyration of the spin-misalignment SANS cross section
$R_\mathrm{H,M}(q)$	Micromagnetic response functions
R_HS	Hard-sphere radius
$\mathbf{r}$	Position vector
$r(\mathbf{q})$	$\frac{d\Sigma}{d\Omega}(\mathbf{q})|_{H_0=0} / \frac{d\Sigma}{d\Omega}(\mathbf{q})|_{\mathbf{q}\|\mathbf{H}_0\to\infty}$ [eqn (5.4)]
r_D	Radial distance on the detector (measured from the beam center)
r_e	Classical electron radius
$\mathrm{rms}(q)$	Root-mean-square uncertainty in q
r_s	Neutron-source aperture radius
S	Spin angular momentum number
$S(\mathbf{q})$	Structure factor
S'	Entropy per unit volume
$S_\mathrm{H}(q)$	Scattering function of the magnetic anisotropy field
$S_\mathrm{M}(q)$	Scattering function of the longitudinal magnetization
$S_\mathrm{mag}(q)$	Magnetic scattering function for paramagnets with random magnetic susceptibility
$\mathcal{S}/V$	Surface-to-volume ratio
s	Size of inhomogeneity
$\widetilde{s}_{ij}$	Fourier-transformed components of the displacement field
T	Absolute temperature
T	Sample transmission
T_B	Superparamagnetic blocking temperature
T_C	Curie temperature
T_f	Spin-glass freezing temperature
t	Sample thickness
t_tof	Neutron time of flight

$U(\mathbf{r})$	Magnetostatic potential
U'	Internal energy density
V	Sample volume
V_{c}	Coherence or correlation volume
$V_{\mathrm{HS}}(r)$	Hard-sphere potential
V_{int}	Interaction potential between neutron and scatterer
$V_{\mathrm{int}}^{\mathrm{mag}}$	Magnetic interaction potential between neutron and scatterer
$V_{\mathrm{int}}^{\mathrm{nuc}}$	Nuclear interaction potential between neutron and scatterer
V_{p}	Particle volume
V_{s}	Sphere volume
v_{n}	Velocity of neutron
v_{s}	Velocity of sound
$\mathbf{W}_{\mathrm{exc}}$	Wave vector of magnon or phonon mode
$W_{\mathbf{k}_0 \to \mathbf{k}_1}$	Number of transitions per second from the state $\mathbf{k}_0$ to the state $\mathbf{k}_1$
$W^{\pm}$	$= (2p-1)(2\epsilon^{\pm}-1) = P(2\epsilon^{\pm}-1)$ [eqn (2.96)]
$w_0, \ldots, w_4$	Arbitrary constants
x_{p}	Volume fraction of particles
$Y_{l,m}(\alpha, \beta)$	Spherical harmonic function of degree l and order m
$Z^{+-}(q, \omega)$	Dynamical chiral function
α	Arbitrary angle, Cartesian-coordinate index
$\alpha_1, \alpha_2, \alpha_3$	Coefficients in the hard-sphere structure factor
β	Arbitrary angle, Cartesian-coordinate index
β	Critical exponent for the magnetization
β_{L}	Argument of Langevin function
γ	Arbitrary angle
γ	Critical exponent for the susceptibility
γ_0	Gyromagnetic ratio
γ_{n}	Magnetic moment of neutron expressed in units of the nuclear magneton
Δ	Gap energy in the dispersion relation for long-wavelength magnons
$\Delta = \nabla^2$	$= \partial^2/\partial x^2 + \partial^2/\partial y^2 + \partial^2/\partial z^2$ Laplace operator
$(\Delta\rho)^2_{\mathrm{mag}}$	Magnetic scattering-length density contrast
$(\Delta\rho)^2_{\mathrm{nuc}}$	Nuclear scattering-length density contrast
$\delta(\mathbf{r})$	Dirac delta function
δ_{ij}	Kronecker delta function
$\delta N(\mathbf{r})$	Excess nuclear scattering-length density function

ϵ	Spin flipper efficiency
ϵ_0	Permittivity of free space
ζ	Arbitrary angle
η	Arbitrary angle
η_{fl}	Viscosity of a fluid
η_{HS}	Volume fraction in the hard-sphere structure factor
$\widetilde{\eta}_{ij}$	Fourier-transformed components of the strain field
$\eta_{\text{M}}(H_0)$	$\frac{d\Sigma_{\text{M}}}{d\Omega}(q^*, H_0) / \frac{d\Sigma_{\text{M}}}{d\Omega}(q^*, H_0 \to \infty)$ [eqn (7.12)]
Θ_{p}	Paramagnetic Curie temperature
θ	Arbitrary angle, angle specifying the orientation of $\mathbf{q}$ on the two-dimensional detector
ϑ	Arbitrary angle
ι	Ratio of $\widetilde{H}_{\text{p}}^2$ and $\widetilde{M}_z^2$
κ^{-1}	Correlation length ξ
κ_{T}	Isothermal compressibility of a liquid
Λ	Exchange anisotropy tensor
λ	Neutron wavelength
λ_{hkl}	Magnetostriction constant along the $\langle hkl \rangle$ direction
$\lambda_{\text{s}} = l_{\text{M}}/s$	Scaled magnetostatic exchange length in Section 4.2.5
μ	Attenuation coefficient
μ_{a}	Atomic magnetic moment
μ_{B}	Bohr magneton
μ_{e}	Electron magnetic moment
$\boldsymbol{\mu}_{\text{I}}$	Nuclear magnetic moment with spin $\mathbf{I}$
μ_{N}	Nuclear magneton
μ_{n}	Neutron magnetic moment
$\boldsymbol{\mu}_{\text{S}}$	Electronic magnetic moment with spin $\mathbf{S}$
μ_0	Magnetic permeability of free space
ν	Critical exponent for the correlation length
$\nu_{\text{d,s,e}}$	Characteristic frequencies $T_{\text{d,s,e}}^{-1}$
ξ	Correlation length
ξ_{H}	Correlation length of the magnetic anisotropy field
ξ_{M}	Correlation length of the saturation-magnetization profile
ξ_{res}	Correlation length of the residual SANS cross section

ρ_a	Atomic density
$\rho_{a,\alpha}$	Atomic density of species α
ρ_d	Defect density
$\rho_{\mathbf{k}_1}$	Number of final neutron momentum states in $d\Omega$ per unit energy range
ρ_m	Mass density
ρ_{mag}	Magnetic scattering-length density
ρ_{nuc}	Nuclear scattering-length density
ρ_p	Particle density
$\rho(\mathbf{r})$	Particle density function
ρ_S	Magnetic surface charges
ρ_V	Magnetic volume charges

$\boldsymbol{\sigma}$	Stress tensor
$\boldsymbol{\sigma}_P$	Pauli spin operator
σ	Total scattering cross section
σ_a	Absorption cross section
σ_C	Width of instrumental resolution function due to finite collimation
σ_{coh}	Nuclear coherent scattering cross section
σ_{inc}	Nuclear incoherent scattering cross section
σ_{tot}	Total cross section
σ_f	Width of particle-size distribution function
σ_{LN}	Variance of log-normal distribution function
σ_R	Width of instrumental resolution function
σ_W	Width of instrumental resolution function due to wavelength spread

τ_B	Brownian relaxation time
τ_N	Néel relaxation time

Φ	Incident neutron flux
ϕ	Arbitrary angle
ϕ_{ps}	Neutron phase shift

$\chi(\mathbf{q})$	Chiral function
$\chi^2(A, H_d^s)$	χ^2-function of Section 4.4.2
χ_{ij}	Components of the magnetic susceptibility tensor
$\chi_{st}(\mathbf{q})$	Wave-vector-dependent static susceptibility

ψ	Scattering angle
$\psi_{C0}^{\Delta'}, \psi_C$	Critical angles for the suppression of one-magnon scattering in the small-angle regime

ω	Angular frequency
ω_{dmi}^{dis}	Dzyaloshinskii–Moriya energy density for torsional magnetoelastic couplings

ω_I	Larmor frequency of nuclear spin
ω_{mc}	Magnetocrystalline anisotropy energy density
ω_{me}	Magnetoelastic anisotropy energy density
ω_S	Larmor frequency of electronic spin
ω_s	Magnetic surface anisotropy energy density
ω_{tot}	Total magnetic energy density

References

[1] H. Kronmüller, A. Seeger, and M. Wilkens. *Z. Phys.*, 171:291–311, 1963.

[2] W. Schmatz, T. Springer, J. Schelten, and K. Ibel. *J. Appl. Cryst.*, 7:96–116, 1974.

[3] K. Ibel. *J. Appl. Cryst.*, 9:296–309, 1976.

[4] W. Schmatz, P. H. Dederichs, and H. Scheuer. *Z. Physik*, 270:337–341, 1974.

[5] W. Schmatz. *Rivista del Nuovo Cimento*, 5:398–422, 1975.

[6] G. Göltz, H. Kronmüller, A. Seeger, H. Scheuer, and W. Schmatz. *Philos. Mag. A*, 54:213–235, 1986.

[7] W. F. Brown Jr. *Phys. Rev.*, 58:736–743, 1940.

[8] T. Springer and W. Schmatz. *Bulletin de la Société Française de Minéralogie et de Cristallographie*, 90:428–435, 1967.

[9] A. Guinier and G. Fournet. *Small-Angle Scattering of X-rays*. Wiley, New York, 1955.

[10] O. Glatter and O. Kratky (editors). *Small Angle X-ray Scattering*. Academic Press, London, 1982.

[11] L. A. Feigin and D. I. Svergun. *Structure Analysis by Small-Angle X-Ray and Neutron Scattering*. Plenum Press, New York, 1987.

[12] D. I. Svergun, M. H. J. Koch, P. A. Timmins, and R. P. May. *Small Angle X-Ray and Neutron Scattering from Solutions of Biological Macromolecules*. Oxford University Press, Oxford, 2013.

[13] W. Gille. *Particle and Particle Systems Characterization: Small-Angle Scattering (SAS) Applications*. CRC Press, Boca Raton, 2014.

[14] E. Jericha, G. Badurek, C. Gösselsberger, and D. Süss. *J. Phys.: Conf. Ser.*, 340:012007, 2012.

[15] E. Jericha, G. Badurek, and C. Gösselsberger. *Phys. Procedia*, 42:58–65, 2013.

[16] B. Jacrot. *Rep. Prog. Phys.*, 39:911–953, 1976.

[17] J. Schelten and R. W. Hendricks. *J. Appl. Cryst.*, 11:297–324, 1978.

[18] J. S. Higgins and R. S. Stein. *J. Appl. Cryst.*, 11:346–375, 1978.

[19] V. Gerold and G. Kostorz. *J. Appl. Cryst.*, 11:376–404, 1978.

[20] J. Schelten. *Colloid & Polym. Sci.*, 259:659–665, 1981.

[21] K. Hardman-Rhyne, N. F. Berk, and E. R. Fuller, Jr. *J. Res. Natl. Inst. Stand. Technol.*, 89:17–34, 1984.

[22] S.-H. Chen and T.-L. Lin. Colloidal Solutions. In D. L. Price and K. Sköld, editors, *Methods of Experimental Physics–Neutron Scattering*, volume 23-Part B, pages 489–543. Academic Press, San Diego, 1987.

[23] J. E. Martin and A. J. Hurd. *J. Appl. Cryst.*, 20:61–78, 1987.

[24] F. S. Bates. *J. Appl. Cryst.*, 21:681–691, 1988.

[25] J. B. Hayter. *J. Appl. Cryst.*, 21:737–743, 1988.

[26] S. H. Chen, E. Y. Sheu, J. Kalus, and H. Hoffmann. *J. Appl. Cryst.*, 21:751–769, 1988.

[27] R. A. Page. *J. Appl. Cryst.*, 21:795–804, 1988.

[28] P. W. Schmidt. *J. Appl. Cryst.*, 24:414–435, 1991.

[29] G. Kostorz. *J. Appl. Cryst.*, 24:444–456, 1991.

[30] B. Boucher and P. Chieux. *J. Phys.: Condens. Matter*, 3:2207–2229, 1991.

[31] P. Lindner and Th. Zemb (editors). *Neutron, X-Ray and Light Scattering: Introduction to an Investigative Tool for Colloidal and Polymeric Systems*. North Holland, Amsterdam, 1991.

[32] S. Mazumder and A. Sequeira. *Pramana J. Phys.*, 38:95–159, 1992.

[33] G. Albertini and R. Coppola. *Nucl. Instrum. Methods Phys. Res. A*, 314:352–365, 1992.

[34] G. Albertini, R. Coppola, and F. Rustichelli. *Phys. Rep.*, 233:137–193, 1993.

[35] J. S. Pedersen. *Adv. Colloid Interface Sci.*, 70:171–210, 1997.

[36] A. Wiedenmann. Magnetic and Crystalline Nanostructures in Ferrofluids as Probed by Small Angle Neutron Scattering. In S. Odenbach, editor, *Ferrofluids: Magnetically Controllable Fluids and Their Applications, Lecture Notes in Physics*, pages 33–58. Springer-Verlag, Berlin, 2002.

[37] P. Thiyagarajan. *J. Appl. Cryst.*, 36:373–380, 2003.

[38] P. Fratzl. *J. Appl. Cryst.*, 36:397–404, 2003.

[39] D. I. Svergun and M. H. J. Koch. *Rep. Prog. Phys.*, 66:1735–1782, 2003.

[40] H. B. Stuhrmann. *Rep. Prog. Phys.*, 67:1073–1115, 2004.

[41] A. P. Radlinski, M. Mastalerz, A. L. Hinde, M. Hainbuchner, H. Rauch, M. Baron, J. S. Lin, L. Fan, and P. Thiyagarajan. *Int. J. Coal Geo.*, 59:245–271, 2004.

[42] M. R. Fitzsimmons, S. D. Bader, J. A. Borchers, G. P. Felcher, J. K. Furdyna, A. Hoffmann, J. B. Kortright, I. K. Schuller, T. C. Schulthess, S. K. Sinha, M. F. Toney, D. Weller, and S. Wolf. *J. Magn. Magn. Mater.*, 271:103–146, 2004.

[43] W. Wagner and J. Kohlbrecher. Small-Angle Neutron Scattering. In Y. Zhu, editor, *Modern Techniques for Characterizing Magnetic Materials*, pages 65–105. Kluwer Academic Publishers, Boston, 2005.

[44] A. J. Allen. *J. Am. Ceram. Soc.*, 88:1367–1381, 2005.

[45] G. D. Wignall and Y. B. Melnichenko. *Rep. Prog. Phys.*, 68:1761–1810, 2005.

[46] G. Fritz and O. Glatter. *J. Phys.: Condens. Matter*, 18:S2403–S2419, 2006.

[47] Y. B. Melnichenko and G. D. Wignall. *J. Appl. Phys.*, 102:021101, 2007.

[48] I. Grillo. Small-Angle Neutron Scattering and Applications in Soft Condensed Matter. In R. Borsali and R. Pecora, editors, *Soft Matter Characterization*, pages 723–782. Springer-Verlag, Dordrecht, 2008.

[49] A. Michels and J. Weissmüller. *Rep. Prog. Phys.*, 71:066501, 2008.

[50] A. Wiedenmann. *Collection de la Société Française de la Neutronique*, 11:219–242, 2010.

[51] M. V. Avdeev and V. L. Aksenov. *Physics–Uspekhi*, 53:971–993, 2010.

[52] B. Hammouda. *Polym. Rev.*, 50:14–39, 2010.

[53] M. R. Eskildsen, E. M. Forgan, and H. Kawano-Furukawa. *Rep. Prog. Phys.*, 74:124504, 2011.

[54] M. Laver. Small-Angle Scattering of Nanostructures and Nanomaterials. In B. Bhushan, editor, *Encyclopedia of Nanotechnology*, pages 2437–2450. Springer Netherlands, Dordrecht, 2012.

[55] B. R. Pauw. *J. Phys.: Condens. Matter*, 25:383201, 2013.

[56] M. J. Hollamby. *Phys. Chem. Chem. Phys.*, 15:10566–10579, 2013.

[57] F. Ott. *Collection de la Société Française de la Neutronique*, 13:02005, 2014.

[58] A. Michels, S. Erokhin, D. Berkov, and N. Gorn. *J. Magn. Magn. Mater.*, 350:55–68, 2014.

[59] A. Michels. *J. Phys.: Condens. Matter*, 26:383201, 2014.

[60] G. Kostorz. X-ray and Neutron Scattering. In D. E. Laughlin and K. Hono, editors, *Physical Metallurgy*, pages 1227–1316. Elsevier, Oxford, 5th edition, 2014.

[61] M. V. Avdeev, V. I. Petrenko, A.V. Feoktystov, I. V. Gapon, V. L. Aksenov, L. Vékás, and P. Kopčanský. *Ukr. J. Phys.*, 60:728–736, 2015.

[62] F. Cousin. *EPJ Web of Conferences*, 104:723–782, 2015.

[63] T. Li, A. J. Senesi, and B. Lee. *Chem. Rev.*, 116:11128–11180, 2016.

[64] S. Mühlbauer, D. Honecker, E. A. Périgo, F. Bergner, S. Disch, A. Heinemann, S. Erokhin, D. Berkov, C. Leighton, M. R. Eskildsen, and A. Michels. *Rev. Mod. Phys.*, 91:015004, 2019.

[65] E. M. Anitas. *Symmetry*, 12:65, 2020.

[66] Cy M. Jeffries, Z. Pietras, and D. I. Svergun. *EPJ Web of Conferences*, 236:03001, 2020.

[67] E. Mahieu, Z. Ibrahim, M. Moulin, M. Härtlein, B. Franzetti, A. Martel, and F. Gabel. *EPJ Web of Conferences*, 236:03002, 2020.

[68] J. S. Bhatt. *EPJ Web of Conferences*, 236:03003, 2020.

[69] R. Pynn. *Los Alamos Science*, 19:1–31, 1990.

[70] J. Chadwick. *Nature*, 129:312, 1932.

[71] H. Abele. *Prog. Part. Nuc. Phys.*, 60:1–81, 2008.

[72] A. Messiah. *Quantenmechanik, Bd. 2*. Walter de Gruyter, Berlin, 1990.

[73] B. Hammouda. *Probing Nanoscale Structures—The SANS Toolbox*. Available from: https://www.ncnr.nist.gov/staff/hammouda [Accessed February 2021].

[74] B. T. M. Willis and C. J. Carlile. *Experimental Neutron Scattering*. Oxford University Press, Oxford, 2009.

[75] J. Byrne. *Neutrons, Nuclei and Matter*. Institute of Physics Publishing, Bristol, 1994.

[76] G. L. Squires. *Introduction to the Theory of Thermal Neutron Scattering*. Cambridge University Press, Cambridge, 2012.

[77] H. Schober. *J. Neutron Res.*, 17:109–357, 2014.

[78] M. Born and E. Wolf. *Principles of Optics*. Pergamon Press, Oxford, 4th edition, 1970.

[79] R. G. Newton. *Scattering Theory of Waves and Particles*. Springer-Verlag, New York, 2nd edition, 1982.

[80] R. G. Newton. *Am. J. Phys.*, 44:639–642, 1976.

[81] B. P. Toperverg, A. Vorobyev, G. Gordeyev, A. Lazebnik, Th. Rekveldt, and W. Kraan. *Physica B*, 267–268:203–206, 1999.

[82] V. F. Sears. *Neutron Optics: An Introduction to the Theory of Neutron Optical Phenomena and their Applications.* Oxford University Press, Oxford, 1989.

[83] P. Rinard. Neutron Interactions with Matter. In D. Reilly, N. Ensslin, H. Smith Jr., and S. Kreiner, editors, *Passive Nondestructive Assay of Nuclear Materials*, pages 357–377. Los Alamos Technical Report, United States Nuclear Regulatory Commission, 1991.

[84] J. Schelten and W. Schmatz. *J. Appl. Cryst.*, 13:385–390, 1980.

[85] P. Staron and D. Bellmann. *J. Appl. Cryst.*, 35:75–81, 2002.

[86] S. W. Lovesey. *Theory of Neutron Scattering from Condensed Matter*, volume I and II. Clarendon Press, Oxford, 1984.

[87] E. Fermi. *Ric. Sci.*, 7:13–52, 1936.

[88] V. F. Sears. Neutron Scattering Lengths and Cross Sections. In K. Sköld and D. L. Price, editors, *Methods of Experimental Physics—Neutron Scattering*, volume 23-Part A, pages 521–550. Academic Press, Orlando, 1986.

[89] E. Fermi and W. H. Zinn. *Reflection of Neutrons on Mirrors.* War Department, Corps of Engineers, Office of the District Engineer, Manhattan District, Oak Ridge, Tennessee, 1946.

[90] E. Fermi and L. Marshall. *Phys. Rev.*, 71:666–677, 1947.

[91] D. J. Hughes, M. T. Burgy, and G. R. Ringo. *Phys. Rev.*, 77:291–292, 1950.

[92] M. T. Burgy, G. R. Ringo, and D. J. Hughes. *Phys. Rev.*, 84:1160–1164, 1951.

[93] W. C. Dickinson, L. Passell, and O. Halpern. *Phys. Rev.*, 126:632–642, 1962.

[94] H. Maier-Leibnitz. *Z. Angew. Phys.*, 14:738–740, 1962.

[95] L. Koester. *Z. Phys.*, 182:328–336, 1965.

[96] L. Koester. *Z. Phys.*, 198:187–200, 1967.

[97] L. Koester and H. Ungerer. *Z. Phys.*, 219:300–310, 1969.

[98] H. Rauch, W. Treimer, and U. Bonse. *Phys. Lett. A*, 47:369–371, 1974.

[99] H. Kaiser, H. Rauch, G. Badurek, W. Bauspiess, and U. Bonse. *Z. Phys. A*, 291:231–238, 1979.

[100] W. Ketter, W. Heil, G. Badurek, M. Baron, R. Loidl, and H. Rauch. *J. Res. Natl. Inst. Stand. Technol.*, 110:241–244, 2005.

[101] H. Rauch and W. Waschkowski. Neutron Scattering Lengths. In H. Schopper, editor, *Low Energy Neutrons and their Interaction with Nuclei and Matter. Part 1. Landolt-Börnstein—Group I Elementary Particles, Nuclei and Atoms (Numerical Data and Functional Relationships in Science and Technology)*, volume 16A1, pages 1–29. Springer-Verlag, Berlin, 2000.

[102] V. F. Sears. *Neutron News*, 3:26–37, 1992.

[103] W. G. Williams. *Polarized Neutrons.* Clarendon Press, Oxford, 1988.

[104] J. Schweizer. Polarized Neutrons and Polarization Analysis. In T. Chatterji, editor, *Neutron Scattering from Magnetic Materials*, pages 153–213. Elsevier, Amsterdam, 2006.

[105] L. L. Foldy. *Rev. Mod. Phys.*, 30:471–481, 1958.

[106] F. Bloch. *Phys. Rev.*, 50:259–260, 1936.

[107] F. Bloch. *Phys. Rev.*, 51:994, 1937.

[108] J. S. Schwinger. *Phys. Rev.*, 51:544–552, 1937.

[109] O. Halpern and M. H. Johnson. *Phys. Rev.*, 55:898–923, 1939.

[110] S. V. Maleev. *Sov. Phys. JETP*, 13:860–862, 1961.

[111] Yu. A. Izyumov and S. V. Maleev. *Sov. Phys. JETP*, 14:1168–1171, 1962.

[112] S. V. Maleev, V. G. Bar'yakhtar, and R. A. Suris. *Sov. Phys. Solid State*, 4:2533–2539, 1963.

[113] M. Blume. *Phys. Rev.*, 130:1670–1676, 1963.

[114] Yu. A. Izyumov. *Sov. Phys. Uspekhi*, 16:359–389, 1963.

[115] P. G. de Gennes. Theory of Neutron Scattering by Magnetic Crystals. In G. T. Rado and H. Suhl, editors, *Magnetism III: Spin Arrangements and Crystal Structure, Domains, and Micromagnetics*, volume 3, pages 115–147. Academic Press, New York, 1963.

[116] S. V. Maleev. *Physics–Uspekhi*, 45:569–596, 2002.

[117] R. M. Moon, T. Riste, and W. C. Koehler. *Phys. Rev.*, 181:920–931, 1969.

[118] A. G. Klein and S. A. Werner. *Rep. Prog. Phys.*, 46:259–335, 1983.

[119] H. Kronmüller and S. Parkin, editors. *Handbook of Magnetism and Advanced Magnetic Materials*. Wiley, Chichester, volume 2: Micromagnetism edition, 2007.

[120] S. Legvold. Rare Earth Metals and Alloys. In E. P. Wohlfarth, editor, *Handbook of Magnetic Materials*, volume 1, pages 183–295. North-Holland Publishing Company, Amsterdam, 1980.

[121] M. Bischof, P. Staron, A. Michels, P. Granitzer, K. Rumpf, H. Leitner, C. Scheu, and H. Clemens. *Acta Mater.*, 55:2637–2646, 2007.

[122] M. Hennion, I. Mirebeau, B. Hennion, S. Lequien, and F. Hippert. *EPL (Europhysics Letters)*, 2:393–399, 1986.

[123] M. Th. Rekveldt. *J. Phys. Colloq. C1*, 32:579–581, 1971.

[124] G. M. Drabkin, A. I. Okorokov, and V. V. Runov. *JETP Lett.*, 15:324–326, 1972.

[125] A. I. Okorokov, V. V. Runov, and A. G. Gukasov. *Nucl. Instrum. Methods*, 157:487–493, 1978.

[126] A. I. Okorokov and V. V. Runov. *Physica B*, 297:239–244, 2001.

[127] M. Batz, S. Baeßler, W. Heil, E. W. Otten, D. Rudersdorf, J. Schmiedeskamp, Y. Sobolev, and M. Wolf. *J. Res. Natl. Inst. Stand. Technol.*, 110:293–298, 2005.

[128] R. J. Weiss. *Phys. Rev.*, 83:379–389, 1951.

[129] H. C. van de Hulst. *Optics of Spherical Particles*. Recherches Astronomiques de l'Observatoire d'Utrecht. N. V. Drukkerij J. F. Duwaer & Zonen, volume XI, Part I, Amsterdam, 1946.

[130] Lord Rayleigh. *Proc. Roy. Soc. A*, 84:25–46, 1910.

[131] R. Gans. *Ann. Phys.*, 381:29–38, 1925.

[132] R. von Nardroff. *Phys. Rev.*, 28:240–246, 1926.

[133] O. Schaerpf. *J. Appl. Cryst.*, 11:626–630, 1978.

[134] H. Kronmüller and M. Fähnle. *Micromagnetism and the Microstructure of Ferromagnetic Solids*. Cambridge University Press, Cambridge, 2003.

[135] S. Flohrer, R. Schäfer, C. Polak, and G. Herzer. *Acta Mater.*, 53:2937–2942, 2005.

[136] D. J. Hughes, J. R. Wallace, and R. H. Holtzman. *Phys. Rev.*, 73:1277–1290, 1948.

[137] D. J. Hughes, M. T. Burgy, R. B. Heller, and J. W. Wallace. *Phys. Rev.*, 75:565–569, 1949.

[138] O. Halpern and T. Holstein. *Phys. Rev.*, 59:960–981, 1941.

[139] S. V. Maleev and V. A. Ruban. *Sov. Phys. JETP*, 31:111–116, 1970.

[140] H. W. Weber, K. Pfeiffer, and H. Rauch. *Z. Phys.*, 244:383–394, 1971.

[141] S. V. Maleev and V. A. Ruban. *Sov. Phys. JETP*, 35:222–226, 1972.

[142] M. Th. Rekveldt. *Z. Phys.*, 259:391–410, 1973.

[143] M. Th. Rekveldt. *J. Magn. Magn. Mater.*, 1:342–350, 1976.

[144] S. Mitsuda and Y. Endoh. *J. Phys. Soc. Jpn.*, 54:1570–1580, 1985.

[145] R. Rosman and M. Th. Rekveldt. *Phys. Rev. B*, 43:8437–8449, 1991.

[146] I. Mirebeau, S. Itoh, S. Mitsuda, T. Watanabe, Y. Endoh, M. Hennion, and P. Calmettes. *Phys. Rev. B*, 44:5120–5128, 1991.

[147] S. Mitsuda, H. Yoshizawa, T. Watanabe, S. Itoh, Y. Endoh, and I. Mirebeau. *J. Phys. Soc. Jpn.*, 60:1721–1729, 1991.

[148] S. Mitsuda, H. Yoshizawa, and Y. Endoh. *Phys. Rev. B*, 45:9788–9797, 1992.

[149] S. V. Grigoriev, S. V. Maleyev, A. I. Okorokov, and V. V. Runov. *Phys. Rev. B*, 58:3206–3211, 1998.

[150] S. V. Grigoriev, S. A. Klimko, W. H. Kraan, S. V. Maleyev, A. I. Okorokov, M. Th. Rekveldt, and V. V. Runov. *Phys. Rev. B*, 64:094426, 2001.

[151] M. Seifert, M. Schulz, G. Benka, C. Pfleiderer, and S. Gilder. *J. Phys.: Conf. Ser.*, 862:012024, 2017.

[152] S. Chikazumi. *Physics of Ferromagnetism*. Oxford University Press, Oxford, 2nd edition, 1997.

[153] G. Bertotti. *Hysteresis in Magnetism*. Academic Press, San Diego, 1998.

[154] A. Aharoni. *Introduction to the Theory of Ferromagnetism*. Oxford University Press, Oxford, 2nd edition, 2000.

[155] S. Blundell. *Magnetism in Condensed Matter*. Oxford University Press, Oxford, 2001.

[156] D. Goll. Micromagnetism–Microstructure Relations and the Hysteresis Loop. In H. Kronmüller and S. Parkin, editors, *Handbook of Magnetism and Advanced Magnetic Materials*, pages 1023–1058. Wiley, Chichester, volume 2: Micromagnetism edition, 2007.

[157] H. Kronmüller. *Z. Phys.*, 168:478–494, 1962.

[158] H.-D. Dietze and K. Schröder. *Phys. Status Solidi*, 27:601–610, 1968.

[159] K. Schröder and H.-D. Dietze. *Phys. Status Solidi*, 27:611–622, 1968.

[160] K. Schröder. *Phys. Status Solidi*, 33:819–830, 1969.

[161] J. D. Jackson. *Classical Electrodynamics*. Wiley, Hoboken, 3rd edition, 1999.

[162] L. G. Vivas, R. Yanes, and A. Michels. *Sci. Rep.*, 7:13060, 2017.

[163] L. G. Vivas, R. Yanes, D. Berkov, S. Erokhin, M. Bersweiler, D. Honecker, P. Bender, and A. Michels. *Phys. Rev. Lett.*, 125:117201, 2020.

[164] E. C. Stoner and E. P. Wohlfarth. *Phil. Trans. Roy. Soc. London*, A240:599–642, 1948.

[165] R. Skomski. *Simple Models of Magnetism*. Oxford University Press, Oxford, 2008.

[166] A. Vansteenkiste, J. Leliaert, M. Dvornik, M. Helsen, Felipe Garcia-Sanchez, and B. Van Waeyenberge. *AIP Advances*, 4:107133, 2014.

[167] J. Leliaert and J. Mulkers. *J. Appl. Phys.*, 125:180901, 2019.

[168] R. H. Kodama, A. E. Berkowitz, E. J. McNiff, and S. Foner. *J. Appl. Phys.*, 81:5552–5557, 1997.

[169] M. Respaud, J. M. Broto, H. Rakoto, A. R. Fert, L. Thomas, B. Barbara, M. Verelst, E. Snoeck, P. Lecante, A. Mosset, J. Osuna, T. Ould Ely, C. Amiens, and B. Chaudret. *Phys. Rev. B*, 57:2925–2935, 1998.

[170] L. Tauxe, H. N. Bertram, and C. Seberino. *Geochem. Geophys. Geosyst.*, 3:1055, 2002.

[171] D. A. Garanin and H. Kachkachi. *Phys. Rev. Lett.*, 90:065504, 2003.

[172] A. Witt, K. Fabian, and U. Bleil. *Earth Planet. Sci. Lett.*, 233:311–324, 2005.

[173] H. Kachkachi and E. Bonet. *Phys. Rev. B*, 73:224402, 2006.

[174] L. Berger, Y. Labaye, M. Tamine, and J. M. D. Coey. *Phys. Rev. B*, 77:104431, 2008.

[175] J. Mazo-Zuluaga, J. Restrepo, F. Muñoz, and J. Mejía-López. *J. Appl. Phys.*, 105:123907, 2009.

[176] C. Gatel, F. J. Bonilla, A. Meffre, E. Snoeck, B. Warot-Fonrose, B. Chaudret, L.-M. Lacroix, and T. Blon. *Nano Letters*, 15:6952–6957, 2015.

[177] Z. Nedelkoski, D. Kepaptsoglou, L. Lari, T. Wen, R. A. Booth, S. D. Oberdick, P. L. Galindo, Q. M. Ramasse, R. F. L. Evans, S. Majetich, and V. K. Lazarov. *Sci. Rep.*, 7:45997, 2017.

[178] D. Aurélio and J. Vejpravova. *Nanomaterials*, 10:1149, 2020.

[179] S. A. Pathak and R. Hertel, arXiv:2007.05939v1 [cond-mat.mes-hall], 2020.

[180] C. D. Dewhurst, I. Grillo, D. Honecker, M. Bonnaud, M. Jacques, C. Amrouni, A. Perillo-Marcone, G. Manzin, and R. Cubitt. *J. Appl. Cryst.*, 49:1–14, 2016.

[181] A. Sokolova, A. E. Whitten, L. de Campo, J. Christoforidis, A. Eltobaji, J. Barnes, F. Darmann, and A. Berry. *J. Appl. Cryst.*, 52:1–12, 2019.

[182] S. Mühlbauer, A. Heinemann, A. Wilhelm, L. Karge, A. Ostermann, I. Defendi, A. Schreyer, W. Petry, and R. Gilles. *Nucl. Instrum. Methods Phys. Res. A*, 832:297–305, 2016.

[183] C. J. Glinka, J. G. Barker, B. Hammouda, S. Krueger, J. J. Moyer, and W. J. Orts. *J. Appl. Cryst.*, 31:430–445, 1998.

[184] A. R. Rennie, M. S. Hellsing, K. Wood, E. P. Gilbert, L. Porcar, R. Schweins, C. D. Dewhurst, P. Lindner, R. K. Heenan, S. E. Rogers, P. D. Butler, J. R. Krzywon, R. E. Ghosh, A. J. Jackson, and M. Malfois. *J. Appl. Cryst.*, 46:1289–1297, 2013.

[185] G. Fritz-Popovski. *J. Appl. Cryst.*, 46:1447–1454, 2013.

[186] T. Keller, T. Krist, A. Danzig, U. Keiderling, F. Mezei, and A. Wiedenmann. *Nucl. Instrum. Methods Phys. Res. A*, 451:474–479, 2000.

[187] A. N. Bazhenov, V. M . Lobashev, A. N. Pirozhkov, and V. N. Slusar. *Nucl. Instrum. Methods Phys. Res. A*, 332:534–536, 1993.

[188] A. K. Petoukhov, V. Guillard, K. H. Andersen, E. Bourgeat-Lami, R. Chung, H. Humblot, D. Jullien, E. Lelievre-Berna, T. Soldner, F. Tasset, and M. Thomas. *Nucl. Instrum. Methods Phys. Res. A*, 560:480–484, 2006.

[189] Y. Quan, B. van den Brandt, J. Kohlbrecher, W. Th. Wenckebach, and P. Hautle. *Nucl. Instrum. Methods Phys. Res. A*, 921:22–26, 2019.

[190] Y. Quan, B. van den Brandt, J. Kohlbrecher, and P. Hautle. *J. Phys.: Conf. Ser.*, 1316:012010, 2019.

[191] B. G. Yerozolimsky. *Nucl. Instrum. Methods Phys. Res. A*, 420:232–242, 1999.

[192] O. Zimmer. *Phys. Lett. B*, 461:307–314, 1999.

[193] V. K. Aswal, B. van den Brandt, P. Hautle, J. Kohlbrecher, J. A. Konter, A. Michels, F. M. Piegsa, J. Stahn, S. Van Petegem, and O. Zimmer. *Nucl. Instrum. Methods Phys. Res. A*, 586:86–89, 2008.

[194] A. Wiedenmann. *J. Appl. Cryst.*, 33:428–432, 2000.

[195] C. G. Shull, E. O. Wollan, and W. C. Koehler. *Phys. Rev.*, 84:912–921, 1951.

[196] W. Marshall and R. D. Lowde. *Rep. Prog. Phys.*, 31:705–775, 1968.

[197] F. Mezei. *Physica*, 137B:295–308, 1986.

[198] F. Tasset. *Physica B*, 156-157:627–630, 1989.

[199] O. Schärpf and H. Capellmann. *Phys. Status Solidi A*, 135:359–379, 1993.

[200] P. J. Brown. Spherical Neutron Polarimetry. In T. Chatterji, editor, *Neutron Scattering from Magnetic Materials*, pages 215–244. Elsevier, Amsterdam, 2006.

[201] T. J. Hicks. *Magnetism in Disorder*. Clarendon Press, Oxford, 1995.

[202] D. F. R. Mildner and J. M. Carpenter. *J. Appl. Cryst.*, 17:249–256, 1984.

[203] J. S. Pedersen, D. Posselt, and K. Mortensen. *J. Appl. Cryst.*, 23:321–333, 1990.

[204] J. S. Pedersen. *J. Phys. Colloques C8*, 3:491–498, 1993.

[205] J. G. Barker and J. S. Pedersen. *J. Appl. Cryst.*, 28:105–114, 1995.

[206] B. Dorner and A. R. Wildes. *Langmuir*, 19:7823–7828, 2003.

[207] T. Freltoft, J. K. Kjems, and S. K. Sinha. *Phys. Rev. B*, 33:269–275, 1986.

[208] M. Kotlarchyk and S.-H. Chen. *J. Chem. Phys.*, 79:2461–2469, 1983.

[209] L. Van Hove. *Phys. Rev.*, 95:249–262, 1954.

[210] R. Gähler, J. Felber, F. Mezei, and R. Golub. *Phys. Rev. A*, 58:280–295, 1998.

[211] R. Gähler, J. Felber, F. Mezei, and R. Golub. *Phys. Rev. A*, 58:4249, 1998.

[212] A. S. Marathay. *Elements of Optical Coherence Theory*. Wiley, New York, 1982.

[213] T. Keller, W. Besenböck, J. Felber, R. Gähler, R. Golub, P. Hank, and M. Köppe. *Physica B*, 234-236:1120–1125, 1997.

[214] J. Felber, R. Gähler, R. Golub, and K. Prechtel. *Physica B*, 252:34–43, 1998.

[215] S. V. Grigoriev, A. V. Syromyatnikov, A. P. Chumakov, N. A. Grigoryeva, K. S. Napolskii, I. V. Roslyakov, A. A. Eliseev, A. V. Petukhov, and H. Eckerlebe. *Phys. Rev. B*, 81:125405, 2010.

[216] S. Demirdiş, C. J. van der Beek, S. Mühlbauer, Y. Su, and Th. Wolf. *J. Phys.: Condens. Matter*, 28:425701, 2016.

[217] T. Adams, S. Mühlbauer, C. Pfleiderer, F. Jonietz, A. Bauer, A. Neubauer, R. Georgii, P. Böni, U. Keiderling, K. Everschor, M. Garst, and A. Rosch. *Phys. Rev. Lett.*, 107:217206, 2011.

[218] R. Lermer and A. Steyerl. *Phys. Status Solidi A*, 33:531–541, 1976.

[219] G. D. Wignall, D. K. Christen, and V. Ramakrishnan. *J. Appl. Cryst.*, 21:438–451, 1988.

[220] A. T. Boothroyd. *J. Appl. Cryst.*, 22:252–255, 1989.

[221] R. P. May. *J. Appl. Cryst.*, 27:298–301, 1994.

[222] A. J. Allen and N. F. Berk. *J. Appl. Cryst.*, 27:878–891, 1994.

[223] S. V. Maleyev. *Phys. Rev. B*, 52:13163–13168, 1995.

[224] G. Long, S. Krueger, and A. Allen. *J. Neutron Res.*, 7:195–210, 1999.

[225] J. Kohlbrecher and W. Wagner. *J. Appl. Cryst.*, 33:804–806, 2000.

[226] J. Šaroun. *J. Appl. Cryst.*, 33:824–828, 2000.

[227] S. Mazumder, D. Sen, S. K. Roy, M. Hainbuchner, M. Baron, and H. Rauch. *J. Phys.: Condens. Matter.*, 13:5089–5102, 2001.

[228] J. Šaroun. *J. Appl. Cryst.*, 40:s701–s705, 2007.

[229] D. F. R. Mildner and R. Cubitt. *J. Appl. Cryst.*, 45:124–126, 2012.

[230] J. G. Barker and D. F. R. Mildner. *J. Appl. Cryst.*, 48:1055–1071, 2015.

[231] H. Frielinghaus. *Nucl. Instrum. Methods Phys. Res. A*, 904:9–14, 2018.

[232] C. Herring and C. Kittel. *Phys. Rev.*, 81:869–880, 1951.

[233] S. V. Maleev. *Sov. Phys. JETP*, 6:776–784, 1958.

[234] S. V. Maleev. *Sov. Phys. JETP*, 21:969–975, 1965.

[235] R. D. Lowde and N. Umakantha. *Phys. Rev. Lett.*, 4:452–454, 1960.

[236] M. Hatherly, K. Hirakawa, R. D. Lowde, J. F. Mallett, M. W. Stringfellow, and B. H. Torrie. *Proc. Phys. Soc.*, 84:55–62, 1964.

[237] M. W. Stringfellow. *J. Phys. C: Solid State Phys.*, 1:950–965, 1968.

[238] G. Maier and N. Stump. *Z. Phys.*, 238:389–398, 1970.

[239] S. V. Grigoriev, K. A. Pshenichnyi, I. A. Baraban, V. V. Rodionova, K. A. Chichai, and A. Heinemann. *JETP Lett.*, 110:793–798, 2019.

[240] G. Shirane, V. J. Minkiewicz, and R. Nathans. *J. Appl. Phys.*, 39:383–390, 1968.

[241] J. W. Lynn. *Phys. Rev. B*, 11:2624–2637, 1975.

[242] J. W. Lynn and H. A. Mook. *Phys. Rev. B*, 23:198–206, 1981.

[243] H. A. Alperin, O. Steinsvoll, G. Shirane, and R. Nathans. *J. Appl. Phys.*, 37:1052–1053, 1966.

[244] A. I. Okorokov, A. G. Gukasov, V. V. Runov, V. E. Mikhaĭlova, and M. Roth. *Sov. Phys. JETP*, 54:775–781, 1981.

[245] A. V. Lazuta, S. V. Maleev, and B. P. Toperverg. *Sov. Phys. JETP*, 54:782–788, 1981.

[246] A. G. Gukasov, A. I. Okorokov, F. Fuzhara, and O. Sherp. *JETP Lett.*, 37:513–516, 1983.

[247] A. I. Okorokov, V. V. Runov, B. P. Toperverg, A. D. Tret'yakov, E. I. Mal'tsev, I. M. Puzeĭ, and V. E. Mikhaĭlova. *JETP Lett.*, 43:503–507, 1986.

[248] V. Deriglazov, A. Okorokov, V. Runov, B. Toperverg, R. Kampmann, H. Eckerlebe, W. Schmidt, and W. Löbner. *Physica B*, 180-181:262–264, 1992.

[249] B. P. Toperverg, V. V. Deriglazov, and V. E. Mikhailova. *Physica B*, 183:326–330, 1993.

[250] S. V. Maleyev. *Phys. Rev. Lett.*, 75:4682–4685, 1995.

[251] S. V. Grigoriev, A. S. Sukhanov, E. V. Altynbaev, S.-A. Siegfried, A. Heinemann, P. Kizhe, and S. V. Maleyev. *Phys. Rev. B*, 92:220415(R), 2015.

[252] S.-A. Siegfried, A. S. Sukhanov, E. V. Altynbaev, D. Honecker, A. Heinemann, A. V. Tsvyashchenko, and S. V. Grigoriev. *Phys. Rev. B*, 95:134415, 2017.

[253] S. V. Grigoriev, E. V. Altynbaev, S.-A. Siegfried, K. A. Pschenichnyi, D. Menzel, A. Heinemann, and G. Chaboussant. *Phys. Rev. B*, 97:024409, 2018.

[254] S. V. Grigoriev, K. A. Pshenichnyi, E. V. Altynbaev, S.-A. Siegfried, A. Heinemann, D. Honecker, and D. Menzel. *JETP Lett.*, 107:640–645, 2018.

[255] S. V. Grigoriev, K. A. Pschenichnyi, E. V. Altynbaev, A. Heinemann, and A. Magrez. *Phys. Rev. B*, 99:054427, 2019.

[256] S. V. Grigoriev, K. A. Pschenichnyi, E. V. Altynbaev, S.-A. Siegfried, A. Heinemann, D. Honecker, and D. Menzel. *Phys. Rev. B*, 100:094409, 2019.

[257] C. Pfleiderer, T. Adams, A. Bauer, W. Biberacher, B. Binz, F. Birkelbach, P. Böni, C. Franz, R. Georgii, M. Janoschek, F. Jonietz, T. Keller, R. Ritz, S. Mühlbauer, W. Münzer, A. Neubauer, B. Pedersen, and A. Rosch. *J. Phys.: Condens. Matter*, 22:164207, 2010.

[258] M. Kataoka. *J. Phys. Soc. Jpn.*, 56:3635–3647, 1987.

[259] P. Bak and M. H. Jensen. *J. Phys. C: Solid State Phys.*, 13:L881–L885, 1980.

[260] A. N. Bogdanov, U. K. Rössler, and C. Pfleiderer. *Physica B*, 359-361:1162–1164, 2005.

[261] S. V. Maleyev. *Phys. Rev. B*, 73:174402, 2006.

[262] S. V. Maleyev. *J. Phys.: Condens. Matter*, 16:S899–S903, 2004.

[263] A. Gukasov. *Physica B*, 267-268:97–105, 1999.

[264] S.-H. Chen and P. Tartaglia. *Scattering Methods in Complex Fluids*. Cambridge University Press, Cambridge, 2015.

[265] A.-J. Dianoux and G. Lander (editors). *ILL Neutron Data Booklet*. OCP Science Imprint, Philadelphia, 2nd edition, 2003.

[266] H. Glättli and M. Goldman. Nuclear Magnetism. In K. Sköld and D. L. Price, editors, *Methods of Experimental Physics–Neutron Scattering*, volume 23-Part C, pages 241–286. Academic Press, San Diego, 1987.

[267] M. Steiner. *J. Low Temp. Phys.*, 135:545–578, 2004.

[268] A. Abragam. *Principles of Nuclear Magnetism*. Oxford University Press, Oxford, 2002.

[269] A. Abragam and M. Goldman. *Rep. Prog. Phys.*, 41:395–467, 1978.

[270] T. Wenckebach. *Essentials of Dynamic Nuclear Polarization*. Spindrift Publications, The Netherlands, 2016.

[271] A. Abragam and M. Goldman. *Nuclear Magnetism: Order and Disorder*. Clarendon Press, Oxford, 1982.

[272] C. G. Shull and R. P. Ferrier. *Phys. Rev. Lett.*, 10:295–297, 1963.

[273] J. B. Hayter, G. T. Jenkin, and J. W. White. *Phys. Rev. Lett.*, 33:696–699, 1974.

[274] M. Kohgi, M. Ishida, Y. Ishikawa, S. Ishimoto, Y. Kanno, A. Masaike, Y. Masuda, and K. Morimoto. *J. Phys. Soc. Jpn.*, 56:2681–2688, 1987.

[275] H. Glättli, C. Fermon, M. Eisenkremer, and M. Pinot. *J. Phys. France*, 50:2375–2387, 1989.

[276] W. Knop, H.-J. Schink, H. B. Stuhrmann, R. Wagner, M. Wenkow-Es-Souni, O. Schärpf, M. Krumpolc, T. O. Niinikoski, M. Rieubland, and A. Rijllart. *J. Appl. Cryst.*, 22:352–362, 1989.

[277] M. G. D. van der Grinten and H. Glättli. *J. Phys. IV France*, 3:427–430, 1993.

[278] M. G. D. van der Grinten, H. Glättli, C. Fermon, M. Eisenkremer, and M. Pinot. *Nucl. Instrum. Methods Phys. Res. A*, 356:422–431, 1995.

[279] R. Willumeit, N. Burkhardt, R. Jünemann, J. Wadzack, K. H. Nierhaus, and H. B. Stuhrmann. *J. Appl. Cryst.*, 30:1125–1131, 1997.

[280] B. van den Brandt, H. Glättli, I. Grillo, P. Hautle, H. Jouve, J. Kohlbrecher, J. A. Konter, E. Leymarie, S. Mango, R. P. May, H. B. Stuhrmann, and O. Zimmer. *EPL (Europhysics Letters)*, 59:62–67, 2002.

[281] B. van den Brandt, H. Glättli, I. Grillo, P. Hautle, H. Jouve, J. Kohlbrecher, J. A. Konter, E. Leymarie, S. Mango, R. P. May, H. B. Stuhrmann, and O. Zimmer. *Physica B*, 335:193–195, 2003.

[282] B. van den Brandt, H. Glättli, I. Grillo, P. Hautle, H. Jouve, J. Kohlbrecher, J. A. Konter, E. Leymarie, S. Mango, R. May, A. Michels, H. B. Stuhrmann, and O. Zimmer. *Nucl. Instrum. Methods Phys. Res. A*, 526:81–90, 2004.

[283] B. van den Brandt, H. Glättli, I. Grillo, P. Hautle, H. Jouve, J. Kohlbrecher, J. A. Konter, E. Leymarie, S. Mango, R. P. May, A. Michels, H. B. Stuhrmann, and O. Zimmer. *Eur. Phys. J. B*, 49:157–165, 2006.

[284] T. Kumada, Y. Noda, S. Koizumi, and T. Hashimoto. *J. Chem. Phys.*, 133:054504, 2010.

[285] Y. Noda, T. Kumada, T. Hashimoto, and S. Koizumi. *J. Appl. Cryst.*, 44:503–513, 2011.

[286] Y. Noda, D. Yamaguchi, T. Hashimoto, S. Shamoto, S. Koizumi, T. Yuasa, T. Tominaga, and T. Sone. *Phys. Procedia*, 42:52–57, 2013.

[287] N. Niketic, B. van den Brandt, W. Th. Wenckebach, J. Kohlbrecher, and P. Hautle. *J. Appl. Cryst.*, 48:1514–1521, 2015.

[288] Y. Noda, S. Koizumi, T. Masui, R. Mashita, H. Kishimoto, D. Yamaguchi, T. Kumada, Shin-ichi Takata, K. Ohishi, and J. Suzuki. *J. Appl. Cryst.*, 49:2036–2045, 2016.

[289] R. Andersson, L. F. van Heijkamp, I. M. de Schepper, and W. G. Bouwman. *J. Appl. Cryst.*, 41:868–885, 2008.

[290] G. Porod. General Theory. In O. Glatter and O. Kratky, editors, *Small Angle X-ray Scattering*, pages 17–51. Academic Press, London, 1982.

[291] O. Glatter. Interpretation. In O. Glatter and O. Kratky, editors, *Small Angle X-ray Scattering*, pages 167–196. Academic Press, London, 1982.

[292] K. Mortensen and J. S. Pedersen. *Macromolecules*, 26:805–812, 1993.

[293] P. Lang and O. Glatter. *Langmuir*, 12:1193–1198, 1996.

[294] B. A. Frandsen, X. Yang, and S. J. L. Billinge. *Acta Cryst. A*, 70:3–11, 2014.

[295] B. A. Frandsen and S. J. L. Billinge. *Acta Cryst. A*, 71:325–334, 2015.

[296] N. Roth, A. F. May, F. Ye, B. C. Chakoumakos, and Bo B. Iversen. *IUCrJ*, 5:410–416, 2018.

[297] N. Roth, F. Ye, A. F. May, B. C. Chakoumakos, and Bo B. Iversen. *Phys. Rev. B*, 100:144404, 2019.

[298] G. Benacchio, I. Titov, A. Malyeyev, I. Peral, M. Bersweiler, P. Bender, D. Mettus, D. Honecker, E. P. Gilbert, M. Coduri, A. Heinemann, S. Mühlbauer, A. Çakır, M. Acet, and A. Michels. *Phys. Rev. B*, 99:184422, 2019.

[299] A. Guinier. *X-Ray Diffraction in Crystals, Imperfect Crystals, and Amorphous Bodies*. Dover Publications, New York, 1994.

[300] P. Debye and A. M. Bueche. *J. Appl. Phys.*, 20:518–525, 1949.

[301] L. S. Ornstein and F. Zernike. *Proceedings of the Huygens Institute—Royal Netherlands Academy of Arts and Sciences*, 17II:793–806, 1914.

[302] J. Löffler and J. Weissmüller. *Phys. Rev. B*, 52:7076–7093, 1995.

[303] J. Weissmüller, A. Michels, D. Michels, A. Wiedenmann, C. E. Krill III, H. M. Sauer, and R. Birringer. *Phys. Rev. B*, 69:054402, 2004.

[304] P. Debye, H. R. Anderson, and H. Brumberger. *J. Appl. Phys.*, 28:679–683, 1957.

[305] S. Ciccariello, J. Goodisman, and H. Brumberger. *J. Appl. Cryst.*, 21:117–128, 1988.

[306] B. Hammouda. *J. Appl. Cryst.*, 43:716–719, 2010.

[307] W. S. Rothwell. *J. Appl. Phys.*, 39:1840–1845, 1968.

[308] G. Kostorz. Inorganic Substances. In O. Glatter and O. Kratky, editors, *Small Angle X-ray Scattering*, pages 467–498. Academic Press, London, 1982.

[309] C. E. Krill and R. Birringer. *Philos. Mag. A*, 77:621–640, 1998.

[310] C. G. Granqvist and R. A. Buhrman. *J. Appl. Phys.*, 47:2200–2219, 1976.

[311] V. Haas and R. Birringer. *Nanostructured Mater.*, 1:491–504, 1992.

[312] D. Mettus and A. Michels. *J. Appl. Cryst.*, 48:1437–1450, 2015.

[313] K. Kranjc. *J. Appl. Cryst.*, 7:211–218, 1974.

[314] P. W. Schmidt, D. Avnir, D. Levy, A. Höhr, M. Steiner, and A. Röll. *J. Chem. Phys.*, 94:1474–1479, 1991.

[315] H. Hermann, A. Wiedenmann, and P. Uebele. *J. Phys.: Condens. Matter*, 9:L509–L516, 1997.

[316] A. Heinemann, H. Hermann, A. Wiedenmann, N. Mattern, and K. Wetzig. *J. Appl. Cryst.*, 33:1386–1392, 2000.

[317] H. Hermann, A. Heinemann, N. Mattern, and A. Wiedenmann. *EPL (Europhysics Letters)*, 51:127–132, 2000.

[318] P. M. Bentley and R. Cywinski. *Phys. Rev. Lett.*, 101:227202, 2008.

[319] R. Takagi, J. S. White, S. Hayami, R. Arita, D. Honecker, H. M. Rønnow, Y. Tokura, and S. Seki. *Sci. Adv.*, 4:eaau3402, 2018.

[320] J. S. White, Á. Butykai, R. Cubitt, D. Honecker, C. D. Dewhurst, L. F. Kiss, V. Tsurkan, and S. Bordács. *Phys. Rev. B*, 97:020401(R), 2018.

[321] A. R. Wildes. *Neutron News*, 17:17–25, 2006.

[322] E. Oran Brigham. *The Fast Fourier Transform*. Prentice-Hall, Englewood Cliffs, 1974.

[323] U. Keiderling. *Appl. Phys. A*, 74:S1455–S1457, 2002.

[324] U. Keiderling, A. Wiedenmann, A. Rupp, J. Klenke, and W. Heil. *Meas. Sci. Technol.*, 19:034009, 2008.

[325] K. Krycka, W. Chen, J. Borchers, B. Maranville, and S. Watson. *J. Appl. Cryst.*, 45:546–553, 2012.

[326] K. Krycka, J. Borchers, Y. Ijiri, R. Booth, and S. Majetich. *J. Appl. Cryst.*, 45:554–565, 2012.

[327] C. D. Dewhurst. Graphical Reduction and Analysis SANS Program (GRASP). Institut Laue-Langevin, Grenoble, France (2019).

[328] A. R. Wildes. *Rev. Sci. Instr.*, 70:4241–4245, 1999.

[329] K. H. Andersen, R. Cubitt, H. Humblot, D. Jullien, A. Petoukhov, F. Tasset, C. Schanzer, V. R. Shah, and A. R. Wildes. *Physica B*, 385-386:1134–1137, 2006.

[330] K. H. Andersen, R. Cubitt, H. Humblot, D. Jullien, A. Petoukhov, F. Tasset, C. Schanzer, V.R. Shah, and A. R. Wildes. *Physica B*, 531:231, 2018.

[331] G. L. Jones, F. Dias, B. Collett, W. C. Chen, T. R. Gentile, P. M. B. Piccoli, M. E. Miller, A. J. Schultz, H. Yan, X. Tong, W. M. Snow, W. T. Lee, C. Hoffmann, and J. Thomison. *Physica B*, 385-386:1131–1133, 2006.

[332] D. Honecker, A. Ferdinand, F. Döbrich, C. D. Dewhurst, A. Wiedenmann, C. Gómez-Polo, K. Suzuki, and A. Michels. *Eur. Phys. J. B*, 76:209–213, 2010.

[333] A. Michels, D. Honecker, F. Döbrich, C. D. Dewhurst, K. Suzuki, and A. Heinemann. *Phys. Rev. B*, 85:184417, 2012.

[334] K. Suzuki, A. Makino, A. Inoue, and T. Masumoto. *J. Magn. Soc. Jpn.*, 18:800–804, 1994.

[335] E. Schlömann. *J. Appl. Phys.*, 38:5027–5034, 1967.

[336] F. Zernike and J. A. Prins. *Z. Phys.*, 41:184–194, 1927.

[337] J. S. Pedersen. *J. Appl. Cryst.*, 27:595–608, 1994.

[338] J. P. Hansen and I. R. McDonald. *Theory of Simple Liquids*. Academic Press, London, 2nd edition, 1986.

[339] D. J. Kinning and E. L. Thomas. *Macromolecules*, 17:1712–1718, 1984.

[340] S. R. Kline. *J. Appl. Cryst.*, 39:895–900, 2006.

[341] I. Bressler, J. Kohlbrecher, and A. F. Thünemann. *J. Appl. Cryst.*, 48:1587–1598, 2015.

[342] SasView for Small Angle Scattering Analysis. Available from: https://www.sasview.org [Accessed February 2021].

[343] J. Kohlbrecher and A. Studer. *J. Appl. Cryst.*, 50:1395–1403, 2017.

[344] W. F. Brown Jr. *Micromagnetics*. Interscience Publishers, New York, 1963.

[345] E. Schlömann. *J. Appl. Phys.*, 42:5798–5807, 1971.

[346] H. Kronmüller and J. Ulner. *J. Magn. Magn. Mater.*, 6:52–56, 1977.

[347] M. Fähnle and H. Kronmüller. *J. Magn. Magn. Mater.*, 8:149–156, 1978.

[348] A. Hubert and R. Schäfer. *Magnetic Domains: The Analysis of Magnetic Microstructures*. Springer-Verlag, Berlin, 1998.

[349] H. B. Callen. *Thermodynamics and an Introduction to Thermostatistics*. Wiley, New York, 2nd edition, 1985.

[350] L. D. Landau and E. M. Lifshitz. *Electrodynamics of Continuous Media*, volume 8. Pergamon Press, Oxford, 2nd edition, 1984. Revised and Enlarged by E. M. Lifshitz and L. P. Pitaevskii.

[351] C. Kittel. *Rev. Mod. Phys.*, 21:541–583, 1949.

[352] T. Moriya. *Phys. Rev.*, 120:91–98, 1960.

[353] P. W. Anderson. Theory of Magnetic Exchange Interactions: Exchange in Insulators and Semiconductors. Volume 14 of *Solid State Physics*, pages 99–214. Academic Press, 1963.

[354] K. Yosida. *Theory of Magnetism*. Springer-Verlag, Berlin, 1996.

[355] J. Curély. *Monatshefte für Chemie/Chemical Monthly*, 136:987–1011, 2005.

[356] J. Curély. *Monatshefte für Chemie/Chemical Monthly*, 136:1013–1036, 2005.

[357] R. M. White. *Quantum Theory of Magnetism*. Springer-Verlag, Berlin, 3rd edition, 2007.

[358] D. Zakharov, H.-A. Krug von Nidda, M. Eremin, J. Deisenhofer, R. Eremina, and A. Loidl. Anisotropic Exchange in Spin Chains. In B. Barbara, Y. Imry, G. Sawatzky, and P. C. E. Stamp, editors, *Quantum Magnetism*, pages 193–238. Springer-Verlag, Dordrecht, 2008. Nato Science for Peace and Security, Series B: Physics and Biophysics.

[359] L. Landau and E. Lifshitz. *Phys. Z. Sowjetunion*, 8:153–169, 1935.

[360] A. S. Arrott. *IEEE Magn. Lett.*, 7:1108505, 2016.

[361] R. Skomski. Nanomagnetics. *J. Phys.: Condens. Matter*, 15:R841–R896, 2003.

[362] R. Skomski and J. M. D. Coey. *Phys. Rev. B*, 48:15812–15816, 1993.

[363] J.-S. Yang and C.-R. Chang. *IEEE Trans. Magn.*, 31:3602–3604, 1995.

[364] T. Leineweber and H. Kronmüller. *Phys. Status Solidi B*, 201:291–301, 1997.

[365] M. Bachmann, R. Fischer, and H. Kronmüller. Simulation of Magnetization Processes in Real Microstructures. In L. Schultz and K.-H. Müller, editors, *Magnetic Anisotropy and Coercivity in Rare-Earth Transition Metal Alloys*, pages 217–236. Werkstoff-Informationsgesellschaft, Frankfurt, 1998.

[366] I. Dzyaloshinsky. *J. Phys. Chem. Solids*, 4:241–255, 1958.

[367] A. N. Bogdanov and D. A. Yablonskiĭ. *Sov. Phys. JETP*, 68:101–103, 1989.

[368] A. Bogdanov and A. Hubert. *J. Magn. Magn. Mater.*, 138:255–269, 1994.

[369] A. N. Bogdanov and U. K. Rößler. *Phys. Rev. Lett.*, 87:037203, 2001.

[370] U. K. Rößler, A. N. Bogdanov, and C. Pfleiderer. *Nature*, 442:797–801, 2006.

[371] B. Binz, A. Vishwanath, and V. Aji. *Phys. Rev. Lett.*, 96:207202, 2006.

[372] M. Bode, M. Heide, K. von Bergmann, P. Ferriani, S. Heinze, G. Bihlmayer, A. Kubetzka, O. Pietzsch, S. Blügel, and R. Wiesendanger. *Nature*, 447:190–193, 2007.

[373] Y. Yamasaki, H. Sagayama, T. Goto, M. Matsuura, K. Hirota, T. Arima, and Y. Tokura. *Phys. Rev. Lett.*, 98:147204, 2007.

[374] S. Mühlbauer, B. Binz, F. Jonietz, C. Pfleiderer, A. Rosch, A. Neubauer, R. Georgii, and P. Böni. *Science*, 323:915–919, 2009.

[375] A. Bauer and C. Pfleiderer. Generic Aspects of Skyrmion Lattices in Chiral Magnets. In J. Seidel, editor, *Topological Structures in Ferroic Materials*, pages 1–28. Springer International Publishing, Cham, 2010.

[376] S. Heinze, K. von Bergmann, M. Menzel, J. Brede, A. Kubetzka, R. Wiesendanger, G. Bihlmayer, and S. Blügel. *Nat. Phys.*, 7:713–718, 2011.

[377] N. Kanazawa, J.-H. Kim, D. S. Inosov, J. S. White, N. Egetenmeyer, J. L. Gavilano, S. Ishiwata, Y. Onose, T. Arima, B. Keimer, and Y. Tokura. *Phys. Rev. B*, 86:134425, 2012.

[378] H. Wilhelm, M. Baenitz, M. Schmidt, C. Naylor, R. Lortz, U. K. Rößler, A. A. Leonov, and A. N. Bogdanov. *J. Phys.: Condens. Matter*, 24:294204, 2012.

[379] P. Milde, D. Köhler, J. Seidel, L. M. Eng, A. Bauer, A. Chacon, J. Kindervater, S. Mühlbauer, C. Pfleiderer, S. Buhrandt, C. Schütte, and A. Rosch. *Science*, 340:1076–1080, 2013.

[380] A. Fert, V. Cros, and J. Sampaio. *Nat. Nanotech.*, 8:152–156, 2013.

[381] S. Rohart and A. Thiaville. *Phys. Rev. B*, 88:184422, 2013.

[382] N. Nagaosa and Y. Tokura. *Nat. Nanotech.*, 8:899–911, 2013.

[383] M. Kostylev. *J. Appl. Phys.*, 115:233902, 2014.

[384] M. N. Wilson, A. B. Butenko, A. N. Bogdanov, and T. L. Monchesky. *Phys. Rev. B*, 89:094411, 2014.

[385] J. S. White, K. Prša, P. Huang, A. A. Omrani, I. Živković, M. Bartkowiak, H. Berger, A. Magrez, J. L. Gavilano, G. Nagy, J. Zang, and H. M. Rønnow. *Phys. Rev. Lett.*, 113:107203, 2014.

[386] Y. Tokunaga, X. Z. Yu, J. S. White, H. M. Rønnow, D. Morikawa, Y. Taguchi, and Y. Tokura. *Nat. Commun.*, 6:7638, 2015.

[387] I. Kézsmárki, S. Bordács, P. Milde, E. Neuber, L. M. Eng, J. S. White, H. M. Rønnow, C. D. Dewhurst, M. Mochizuki, K. Yanai, H. Nakamura, D. Ehlers, V. Tsurkan, and A. Loidl. *Nat. Mater.*, 14:1116–1122, 2015.

[388] N. Romming, A. Kubetzka, C. Hanneken, K. von Bergmann, and R. Wiesendanger. *Phys. Rev. Lett.*, 114:177203, 2015.

[389] F. N. Rybakov, A. B. Borisov, S. Blügel, and N. S. Kiselev. *Phys. Rev. Lett.*, 115:117201, 2015.

[390] K. Karube, J. S. White, N. Reynolds, J. L. Gavilano, H. Oike, A. Kikkawa, F. Kagawa, Y. Tokunaga, H. M. Rønnow, Y. Tokura, and Y. Taguchi. *Nat. Mater.*, 15:1237–1242, 2016.

[391] R. Wiesendanger. *Nat. Rev. Mater.*, 1:16044, 2016.

[392] A. K. Nayak, V. Kumar, T. Ma, P. Werner, E. Pippel, R. Sahoo, F. Damay, U. K. Rößler, C. Felser, and S. S. P. Parkin. *Nature*, 548:561–566, 2017.

[393] N. Kanazawa, S. Seki, and Y. Tokura. *Adv. Mater.*, 29:1603227, 2017.

[394] A. Kovács, J. Caron, A. S. Savchenko, N. S. Kiselev, K. Shibata, Z.-A. Li, N. Kanazawa, Y. Tokura, S. Blügel, and R. E. Dunin-Borkowski. *Appl. Phys. Lett.*, 111:192410, 2017.

[395] A. Chacon, L. Heinen, M. Halder, A. Bauer, W. Simeth, S. Mühlbauer, H. Berger, M. Garst, A. Rosch, and C. Pfleiderer. *Nat. Phys.*, 14:936–941, 2018.

[396] J. S. White, I. Živković, A. J. Kruchkov, M. Bartkowiak, A. Magrez, and H. M. Rønnow. *Phys. Rev. Applied*, 10:014021, 2018.

[397] J. Kindervater, I. Stasinopoulos, A. Bauer, F. X. Haslbeck, F. Rucker, A. Chacon, S. Mühlbauer, C. Franz, M. Garst, D. Grundler, and C. Pfleiderer. *Phys. Rev. X*, 9:041059, 2019.

[398] C. Back, V. Cros, H. Ebert, K. Everschor-Sitte, A. Fert, M. Garst, T. Ma, S. Mankovsky, T. L. Monchesky, M. Mostovoy, N. Nagaosa, S. S. P. Parkin, C. Pfleiderer, N. Reyren, A. Rosch, Y. Taguchi, Y. Tokura, K. von Bergmann, and J. Zang. *J. Phys. D: Appl. Phys.*, 53:363001, 2020.

[399] F. Keffer. *Phys. Rev.*, 126:896–900, 1962.

[400] A. Bogdanov. *J. Exp. Theo. Phys. Lett.*, 68:317–319, 1998.

[401] D. Cortés-Ortuño and P. Landeros. *J. Phys.: Condens. Matter*, 25:156001, 2013.

[402] Y. Iguchi, S. Uemura, K. Ueno, and Y. Onose. *Phys. Rev. B*, 92:184419, 2015.

[403] S. Seki, Y. Okamura, K. Kondou, K. Shibata, M. Kubota, R. Takagi, F. Kagawa, M. Kawasaki, G. Tatara, Y. Otani, and Y. Tokura. *Phys. Rev. B*, 93:235131, 2016.

[404] D.-H. Kim, M. Haruta, H.-W. Ko, G. Go, H.-J. Park, T. Nishimura, D.-Y. Kim, T. Okuno, Y. Hirata, Y. Futakawa, H. Yoshikawa, W. Ham, S. Kim, H. Kurata, A. Tsukamoto, Y. Shiota, T. Moriyama, S.-B. Choe, K.-J. Lee, and T. Ono. *Nat. Mater.*, 18:685–690, 2019.

[405] D. C. Wright and N. D. Mermin. *Rev. Mod. Phys.*, 61:385–432, 1989.

[406] P. G. de Gennes and J. Prost. *The Physics of Liquid Crystals*. Clarendon Press, Oxford, 2nd edition, 1993.

[407] A. Sparavigna. *Materials*, 2:674–698, 2009.

[408] M. Fiebig. *J. Phys. D: Appl. Phys.*, 38:R123–R152, 2005.

[409] Y. Tokura and S. Seki. *Adv. Mater.*, 22:1554–1565, 2010.

[410] M. Fiebig, T. Lottermoser, D. Meier, and M. Trassin. *Nat. Rev. Mater.*, 1:16046, 2016.

[411] A. Fert and P. M. Levy. *Phys. Rev. Lett.*, 44:1538–1541, 1980.

[412] P. M. Levy and A. Fert. *J. Appl. Phys.*, 52:1718–1719, 1981.

[413] P. M. Levy, C. Morgan-Pond, and A. Fert. *J. Appl. Phys.*, 53:2168–2173, 1982.

[414] A. Michels, D. Mettus, D. Honecker, and K. L. Metlov. *Phys. Rev. B*, 94:054424, 2016.

[415] R. L. Melcher. *Phys. Rev. Lett.*, 30:125–128, 1973.

[416] H. Puszkarski and P. E. Wigen. *Phys. Rev. Lett.*, 35:1017–1018, 1975.

[417] J.-H. Moon, S.-M. Seo, K.-J. Lee, K.-W. Kim, J. Ryu, H.-W. Lee, R. D. McMichael, and M. D. Stiles. *Phys. Rev. B*, 88:184404, 2013.

[418] B. W. Zingsem, M. Farle, R. L. Stamps, and R. E. Camley. *Phys. Rev. B*, 99:214429, 2019.

[419] N. Josten, T. Feggeler, R. Meckenstock, D. Spoddig, M. Spasova, K. Chai, I. Radulov, Z.-A. Li, O. Gutfleisch, M. Farle, and B. Zingsem. *Sci. Rep.*, 10:2861, 2020.

[420] A. Arrott. *J. Appl. Phys.*, 34:1108–1109, 1963.

[421] V. I. Fedorov, A. G. Gukasov, V. Kozlov, S. V. Maleyev, V. P. Plakhty, and I. A. Zobkalo. *Phys. Lett. A*, 224:372–378, 1997.

[422] S. V. Grigoriev, Yu. O. Chetverikov, D. Lott, and A. Schreyer. *Phys. Rev. Lett.*, 100:197203, 2008.

[423] V. V. Tarnavich, D. Lott, S. Mattauch, A. Oleshkevych, V. Kapaklis, and S. V. Grigoriev. *Phys. Rev. B*, 89:054406, 2014.

[424] P. Beck and M. Fähnle. *J. Magn. Magn. Mater.*, 322:3701–3703, 2010.

[425] D. A. Kitchaev, I. J. Beyerlein, and A. Van der Ven. *Phys. Rev. B*, 98:214414, 2018.

[426] A. B. Butenko and U. K. Rößler. *EPJ Web of Conferences*, 40:08006, 2013.

[427] M. S. S. Brooks and D. A. Goodings. *J. Phys. C: Solid State Phys.*, 1:1279–1287, 1968.

[428] M. Colarieti-Tosti, S. I. Simak, R. Ahuja, L. Nordström, O. Eriksson, D. Åberg, S. Edvardsson, and M. S. S. Brooks. *Phys. Rev. Lett.*, 91:157201, 2003.

[429] S. Chikazumi. *Physics of Ferromagnetism*. Clarendon Press, Oxford, 1997. chapter 12.

[430] W. F. Brown Jr. *Magnetoelastic Interactions*. Springer-Verlag, Berlin, 1966.

[431] W. F. Brown Jr. *Phys. Rev.*, 60:139–147, 1941.

[432] H. Kronmüller. Magnetisierungskurve der Ferromagnetika. In A. Seeger, editor, *Moderne Probleme der Metallphysik*, volume 2, pages 24–156. Springer-Verlag, Berlin, 1966.

[433] J. A. Osborn. *Phys. Rev.*, 67:351–357, 1945.

[434] D.-X. Chen, J. A. Brug, and R. B. Goldfarb. *IEEE Trans. Magn.*, 27:3601–3619, 1991.

[435] A. Aharoni. *J. Appl. Phys.*, 83:3432–3434, 1998.

[436] D.-X. Chen, E. Pardo, and A. Sanchez. *IEEE Trans. Magn.*, 41:2077–2088, 2005.

[437] W. F. Brown, Jr. and A. H. Morrish. *Phys. Rev.*, 105:1198–1201, 1957.

[438] W. F. Brown Jr. *Magnetostatic Principles in Ferromagnetism*. North-Holland Publishing Company, Amsterdam, 1962.

[439] K. L. Metlov and A. Michels. *Phys. Rev. B*, 91:054404, 2015.

[440] K. L. Metlov, K. Suzuki, D. Honecker, and A. Michels. *Phys. Rev. B*, 101:214410, 2020.

[441] J. Weissmüller, A. Michels, J. G. Barker, A. Wiedenmann, U. Erb, and R. D. Shull. *Phys. Rev. B*, 63:214414, 2001.

[442] A. Michels, R. N. Viswanath, J. G. Barker, R. Birringer, and J. Weissmüller. *Phys. Rev. Lett.*, 91:267204, 2003.

[443] A. Michels. *Phys. Rev. B*, 82:024433, 2010.

[444] D. Honecker and A. Michels. *Phys. Rev. B*, 87:224426, 2013.

[445] A. Michels, C. Vecchini, O. Moze, K. Suzuki, J. M. Cadogan, P. K. Pranzas, and J. Weissmüller. *EPL (Europhysics Letters)*, 72:249–255, 2005.

[446] A. Michels, C. Vecchini, O. Moze, K. Suzuki, P. K. Pranzas, J. Kohlbrecher, and J. Weissmüller. *Phys. Rev. B*, 74:134407, 2006.

[447] A. Michels, F. Döbrich, M. Elmas, A. Ferdinand, J. Markmann, M. Sharp, H. Eckerlebe, J. Kohlbrecher, and R. Birringer. *EPL (Europhysics Letters)*, 81:66003, 2008.

[448] A. Michels, M. Elmas, F. Döbrich, M. Ames, J. Markmann, M. Sharp, H. Eckerlebe, J. Kohlbrecher, and R. Birringer. *EPL (Europhysics Letters)*, 85:47003, 2009.

[449] F. Döbrich, J. Kohlbrecher, M. Sharp, H. Eckerlebe, R. Birringer, and A. Michels. *Phys. Rev. B*, 85:094411, 2012.

[450] S. V. Maleev and B. P. Toperverg. *Sov. Phys. JETP*, 51:158–165, 1980.

[451] M. Ernst, J. Schelten, and W. Schmatz. *Phys. Status Solidi A*, 7:469–476, 1971.

[452] M. Ernst, J. Schelten, and W. Schmatz. *Phys. Status Solidi A*, 7:477–483, 1971.

[453] O. Schärpf and H. Strothmann. *Physica Scripta*, T24:58–70, 1988.

[454] A. del Moral and J. Cullen. *J. Magn. Magn. Mater.*, 139:39–58, 1995.

[455] J. Weissmüller, R. D. McMichael, J. G. Barker, H. J. Brown, U. Erb, and R. D. Shull. *Mater. Res. Soc. Symp. Proc.*, 457:231–236, 1997.

[456] J. Weissmüller, R. D. McMichael, A. Michels, and R. D. Shull. *J. Res. Natl. Inst. Stand. Technol.*, 104:261–275, 1999.

[457] J. F. Löffler, H. B. Braun, W. Wagner, G. Kostorz, and A. Wiedenmann. *Phys. Rev. B*, 71:134410, 2005.

[458] R. Cywinski, J. G. Booth, and B. D. Rainford. *J. Phys. F: Met. Phys.*, 7:2567–2581, 1977.

[459] R. Pynn, J. B. Hayter, and S. W. Charles. *Phys. Rev. Lett.*, 51:710–713, 1983.

[460] C. Bellouard, I. Mirebeau, and M. Hennion. *J. Magn. Magn. Mater.*, 140-144:431–432, 1995.

[461] M. Hennion and I. Mirebeau. *J. Phys. IV France*, 9:51–66, 1999.

[462] J. Kohlbrecher, A. Wiedenmann, and H. Wollenberger. *Z. Physik B*, 104:1–4, 1997.

[463] A. Wiedenmann. *Physica B*, 297:226–233, 2001.

[464] A. Heinemann and A. Wiedenmann. *J. Appl. Cryst.*, 36:845–849, 2003.

[465] A. Heinemann, A. Wiedenmann, and M. Kammel. *Physica B*, 350:e207–e210, 2004.

[466] A. Wiedenmann. *Physica B*, 356:246–253, 2005.

[467] K. L. Metlov and A. Michels. *Sci. Rep.*, 6:25055, 2016.

[468] I. Mirebeau, N. Martin, M. Deutsch, L. J. Bannenberg, C. Pappas, G. Chaboussant, R. Cubitt, C. Decorse, and A. O. Leonov. *Phys. Rev. B*, 98:014420, 2018.

[469] D. Honecker, L. F. Barquín, and P. Bender. *Phys. Rev. B*, 101:134401, 2020.

[470] F. Y. Ogrin, S. L. Lee, M. Wismayer, T. Thomson, C. D. Dewhurst, R. Cubitt, and S. M. Weekes. *J. Appl. Phys.*, 99:08G912, 2006.

[471] S. Saranu, A. Grob, J. Weissmüller, and U. Herr. *Phys. Status Solidi A*, 205:1774–1778, 2008.

[472] F. Zighem, F. Ott, T. Maurer, G. Chaboussant, J.-Y. Piquemal, and G. Viau. *Phys. Procedia*, 42:66–73, 2013.

[473] S. Erokhin, D. Berkov, N. Gorn, and A. Michels. *IEEE Trans. Magn.*, 47:3044–3047, 2011.

[474] S. Erokhin, D. Berkov, N. Gorn, and A. Michels. *Phys. Rev. B*, 85:024410, 2012.

[475] S. Erokhin, D. Berkov, N. Gorn, and A. Michels. *Phys. Rev. B*, 85:134418, 2012.

[476] S. Erokhin, D. Berkov, and A. Michels. *Phys. Rev. B*, 92:014427, 2015.

[477] S. Erokhin and D. Berkov. *Phys. Rev. Applied*, 7:014011, 2017.

[478] S. Erokhin, D. Berkov, M. Ito, A. Kato, M. Yano, and A. Michels. *J. Phys.: Condens. Matter*, 30:125802, 2018.

[479] S. V. Grigoriev, K. S. Napolskii, N. A. Grigoryeva, A. V. Vasilieva, A. A. Mistonov, D. Yu. Chernyshov, A. V. Petukhov, D. V. Belov, A. A. Eliseev, A. V. Lukashin, Yu. D. Tretyakov, A. S. Sinitskii, and H. Eckerlebe. *Phys. Rev. B*, 79:045123, 2009.

[480] N. A. Grigoryeva, A. A. Mistonov, K. S. Napolskii, N. A. Sapoletova, A. A. Eliseev, W. Bouwman, D. V. Byelov, A. V. Petukhov, D. Yu. Chernyshov, H. Eckerlebe, A. V. Vasilieva, and S. V. Grigoriev. *Phys. Rev. B*, 84:064405, 2011.

[481] A. A. Mistonov, N. A. Grigoryeva, A. V. Chumakova, H. Eckerlebe, N. A. Sapoletova, K. S. Napolskii, A. A. Eliseev, D. Menzel, and S. V. Grigoriev. *Phys. Rev. B*, 87:220408(R), 2013.

[482] A. A. Mistonov, I. S. Shishkin, I. S. Dubitskiy, N. A. Grigoryeva, H. Eckerlebe, and S. V. Grigoriev. *J. Exp. Theo. Phys.*, 120:844–850, 2015.

[483] I. S. Dubitskiy, A. V. Syromyatnikov, N. A. Grigoryeva, A. A. Mistonov, N. A. Sapoletova, and S. V. Grigoriev. *J. Magn. Magn. Mater.*, 441:609–619, 2017.

[484] I. S. Dubitskiy, A. A. Mistonov, N. A. Grigoryeva, and S. V. Grigoriev. *Physica B*, 549:107–112, 2018.

[485] A. A. Mistonov, I. S. Dubitskiy, I. S. Shishkin, N. A. Grigoryeva, A. Heinemann, N. A. Sapoletova, G. A. Valkovskiy, and S. V. Grigoriev. *J. Magn. Magn. Mater.*, 477:99–108, 2019.

[486] J. McCord and A. Hubert. *Phys. Status Solidi A*, 171:555–562, 1999.

[487] E. A. Périgo, E. P. Gilbert, K. L. Metlov, and A. Michels. *New. J. Phys.*, 16:123031, 2014.

[488] D. Honecker, C. D. Dewhurst, K. Suzuki, S. Erokhin, and A. Michels. *Phys. Rev. B*, 88:094428, 2013.

[489] Y. Oba, N. Adachi, Y. Todaka, E. P. Gilbert, and H. Mamiya. *Phys. Rev. Research*, 2:033473, 2020.

[490] A. Michels, J. Weissmüller, and R. Birringer. *Eur. Phys. J. B*, 29:533–540, 2002.

[491] A. Michels, D. Mettus, I. Titov, A. Malyeyev, M. Bersweiler, P. Bender, I. Peral, R. Birringer, Y. Quan, P. Hautle, J. Kohlbrecher, D. Honecker, J. R. Fernández, L. F. Barquín, and K. L. Metlov. *Phys. Rev. B*, 99:014416, 2019.

[492] D. A. Tatarskiy. *J. Magn. Magn. Mater.*, 509:166899, 2020.

[493] J. Kindervater, W. Häußler, M. Janoschek, C. Pfleiderer, P. Böni, and M. Garst. *Phys. Rev. B*, 89:180408, 2014.

[494] Y. Quan, J. Kohlbrecher, P. Hautle, and A. Michels. *J. Phys.: Condens. Matter*, 32:285804, 2020.

[495] A. Michels, A. Malyeyev, I. Titov, D. Honecker, R. Cubitt, E. Blackburn, and K. Suzuki. *IUCrJ*, 7:136–142, 2020.

[496] S. K. Burke. *J. Phys. F: Met. Phys.*, 11:L53–L58, 1981.

[497] A. Michels, J. Weissmüller, A. Wiedenmann, and J. G. Barker. *J. Appl. Phys.*, 87:5953–5955, 2000.

[498] D. F. R. Mildner and P. L. Hall. *J. Phys. D: Appl. Phys.*, 19:1535–1545, 1986.

[499] D. L. Dexter. *Phys. Rev.*, 90:1007–1012, 1953.

[500] A. Seeger. *Z. Naturf.*, 11a:724–730, 1956.

[501] A. Seeger. *Acta Metall.*, 5:24–28, 1957.

[502] J. Blin. *Acta Metall.*, 5:528–533, 1957.

[503] H. H. Atkinson and R. D. Lowde. *Philos. Mag.*, 2:589–590, 1957.

[504] H. H. Atkinson and P. B. Hirsch. *Philos. Mag.*, 3:213–228, 1958.

[505] H. H. Atkinson. *Philos. Mag.*, 3:476–488, 1958.

[506] H. H. Atkinson. *J. Appl. Phys.*, 30:637–645, 1959.

[507] A. Seeger and E. Kröner. *Z. Naturf.*, 14a:74–80, 1959.

[508] A. K. Seeger. *J. Appl. Phys.*, 30:629–637, 1959.

[509] A. Seeger and M. Rühle. *Ann. Phys.*, 466:216–229, 1963.

[510] A. Stork. *Z. Phys.*, 198:386–408, 1967.

[511] W. Vorbrugg and O. Schärpf. *Philos. Mag.*, 32:629–641, 1975.

[512] W. Vorbrugg. *Philos. Mag.*, 32:643–662, 1975.

[513] G. Göltz and H. Kronmüller. *Phys. Lett.*, 77A:70–72, 1980.

[514] R. Anders, G. Göltz, H. Scheuer, and K. Stierstadt. *Solid State Commun.*, 35:423–427, 1980.

[515] R. Anders, M. Giehrl, E. Röber, K. Stierstadt, and D. Schwahn. *Solid State Commun.*, 51:111–113, 1984.

[516] B. J. Heuser. *J. Appl. Cryst.*, 27:1020–1029, 1994.

[517] B. J. Heuser and J. S. King. *J. Alloys Comp.*, 261:225–230, 1997.

[518] V. B. Shenoy, H. H. M. Cleveringa, R. Phillips, E. Van der Giessen, and A. Needleman. *Modelling Simul. Mater. Sci. Eng.*, 8:557–581, 2000.

[519] M. Maxelon, A. Pundt, W. Pyckhout-Hintzen, J. Barker, and R. Kirchheim. *Acta Mater.*, 49:2625–2634, 2001.

[520] S. M. He, N. H. van Dijk, M. Paladugu, H. Schut, J. Kohlbrecher, F. D. Tichelaar, and S. van der Zwaag. *Phys. Rev. B*, 82:174111, 2010.

[521] B. J. Heuser and H. Ju. *Phys. Rev. B*, 83:094103, 2011.

[522] S. Zhang, J. Kohlbrecher, F. D. Tichelaar, G. Langelaan, E. Brück, S. van der Zwaag, and N. H. van Dijk. *Acta Mater.*, 61:7009–7019, 2013.

[523] N. Kalanda, V. Garamus, M. Avdeev, M. Zheludkevich, M. Yarmolich, M. Serdechnova, D. C. Florian Wieland, A. Petrov, A. Zhaludkevich, and N. Sobolev. *Phys. Status Solidi B*, 256:1800428, 2019.

[524] R. Thomson, L. E. Levine, and G. G. Long. *Acta Cryst.*, A55:433–447, 1999.

[525] G. G. Long and L. E. Levine. *Acta Cryst.*, A61:557–567, 2005.

[526] J.-P. Bick, D. Honecker, F. Döbrich, K. Suzuki, E. P. Gilbert, H. Frielinghaus, J. Kohlbrecher, J. Gavilano, E. M. Forgan, R. Schweins, P. Lindner, R. Birringer, and A. Michels. *Appl. Phys. Lett.*, 102:022415, 2013.

[527] R. Przenioslo, R. Winter, H. Natter, M. Schmelzer, R. Hempelmann, and W. Wagner. *Phys. Rev. B*, 63:054408, 2001.

[528] D. Mettus, M. Deckarm, A. Leibner, R. Birringer, M. Stolpe, R. Busch, D. Honecker, J. Kohlbrecher, P. Hautle, N. Niketic, J. R. Fernández, L. F. Barquín, and A. Michels. *Phys. Rev. Materials*, 1:074403, 2017.

[529] S. Disch, E. Wetterskog, R. P. Hermann, A. Wiedenmann, U. Vainio, G. Salazar-Alvarez, L. Bergström, and Th. Brückel. *New J. Phys.*, 14:013025, 2012.

[530] J. M. D. Coey. *Magnetism and Magnetic Materials*. Cambridge University Press, Cambridge, 2009.

[531] H. Naser, C. Rado, G. Lapertot, and S. Raymond. *Phys. Rev. B*, 102:014443, 2020.

[532] A. Michels, J. Weissmüller, A. Wiedenmann, J. S. Pedersen, and J. G. Barker. *Philos. Mag. Lett.*, 80:785–792, 2000.

[533] J.-P. Bick, K. Suzuki, E. P. Gilbert, E. M. Forgan, R. Schweins, P. Lindner, C. Kübel, and A. Michels. *Appl. Phys. Lett.*, 103:122402, 2013.

[534] G. B. Arfken and H. J. Weber. *Mathematical Methods for Physicists*. Elsevier, Amsterdam, 6th edition, 2005.

[535] A. Michels, R. N. Viswanath, and J. Weissmüller. *EPL (Europhysics Letters)*, 64:43–49, 2003.

[536] A. Grob, S. Saranu, U. Herr, A. Michels, R. N. Viswanath, and J. Weissmüller. *Phys. Status Solidi A*, 201:3354–3360, 2004.

[537] N. Ito, A. Michels, J. Kohlbrecher, J. S. Garitaonandia, K. Suzuki, and J. D. Cashion. *J. Magn. Magn. Mater.*, 316:458–461, 2007.

[538] T. Schrefl and J. Fidler. *J. Magn. Magn. Mater.*, 177-181:970–975, 1998.

[539] D. Honecker, F. Döbrich, C. D. Dewhurst, A. Wiedenmann, and A. Michels. *J. Phys.: Condens. Matter*, 23:016003, 2011.

[540] R. A. Cowley. Phase Transitions. In K. Sköld and D. L. Price, editors, *Methods of Experimental Physics–Neutron Scattering*, volume 23-Part C, pages 1–67. Academic Press, San Diego, 1987.

[541] M. F. Collins. *Magnetic Critical Scattering*. Oxford University Press, Oxford, 1989.

[542] T. Oguchi and I. Ono. *J. Phys. Soc. Jpn.*, 21:2178–2193, 1966.

[543] R. Birringer, H. Gleiter, H.-P. Klein, and P. Marquardt. *Phys. Lett. A*, 102:365–369, 1984.

[544] H. Gleiter. *Prog. Mater. Sci.*, 33:223–315, 1989.

[545] D. Michels, C. E. Krill III, and R. Birringer. *J. Magn. Magn. Mater.*, 250:203–211, 2002.

[546] G. Balaji, S. Ghosh, F. Döbrich, H. Eckerlebe, and J. Weissmüller. *Phys. Rev. Lett.*, 100:227202, 2008.

[547] F. Döbrich, J.-P. Bick, R. Birringer, M. Wolff, J. Kohlbrecher, and A. Michels. *J. Phys.: Condens. Matter*, 27:046001, 2015.

[548] G. E. Bacon. *Neutron Diffraction*. Clarendon Press, Oxford, 1955.

[549] D. L. Strandburg, S. Legvold, and F. H. Spedding. *Phys. Rev.*, 127:2046–2051, 1962.

[550] C. D. Graham Jr. *J. Appl. Phys.*, 34:1341–1342, 1963.

[551] A. Ferdinand, A.-C. Probst, A. Michels, R. Birringer, and S. N. Kaul. *J. Phys.: Condens. Matter*, 26:056003, 2014.

[552] J. Blin and A. Guinier. *C. R. Acad. Sci. Paris*, 233:1288–1290, 1951.

[553] J. Blin and A. Guinier. *C. R. Acad. Sci. Paris*, 236:2150–2152, 1953.

[554] A. Molinari, R. Witte, K. K. Neelisetty, S. Gorji, C. Kübel, I. Münch, F. Wöhler, L. Hahn, S. Hengsbach, K. Bade, H. Hahn, and R. Kruk. *Adv. Mater.*, 32:1907541, 2020.

[555] A. Seeger and H. Kronmüller. *J. Phys. Chem. Solids*, 12:298–313, 1960.

[556] H. Kronmüller and A. Seeger. *J. Phys. Chem. Solids*, 18:93–115, 1961.

[557] P. H. Dederichs and G. Leibfried. *Phys. Rev.*, 188:1175–1183, 1969.

[558] G. Leibfried. *Z. Phys.*, 135:23–43, 1953.

[559] A. Michels and J. Weissmüller. *Eur. Phys. J. B*, 26:57–65, 2002.

[560] M. Bersweiler, P. Bender, L. G. Vivas, M. Albino, M. Petrecca, S. Mühlbauer, S. Erokhin, D. Berkov, C. Sangregorio, and A. Michels. *Phys. Rev. B*, 100:144434, 2019.

[561] G. Abersfelder, K. Noack, K. Stierstadt, J. Schelten, and W. Schmatz. *Philos. Mag. B*, 41:519–534, 1980.

[562] P. Kournettas, K. Stierstadt, and D. Schwahn. *Philos. Mag. B*, 51:381–388, 1985.

[563] M. Sato and K. Hirakawa. *J. Phys. Soc. Jpn.*, 39:1467–1472, 1975.

[564] C. Bellouard, I. Mirebeau, and M. Hennion. *Phys. Rev. B*, 53:5570–5578, 1996.

[565] T. Thomson, S. L. Lee, M. F. Toney, C. D. Dewhurst, F. Y. Ogrin, C. J. Oates, and S. Sun. *Phys. Rev. B*, 72:064441, 2005.

[566] Y. Ijiri, C. V. Kelly, J. A. Borchers, J. J. Rhyne, D. F. Farrell, and S. A. Majetich. *Appl. Phys. Lett.*, 86:243102, 2005.

[567] I. Bergenti, A. Deriu, L. Savini, E. Bonetti, and A. Hoell. *J. Appl. Cryst.*, 36:450–453, 2003.

[568] Y. Oba, T. Shinohara, T. Oku, J. Suzuki, M. Ohnuma, and T. Sato. *J. Phys. Soc. Jpn.*, 78:044711, 2009.

[569] T. Oku, T. Kikuchi, T. Shinohara, J. Suzuki, Y. Ishii, M. Takeda, K. Kakurai, Y. Sasaki, M. Kishimoto, M. Yokoyama, and Y. Nishihara. *Physica B*, 404:2575–2577, 2009.

[570] K. L. Krycka, R. A. Booth, C. R. Hogg, Y. Ijiri, J. A. Borchers, W. C. Chen, S. M. Watson, M. Laver, T. R. Gentile, L. R. Dedon, S. Harris, J. J. Rhyne, and S. A. Majetich. *Phys. Rev. Lett.*, 104:207203, 2010.

[571] K. L. Krycka, J. A. Borchers, R. A. Booth, Y. Ijiri, K. Hasz, J. J. Rhyne, and S. A. Majetich. *Phys. Rev. Lett.*, 113:147203, 2014.

[572] K. Hasz, Y. Ijiri, K. L. Krycka, J. A. Borchers, R. A. Booth, S. Oberdick, and S. A. Majetich. *Phys. Rev. B*, 90:180405(R), 2014.

[573] V. S. Raghuwanshi, R. Harizanova, S. Haas, D. Tatchev, I. Gugov, C. Dewhurst, C. Rüssel, and A. Hoell. *J. Non. Cryst. Solids*, 385:24–29, 2014.

[574] V. S. Raghuwanshi, R. Harizanova, D. Tatchev, A. Hoell, and C. Rüssel. *J. Solid State Chem.*, 222:103–110, 2015.

[575] S. H. Lee, D. H. Lee, H. Jung, Y.-S. Han, T.-H. Kim, and W. Yang. *Curr. Appl. Phys.*, 15:915–919, 2015.

[576] M. Herlitschke, S. Disch, I. Sergueev, K. Schlage, E. Wetterskog, L. Bergström, and R. P. Hermann. *J. Phys.: Conf. Ser.*, 711:012002, 2016.

[577] P. Bender, J. Fock, C. Frandsen, M. F. Hansen, C. Balceris, F. Ludwig, O. Posth, E. Wetterskog, L. K. Bogart, P. Southern, W. Szczerba, L. Zeng, K. Witte, C. Grüttner, F. Westphal, D. Honecker, D. González-Alonso, L. Fernández Barquín, and C. Johansson. *J. Phys. Chem. C*, 122:3068–3077, 2018.

[578] D. Zákutná, J. Vlček, P. Fitl, K. Nemkovski, D. Honecker, D. Nižňanský, and S. Disch. *Phys. Rev. B*, 98:064407, 2018.

[579] S. D. Oberdick, A. Abdelgawad, C. Moya, S. Mesbahi-Vasey, D. Kepaptsoglou, V. K. Lazarov, R. F. L. Evans, D. Meilak, E. Skoropata, J. van Lierop, I. Hunt-Isaak, H. Pan, Y. Ijiri, K. L. Krycka, J. A. Borchers, and S. A. Majetich. *Sci. Rep.*, 8:3425, 2018.

[580] M.-D. Yang, C.-H. Ho, S. Ruta, R. Chantrell, K. Krycka, O. Hovorka, F.-R. Chen, P.-S. Lai, and C.-H. Lai. *Adv. Mater.*, 30:1802444, 2018.

[581] Y. Ijiri, K. L. Krycka, I. Hunt-Isaak, H. Pan, J. Hsieh, J. A. Borchers, J. J. Rhyne, S. D. Oberdick, A. Abdelgawad, and S. A. Majetich. *Phys. Rev. B*, 99:094421, 2019.

[582] T. O. Farmer, E.-J. Guo, R. D. Desautels, L. DeBeer-Schmitt, A. Chen, Z. Wang, Q. Jia, J. A. Borchers, D. A. Gilbert, B. Holladay, S. K. Sinha, and M. R. Fitzsimmons. *Phys. Rev. Materials*, 3:081401(R), 2019.

[583] C. Kons, M.-H. Phan, H. Srikanth, D. A. Arena, Z. Nemati, J. A. Borchers, and K. L. Krycka. *Phys. Rev. Materials*, 4:034408, 2020.

[584] P. Bender, J. Leliaert, M. Bersweiler, D. Honecker, and A. Michels. *Small Sci.*, 1:2000003, 2021.

[585] D. Zákutná, D. Nižňanský, L. C. Barnsley, E. Babcock, Z. Salhi, A. Feoktystov, D. Honecker, and S. Disch. *Phys. Rev. X*, 10:031019, 2020.

[586] A. Günther, D. Honecker, J.-P. Bick, P. Szary, C. D. Dewhurst, U. Keiderling, A. V. Feoktystov, A. Tschöpe, R. Birringer, and A. Michels. *J. Appl. Cryst.*, 47:992–998, 2014.

[587] K. S. Napolskii, A. P. Chumakov, S. V. Grigoriev, N. A. Grigoryeva, H. Eckerlebe, A. A. Eliseev, A. V. Lukashin, and Yu. D. Tretyakov. *Physica B*, 404:2568–2571, 2009.

[588] T. Maurer, F. Zighem, S. Gautrot, F. Ott, G. Chaboussant, L. Cagnon, and O. Fruchart. *Phys. Procedia*, 42:74–79, 2013.

[589] T. Maurer, S. Gautrot, F. Ott, G. Chaboussant, F. Zighem, L. Cagnon, and O. Fruchart. *Phys. Rev. B*, 89:184423, 2014.

[590] A. J. Grutter, K. L. Krycka, E. V. Tartakovskaya, J. A. Borchers, K. S. M. Reddy, E. Ortega, A. Ponce, and B. J. H. Stadler. *ACS Nano*, 11:8311–8319, 2017.

[591] S. Bedanta and W. Kleemann. *J. Phys. D: Appl. Phys.*, 42:013001, 2009.

[592] S. Bedanta, O. Petracic, and W. Kleemann. Supermagnetism. In K. H. J. Buschow, editor, *Handbook of Magnetic Materials*, volume 23, pages 1–83. Elsevier, Amsterdam, 2015.

[593] C. V. Topping and S. J. Blundell. *J. Phys.: Condens. Matter*, 31:013001, 2018.

[594] K. P. Bhatti, S. El-Khatib, V. Srivastava, R. D. James, and C. Leighton. *Phys. Rev. B*, 85:134450, 2012.

[595] P. Bender, E. Wetterskog, D. Honecker, J. Fock, C. Frandsen, C. Moerland, L. K Bogart, O. Posth, W. Szczerba, H. Gavilán, R. Costo, M. T. Fernández-Díaz, D. González-Alonso, L. Fernández Barquín, and C. Johansson. *Phys. Rev. B*, 98:224420, 2018.

[596] R. E. Rosensweig. *J. Magn. Magn. Mater.*, 252:370–374, 2002.

[597] P. C. Fannin, A. Slawska-Waniewska, P. Didukh, A. T. Giannitsis, and S. W. Charles. *Eur. Phys. J. AP*, 17:3–9, 2002.

[598] K. Ridier, B. Gillon, G. Chaboussant, L. Catala, S. Mazérat, E. Rivière, and T. Mallah. *Eur. Phys. J. B*, 90:77, 2017.

[599] G. T. Rado and J. R. Weertman. *J. Phys. Chem. Solids*, 11:315–333, 1959.

[600] K. L. Metlov. *Phys. Rev. Lett.*, 105:107201, 2010.

[601] N. A. Usov and S. E. Peschany. *J. Magn. Magn. Mater.*, 118:L290–L294, 1993.

[602] Yu. K. Guslienko and K. L. Metlov. *Phys. Rev. B*, 63:100403R, 2001.

[603] K. L. Metlov and Yu. K. Guslienko. *J. Magn. Magn. Mater.*, 242-245:1015–1017, 2002.

[604] K. L. Metlov and Yu. K. Guslienko. *Phys. Rev. B*, 70:052406, 2004.

[605] K. L. Metlov. *Phys. Rev. Lett.*, 97:127205, 2006.

[606] K. L. Metlov. *Phys. Rev. B*, 88:014427, 2013.

[607] A. B. Bogatyrev and K. L. Metlov. *Low Temp. Phys.*, 41:767–771, 2015.

[608] A. B. Bogatyrëv and K. L. Metlov. *Phys. Rev. B*, 95:024403, 2017.

[609] G. Gubbiotti, G. Carlotti, F. Nizzoli, R. Zivieri, T. Okuno, and T. Shinjo. *IEEE Trans. Magn.*, 38:2532–2534, 2002.

[610] I. V. Roshchin, C.-P. Li, H. Suhl, X. Batlle, S. Roy, S. K. Sinha, S. Park, R. Pynn, M. R. Fitzsimmons, J. Mejía-López, D. Altbir, A. H. Romero, and I. K. Schuller. *EPL (Europhysics Letters)*, 86:67008, 2009.

[611] N. A. Usov and S. E. Peschany. *Fiz. Met. Metalloved.*, 12:13–24, 1994 (in Russian).

[612] P. Böni, S. M. Shapiro, and K. Motoya. *Solid State Commun.*, 60:881–884, 1986.

[613] S. Lequien, I. Mirebeau, M. Hennion, B. Hennion, F. Hippert, and A. P. Murani. *Phys. Rev. B*, 35:7279–7282, 1987.

[614] M. Hennion, B. Hennion, I. Mirebeau, S. Lequien, and F. Hippert. *J. Appl. Phys.*, 63:4071–4076, 1988.

[615] R. E. Rosensweig. *Ferrohydrodynamics*. Cambridge University Press, Cambridge, 1985.

[616] S. Odenbach. *J. Phys.: Condens. Matter*, 16:R1135–R1150, 2004.

[617] S. Odenbach, editor. *Colloidal Magnetic Fluids: Basics, Development and Application of Ferrofluids*. Springer-Verlag, Berlin, 2009.

[618] P. G. de Gennes and P. A. Pincus. *Phys. kondens. Materie*, 11:189–198, 1970.

[619] P. C. Jordan. *Mol. Phys.*, 25:961–973, 1973.

[620] D. J. Cebula, S. W. Charles, and J. Popplewell. *Colloid & Polym. Sci.*, 259:395–397, 1981.

[621] D. J. Cebula, S. W. Charles, and J. Popplewell. *J. Phys. F: Met. Phys.*, 12:L229–L234, 1982.

[622] J. B. Hayter and R. Pynn. *Phys. Rev. Lett.*, 49:1103–1106, 1982.

[623] R. Rosman, J. J. M. Janssen, and M. Th. Rekveldt. *J. Appl. Phys.*, 67:3072–3080, 1990.

[624] F. Boué, V. Cabuil, J.-C. Bacri, and R. Perzynski. *J. Magn. Magn. Mater*, 122:78–82, 1993.

[625] S. Odenbach, D. Schwahn, and K. Stierstadt. *Z. Phys. B*, 96:567–569, 1995.

[626] T. Upadhyay, R. V. Upadhyay, R. V. Mehta, V. K. Aswal, and P. S. Goyal. *Phys. Rev. B*, 55:5585, 1997.

[627] E. Dubois, V. Cabuil, F. Boué, and R. Perzynski. *J. Chem. Phys.*, 111:7147–7160, 1999.

[628] E. Dubois, R. Perzynski, F. Boué, and V. Cabuil. *Langmuir*, 16:5617–5625, 2000.

[629] M. Avdeev, M. Balasoiu, Gy. Torok, D. Bica, L. Rosta, V. L. Aksenov, L. Vekas. *J. Magn. Magn. Mater.*, 252:86–88, 2002.

[630] F. Cousin, V. Cabuil, and P. Levitz. *Langmuir*, 18:1466–1473, 2002.

[631] F. Gazeau, E. Dubois, J.-C. Bacri, F. Boué, A. Cebers, and R. Perzynski. *Phys. Rev. E*, 65:031403, 2002.

[632] A. Wiedenmann, A. Hoell, and M. Kammel. *J. Magn. Magn. Mater.*, 252:83–85, 2002.

[633] A. Hoell, R. Müller, A. Wiedenmann, and W. Gawalek. *J. Magn. Magn. Mater.*, 252:92–94, 2002.

[634] F. Gazeau, F. Boué, E. Dubois, and R. Perzynski. *J. Phys.: Condens. Matter*, 15:S1305–S1334, 2003.

[635] A. Wiedenmann, A. Hoell, M. Kammel, and P. Boesecke. *Phys. Rev. E*, 68:031203, 2003.

[636] M. Avdeev, M. V. Balasoiu, V. L. Aksenov, V. M. Garamus, J. Kohlbrecher, D. Bica, and L. Vekas. *J. Magn. Magn. Mater.*, 270:371–379, 2004.

[637] A. Hoell, A. Wiedenmann, U. Heyen, and D. Schüler. *Physica B*, 350:e309–e313, 2004.

[638] K. Butter, A. Hoell, A. Wiedenmann, A. V. Petukhov, and G. J. Vroege. *J. Appl. Cryst.*, 37:847–856, 2004.

[639] L. M. Pop, J. Hilljegerdes, S. Odenbach, and A. Wiedenmann. *Appl. Organometal. Chem.*, 18:523–528, 2004.

[640] M. Bonini, A. Wiedenmann, and P. Baglioni. *J. Phys. Chem. B*, 108:14901–14906, 2004.

[641] A. Wiedenmann, M. Kammel, and A. Hoell. *J. Magn. Magn. Mater.*, 272-276:1487–1489, 2004.

[642] A. Wiedenmann and A. Heinemann. *J. Magn. Magn. Mater.*, 289:58–61, 2005.

[643] A. Heinemann and A. Wiedenmann. *J. Magn. Magn. Mater.*, 289:149–151, 2005.

[644] G. Mériguet, E. Dubois, A. Bourdon, G. Demouchy, V. Dupuis, and R. Perzynski. *J. Magn. Magn. Mater.*, 289:39–42, 2005.

[645] L. M. Pop, S. Odenbach, A. Wiedenmann, N. Matoussevitch, and H. Bönnemann. *J. Magn. Magn. Mater.*, 289:303–306, 2005.

[646] A. Wiedenmann, U. Keiderling, K. Habicht, M. Russina, and R. Gähler. *Phys. Rev. Lett.*, 97:057202, 2006.

[647] A. Wiedenmann, M. Kammel, A. Heinemann, and U. Keiderling. *J. Phys.: Condens. Matter*, 18:S2713–S2736, 2006.

[648] M. Bonini, A. Wiedenmann, and P. Baglioni. *Mater. Sci. Eng. C*, 26:745–750, 2006.

[649] L. M. Pop and S. Odenbach. *J. Phys.: Condens. Matter*, 18:S2785–S2802, 2006.

[650] G. Mériguet, E. Dubois, M. Jardat, A. Bourdon, G. Demouchy, V. Dupuis, B. Farago, R. Perzynski, and P. Turq. *J. Phys.: Condens. Matter*, 18:S2685–S2696, 2006.

[651] G. Mériguet, F. Cousin, E. Dubois, F. Boué, A. Cebers, B. Farago, and R. Perzynski. *J. Phys. Chem. B*, 110:4378–4386, 2006.

[652] M. V. Avdeev, V. L. Aksenov, M. Balasoiu, V. M. Garamus, A. Schreyer, Gy. Török, L. Rosta, D. Bica, and L. Vékás. *J. Colloid Interface Sci.*, 295:100–107, 2006.

[653] A. Heinemann, A. Wiedenmann, and M. Kammel. *J. Appl. Cryst.*, 40:s57–s61, 2007.

[654] M. V. Avdeev, D. Bica, L. Vékás, O. Marinica, M. Balasoiu, V. L. Aksenov, L. Rosta, V. M. Garamus, and A. Schreyer. *J. Magn. Magn. Mater.*, 311:6–9, 2007.

[655] M. V. Avdeev. *J. Appl. Cryst.*, 40:56–70, 2007.

[656] M. Bonini, A. Wiedenmann, and P. Baglioni. *J. Appl. Cryst.*, 40:s254–s258, 2007.

[657] M. Klokkenburg, B. H. Erné, A. Wiedenmann, A. V. Petukhov, and A. P. Philipse. *Phys. Rev. E*, 75:051408, 2007.

[658] M. Bonini, S. Lenz, E. Falletta, F. Ridi, E. Carretti, E. Fratini, A. Wiedenmann, and P. Baglioni. *Langmuir*, 24:12644–12650, 2008.

[659] A. Wiedenmann, U. Keiderling, M. Meissner, D. Wallacher, R. Gähler, R. P. May, S. Prévost, M. Klokkenburg, B. H. Erné, and J. Kohlbrecher. *Phys. Rev. B*, 77:184417, 2008.

[660] M. V. Avdeev, E. Dubois, G. Mériguet, E. Wandersman, V. M. Garamus, A. V. Feoktystov, and R. Perzynski. *J. Appl. Cryst.*, 42:1009–1019, 2009.

[661] E. Pyanzina, S. Kantorovich, J. J. Cerdà, A. Ivanov, and C. Holm. *Mol. Phys.*, 107:571–590, 2009.

[662] J. J. Cerdà, E. Elfimova, V. Ballenegger, E. Krutikova, A. Ivanov, and C. Holm. *Phys. Rev. E*, 81:011501, 2010.

[663] A. Nagornyi, V. Petrenko, M. Avdeev, L. Bulavin, and V. Aksenov. *Journal of Surface Investigation: X-ray, Synchrotron and Neutron Techniques*, 4:976–981, 2010.

[664] M. Barrett, A. Deschner, J. P. Embs, and M. C. Rheinstädter. *Soft Matter*, 7:6678–6683, 2011.

[665] A. Wiedenmann, R. Gähler, C. D. Dewhurst, U. Keiderling, S. Prévost, and J. Kohlbrecher. *Phys. Rev. B*, 84:214303, 2011.

[666] G. Mériguet, E. Wandersman, E. Dubois, A. Cébers, J. de Andrade Gomes, G. Demouchy, J. Depeyrot, A. Robert, and R. Perzynski. *Magnetohydrodynamics*, 48:415–425, 2012.

[667] B. Frka-Petesic, E. Dubois, L. Almasy, V. Dupuis, F. Cousin, and R. Perzynski. *Magnetohydrodynamics*, 49:328–338, 2013.

[668] N. Jain, C. K. Liu, B. S. Hawkett, G. G. Warr, and W. A. Hamilton. *J. Appl. Cryst.*, 47:41–52, 2014.

[669] L. Melníková, V. I. Petrenko, M. V. Avdeev, V. M. Garamus, L. Almásy, O. I. Ivankov, L. A. Bulavin, Z. Mitróová, and P. Kopčanský. *Colloids Surf. B*, 123:82–88, 2014.

[670] L. Melnikova, V. I. Petrenko, M. V. Avdeev, O. I. Ivankov, L. A. Bulavin, V. M. Garamus, L. Almásy, Z. Mitroova, and P. Kopcansky. *J. Magn. Magn. Mater.*, 377:77–80, 2015.

[671] M. Rajnak, V. I. Petrenko, M. V. Avdeev, O. I. Ivankov, A. Feoktystov, B. Dolnik, J. Kurimsky, P. Kopcansky, and M. Timko. *Appl. Phys. Lett.*, 107:073108, 2015.

[672] P. Bender, A. Günther, D. Honecker, A. Wiedenmann, S. Disch, A. Tschöpe, A. Michels, and R. Birringer. *Nanoscale*, 7:17122–17130, 2015.

[673] V. I. Petrenko, M. V. Avdeev, L. A. Bulavin, L. Almasy, N. A. Grigoryeva, and V. L. Aksenov. *Cryst. Rep.*, 61:121–125, 2016.

[674] Z. Fu, Y. Xiao, A. Feoktystov, V. Pipich, M.-S. Appavou, Y. Su, E. Feng, W. Jin, and T. Brückel. *Nanoscale*, 8:18541–18550, 2016.

[675] J. Zhao, S. Bolisetty, S. Isabettini, J. Kohlbrecher, J. Adamcik, P. Fischer, and R. Mezzenga. *Biomacromolecules*, 17:2555–2561, 2016.

[676] M. Rajnak, M. Timko, P. Kopcansky, K. Paulovicova, J. Tothova, J. Kurimsky, B. Dolnik, R. Cimbala, M. V. Avdeev, V. I. Petrenko, and A. Feoktystov. *J. Magn. Magn. Mater.*, 431:99–102, 2017.

[677] V. I. Petrenko, A. V. Nagornyi, I. V. Gapon, L. Vekas, V. M. Garamus, L. Almasy, A. V. Feoktystov, and M. V. Avdeev. Magnetic Fluids: Structural Aspects by Scattering Techniques. In L. A. Bulavin and A. V. Chalyi, editors, *Modern Problems of Molecular Physics*, pages 205–226. Springer International Publishing, Cham, 2018.

[678] D. V. Berkov, N. L. Gorn, R. Schmitz, and D. Stock. *J. Phys.: Condens. Matter*, 18:S2595–S2621, 2006.

[679] H. B. Stuhrmann. Contrast Variation. In O. Glatter and O. Kratky, editors, *Small Angle X-ray Scattering*, pages 197–213. Academic Press, London, 1982.

[680] H. B. Stuhrmann. Contrast Variation. In H. Brumberger, editor, *Modern Aspects of Small-Angle Scattering*, pages 221–253. Kluwer Academic Publishers, Dordrecht, 1995.

[681] D. Zákutná, K. Graef, D. Dresen, L. Porcar, D. Honecker, and S. Disch. *Colloid & Polym. Sci.*, 299:281–288, 2021.

[682] M. Rajnak, V. M. Garamus, M. Timko, P. Kopcansky, K. Paulovicova, J. Kurimsky, B. Dolnik, and R. Cimbala. *Acta Phys. Pol. A*, 137:942–944, 2020.

[683] R. Gähler and R. Golub. *Z. Phys.*, 56:5–12, 1984.

[684] D. Kipping, R. Gähler, and K. Habicht. *Phys. Lett. A*, 372:1541–1546, 2008.

[685] M. Bleuel. *Nucl. Instrum. Methods Phys. Res. A*, 927:184–186, 2019.

[686] C. Glinka, M. Bleuel, P. Tsai, D. Zákutná, D. Honecker, D. Dresen, F. Mees, and S. Disch. *J. Appl. Cryst.*, 53:598–604, 2020.

[687] S. Mühlbauer, C. Pfleiderer, P. Böni, E. M. Forgan, E. H. Brandt, A. Wiedenmann, U. Keiderling, and G. Behr. *Phys. Rev. B*, 83:184502, 2011.

[688] S. Mühlbauer, J. Kindervater, T. Adams, A. Bauer, U. Keiderling, and C. Pfleiderer. *New J. Phys.*, 18:075017, 2016.

[689] A. Müller, Y. Pütz, R. Oberhoffer, N. Becker, R. Strey, A. Wiedenmann, and T. Sottmann. *Phys. Chem. Chem. Phys.*, 16:18092–18097, 2014.

[690] Y. Pütz. Ph.D. dissertation, Universität zu Köln, Germany, 2015, unpublished.

[691] E. Eidenberger, R. Schnitzer, G. A. Zickler, M. Eidenberger, M. Bischof, P. Staron, H. Leitner, A. Schreyer, and H. Clemens. *Adv. Eng. Mater.*, 13:664–673, 2011.

[692] A. Deschamps and F. de Geuser. *Metall. Mater. Trans. A*, 44A:77–86, 2013.

[693] F. Frisius and D. Buenemann. The Measurement of Radiation Defects in Iron Alloys by Means of the Small Angle Neutron Scattering. In J. Poirier and J. M. Dupouy, editors, *Proceedings of the International Conference on Irradiation Behaviour of Metallic Materials for Fast Reactor Core Components*, pages 247–252. CEA Saclay, France, 1979.

[694] W. Wagner. *Acta Metall. Mater.*, 38:2711–2719, 1990.

[695] F. Bley. *Acta Metall. Mater.*, 40:1505–1517, 1992.

[696] P. Staron, R. Kampmann, and R. Wagner. *Physica B*, 213–214:815–817, 1995.

[697] Y. J. Wang, R. Kampmann, and R. Wagner. *Physica B*, 234–236:992–994, 1997.

[698] P. Staron and R. Kampmann. *Acta Mater.*, 48:701–712, 2000.

[699] M. Schober, E. Eidenberger, P. Staron, and H. Leitner. *Microsc. Microanal.*, 17:26–33, 2011.

[700] X. Xu, J. Odqvist, S. M. King, D. Alba Venero, and P. Hedström. *Mater. Char.*, 164:110347, 2020.

[701] R. Coppola, R. Kampmann, M. Magnani, and P. Staron. *Acta Mater.*, 46:5447–5456, 1998.

[702] A. Ulbricht. Ph.D. dissertation, TU Bergakademie Freiberg, Germany, 2006, unpublished.

[703] V. G. Gavriljuk, A. L. Sozinov, A. G. Balanyuk, S. V. Grigoriev, O. A. Gubin, G. P. Kopitsa, A. I. Okorokov, and V. V. Runov. *Metall. Mater. Trans. A*, 28A:2195–2199, 1997.

[704] P. Staron, B. Jamnig, H. Leitner, R. Ebner, and H. Clemens. *J. Appl. Cryst.*, 36:415–419, 2003.

[705] J. B. Wiskel, D. G. Ivey, and H. Henein. *Metall. Mater. Trans. B*, 39B:116–124, 2008.

[706] M. J. Alinger, G. R. Odette, and D. T. Hoelzer. *Acta Mater.*, 57:392–406, 2009.

[707] M. Perrut, M.-H. Mathon, and D. Delagnes. *J. Mater. Sci.*, 47:1920–1929, 2012.

[708] M. H. Mathon, M. Perrut, S. Y. Zhong, and Y. de Carlan. *J. Nucl. Mater.*, 428:147–153, 2012.

[709] X. Boulnat, M. Perez, D. Fabregue, T. Douillard, M.-H. Mathon, and Y. de Carlan. *Metall. Mater. Trans. A*, 45A:1485–1497, 2014.

[710] R. Pareja, P. Parente, A. Muñoz, A. Radulescu, and V. de Castro. *Philos. Mag.*, 95:2450–2465, 2015.

[711] Y. Oba, S. Morooka, K. Ohishi, N. Sato, R. Inoue, N. Adachi, J. Suzuki, T. Tsuchiyama, E. P. Gilbert, and M. Sugiyama. *J. Appl. Cryst.*, 49:1659–1664, 2016.

[712] Y. Q. Wang, S. J. Clark, V. Janik, R. K. Heenan, D. Alba Venero, K. Yan, D. G. McCartney, S. Sridhar, and P. D. Lee. *Acta Mater.*, 145:84–96, 2018.

[713] C. P. Massey, S. N. Dryepondt, P. D. Edmondson, M. G. Frith, K. C. Littrell, A. Kini, B. Gault, K. A. Terrani, and S. J. Zinkle. *Acta Mater.*, 166:1–17, 2019.

[714] Y. Q. Wang, S. J. Clark, B. Cai, D. Alba Venero, K. Yan, M. Gorley, E. Surrey, D. G. McCartney, S. Sridhar, and P. D. Lee. *Scripta Mater.*, 174:24–28, 2020.

[715] R. Coppola, A. Feoktystov, T. Mueller, L. Pilloni, and A. Radulescu. *Nucl. Mater. Energy*, 24:100772, 2020.

[716] G. Albertini, F. Carsughi, R. Coppola, W. Kesternich, G. Mercurio, F. Rustichelli, D. Schwahn, and H. Ullmaier. *J. Nucl. Mater.*, 191-194:1327–1330, 1992.

[717] R. G. Carter, N. Soneda, K. Dohi, J. M. Hyde, C. A. English, and W. L. Server. *J. Nucl. Mater.*, 298:211–224, 2001.

[718] M. K. Miller, B. D. Wirth, and G. R. Odette. *Mater. Sci. Eng.*, A353:133–139, 2003.

[719] M. H. Mathon, Y. de Carlan, G. Geoffroy, X. Averty, A. Alamo, and C. H. de Novion. *J. Nucl. Mater.*, 312:236–248, 2003.

[720] F. Bergner, A. Ulbricht, H. Hein, and M. Kammel. *J. Phys.: Condens. Matter*, 20:104262, 2008.

[721] R. Coppola, M. Klimenkov, R. Lindau, A. Möslang, M. Valli, and A. Wiedenmann. *J. Nucl. Mater.*, 409:100–105, 2011.

[722] J. M. Hyde, M. G. Burke, G. D. W. Smith, P. Styman, H. Swan, and K. Wilford. *J. Nucl. Mater.*, 449:308–314, 2014.

[723] S. Shu, B. D. Wirth, P. B. Wells, D. D. Morgan, and G. R. Odette. *Acta Mater.*, 146:237–252, 2018.

[724] R. Coppola and M. Klimenkov. *Metals*, 9:552, 2019.

[725] M. J. Konstantinović, A. Ulbricht, T. Brodziansky, N. Castin, and L. Malerba. *J. Nucl. Mater.*, 540:152341, 2020.

[726] A. Ulbricht, J. Böhmert, and H.-W. Viehrig. *J. ASTM Inter.*, 2:1–14, 2005.

[727] G. R. Odette and G. E. Lucas. *Radiat. Eff. Def. Sol.*, 144:189–231, 1998.

[728] E. Altstadt, E. Keim, H. Hein, M. Serrano, F. Bergner, H.-W. Viehrig, A. Ballesteros, R. Chaouadi, and K. Wilford. *Nucl. Eng. Des.*, 278:753–757, 2014.

[729] F. Bergner, A. Ulbricht, and C. Heintze. *Scripta Mater.*, 61:1060–1063, 2009.

[730] D. Sherrington and S. Kirkpatrick. *Phys. Rev. Lett.*, 35:1792–1796, 1975.

[731] S. F. Edwards and P. W. Anderson. *J. Phys. F: Met. Phys.*, 5:965–974, 1975.

[732] C. Y. Huang. *J. Magn. Magn. Mater*, 51:1–74, 1985.

[733] K. Binder and A. P. Young. *Rev. Mod. Phys.*, 58:801–976, 1986.

[734] K. H. Fischer and J. A. Hertz. *Spin Glasses.* Cambridge University Press, Cambridge, 1991.

[735] J. A. Mydosh. *Spin Glasses: An Experimental Introduction.* Taylor & Francis, London, 1993.

[736] G. Parisi. *Proc. Natl. Acad. Sci. U.S.A.*, 103:7948–7955, 2006.

[737] E. Bolthausen and A. Bovier, editors. *Spin Glasses.* Springer-Verlag, Berlin, 2007.

[738] J. A. Mydosh. *Rep. Prog. Phys.*, 78:052501, 2015.

[739] C. Dekker, A. F. M. Arts, and H. W. de Wijn. *J. Appl. Phys.*, 63:4334–4336, 1988.

[740] M. A. Ruderman and C. Kittel. *Phys. Rev.*, 96:99–102, 1954.

[741] T. Kasuya. *Prog. Theor. Phys.*, 16:45–57, 1956.

[742] K. Yosida. *Phys. Rev.*, 106:893–898, 1957.

[743] B. R. Coles, B. V. B. Sarkissian, and R. H. Taylor. *Philos. Mag. B*, 37:489–498, 1978.

[744] A. P. Murani. *Phys. Rev. Lett.*, 37:450–453, 1976.

[745] C. M. Soukoulis, G. S. Grest, and K. Levin. *Phys. Rev. Lett.*, 41:568–571, 1978.

[746] A. P. Murani. *Phys. Rev. B*, 22:3495–3499, 1980.

[747] D. Sherrington. Spin Glasses: A Perspective. In E. Bolthausen and A. Bovier, editors, *Spin Glasses*, pages 45–62. Springer-Verlag, Berlin, 2007.

[748] F. E. Luborsky. Amorphous Ferromagnets. In E. P. Wohlfarth, editor, *Handbook of Magnetic Materials*, volume 1, pages 451–529. North-Holland Publishing Company, Amsterdam, 1980.

[749] U. Mizutani. *Prog. Mater. Sci.*, 28:97–228, 1983.

[750] P. Hansen. Magnetic Amorphous Alloys. In K. H. J. Buschow, editor, *Handbook of Magnetic Materials*, volume 6, pages 289–452. Elsevier, Amsterdam, 1991.

[751] R. Harris, M. Plischke, and M. J. Zuckermann. *Phys. Rev. Lett.*, 31:160–162, 1973.

[752] R. Alben, J. J. Becker, and M. C. Chi. *J. Appl. Phys.*, 49:1653–1658, 1978.

[753] D. J. Sellmyer and S. Nafis. *J. Appl. Phys.*, 57:3584–3588, 1985.

[754] E. M. Chudnovsky, W. M. Saslow, and R. A. Serota. *Phys. Rev. B*, 33:251–261, 1986.

[755] G. Herzer. *IEEE Trans. Magn.*, 26:1397–1402, 1990.

[756] G. Herzer. *Acta Mater.*, 61:718–734, 2013.

[757] A. P. Murani, S. Roth, P. Radhakrishna, B. D. Rainford, B. R. Coles, K. Ibel, G. Goeltz, and F. Mezei. *J. Phys. F: Met. Phys.*, 6:425–432, 1976.

[758] S. K. Burke, R. Cywinski, and B. D. Rainford. *J. Appl. Cryst.*, 11:644–648, 1978.

[759] R. Cywinski and S. K. Burke. *J. Magn. Magn. Mater.*, 14:247–249, 1979.

[760] A. P. Murani. *Solid State Commun.*, 34:705–708, 1980.

[761] S. M. Shapiro, G. Shirane, B. H. Verbeek, G. J. Nieuwenhuys, and J. A. Mydosh. *Solid State Commun.*, 36:167–170, 1980.

[762] H. R. Child. *J. Appl. Phys.*, 52:1732–1734, 1981.

[763] G. Aeppli, S. M. Shapiro, R. J. Birgeneau, and H. S. Chen. *Phys. Rev. B*, 28:5160–5172, 1983.

[764] S. K. Burke and B. D. Rainford. *J. Phys. F: Met. Phys.*, 13:441–450, 1983.

[765] S. K. Burke, R. Cywinski, J. R. Davis, and B. D. Rainford. *J. Phys. F: Met. Phys.*, 13:451–470, 1983.

[766] S. K. Burke and B. D. Rainford. *J. Phys. F: Met. Phys.*, 13:471–482, 1983.

[767] J. J. Rhyne and C. J. Glinka. *J. Appl. Phys.*, 55:1691–1693, 1984.

[768] J. J. Rhyne. *IEEE Trans. Magn.*, 21:1990–1995, 1985.

[769] M. Arai, Y. Ishikawa, N. Saito, and H. Takei. *J. Phys. Soc. Jpn.*, 54:781–794, 1985.

[770] M. Arai, Y. Ishikawa, and H. Takei. *J. Phys. Soc. Jpn.*, 54:2279–2286, 1985.

[771] O. Moze, E. J. Lindley, B. D. Rainford, and D. MckPaul. *J. Magn. Magn. Mater.*, 53:167–174, 1985.

[772] J. J. Rhyne. *Physica*, 136B:30–35, 1986.

[773] J. J. Rhyne, R. W. Erwin, J. A. Fernandez-Baca, and G. E. Fish. *J. Appl. Phys.*, 63:4080–4082, 1988.

[774] Ph. Mangin, D. Boumazouza, B. George, J. J. Rhyne, and R. W. Erwin. *Phys. Rev. B*, 40:11123–11139, 1989.

[775] J. Suzuki, Y. Endoh, M. Arai, M. Furusaka, and H. Yoshizawa. *J. Phys. Soc. Jpn.*, 59:718–724, 1990.

[776] T. Sato, T. Ando, T. Watanabe, S. Itoh, Y. Endoh, and M. Furusaka. *Phys. Rev. B*, 48:6074–6086, 1993.

[777] K. Mergia, S. Messoloras, G. Nicolaides, D. Niarchos, and R. J. Stewart. *J. Appl. Phys.*, 76:6380–6382, 1994.

[778] R. I. Bewley and R. Cywinski. *J. Magn. Magn. Mater.*, 140–144:869–870, 1995.

[779] L. Fernández Barquín, J. C. Gómez Sal, S. N. Kaul, J. M. Barandiarán, P. Gorría, J. S. Pedersen, and R. Heenan. *J. Appl. Phys.*, 79:5146–5148, 1996.

[780] F. Hellman, A. L. Shapiro, E. N. Abarra, R. A. Robinson, R. P. Hjelm, P. A. Seeger, J. J. Rhyne, and J. I. Suzuki. *Phys. Rev. B*, 59:11408–11417, 1999.

[781] A. Bracchi, K. Samwer, P. Schaaf, J. F. Löffler, and S. Schneider. *Mater. Sci. Eng. A*, 375-377:1027–1031, 2004.

[782] R. García Calderón, L. Fernández Barquín, S. N. Kaul, J. C. Gómez Sal, P. Gorria, J. S. Pedersen, and R. K. Heenan. *Phys. Rev. B*, 71:134413, 2005.

[783] D. Martín Rodríguez, F. Plazaola, J. J. del Val, J. S. Garitaonandia, G. J. Cuello, and C. Dewhurst. *J. Appl. Phys.*, 99:08H502, 2006.

[784] C. Magen, P. A. Algarabel, L. Morellon, J. P. Araújo, C. Ritter, M. R. Ibarra, A. M. Pereira, and J. B. Sousa. *Phys. Rev. Lett.*, 96:167201, 2006.

[785] N. Marcano, J. C. Gómez Sal, J. I. Espeso, J. M. De Teresa, P. A. Algarabel, C. Paulsen, and J. R. Iglesias. *Phys. Rev. Lett.*, 98:166406, 2007.

[786] K. Mergia and S. Messoloras. *J. Phys.: Condens. Matter*, 20:104219, 2008.

[787] K. Mergia and S. Messoloras. *J. Phys.: Conf. Ser.*, 340:012069, 2012.

[788] I. J. McDonald, M. E. Jamer, K. L. Krycka, E. Anber, D. Foley, A. C. Lang, W. D. Ratcliff, D. Heiman, M. L. Taheri, J. A. Borchers, and L. H. Lewis. *ACS Appl. Nano Mater.*, 2:1940–1950, 2019.

[789] A. Schroeder, S. Bhattarai, A. Gebretsadik, H. Adawi, J.-G. Lussier, and K. L. Krycka. *AIP Advances*, 10:015036, 2020.

[790] M. Bersweiler, P. Bender, I. Peral, L. Eichenberger, M. Hehn, V. Polewczyk, S. Mühlbauer, and A. Michels. *J. Phys. D: Appl. Phys.*, 53:335302, 2020.

[791] E. G. Iashina, E. V. Altynbaev, L. N. Fomicheva, A. V. Tsvyashchenko, and S. V. Grigoriev. *Journal of Surface Investigation: X-ray, Synchrotron and Neutron Techniques*, 14:429–433, 2020.

[792] H. A. Gersch, C. G. Shull, and M. K. Wilkinson. *Phys. Rev.*, 103:525–534, 1956.

[793] M. Ericson and B. Jacrot. *J. Phys. Chem. Solids*, 13:235–243, 1960.

[794] B. Jacrot, J. Konstantinovic, G. Parette, and D. Cribier. Diffusion aux petit angles des neutrons par le fer et le nickel au voisinage du point de Curie. In *Proceedings of the Symposium on Inelastic Scattering of Neutrons in Solids and Liquids*, Chalk River, Canada, 1962, volume II, pages 317–326. International Atomic Energy Agency, Vienna, 1963.

[795] L. Passell, K. Blinowski, T. Brun, and P. Nielsen. *J. Appl. Phys.*, 35:933–934, 1964.

[796] L. Passell, K. Blinowski, T. Brun, and P. Nielsen. *Phys. Rev.*, 139:A1866–A1876, 1965.

[797] G. M. Drabkin, E. I. Zabidarov, Ya. A. Kasman, and A. I. Okorokov. *JETP Lett.*, 2:336–338, 1965.

[798] D. Bally, B. Grabcev, A. M. Lungu, M. Popovici, and M. Totia. *J. Phys. Chem. Solids*, 28:1947–1955, 1967.

[799] N. Stump and G. Maier. *Phys. Lett.*, 24A:625–626, 1967.

[800] C. J. Glinka, V. J. Minkiewicz, and L. Passell. *Phys. Rev. B*, 16:4084–4103, 1977.

[801] H. S. Kogon and D. J. Wallace. *J. Phys. A: Math. Gen.*, 14:L527–L531, 1981.

[802] A. Aharony and E. Pytte. *Phys. Rev. B*, 27:5872–5874, 1983.

[803] R. J. Birgeneau, H. Yoshizawa, R. A. Cowley, G. Shirane, and H. Ikeda. *Phys. Rev. B*, 28:1438–1448, 1983.

[804] G. Placzek. *Phys. Rev.*, 86:377–388, 1952.

[805] J. Als-Nielsen. *Phys. Rev. Lett.*, 25:730–734, 1970.

[806] A. Tucciarone, H. Y. Lau, L. M. Corliss, A. Delapalme, and J. M. Hastings. *Phys. Rev. B*, 4:3206–3245, 1971.

[807] P. A. Egelstaff. Classical Fluids. In K. Sköld and D. L. Price, editors, *Methods of Experimental Physics–Neutron Scattering*, volume 23-Part B, pages 405–470. Academic Press, San Diego, 1987.

[808] A. K. Soper. *Mol. Phys.*, 107:1667–1684, 2009.

[809] M. Ohnuma, K. Hono, S. Linderoth, J. S. Pedersen, Y. Yoshizawa, and H. Onodera. *Acta Mater.*, 48:4783–4790, 2000.

[810] J. R. Childress, C. L. Chien, J. J. Rhyne, and R. W. Erwin. *J. Magn. Magn. Mater.*, 104-107:1585–1586, 1992.

[811] T. Oku, E. Ohta, T. Sato, M. Furusaka, M. Imai, and Y. Matsushita. *J. Magn. Magn. Mater.*, 188:291–300, 1998.

[812] D. F. Farrell, Y. Ijiri, C. V. Kelly, J. A. Borchers, J. J. Rhyne, Y. Ding, and S. A. Majetich. *J. Magn. Magn. Mater.*, 303:318–322, 2006.

[813] M. Sachan, C. Bonnoit, S. A. Majetich, Y. Ijiri, P. O. Mensah-Bonsu, J. A. Borchers, and J. J. Rhyne. *Appl. Phys. Lett.*, 92:152503, 2008.

[814] W. C. Koehler. *J. Appl. Phys.*, 36:1078–1087, 1965.

[815] J. Jensen and A. R. Mackintosh. *Phys. Rev. Lett.*, 64:2699–2702, 1990.

[816] T. Kosugi, S. Kawano, N. Achiwa, A. Onodera, Y. Nakai, and N. Yamamoto. *Physica B*, 334:365–368, 2003.

[817] T. Chatterji, editor. *Neutron Scattering from Magnetic Materials*. Elsevier, Amsterdam, 2006.

[818] M. Ramazanoglu, M. Laver, W. Ratcliff II, S. M. Watson, W. C. Chen, A. Jackson, K. Kothapalli, S. Lee, S.-W. Cheong, and V. Kiryukhin. *Phys. Rev. Lett.*, 107:207206, 2011.

[819] A. Michels, J.-P. Bick, R. Birringer, A. Ferdinand, J. Baller, R. Sanctuary, S. Philippi, D. Lott, S. Balog, E. Rotenberg, G. Kaindl, and K. M. Döbrich. *Phys. Rev. B*, 83:224415, 2011.

[820] P. Szary, D. Kaiser, J.-P. Bick, D. Lott, A. Heinemann, C. Dewhurst, R. Birringer, and A. Michels. *J. Appl. Cryst.*, 49:533–538, 2016.

[821] D. M. Fobes, S.-Z. Lin, N. J. Ghimire, E. D. Bauer, J. D. Thompson, M. Bleuel, L. M. DeBeer-Schmitt, and M. Janoschek. *Phys. Rev. B*, 96:174413, 2017.

[822] Y. Li, Q. Wang, L. DeBeer-Schmitt, Z. Guguchia, R. D. Desautels, J.-X. Yin, Q. Du, W. Ren, X. Zhao, Z. Zhang, I. A. Zaliznyak, C. Petrovic, W. Yin, M. Z. Hasan, H. Lei, and J. M. Tranquada. *Phys. Rev. Lett.*, 123:196604, 2019.

[823] P. Puphal, V. Pomjakushin, N. Kanazawa, V. Ukleev, D. J. Gawryluk, J. Ma, M. Naamneh, N. C. Plumb, L. Keller, R. Cubitt, E. Pomjakushina, and J. S. White. *Phys. Rev. Lett.*, 124:017202, 2020.

[824] P. Puphal, S. Krebber, E. Suard, R. Cubitt, C. Wang, T. Shang, V. Ukleev, J. S. White, and E. Pomjakushina. *Phys. Rev. B*, 101:214416, 2020.

[825] R. L. Dally, W. D. Ratcliff II, L. Zhang, H.-S. Kim, M. Bleuel, J. W. Kim, K. Haule, D. Vanderbilt, S.-W. Cheong, and J. W. Lynn. *Phys. Rev. B*, 102:014410, 2020.

[826] D. Cribier, B. Jacrot, L. Madhav Rao, and B. Farnoux. *Phys. Lett.*, 9:106–107, 1964.

[827] J. Schelten, H. Ullmaier, and W. Schmatz. *Phys. Status Solidi B*, 48:619–628, 1971.

[828] J. Schelten, H. Ullmaier, and G. Lippmann. *Z. Phys.*, 253:219–231, 1972.

[829] D. Cribier, Y. Simon, and P. Thorel. *Phys. Rev. Lett.*, 28:1370–1372, 1972.

[830] H. W. Weber, J. Schelten, and G. Lippmann. *Phys. Status Solidi B*, 57:515–522, 1973.

[831] G. Lippmann, J. Schelten, R. W. Hendricks, and W. Schmatz. *Phys. Status Solidi B*, 58:633–641, 1973.

[832] G. Lippmann and J. Schelten. *J. Appl. Cryst.*, 7:236–239, 1974.

[833] J. Schelten, G. Lippmann, and H. Ullmaier. *J. Low Temp. Phys.*, 14:213–226, 1974.

[834] D. K. Christen, H. R. Kerchner, S. T. Sekula, and P. Thorel. *Phys. Rev. B*, 21:102–117, 1980.

[835] A. A. Abrikosov. *Sov. Phys. JETP*, 5:1174–1182, 1957.

[836] R. Cubitt, E. M. Forgan, G. Yang, S. L. Lee, D. McK. Paul, H. A. Mook, M. Yethiraj, P. H. Kes, T. W. Li, A. A. Menovsky, Z. Tarnawski, and K. Mortensen. *Nature*, 365:407–411, 1993.

[837] M. R. Eskildsen, K. Harada, P. L. Gammel, A. B. Abrahamsen, N. H. Andersen, G. Ernst, A. P. Ramirez, D. J. Bishop, K. Mortensen, D. G. Naugle, K. D. D. Rathnayaka, and P. C. Canfield. *Nature*, 393:242–245, 1998.

[838] T. M. Riseman, P. G. Kealey, E. M. Forgan, A. P. Mackenzie, L. M. Galvin, A. W. Tyler, S. L. Lee, C. Ager, D. McK. Paul, C. M. Aegerter, R. Cubitt, Z. Q. Mao, T. Akima, and Y. Maeno. *Nature*, 396:242–245, 1998.

[839] R. Gilardi, J. Mesot, S. P. Brown, E. M. Forgan, A. Drew, S. L. Lee, R. Cubitt, C. D. Dewhurst, T. Uefuji, and K. Yamada. *Phys. Rev. Lett.*, 93:217001, 2004.

[840] M. Laver, E. M. Forgan, S. P. Brown, D. Charalambous, D. Fort, C. Bowell, S. Ramos, R. J. Lycett, D. K. Christen, J. Kohlbrecher, C. D. Dewhurst, and R. Cubitt. *Phys. Rev. Lett.*, 96:167002, 2006.

[841] I. K. Dimitrov, N. D. Daniilidis, C. Elbaum, J. W. Lynn, and X. S. Ling. *Phys. Rev. Lett.*, 99:047001, 2007.

[842] A. D. Bianchi, M. Kenzelmann, L. DeBeer-Schmitt, J. S. White, E. M. Forgan, J. Mesot, M. Zolliker, J. Kohlbrecher, R. Movshovich, E. D. Bauer, J. L. Sarrao, Z. Fisk, C. Petrović, and M. R. Eskildsen. *Science*, 319:177–180, 2008.

[843] J. S. White, V. Hinkov, R. W. Heslop, R. J. Lycett, E. M. Forgan, C. Bowell, S. Strässle, A. B. Abrahamsen, M. Laver, C. D. Dewhurst, J. Kohlbrecher, J. L. Gavilano, J. Mesot, B. Keimer, and A. Erb. *Phys. Rev. Lett.*, 102:097001, 2009.

[844] D. S. Inosov, J. S. White, D. V. Evtushinsky, I. V. Morozov, A. Cameron, U. Stockert, V. B. Zabolotnyy, T. K. Kim, A. A. Kordyuk, S. V. Borisenko, E. M. Forgan, R. Klingeler, J. T. Park, S. Wurmehl, A. N. Vasiliev, G. Behr, C. D. Dewhurst, and V. Hinkov. *Phys. Rev. Lett.*, 104:187001, 2010.

[845] M. R. Eskildsen. *Front. Phys.*, 6:398–409, 2011.

[846] R. Toft-Petersen, A. B. Abrahamsen, S. Balog, L. Porcar, and M. Laver. *Nat. Commun.*, 9:901, 2018.

[847] E. R. Louden, C. Rastovski, S. J. Kuhn, A. W. D. Leishman, L. DeBeer-Schmitt, C. D. Dewhurst, N. D. Zhigadlo, and M. R. Eskildsen. *Phys. Rev. B*, 99:060502(R), 2019.

[848] E. R. Louden, C. Rastovski, L. DeBeer-Schmitt, C. D. Dewhurst, N. D. Zhigadlo, and M. R. Eskildsen. *Phys. Rev. B*, 99:144515, 2019.

[849] K. E. Avers, W. J. Gannon, S. J. Kuhn, W. P. Halperin, J. A. Sauls, L. DeBeer-Schmitt, C. D. Dewhurst, J. Gavilano, G. Nagy, U. Gasser, and M. R. Eskildsen. *Nat. Phys.*, 16:531–535, 2020.

[850] E. Jellyman, P. Jefferies, S. Pollard, E. M. Forgan, E. Blackburn, E. Campillo, A. T. Holmes, R. Cubitt, J. Gavilano, H. Wang, J. Du, and M. Fang. *Phys. Rev. B*, 101:134523, 2020.

[851] D. K. Christen, F. Tasset, S. Spooner, and H. A. Mook. *Phys. Rev. B*, 15:4506–4509, 1977.

[852] D. Mazzone, J. L. Gavilano, R. Sibille, M. Ramakrishnan, and M. Kenzelmann. *Phys. Rev. B*, 90:020507(R), 2014.

[853] A. Pautrat and A. Brûlet. *J. Phys.: Condens. Matter*, 26:232201, 2014.

[854] D. G. Mazzone, J. L. Gavilano, R. Sibille, M. Ramakrishnan, C. D. Dewhurst, and M. Kenzelmann. *J. Phys.: Condens. Matter*, 27:245701, 2015.

[855] T. Reimann, S. Mühlbauer, M. Schulz, B. Betz, A. Kaestner, V. Pipich, P. Böni, and C. Grünzweig. *Nat. Commun.*, 6:8813, 2015.

[856] T. Reimann, M. Schulz, C. Grünzweig, A. Kaestner, A. Bauer, P. Böni, and S. Mühlbauer. *J. Low Temp. Phys.*, 182:107–116, 2016.

[857] T. Reimann, M. Schulz, D. F. R. Mildner, M. Bleuel, A. Brûlet, R. P. Harti, G. Benka, A. Bauer, P. Böni, and S. Mühlbauer. *Phys. Rev. B*, 96:144506, 2017.

[858] A. Backs, M. Schulz, V. Pipich, M. Kleinhans, P. Böni, and S. Mühlbauer. *Phys. Rev. B*, 100:064503, 2019.

[859] L. J. Bannenberg, F. Qian, R. M. Dalgliesh, N. Martin, G. Chaboussant, M. Schmidt, D. L. Schlagel, T. A. Lograsso, H. Wilhelm, and C. Pappas. *Phys. Rev. B*, 96:184416, 2017.

[860] T. Kurumaji, T. Nakajima, V. Ukleev, A. Feoktystov, T. Arima, K. Kakurai, and Y. Tokura. *Phys. Rev. Lett.*, 119:237201, 2017.

[861] T. Nakajima, Y. Inamura, T. Ito, K. Ohishi, H. Oike, F. Kagawa, A. Kikkawa, Y. Taguchi, K. Kakurai, Y. Tokura, and T. Arima. *Phys. Rev. B*, 98:014424, 2018.

[862] L. J. Bannenberg, R. M. Dalgliesh, T. Wolf, F. Weber, and C. Pappas. *Phys. Rev. B*, 98:184431, 2018.

[863] T. Adams, M. Garst, A. Bauer, R. Georgii, and C. Pfleiderer. *Phys. Rev. Lett.*, 121:187205, 2018.

[864] D. A. Gilbert, A. J. Grutter, P. M. Neves, G.-J. Shu, G. Zimanyi, B. B. Maranville, F.-C. Chou, K. Krycka, N. P. Butch, S. Huang, and J. A. Borchers. *Phys. Rev. Materials*, 3:014408, 2019.

[865] O. I. Utesov and A. V. Syromyatnikov. *Phys. Rev. B*, 99:134412, 2019.

[866] L. J. Bannenberg, R. Sadykov, R. M. Dalgliesh, C. Goodway, D. L. Schlagel, T. A. Lograsso, P. Falus, E. Lelièvre-Berna, A. O. Leonov, and C. Pappas. *Phys. Rev. B*, 100:054447, 2019.

[867] A. S. Sukhanov, P. Vir, A. S. Cameron, H. C. Wu, N. Martin, S. Mühlbauer, A. Heinemann, H. D. Yang, C. Felser, and D. S. Inosov. *Phys. Rev. B*, 100:184408, 2019.

[868] E. Altynbaev, N. Martin, A. Heinemann, L. Fomicheva, A. Tsvyashchenko, I. Mirebeau, and S. Grigoriev. *Phys. Rev. B*, 101:100404(R), 2020.

[869] A. S. Sukhanov, B. E. Zuniga Cespedes, P. Vir, A. S. Cameron, A. Heinemann, N. Martin, G. Chaboussant, V. Kumar, P. Milde, L. M. Eng, C. Felser, and D. S. Inosov. *Phys. Rev. B*, 102:174447, 2020.

[870] R. J. Elliott. *Proc. Roy. Soc. A*, 235:289–304, 1956.

[871] S. L. Ginzburg and S. V. Maleyev. *Fiz. Tverd. Tela* (Leningrad), 7:3063–3069, 1965. (*Sov. Phys. Solid State*, 7:2477, 1965.).

[872] G. P. Kopitsa, S. V. Grigoriev, V. V. Runov, S. V. Maleyev, V. M. Garamus, and A. G. Yashenkin. *JETP Letters*, 81:556–560, 2005.

[873] S. Komura, G. Lippmann, and W. Schmatz. *J. Appl. Cryst.*, 7:233–236, 1974.

[874] S. Komura, G. Lippmann, and W. Schmatz. *J. Magn. Magn. Mater.*, 5:123–128, 1977.

[875] A. Chamberod, M. Roth, and L. Billard. *J. Magn. Magn. Mater.*, 7:101–103, 1978.

[876] A. Z. Menshikov and J. Schweizer. *Solid State Commun.*, 100:251–255, 1996.

[877] S. V. Grigoriev, S. V. Maleyev, A. I. Okorokov, and V. V. Runov. *Phys. Rev. B*, 58:3206–3211, 1998.

[878] S. V. Grigoriev, S. A. Klimko, W. H. Kraan, S. V. Maleyev, A. I. Okorokov, M. Th. Rekveldt, and V. V. Runov. *Phys. Rev. B*, 64:094426, 2001.

[879] S. V. Grigoriev, S. V. Maleyev, A. I. Okorokov, H. Eckerlebe, and N. H. van Dijk. *Phys. Rev. B*, 69:134417, 2004.

[880] J. Ross Stewart, S. R. Giblin, D. Honecker, P. Fouquet, D. Prabhakaran, and J. W Taylor. *J. Phys.: Condens. Matter*, 31:025802, 2019.

[881] M. Th. Rekveldt, C. M. E. Zeyen, J. C. Lodder, and W. H. Kraan. *J. Magn. Magn. Mater.*, 78:110–117, 1989.

[882] K. Takei, J. Suzuki, Y. Maeda, and S. Funahashi. *Jap. J. Appl. Phys.*, 32:2665–2666, 1993.

[883] K. Takei, J. Suzuki, Y. Maeda, and Y. Morii. *IEEE Trans. Magn.*, 30:4029–4031, 1994.

[884] W. H. Kraan, P. T. Por, and M. Th. Rekveldt. *J. Magn. Magn. Mater.*, 155:219–221, 1996.

[885] Y. Maeda, D. J. Rogers, O. Song, J. Suzuki, Y. Morii, and K. Takei. *J. Appl. Phys.*, 79:1819–1821, 1996.

[886] J. Suzuki, K. Takei, Y. Maeda, and Y. Morii. *J. Magn. Magn. Mater.*, 184:116–125, 1998.

[887] S. Sankar, A. E. Berkowitz, D. Dender, J. A. Borchers, R. W. Erwin, S. R. Kline, and David J. Smith. *J. Magn. Magn. Mater.*, 221:1–9, 2000.

[888] M. F. Toney, K. A. Rubin, S.-M. Choi, and C. J. Glinka. *Appl. Phys. Lett.*, 82:3050–3052, 2003.

[889] M. P. Wismayer, S. L. Lee, T. Thomson, F. Y. Ogrin, C. D. Dewhurst, S. M. Weekes, and R. Cubitt. *J. Appl. Phys.*, 99:08E707, 2006.

[890] S. J. Lister, M. P. Wismayer, V. Venkataramana, M. A. De Vries, S. J. Ray, S. L. Lee, T. Thomson, J. Kohlbrecher, H. Do, Y. Ikeda, K. Takano, and C. Dewhurst. *J. Appl. Phys.*, 106:063908, 2009.

[891] S. J. Lister, T. Thomson, J. Kohlbrecher, K. Takano, V. Venkataramana, S. J. Ray, M. P. Wismayer, M. A. de Vries, H. Do, Y. Ikeda, and S. L. Lee. *Appl. Phys. Lett.*, 97:112503, 2010.

[892] S. J. Lister, J. Kohlbrecher, V. Venkataramana, T. Thomson, K. Takano, and S. L. Lee. *Int. J. Mater. Res.*, 102:1142–1146, 2011.

[893] M. Feygenson, J. C. Bauer, Z. Gai, C. Marques, M. C. Aronson, X. Teng, D. Su, V. Stanic, V. S. Urban, K. A. Beyer, and S. Dai. *Phys. Rev. B*, 92:054416, 2015.

[894] R. Maruyama, T. Bigault, T. Saerbeck, D. Honecker, K. Soyama, and P. Courtois. *Crystals*, 9:383, 2019.

[895] C. Dufour, M. R. Fitzsimmons, J. A. Borchers, M. Laver, K. L. Krycka, K. Dumesnil, S. M. Watson, W. C. Chen, J. Won, and S. Singh. *Phys. Rev. B*, 84:064420, 2011.

[896] A. Planes, L. Mañosa, and M. Acet. *J. Phys.: Condens. Matter*, 21:233201, 2009.

[897] M. Acet, Ll. Mañosa, and A. Planes. Magnetic-Field-Induced Effects in Martensitic Heusler-Based Magnetic Shape Memory Alloys. In K. H. J. Buschow, editor, *Handbook of Magnetic Materials*, volume 19, pages 231–289. Elsevier, Amsterdam, 2011.

[898] O. Gutfleisch, M. A. Willard, E. Brück, C. H. Chen, S. G. Sankar, and J. Ping Liu. *Adv. Mater.*, 23:821–842, 2011.

[899] V. V. Runov, Yu. P. Chernenkov, M. K. Runova, V. G. Gavriljuk, and N. I. Glavatska. *JETP Letters*, 74:590–595, 2001.

[900] V. V. Runov, Yu. P. Chernenkov, M. K. Runova, V. G. Gavriljuk, and N. I. Glavatska. *Physica B*, 335:109–113, 2003.

[901] G. P. Kopitsa, V. V. Runov, S. V. Grigoriev, V. V. Bliznuk, V. G. Gavriljuk, and N. I. Glavatska. *Physica B*, 335:134–139, 2003.

[902] V. Runov, M. Runova, V. Gavriljuk, and N. Glavatska. *Physica B*, 350:e87–e89, 2004.

[903] V. V. Bliznuk, V. G. Gavriljuk, G. P. Kopitsa, S. V. Grigoriev, and V. V. Runov. *Acta Mater.*, 52:4791–4799, 2004.

[904] V. V. Runov, Yu. P. Chernenkov, M. K. Runova, V. G. Gavrilyuk, N. I. Glavatska, A. G. Goukasov, V. V. Koledov, V. G. Shavrov, and V. V. Khovaĭlo. *J. Exp. Theo. Phys.*, 102:102–113, 2006.

[905] K. P. Bhatti, V. Srivastava, D. P. Phelan, S. El-Khatib, R. D. James, and C. Leighton. Magnetic Phase Competition in Off-Stoichiometric Martensitic Heusler Alloys: The $Ni_{50-x}Co_xMn_{25+y}Sn_{25-y}$ System. In C. Felser and A. Hirohata, editors, *Heusler Alloys: Properties, Growth, Applications*, pages 193–216. Springer International Publishing, Switzerland, 2016.

[906] S. El-Khatib, K. P. Bhatti, V. Srivastava, R. D. James, and C. Leighton. *Phys. Rev. Materials*, 3:104413, 2019.

[907] S. K. Sarkar, S. Ahlawat, S. D. Kaushik, P. D. Babu, D. Sen, D. Honecker, and A. Biswas. *J. Phys.: Condens. Matter*, 32:115801, 2020.

[908] A. Çakır, M. Acet, and M. Farle. *Sci. Rep.*, 6:28931, 2016.

[909] A. Çakır, M. Acet, U. Wiedwald, and M. Farle. *Acta Mater.*, 127:117–123, 2017.

[910] C. Mudivarthi, M. Laver, J. Cullen, A. B. Flatau, and M. Wuttig. *J. Appl. Phys.*, 107:09A957, 2010.

[911] M. Laver, C. Mudivarthi, J. R. Cullen, A. B. Flatau, W.-C. Chen, S. M. Watson, and M. Wuttig. *Phys. Rev. Lett.*, 105:027202, 2010.

[912] C. Zhang, J. Chen, L. Sun, P. Zhang, Y. Liu, L. Chen, Z. Yang, B. Pang, Y. Huang, G. Sun, and C. Huang. *J. Magn. Magn. Mater.*, 490:165495, 2019.

[913] Y. Ke, H.-H. Wu, S. Lan, H. Jiang, Y. Ren, S. Liu, and C. Jiang. *J. Alloys Comp.*, 822:153687, 2020.

[914] H. Fujii, M. Saga, T. Takeda, S. Komura, T. Okamoto, S. Hirosawa, and M. Sagawa. *IEEE Trans. Magn.*, 23:3119–3121, 1987.

[915] M. Takeda, J. Suzuki, T. Akiya, and H. Kato. *J. Japan Inst. Metals*, 76:165–176, 2012.

[916] A. Kreyssig, R. Prozorov, C. D. Dewhurst, P. C. Canfield, R. W. McCallum, and A. I. Goldman. *Phys. Rev. Lett.*, 102:047204, 2009.

[917] T. Ueno, K. Saito, M. Yano, M. Ito, T. Shoji, N. Sakuma, A. Kato, A. Manabe, A. Hashimoto, E. P. Gilbert, U. Keiderling, and K. Ono. *Sci. Rep.*, 6:28167, 2016.

[918] M. Yano, K. Ono, A. Manabe, N. Miyamoto, T. Shoji, A. Kato, Y. Kaneko, M. Harada, H. Nozaki, and J. Kohlbrecher. *IEEE Trans. Magn.*, 48:2804–2807, 2012.

[919] T. Ueno, K. Saito, M. Yano, M. Harada, T. Shoji, N. Sakuma, A. Manabe, A. Kato, U. Keiderling, and K. Ono. *IEEE Trans. Magn.*, 50:2103104, 2014.

[920] M. Yano, K. Ono, M. Harada, A. Manabe, T. Shoji, A. Kato, and J. Kohlbrecher. *J. Appl. Phys.*, 115:17A730, 2014.

[921] E. A. Périgo, E. P. Gilbert, and A. Michels. *Acta Mater.*, 87:142–149, 2015.

[922] K. Saito, T. Ueno, M. Yano, M. Harada, T. Shoji, N. Sakuma, A. Manabe, A. Kato, U. Keiderling, and K. Ono. *J. Appl. Phys.*, 117:17B302, 2015.

[923] E. A. Périgo, D. Mettus, E. P. Gilbert, P. Hautle, N. Niketic, B. van den Brandt, J. Kohlbrecher, P. McGuiness, Z. Fu, and A. Michels. *J. Alloys Comp.*, 661:110–114, 2016.

[924] E. A. Périgo, I. Titov, R. Weber, D. Honecker, E. P. Gilbert, M. F. De Campos, and A. Michels. *J. Alloys Comp.*, 677:139–142, 2016.

[925] A. Michels, R. Weber, I. Titov, D. Mettus, É. A. Périgo, I. Peral, O. Vallcorba, J. Kohlbrecher, K. Suzuki, M. Ito, A. Kato, and M. Yano. *Phys. Rev. Applied*, 7:024009, 2017.

[926] É. A. Périgo, I. Titov, R. Weber, D. Mettus, I. Peral, O. Vallcorba, D. Honecker, A. Feoktystov, and A. Michels. *Mater. Res. Express*, 5:036110, 2018.

[927] I. Titov, M. Barbieri, P. Bender, I. Peral, J. Kohlbrecher, K. Saito, V. Pipich, M. Yano, and A. Michels. *Phys. Rev. Materials*, 3:084410, 2019.

[928] I. Titov, D. Honecker, D. Mettus, A. Feoktystov, J. Kohlbrecher, P. Strunz, and A. Michels. *Phys. Rev. Materials*, 4:054419, 2020.

[929] J. M. D. Coey, M. Viret, and S. von Molnár. *Adv. Phys.*, 48:167–293, 1999.

[930] Y. Tokura and Y. Tomioka. *J. Magn. Magn. Mater.*, 200:1–23, 1999.

[931] E. Dagotto, T. Hotta, and A. Moreo. *Phys. Rep.*, 344:1–153, 2001.

[932] E. Dagotto, editor. *Nanoscale Phase Separation and Colossal Magnetoresistance: The Physics of Manganites and Related Compounds*. Springer-Verlag, Berlin, 2003.

[933] J. M. De Teresa, M. R. Ibarra, P. A. Algarabel, C. Ritter, C. Marquina, J. Blasco, J. Garcia, A. del Moral, and Z. Arnold. *Nature*, 386:256–259, 1997.

[934] C. Ritter, M. R. Ibarra, J. M. De Teresa, P. A. Algarabel, C. Marquina, J. Blasco, J. García, S. Oseroff, and S.-W. Cheong. *Phys. Rev. B*, 56:8902–8911, 1997.

[935] M. Viret, H. Glättli, C. Fermon, A. M. de Leon-Guevara, and A. Revcolevschi. *EPL (Europhysics Letters)*, 42:301–306, 1998.

[936] J. W. Lynn, L. Vasiliu-Doloc, and M. A. Subramanian. *Phys. Rev. Lett.*, 80:4582–4585, 1998.

[937] V. V. Runov, G. P. Kopitsa, A. I. Okorokov, M. K. Runova, and H. Glattli. *JETP Letters*, 69:353–360, 1999.

[938] V. Runov, H. Glattli, G. Kopitsa, A. Okorokov, and M. Runova. *Physica B*, 276-278:795–796, 2000.

[939] P. G. Radaelli, R. M. Ibberson, D. N. Argyriou, H. Casalta, K. H. Andersen, S.-W. Cheong, and J. F. Mitchell. *Phys. Rev. B*, 63:172419, 2001.

[940] Y. Yamada, T. Iwase, M. Watahiki, and J. Suzuki. *J. Phys. Soc. Jpn.*, 70:1593–1597, 2001.

[941] G. Biotteau, M. Hennion, F. Moussa, J. Rodríguez-Carvajal, L. Pinsard, A. Revcolevschi, Y. M. Mukovskii, and D. Shulyatev. *Phys. Rev. B*, 64:104421, 2001.

[942] Ch. Simon, S. Mercone, N. Guiblin, C. Martin, A. Brûlet, and G. André. *Phys. Rev. Lett.*, 89:207202, 2002.

[943] J. M. De Teresa, M. R. Ibarra, P. Algarabel, L. Morellon, B. García-Landa, C. Marquina, C. Ritter, A. Maignan, C. Martin, B. Raveau, A. Kurbakov, and V. Trounov. *Phys. Rev. B*, 65:100403, 2002.

[944] S. Mercone, V. Hardy, C. Martin, C. Simon, D. Saurel, and A. Brûlet. *Phys. Rev. B*, 68:094422, 2003.

[945] C. Yaicle, C. Martin, Z. Jirak, F. Fauth, G. André, E. Suard, A. Maignan, V. Hardy, R. Retoux, M. Hervieu, S. Hébert, B. Raveau, Ch. Simon, D. Saurel, A. Brûlet, and F. Bourée. *Phys. Rev. B*, 68:224412, 2003.

[946] F. M. Woodward, J. W. Lynn, M. B. Stone, R. Mahendiran, P. Schiffer, J. F. Mitchell, D. N. Argyriou, and L. C. Chapon. *Phys. Rev. B*, 70:174433, 2004.

[947] M. Viret, F. Ott, J. P. Renard, H. Glättli, L. Pinsard-Gaudart, and A. Revcolevschi. *Phys. Rev. Lett.*, 93:217402, 2004.

[948] J. M. De Teresa, P. A. Algarabel, C. Ritter, J. Blasco, M. R. Ibarra, L. Morellon, J. I. Espeso, and J. C. Gómez-Sal. *Phys. Rev. Lett.*, 94:207205, 2005.

[949] J. Wu, J. W. Lynn, C. J. Glinka, J. Burley, H. Zheng, J. F. Mitchell, and C. Leighton. *Phys. Rev. Lett.*, 94:037201, 2005.

[950] D. Saurel, A. Brûlet, A. Heinemann, C. Martin, S. Mercone, and C. Simon. *Phys. Rev. B*, 73:094438, 2006.

[951] D. Saurel, C. Simon, A. Brûlet, A. Heinemann, and C. Martin. *Phys. Rev. B*, 75:184442, 2007.

[952] C. He, M. A. Torija, J. Wu, J. W. Lynn, H. Zheng, J. F. Mitchell, and C. Leighton. *Phys. Rev. B*, 76:014401, 2007.

[953] Y. Qin, T. A Tyson, K. Pranzas, and H. Eckerlebe. *J. Phys.: Condens. Matter*, 20:195209, 2008.

[954] C. Leighton, D. D. Stauffer, Q. Huang, Y. Ren, S. El-Khatib, M. A. Torija, J. Wu, J. W. Lynn, L. Wang, N. A. Frey, H. Srikanth, J. E. Davies, Kai Liu, and J. F. Mitchell. *Phys. Rev. B*, 79:214420, 2009.

[955] C. He, S. El-Khatib, J. Wu, J. W. Lynn, H. Zheng, J. F. Mitchell, and C. Leighton. *EPL (Europhysics Letters)*, 87:27006, 2009.

[956] D. Saurel, Ch. Simon, A. Pautrat, C. Martin, C. Dewhurst, and A. Brûlet. *Phys. Rev. B*, 82:054427, 2010.

[957] B. G. Ueland, J. W. Lynn, M. Laver, Y. J. Choi, and S.-W. Cheong. *Phys. Rev. Lett.*, 104:147204, 2010.

[958] D. Phelan, Y. Suzuki, S. Wang, A. Huq, and C. Leighton. *Phys. Rev. B*, 88:075119, 2013.

[959] D. Phelan, K. P. Bhatti, M. Taylor, S. Wang, and C. Leighton. *Phys. Rev. B*, 89:184427, 2014.

[960] S. El-Khatib, D. Phelan, J. G. Barker, H. Zheng, J. F. Mitchell, and C. Leighton. *Phys. Rev. B*, 92:060404, 2015.

[961] P. Anil Kumar, A. Nag, R. Mathieu, R. Das, S. Ray, P. Nordblad, A. Hossain, D. Cherian, D. Alba Venero, L. DeBeer-Schmitt, O. Karis, and D. D. Sarma. *Phys. Rev. Research*, 2:043344, 2020.

[962] J. S. Gardner, M. J. P. Gingras, and J. E. Greedan. *Rev. Mod. Phys.*, 82:53–107, 2010.

[963] B. D. Gaulin, J. N. Reimers, T. E. Mason, J. E. Greedan, and Z. Tun. *Phys. Rev. Lett.*, 69:3244–3247, 1992.

[964] J. E. Greedan, J. N. Reimers, C. V. Stager, and S. L. Penny. *Phys. Rev. B*, 43:5682–5691, 1991.

[965] J. E. Greedan, J. Avelar, and M. A. Subramanian. *Solid State Commun.*, 82:797–799, 1992.

[966] N. P. Raju, J. E. Greedan, and M. A. Subramanian. *Phys. Rev. B*, 49:1086–1091, 1994.

[967] J. E. Greedan, N. P. Raju, A. Maignan, Ch. Simon, J. S. Pedersen, A. M. Niraimathi, E. Gmelin, and M. A. Subramanian. *Phys. Rev. B*, 54:7189–7200, 1996.

[968] C. R. C. Buhariwalla, Q. Ma, L. DeBeer-Schmitt, K. G. S. Xie, D. Pomaranski, J. Gaudet, T. J. Munsie, H. A. Dabkowska, J. B. Kycia, and B. D. Gaulin. *Phys. Rev. B*, 97:224401, 2018.

[969] A. Scheie, J. Kindervater, S. Zhang, H. J. Changlani, G. Sala, G. Ehlers, A. Heinemann, G. S. Tucker, S. M. Koohpayeh, and C. Broholm. *Proc. Natl. Acad. Sci. U.S.A.*, 117:27245–27254, 2020.

[970] N. Baddour. *J. Opt. Soc. Am.*, 26:1767–1777, 2009.

[971] N. Baddour. *J. Opt. Soc. Am.*, 27:2144–2155, 2010.

[972] G. Fritz-Popovski. *J. Appl. Cryst.*, 48:44–51, 2015.

[973] G. N. Watson. *A Treatise on the Theory of Bessel Functions*. Cambridge University Press, Cambridge, 2nd edition, 1966.

[974] T. G. Woodcock, Y. Zhang, G. Hrkac, G. Ciuta, N. M. Dempsey, T. Schrefl, O. Gutfleisch, and D. Givord. *Scripta Mater.*, 67:536–541, 2012.

[975] R. Skomski, H. Zeng, and D. J. Sellmyer. *IEEE Trans. Magn.*, 37:2549–2551, 2001.

[976] G. Herzer. Nanocrystalline Soft Magnetic Alloys. In K. H. J. Buschow, editor, *Handbook of Magnetic Materials*, volume 10, pages 415–462. Elsevier, Amsterdam, 1997.

[977] D. E. Hegland, S. Legvold, and F. H. Spedding. *Phys. Rev.*, 131:158–162, 1963.

[978] P. G. Sanders, J. A. Eastman, and J. R. Weertman. *Acta Mater.*, 46:4195–4202, 1998.

[979] J. J. Rhyne and A. E. Clark. *J. Appl. Phys.*, 38:1379–1380, 1967.

[980] F. Döbrich, M. Elmas, A. Ferdinand, J. Markmann, M. Sharp, H. Eckerlebe, J. Kohlbrecher, R. Birringer, and A. Michels. *J. Phys.: Condens. Matter*, 21:156003, 2009.

[981] E. E Underwood. *Quantitative Stereology*. Addison-Wesley Publishing Company, Reading, 1970.

[982] A. P. Sutton and R. W. Balluffi. *Interfaces in Crystalline Materials*. Clarendon Press, Oxford, 1995.

[983] Ph. Kurz, G. Bihlmayer, and S. Blügel. *J. Phys.: Condens. Matter*, 14:6353–6371, 2002.

[984] I. Turek, J. Kudrnovský, G. Bihlmayer, and S. Blügel. *J. Phys.: Condens. Matter*, 15:2771–2782, 2003.

[985] M. E. Fisher and A. E. Ferdinand. *Phys. Rev. Lett.*, 19:169–172, 1967.

[986] M. Donahue and D. Porter. The Object Oriented MicroMagnetic Framework (OOMMF). Available from: https://math.nist.gov/oommf [Accessed February 2021].

[987] T. Fischbacher, M. Franchin, G. Bordignon, and H. Fangohr. *IEEE Trans. Magn.*, 43:2896–2898, 2007.

[988] T. Fischbacher, M. Franchin, G. Bordignon, A. Knittel, and H. Fangohr. *J. Appl. Phys.*, 105:07D527, 2009.

[989] D. V. Berkov and N. L. Gorn. Numerical Simulation of Quasistatic and Dynamic Remagnetization Processes with Special Applications to Thin Films and Nanoparticles. In Y. Liu, D. J. Sellmyer, and D. Shindo, editors, *Handbook of Advanced Magnetic Materials*, pages 421–507. Springer-Verlag, New York, 2005.

[990] D. V. Berkov and N. L. Gorn. *Phys. Rev. B*, 57:14332, 1998.

[991] N. L. Gorn, D. V. Berkov, P. Görnert, and D. Stock. *J. Magn. Magn. Mater.*, 310:2829, 2007.

[992] D. V. Berkov. Magnetization Dynamics Including Thermal Fluctuations: Basic Phenomenology, Fast Remagnetization Processes and Transitions Over High-Energy Barriers. In H. Kronmüller and S. Parkin, editors, *Handbook of Magnetism and Advanced Magnetic Materials*, pages 795–823. Wiley, Chichester, volume 2: Micromagnetism edition, 2007.

[993] K. Suzuki and G. Herzer. Soft Magnetic Nanostructures and Applications. In D. Sellmyer and R. Skomski, editors, *Advanced Magnetic Nanostructures*, pages 365–401. Springer-Verlag, New York, 2006.

[994] R. Hertel. *J. Appl. Phys.*, 90:5752–5758, 2001.

[995] J. Escrig, R. Lavín, J. L. Palma, J. C. Denardin, D. Altbir, A. Cortés, and H. Gómez. *Nanotechnology*, 19:075713, 2008.

[996] F. Zighem, T. Maurer, F. Ott, and G. Chaboussant. *J. Appl. Phys.*, 109:013910, 2011.

[997] Yu. P. Ivanov, L. G. Vivas, A. Asenjo, A. Chuvilin, O. Chubykalo-Fesenko, and M. Vázquez. *EPL (Europhysics Letters)*, 102:17009, 2013.

[998] J. A. Fernandez-Roldan, Yu. P. Ivanov, and O. Chubykalo-Fesenko. Micromagnetic modeling of magnetic domain walls and domains in cylindrical nanowires. In M. Vázquez, editor, *Magnetic Nano- and Microwires*, Woodhead Publishing Series in Electronic and Optical Materials, pages 403–426. Woodhead Publishing, 2nd edition, 2020.

[999] L. G. Vivas, Yu. P. Ivanov, D. G. Trabada, M. P. Proenca, O. Chubykalo-Fesenko, and M. Vázquez. *Nanotechnology*, 24:105703, 2013.

[1000] Yu. P. Ivanov, A. Chuvilin, L. G. Vivas, J. Kosel, O. Chubykalo-Fesenko, and M. Vázquez. *Sci. Rep.*, 6:23844, 2016.

[1001] S. M. Yusuf, J. M. De Teresa, M. D. Mukadam, J. Kohlbrecher, M. R. Ibarra, J. Arbiol, P. Sharma, and S. K. Kulshreshtha. *Phys. Rev. B*, 74:224428, 2006.

[1002] K. L. Krycka, R. Booth, J. A. Borchers, W. C. Chen, C. Conlon, T. R. Gentile, C. Hogg, Y. Ijiri, M. Laver, B. B. Maranville, S. A. Majetich, J. J. Rhyne, and S. M. Watson. *Physica B*, 404:2561–2564, 2009.

[1003] C. L. Dennis, K. L. Krycka, J. A. Borchers, R. D. Desautels, J. van Lierop, N. F. Huls, A. J. Jackson, C. Gruettner, and R. Ivkov. *Adv. Funct. Mater.*, 25:4300–4311, 2015.

[1004] I. Orue, L. Marcano, P. Bender, A. García-Prieto, S. Valencia, M. A. Mawass, D. Gil-Cartón, D. Alba Venero, D. Honecker, A. García-Arribas, L. Fernández Barquín, A. Muela, and M. L. Fdez-Gubieda. *Nanoscale*, 10:7407–7419, 2018.

[1005] P. Bender, D. Honecker, and L. F. Barquín. *Appl. Phys. Lett.*, 115:132406, 2019.

[1006] H. Carlton, K. Krycka, M. Bleuel, and D. Huitink. *Part. Part. Syst. Charact.*, 37:1900358, 2019.

[1007] P. Bender, L. Marcano, I. Orue, D. Alba Venero, D. Honecker, L. Fernández Barquín, A. Muela, and M. L. Fdez-Gubieda. *Nanoscale Adv.*, 2:1115–1121, 2020.

[1008] M. Bersweiler, H. Gavilan Rubio, D. Honecker, A. Michels, and P. Bender. *Nanotechnology*, 31:435704, 2020.

[1009] T. Mekuria, S. Khalid, K. Krycka, M. Bleuel, H. Verma, H. Hong, S. P. Karna, and D. Seifu. *AIP Advances*, 10:065134, 2020.

[1010] P. Hrubovčák, N. Kučerka, A. Zeleňáková, and V. Zeleňák. *Acta Phys. Pol. A*, 137:730–732, 2020.

[1011] R. F. L. Evans, W. J. Fan, P. Chureemart, T. A. Ostler, M. O. A. Ellis, and R. W. Chantrell. *J. Phys.: Condens. Matter*, 26:103202, 2014.

[1012] G. Bertotti, C. Serpico, and I. D. Mayergoyz. *Phys. Rev. Lett.*, 86:724, 2001.

[1013] G. Bertotti, I. D. Mayergoyz, and C. Serpico. *Phys. Rev. Lett.*, 87:217203, 2001.

[1014] J. M. D. Coey. *J. Phys.: Condens. Matter*, 26:064211, 2014.

[1015] R. N. Bracewell. *Aust. J. Phys.*, 9:198–217, 1956.

[1016] S. Ciccariello, personnel communication, 2020.

[1017] I. S. Gradshteyn and I. M. Ryzhik. *Table of Integrals, Series, and Products*. Academic Press, San Diego, 7th edition, 2007.

[1018] S. Ciccariello, G. Cocco, A. Benedetti, and S. Enzo. *Phys. Rev. B*, 23:6474–6485, 1981.

[1019] R. N. Bracewell. *The Fourier Transform and its Applications*. McGraw-Hill, Boston, 3rd edition, 2000.

[1020] W. G. Bouwman, W. Stam, T. V. Krouglov, J. Plomp, S. V. Grigoriev, W. H. Kraan, and M. Th. Rekveldt. *Nucl. Instrum. Methods Phys. Res. A*, 529:16–21, 2004.

[1021] M. Th. Rekveldt, N. H. van Dijk, S. V. Grigoriev, W. H. Kraan, and W. G. Bouwman. *Rev. Sci. Instr.*, 77:073902, 2006.

[1022] W. G. Bouwman, J. Plomp, V. O. de Haan, W. H. Kraan, A. A. van Well, K. Habicht, T. Keller, and M. Th. Rekveldt. *Nucl. Instrum. Methods Phys. Res. A*, 586:9–14, 2008.

[1023] M. Strobl, A. S. Tremsin, A. Hilger, F. Wieder, N. Kardjilov, I. Manke, W. G. Bouwman, and J. Plomp. *J. Appl. Phys.*, 112:014503, 2012.

[1024] E. G. Iashina, W. G. Bouwman, C. P. Duif, M. V. Filatov, and S. V. Grigoriev. *J. Phys.: Conf. Series*, 862:012010, 2017.

[1025] M. Bacak, J. Valsecchi, J. Čapek, E. Polatidis, A. Kaestner, A. Arabi-Hashemi, I. Kruk, C. Leinenbach, A. M. Long, A. Tremsin, S. C. Vogel, E. B. Watkins, and M. Strobl. *Materials & Design*, 195:109009, 2020.

[1026] E. R. Cohen and P. Giacomo. *Physica*, 146A:1–68, 1987.

Index